AST COT 1977   c.08

The Elements of navigation and
nautical astronomy: C. H. Cotter

The Elements of Navigation
and Nautical Astronomy

The Elements of Navigation and Nautical Astronomy

A TEXT-BOOK FOR CADETS

BY

CHARLES H. COTTER

Ex.C., B.Sc. (London), M.Sc. (Wales), Ph.D. (London), F.R.I.N.

*Senior Lecturer in the Department of Maritime Studies
University of Wales Institute of Science and Technology
Cardiff*

GLASGOW
BROWN, SON & FERGUSON, Ltd., NAUTICAL PUBLISHERS
52 DARNLEY STREET, G41 2SG

Copyright in all countries signatory to the Berne Convention
All rights reserved

First Edition 1977

STUDENT TEXT BOOK

No. AST COT 1977 C.08

ISBN 0 85174 270 X
© 1977 BROWN, SON & FERGUSON, LTD., GLASGOW G41 2SG
Made and Printed in Great Britain

What is beyond the horizon?
Let us discover

FOREWORD

THE present work is a completely revised version of my earlier book, *The Elements of Navigation*, written a quarter of a century ago. Since its first publication, two editions in French for the Government of Quebec have appeared and several reprints have been issued. The time, however, is long overdue for an entirely new edition, and Messrs. Brown, Son and Ferguson, the well-known nautical publishers of Glasgow, have very kindly adopted the book in its new guise for inclusion in their comprehensive range of manuals and text-books aimed specifically to assist mariners in their professional studies.

In overhauling *The Elements of Navigation* I particularly had in mind Merchant Naval Cadets preparing for the O.N.C. and O.N.D. examinations in Nautical Science. Having had the privilege and invaluable experience of serving on the Joint Committee for National Certificates and Diplomas for many years—indeed since its inception—to 1974, I am aware of the remarkable advances that have been made in nautical education over the last couple of decades. Two things have given me immense satisfaction during this period: the first is that Cadets are now able to gain a nationally recognized qualification in the form of a National Certificate or Diploma, in consequence of which we have witnessed a profound effect on the teaching of navigation (and other nautical subjects) to the extent that Cadets are prepared for their professional or vocational duties in a manner far superior to that which prevailed before the advent of the O.N.D. scheme; and the second is that Merchant Naval Officers are no longer debarred, on practical grounds, from reading for a degree at a university or polytechnic.

It has been an experience of considerable personal satisfaction and pleasure for me to have come into close contact with a large number of officers from many countries who have read for a B.Sc. degree in Maritime Studies at the University of Wales Institute of Science and Technology at Cardiff. I and my colleagues can bear testimony to the excellence of many of these officers who have gone down from the University, not only with splendid academic qualifications but, and more importantly, with increased knowledge of ships, ports and the sea. And it is this that cannot fail, in the long run, to enhance the quality of the services offered by merchant navies of the world, so to improve their role in ensuring that the material needs of mankind are satisfied.

The safe carriage of goods by sea is closely linked with the processes and techniques of getting vessels safely from port to port, and this is the essence of navigation.

Although the plan of the present book follows closely that of the original, I have made numerous changes so that the work may now be considered to provide a carefully arranged programme of studies in all fields of navigation and nautical astronomy—of particular value to those who study at sea. The book is elementary in character and it covers the groundwork of the subject for the O.N.D. and O.N.C. examinations in theoretical and practical navigation. Readers who may wish to extend their knowledge of navigation and nautical astronomy are referred to my two books of a more advanced nature—*The Complete Coastal Navigator*, and *The Complete Nautical Astronomer*.

I can fairly claim that *The Elements of Navigation*, despite its defects, was a popular text; and, judging by the numerous encouraging letters of tribute which I have received from readers, it helped a large number of students in their studies of navigation. It is a source of deep satisfaction to an author to see his book go into a new edition; for this, surely, indicates that his work has found favour and that it has stood the test of time. It is my sincere wish that this new work will meet with the approval of all those into whose hands it may fall: this applies particularly to Cadets preparing for the O.N.C. or O.N.D. examinations in Nautical Science.

My share of the labour involved in producing this book was greatly reduced by the care with which the publisher and printer, and their respective staffs, followed my original typescript. It is a pleasure for me to record my appreciation to all who have contributed towards its production.

CHARLES H. COTTER
Cardiff 1977.

CONTENTS

PART 1

THE MATHEMATICS OF NAVIGATION AND NAUTICAL ASTRONOMY

Chapter 1 Plane Trigonometry 3

Para. 1 Introduction 3
2 Complementary Angles 3
3 Trigonometrical Functions as Straight Lines 5
4 The Signs of the Trigonometrical Ratios of Angles between 90° and 180° 6
5 The Standard Formulae 7
6 Special Angles 7

Chapter 2 Circular Measure 10

Para. 1 The Radian 10
2 Trigonometrical Ratios of Small Angles 12

Chapter 3 The Traverse Table and the Solution of Plane Right-angled Triangles 15

Para. 1 Introduction 15
2 Plane Right-angled Triangles and their Traverse Table Solutions 15

Chapter 4 Compound Angles 22

Para. 1 Trigonometrical Ratios of the Sum of Two Angles 22
2 Trigonometrical Ratios of the Difference of Two Angles 23
3 Products as Sums and Differences 23
4 Sums and Differences as Products 24

Chapter 5 Oblique-angled Triangles and their Solutions 25

Para. 1 Introduction 25
2 The Sine Formula 25
3 The Cosine Formula 27
4 The Haversine Formula 28
5 The Tangent Formula 30

Chapter 6 Spherical Trigonometry 33

Para. 1 The Geometry of the Sphere 33
2 Spherical Trigonometry 35

CONTENTS

Chapter 7 The Stereographic Projection and the Graphical Solutions of Spherical Triangles 36
Para. 1 Introduction 36
2 The Stereographic Projection 36
3 The Principles of the Stereographic Projection 36
4 To Project a Great Circle about a Given Point as Pole 38
5 To Project a Small Circle about a Given Point as Pole 38
6 To Find the Locus of Centres of all Great Circles which pass through a Given Point 39
7 To Project a Great Circle through Two Given Points 40
8 To Measure a Given Arc of a Projected Great Circle 41
9 To Measure a Projection of a Spherical Angle 41
10 Examples 41
11 Figure Drawing 45

Chapter 8 The Trigonometrical Solutions of Spherical Triangles 46
Para. 1 The Spherical Sine Formula 46
2 The Spherical Cosine Formula 47
3 The Spherical Haversine Formula 48
4 The Four Parts Formula 49

Chapter 9 Napier's Rules 51
Para. 1 Napier's Rules for Solving Right-angled Spherical Triangles 51
2 Napier's Rules for Solving Quadrantal Spherical Triangles 55
3 The Solution of Oblique Spherical Triangles by Napier's Rules 56

PART 2

THE SAILINGS

Chapter 10 The Shape and Size of the Earth 61
Para. 1 The Earth 61
2 Describing a Terrestrial Position 62
3 The True Shape of the Earth 64
4 The Nautical Mile 65
5 Reduction of the Geographical Latitude 66
6 The Geographical Mile 67

Chapter 11 The Rhumb Line 68
Para. 1 Introduction 68
2 The Sailings 68
3 Parallel Sailing 69
4 The Parallel Sailing Formula 69
5 Plane Sailing 71
6 Proof of Plane Sailing Formulae 71
7 Traverse Sailing 73
8 The Departure Position 73
9 Current 74

CONTENTS

Chapter 12 The Mercator Chart 77
- Para. 1 The Navigator's Chart 77
- 2 Features of a Mercator Chart 77
- 3 The Defects of a Mercator Chart 78
- 4 Distortion of the Mercator Projection 78
- 5 Meridional Parts 80
- 6 Meridional Parts for the Terrestrial Spheroid 82
- 7 Constructing Mercator Charts 83

Chapter 13 Mercator Sailing and Middle Latitude Sailing 86
- Para. 1 Introduction 86
- 2 Mercator Sailing 86
- 3 Rhumb Line Sailing when Course Angle is Large 88
- 4 Middle Latitude Sailing 90
- 5 Mean to Middle Latitude Correction Table 90
- 6 Crossing the Equator 94
- 7 The Day's Run 95
- 8 The Day's Work 96

Chapter 14 Great Circle Sailing 100
- Para. 1 Introduction 100
- 2 The Gnomonic Chart 101
- 3 Practical Great Circle Sailing 102
- 4 Composite Great Circle Sailing 106

PART 3
INTRODUCTION TO CHARTWORK

Chapter 15 Introduction to Chartwork 113
- Para. 1 Coastal Navigation 113
- 2 Charts 113
- 3 The Natural Scale of a Chart 113
- 4 Description of a Chart 114
- 5 Chart Abbreviations and Symbols 115
- 6 Hints when using Charts 118

Chapter 16 Correcting the Course 121
- Para. 1 The Three Norths 121
- 2 The Earth's Magnetism 121
- 3 The Three Courses 122
- 4 The Three Bearings 122
- 5 Compass Error 122
- 6 Variation 123
- 7 Deviation 124
- 8 The Deviation Card or Table 125
- 9 Leeway 127

Chapter 17 The Position Line 131

Para. 1 Fixing by Cross-Bearings 131
    2 The Cocked Hat 132
    3 Transits 132
    4 Relative Bearings 132
    5 Fixing by Bearing and Angle 132
    6 Fixing by Bearing and Sounding 133
    7 Fixing by Sector Light 133
    8 Choosing Marks for Fixing 133
    9 Angle of Cut 133

Chapter 18 The Transferred Position Line 135

Para. 1 Introduction 135
    2 Transferring a Position Line 135
    3 The Running Fix 136
    4 Additional Use of a Position Line 137
    5 Doubling the Angle on the Bow 137
    6 The Four-Point Bearing Problem with Leeway and Current 138
    7 Special Angles 139

Chapter 19 Position Line by Vertical Angle: Distance of the Horizon 142

Para. 1 Distance off by Vertical Angle 142
    2 Distance of the Theoretical Horizon 144

Chapter 20 Position Line by Horizontal Angle 147

Para. 1 Geometrical Principles 147
    2 Application to Fixing 148
    3 The Horizontal Danger Angle 149
    4 Fixing by Horizontal Angles 149
    5 Reliability of the Horizontal Angle Fix 149
    6 Method of Recording a Fix by Horizontal Angles 150
    7 Examples of Fixes by Horizontal Angles 150
    8 Use of Tracing Paper for Fixing by Horizontal Angles 151
    9 The Station Pointer 151

Chapter 21 The Three-Bearing Problem 154

Para. 1 Principles 154
    2 Practice 154
    3 Examples 155

Chapter 22 The Three Positions: Current Sailing 157

Para. 1 The Three Positions 157
    2 Examples 158
    3 Current Sailing 158

CONTENTS

Chapter 23 Position Line by Radio Bearing 162

Para. 1 Introduction 162
2 Convergency of the Meridians 162
3 The Half-Convergency Correction 163

Chapter 24 Tides 166

Para. 1 The Tide 166
2 The Equilibrium Theory of the Tide 167
3 Effect of Earth's Rotation on Tides 167
4 The Moon's Effect 168
5 The Sun's Effect 168
6 The Luni-Solar Tide 169
7 The Progressive Wave Theory of the Tide 170
8 The Standing Wave Theory of the Tide 170
9 Priming and Lagging of the Tides 170
10 Tidal Streams 171
11 Practical Tide Problems 171

PART 4

GENERAL ASTRONOMY

Chapter 25 The Universe 175

Para. 1 The Stars 175
2 The Solar System 176
3 Kepler's Laws of Planetary Motion 183

Chapter 26 The Earth's Motions and the Seasons 187

Para. 1 The Earth's Axial Motion 187
2 The Earth's Orbital Motion 187
3 The Celestial Sphere 187
4 The Seasons 189
5 Unequal Lengths of Daylight and Darkness During the Year 190
6 Climatic Zones 192
7 Unequal Lengths of the Seasons 193
8 The Zodiacal Belt 193
9 Planetary Orbits 194
10 Direct and Retrograde Motion 195

Chapter 27 Defining Celestial Positions 196

Para. 1 The Cartesian System of Co-ordinates 196
2 The Ecliptic System of Defining Celestial Positions 196
3 The Horizon System of Defining Celestial Positions 197
4 The Celestial Equatorial System of Defining Celestial Positions 198

CONTENTS

Chapter 28 **The Apparent Diurnal Motion of Celestial Bodies** 201
  Para. 1 Diurnal Circles 201

Chapter 29 **Time** 205
  Para. 1 The Units of Time 205
       2 Time at an Instant 207
       3 The Equation of Time 209
       4 Comparison of Solar and Sidereal Time Units 211
       5 Time and Longitude 212
       6 Time at Sea 212
       7 Standard Time 213
       8 The Years 213
       9 Co-ordinated Universal Time 214
     10 The Calendar 215

Chapter 30 **The Moon** 218
  Para. 1 The Moon and its Motions 218
       2 The Phases of the Moon 219
       3 The Age of the Moon 220
       4 Winter and Summer Full Moons 221
       5 Spring and Autumn Full Moons 221
       6 Earth-Shine 222
       7 Moon's Librations 222
       8 Eclipses 222
       9 Occultations 224

PART 5

NAUTICAL ASTRONOMY

Chapter 31 **Finding the True Altitude** 229
  Para. 1 The Altitude 229
       2 Dip 229
       3 Refraction 230
       4 Effect of Refraction on Dip 231
       5 Semi-Diameter 232
       6 Augmentation of the Moon's Semi-Diameter 232
       7 Parallax 232
       8 Effect of Earth's Shape on Horizontal Parallax 234
       9 Irradiation 235
     10 The Correction of Observed Altitudes 235
     11 Correcting Star Altitudes 235
     12 Correcting Planet Altitudes 236
     13 Correcting Sun Altitudes 237
     14 Correcting Moon Altitudes 238
     15 Back Angles 240
     16 The Artificial Horizon 242
     17 The Bubble Attachment 242

CONTENTS

Chapter 32   **The Astronomical Position Line**   245

Para. 1   The Geographical Position of a Heavenly Body   245
       2   Circles of Equal Altitude   249

Chapter 33   **Meridian Altitude Observations**   253

Para. 1   Latitude by Meridian Altitude   253
       2   Latitude by Meridian Altitude of a Body at Lower Meridian Passage   255
       3   Effect of Observer's Motion on Meridian Altitude Observations   255
       4   G.M.T. of Sun's Meridian Passage   257
       5   G.M.T. of Moon's Meridian Passage   258
       6   G.M.T. of Planet's Meridian Passage   260
       7   G.M.T. of Star's Meridian Passage   261
       8   Position Line from Meridian Altitude Observation   262

Chapter 34   **The Astronomical Triangle and Sight Reduction**   268

Para. 1   Introduction   268
       2   The Longitude Method   269
       3   General Remarks on the Longitude Method   271
       4   The Longitude Method in Practice   273
       5   The Intercept Method   278
       6   The Intercept Method in Practice   279
       7   The Modified Formula   283
       8   The Azimuth   286
       9   Compass Error by Azimuth   287
     10   Compass Error by Amplitude   287

Chapter 35   **Position Lines and Plotting Chart**   291

Para. 1   Introduction   291
       2   Error in Longitude due to Error in Latitude   292
       3   Effects of Errors in Altitude   293
       4   The Plotting Chart   294

Chapter 36   **Celestial Bodies Near the Meridian**   299

Para. 1   General Remarks   299
       2   Position Line from Ex-Meridian Observation   301
       3   Ex-Meridian Observation Using the Intercept Method   302
       4   Ex-Meridian Observation Using Napier's Rules   303
       5   Reduction to the Meridian   304
       6   Concluding Remarks on the Ex-Meridian Problem   306

Chapter 37   **The Pole Star**   308

Para. 1   Latitude from Observation of Polaris   308
       2   The Pole Star Altitude Tables   309
       3   The Pole Star Azimuth Table   311

CONTENTS

Chapter 38   Nautical Astronomical Tables   314
  Para. 1   Introduction   314
       2   The Principles of ABC Tables   314
       3   Uses of the ABC Tables   316
       4   Inspection Tables   318
       5   Short Method Tables   320
       6   Ogura's Table   320

Chapter 39   Rising and Setting Phenomena   323
  Para. 1   Introduction   323
       2   Sunrise and Sunset   323
       3   Moonrise and Moonset   324
       4   Twilight   325
       5   The Midnight Sun   327

Chapter 40   Rate of Change of Hour Angle, Azimuth and Altitude   328
  Para. 1   Rate of Change of Hour Angle   328
       2   Rate of Change of Azimuth   328
       3   Rate of Change of Altitude   330
       4   Interval Between Maximum and Meridian Altitudes of the Sun   332

PART 6
THE INSTRUMENTS OF NAVIGATION

Chapter 41   The Magnetic Compass   337
  Para. 1   Magnetism   337
       2   Terrestrial Magnetism   337
       3   Variation   338
       4   The Magnetic Compass   338
       5   Ship Magnetism   339
       6   Simple Ideas on Compass Adjustment   340
       7   The Azimuth Mirror   341

Chapter 42   The Gyroscopic Compass   343
  Para. 1   Introduction   343
       2   The Principles of the Gyroscope   343
       3   Tilt and Drift   349
       4   Tilting and Drifting   350
       5   The Gravity Controlled Gyroscope   351
       6   The Liquid Gravity Control   352
       7   The Movement of the Axle of a Gravity Controlled Gyroscope   352
       8   Damping   354
       9   The Natural Errors of the Gyro Compass   354
            Latitude Error   354
            Rolling Error   355
            Latitude Course/Speed Error   355
       10   Ballistic Deflection   357
       11   Ballistic Tilt   357

CONTENTS

Chapter 43 The Sextant and Chronometer 359

Para. 1 Description of Sextant 359
2 The Sextant Telescope 359
3 The Principle of the Sextant 360
4 The Errors and Adjustments of the Sextant 361
5 Using the Sextant 363
6 Care of the Sextant 364
7 The Chronometer and its Care 364
8 Use of Chronometer 365

Chapter 44 Sounding Instruments and Logs 367

Para. 1 The Lead Line and Mechanical Sounding Machine 367
2 The Sounding Machine and Sounding Tube 367
3 The Echo-sounder 368
4 Logs 369

Chapter 45 Radio Direction Finding 371

Para. 1 The Simple Direction Finder 371
2 The Bellini-Tosi Direction Finder 372
3 Errors in Radio Direction Finding 373

Chapter 46 Hyperbolic Navigation 375

Para. 1 Introduction 375
2 The Hyperbola 375
3 Principles of Hyperbolic Navigation 376
4 The Nature of Hyperbolic Position Lines 376
5 Consol 377
6 Loran 378
7 Decca Navigator 379
8 Omega 380

Chapter 47 Radar Navigation 382

Para. 1 Principles of Radar 382
2 The Use of Primary Radar in Navigation 382
3 Radar Beacons 383

Chapter 48 Navigational Satellite and Inertial Navigation 385

Para. 1 Introduction 385
2 Navigational Satellites 385
3 Inertial Navigation 387

Extracts from *Admiralty Tide Tables* and *The Nautical Almanac* 389

Answers to Exercises 417

Index 427

PART 1

THE MATHEMATICS OF NAVIGATION AND NAUTICAL ASTRONOMY

Navigation embraces the techniques used by a navigator for finding the distance to travel and the direction to steer from one given position to another. The basic instruments of navigation are the chart, on which positions and course-lines are plotted; the log, by which distances travelled through the water are measured; and the compass, by which horizontal directions are indicated or ascertained. Navigating by log and compass—a technique known as dead reckoning—yields approximate and usually unreliable results, so that the prudent navigator loses no opportunity of checking his ship's position to reassure himself that his vessel is never on a dangerous course.

To determine the position of his vessel when the land is in sight observations of charted land- and sea-marks, such as lighthouses, buoys and beacons, and the application of simple geometrical principles, permit the navigator to fix his ship's position with relative ease. In recent decades a wide range of electronic and radio instruments is available for determining a ship's position even when the land is not in sight. In the absence of such aids the traditional methods of finding latitude and longitude are those which employ astronomical principles: these methods collectively form the science of nautical astronomy.

To understand navigation and nautical astronomy a thorough knowledge of elementary mathematics—particularly geometry and trigonometry—is essential. It is upon these branches of mathematics that the whole structure of navigation and nautical astronomy has been built. Part I deals essentially with elementary trigonometry, the application of which to the work of a navigator is easy and interesting only if it is properly comprehended.

CHAPTER 1

PLANE TRIGONOMETRY

1. Introduction

The branch of mathematics which deals with the calculation of the unknown parts of a triangle is known as trigonometry. The word *trigonometry*—usually frightening to a beginner—means nothing more than triangle-measurement. An important part of a navigator's work involves solving triangles, so that it is essential for a student of navigation to have a full knowledge of the elements of this important branch of mathematics.

Trigonometry is divided into two parts. The part which deals with the mathematics of plane triangles is known as plane trigonometry: that which deals with spherical triangles is known as spherical trigonometry.

Of the six parts—three angles and three sides—of a plane triangle, provided that at least one of three known parts is a side, the unknown parts may be found geometrically by construction, or trigonometrically by calculation. The calculation of the unknown parts of a triangle is facilitated by using tables of trigonometrical functions. A function in mathematics is a quantity whose value depends upon the value of some other quantity. We say, for example, that the speed of a ship through the water is dependent upon her displacement, so that speed is a function of displacement. It is also a function of any of a variety of other factors, such as the rate at which the propeller is revolving, and the rate at which fuel is being consumed.

A trigonometrical ratio is a function of an angle. In a right-angled triangle which contains an acute angle θ, the trigonometrical ratios of the angle θ are the numerical comparisons between the lengths of pairs of sides of the triangle. Because a triangle has three sides there are, accordingly, six trigonometrical ratios. These are known, respectively, as the sine, cosine, tangent, cotangent, secant and cosecant.

In the right-angled triangle illustrated in fig. 1·1, the trigonometrical ratios of the angle θ are:

The ratio $a:c$, that is a/c, is the sine of θ, or $\sin \theta$

The ratio $b:c$, that is b/c, is the cosine of θ, or $\cos \theta$

The ratio $a:b$, that is a/b, is the tangent of θ, or $\tan \theta$

The ratio $b:a$, that is b/a, is the cotangent of θ, or $\cot \theta$

The ratio $c:b$, that is c/b, is the secant of θ, or $\sec \theta$

The ratio $c:a$, that is c/a, is the cosecant of θ, or $\csc \theta$.

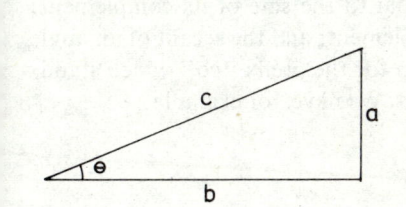

Fig. 1·1

In addition to these six trigonometrical ratios there are two other trigonometrical functions known, respectively, as the versine and the haversine. These are of considerable use in the practice of navigation and nautical astronomy.

versine (or vers) $\theta = 1 - \cos\theta$

haversine (or hav) θ = half vers θ

$$= \tfrac{1}{2}(1 - \cos\theta)$$

The trigonometrical functions of an angle are dependent solely upon the magnitude of the angle.

Referring to fig. 1·2, suppose the line AX to be rotated about the point A so that any acute angle θ is swept out. Perpendiculars dropped from any points B and D on AY onto AX will form two similar or equiangular triangles ABC and ADE respectively. Now the ratio between any two corresponding sides of similar triangles is a constant amount. It follows that:

$$BC/AB = DE/AD = \sin\theta$$
$$AC/AB = AE/AD = \cos\theta$$
$$BC/AC = DE/AE = \tan\theta$$

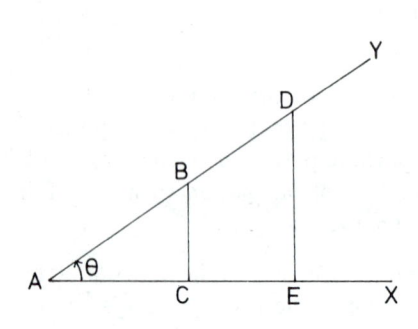

Fig. 1·2

2. Complementary Angles

Two angles are said to be complementary when their sum is 90°. Angles of 30° and 60° are complementary, each being the complement of the other. Because the sum of the three angles of a plane triangle is 180°, it follows that the two non-90° angles of a right-angled triangle are complementary.

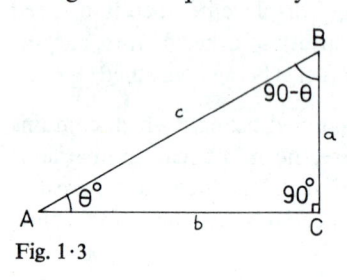

In the right-angled triangle illustrated in fig. 1·3, the angle BAC is denoted by θ. The angle ABC, therefore, is $(90° - \theta)$.

It is readily seen that:

$$\left\{\begin{array}{l} \sin\ \theta = a/c = \cos(90° - \theta) \\ \tan\ \theta = a/b = \cot(90° - \theta) \\ \sec\ \theta = c/b = \operatorname{cosec}(90° - \theta) \end{array}\right\}$$

Fig. 1·3

Thus, the sine of an angle is equal to the cosine of its complement; the tangent of an angle is equal to the cotangent of its complement; and the secant of an angle is equal to the cosecant of its complement. Similarly, the cosine of an angle is equal to the sine of its complement; the cotangent of an angle is equal to the tangent of its complement; and the secant of an angle is equal to the cosecant of its complement. This is the reason for the prefix "co" which stands for complement, in the names cosine, cotangent and cosecant. We have, for example:

$$\sin 55° = \cos 35°$$
$$\tan 60° = \cot 30°$$
$$\sec 24° = \operatorname{cosec} 66°$$
$$\cos 67° = \sin 23°$$
$$\cot 86° = \tan 4°$$
$$\operatorname{cosec} 14° = \sec 76°$$

3. Trigonometrical Functions as Straight Lines

Sine and Cosine. Suppose the radius of the circle illustrated in fig. 1·4 to be of any unit length. Let the radius AB sweep out any acute angle θ. Because the radius of the circle is unity the length of the side DC in the right-angled triangle ACD is the sine of the angle θ. That is:

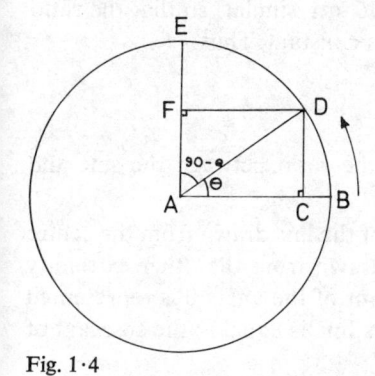

$$\sin \theta = DC$$

The sine of an arc or angle may, therefore, be defined as the length of a perpendicular dropped from one extremity of an arc of unit radius onto the diameter of the circle of which the arc forms part drawn through the other extremity.

If the angle BAE in fig. 1·4 is 90°, the angle EAD is the complement of θ. The sine of the complement of θ, that is to say, $\cos \theta$, is denoted in fig. 1·4 by the line DF. But DF is equal in length to AC, so that:

$$\cos \theta = AC$$

Fig. 1·4

The sine of 0° is zero because when θ is 0° the length of the perpendicular DC is zero. The cosine of 0° is unity because when θ is 0° the length of AC is equal to the radius of the circle, which is unity.

The value of the sine of an angle increases from 0 to 1 as the angle increases from 0 to 90°, but the value of the cosine of an angle decreases from 1 to 0 as the angle increases from 0 to 90°.

If, in fig. 1·4, distances measured to the right of A in the direction of B are designated positive, then distances measured to the left of A will be designated negative. Again, if distances measured from A in the direction of E are designated positive, distances measured in the opposite direction will be designated negative. It follows that the sines and cosines of all acute angles are positive. But consider the situation for angles greater than 90°.

The sine of an angle decreases from $+1$ to 0 as the angle increases from 90° to 180°, and the cosine decreases from 0 to -1 as the angle increases from 90° to 180°. As the angle increases from 180° to 270° the sine decreases from 0 to -1 and the cosine increases from -1 to 0. As the angle increases from 270° to 360°, to complete the circle, the sine increases from -1 to 0 and the cosine increases from 0 to $+1$.

The graphs of the sine and cosine of angles between 0° and 360° are illustrated in fig. 1·5. The two curves are said to be out of phase with each other to the extent of 90°. In other words if the cosine curve is moved 90° to the right it will coincide with the sine curve.

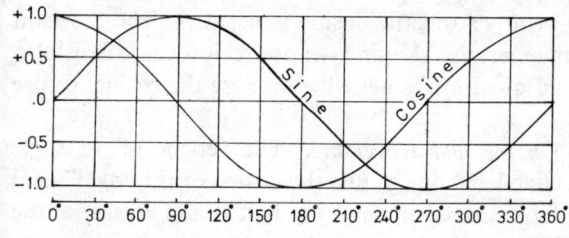

Fig. 1·5

Tangent and Cotangent. Suppose the radius of the arc illustrated in fig. 1·6 is of unit length. Let the radius AB rotate about A to form the acute angle θ. Let a straight line drawn tangentially to the arc at B cut the radius AD produced at G. The radius of the arc is unity, so that the length of the line BG is the tangent of the angle θ.

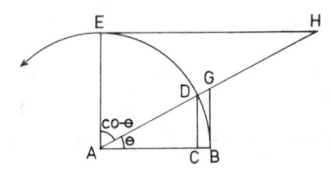

Fig. 1·6

If a straight line be drawn tangentially to the arc at *E*, which is 90° from *B*, to cut the radius *AD* produced at *H*, the length of the line *EH* is the cotangent of the angle θ. In other words *EH* is the tangent of the complement of the angle θ.

The triangles *ACD* and *ABG* are similar, so that the ratio between corresponding sides is constant. Thus:

$$CD/AC = BG/AB$$

or:

$$\sin θ/\cos θ = \tan θ/1$$

The tangent of an angle may, therefore, be defined as the ratio between the sine and cosine of the angle.

Secant and Cosecant. The secant of an angle is the length of the line drawn from the centre of a circle through one extremity of an arc to the tangent drawn from the other extremity of the same arc in a circle of unit radius. In fig. 1·6 the secant of the angle θ is represented by the line *AH*. It should also be noted that the length of this line is equal to the cosecant of the complement of θ.

The triangles *ACD* and *ABG* are similar, so that:

$$AG/AB = AD/AC$$

and:

$$\sec θ/1 = 1/\cos θ$$

The secant of an angle may, therefore, be defined as the reciprocal of the cosine of the angle. Similarly, the cosecant of an angle may be defined as the reciprocal of the sine of the angle.

4. The Signs of the Trigonometrical Ratios of Angles between 90° and 180°

In the practice of navigation, when solving triangles, angles having values of up to 180° only are involved. For this reason angles over 180° need never be considered in the practical work of solving triangles.

Two angles are said to be supplementary when their sum is 180°. It will be noted from the graphs of the sines and cosines of angles illustrated in fig. 1·5, that the sine of an angle is equal to the sine of its supplement, and that the cosine of an angle is equal to minus the cosine of its supplement.

In the second quadrant, which refers to angles between 90° and 180°, the tangent and the cotangent are negative because the sine is positive and the cosine is negative. The cosecant of an angle in the second quadrant is positive because the sine, whose reciprocal it is, is also positive. The secant of an angle in the second quadrant is negative because the cosine, whose reciprocal it is, is also negative.

Versine and Haversine. The versine of an angle is defined as "one minus the cosine of the angle", and the haversine of an angle is a half of the versine of the angle.

It is noticed from fig. 1·7 that when the angle θ increases from 0° to 180°, the versine of θ—which is equal to the length of the line *BC*—increases from 0 to +2.

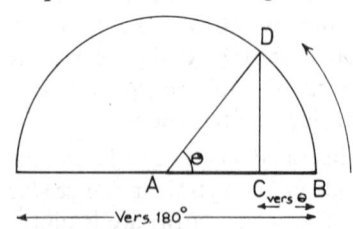

Fig. 1·7

PLANE TRIGONOMETRY

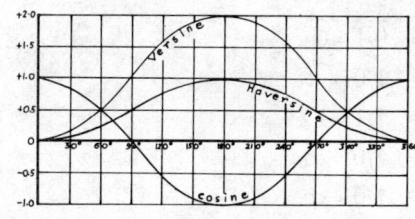

Fig. 1·8 illustrates the graphs of the versines and haversines, and the cosines, of angles between 0° and 360°.

The principal feature of the functions versine and haversine is that they are always positive. In contrast to cosines, which in the second quadrant are negative, versines and haversines are easily handled in computations because they are always positive. It is for this reason that they are used in preference to cosines for solving certain navigational triangles.

Fig. 1·8

5. The Standard Formulae

From the foregoing remarks it will be observed that sines and cosines are lengths of lines *within* a circle; tangents and cotangents are lengths of lines which *touch* a circle; and that secants and cosecants are lengths of lines which *cut* a circle.

Consider the triangles *ACD*, *ABG*, and *AEH*, in fig. 1·9. These are similar triangles, so that:

$$CD/AC = BG/AB = AE/EH$$

or: $\sin \theta / \cos \theta = \tan \theta = 1/\cot \theta$

Also: $CD/AD = BG/AG = AE/AH$

or: $\sin \theta / 1 = \tan \theta / \sec \theta = 1/\operatorname{cosec} \theta$

Also: $AC/AD = AB/AG = EH/AH$

or: $\cos \theta / 1 = 1/\sec \theta = \cot \theta / \operatorname{cosec} \theta$

Fig. 1·9

By Pythagoras' Theorem:

$$CD^2 + AC^2 = AD^2$$

so that: $\sin^2 \theta + \cos^2 \theta = 1$

Also: $$BG^2 + AB^2 = AG^2$$

so that: $\tan^2 \theta + 1 = \sec^2 \theta$

Also: $$EH^2 + AE^2 = AH^2$$

so that: $1 + \cot^2 \theta = \operatorname{cosec}^2 \theta$

The above formulae are known as the Standard Formulae. They are sometimes useful in navigational work, and are worth remembering.

6. Special Angles

It is a comparatively easy matter to derive the trigonometrical ratios of 0°, 30°, 45°, 60° and 90°. It is useful to memorise the ratios of these so-called "special angles".

It has already been established that:

$$\sin 0° = 0$$
$$\cos 0° = 1$$
$$\sin 90° = 1$$
$$\cos 90° = 0$$

8   THE ELEMENTS OF NAVIGATION AND NAUTICAL ASTRONOMY

It follows that:

$$\tan 0° = \sin 0°/\cos 0° = 0/1 = 0$$
$$\tan 90° = \sin 90°/\cos 90° = 1/0 = \infty$$
$$\sec 0° = 1/\cos 0° = 1/1 = 1$$
$$\sec 90° = 1/\cos 90° = 1/0 = \infty$$
$$\operatorname{cosec} 0° = 1/\sin 0° = 1/0 = \infty$$
$$\operatorname{cosec} 90° = 1/\sin 90° = 1/1 = 1$$
$$\cot 0° = 1/\tan 0° = 1/0 = \infty$$
$$\cot 90° = 1/\tan 90° = 1/\infty = 0$$

The trigonometrical ratios of 30° and 60° may be found as follows:

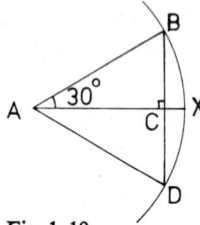

Let the radius of the circular arc illustrated in fig. 1·10 be of any unit length. Let AX be rotated about A through an angle of 30° to AB. From B drop a perpendicular onto AX and produce to D on the arc. Join AD. In the triangles ABC and ACD:

$$AC = AC \text{ (common)}$$
$$AB = AD \text{ (radii)}$$
$$ACB = ACD \text{ (90°)}$$

Fig. 1·10

The triangles ABC and ACD are, therefore, congruent triangles. Also the triangle ABD is equilateral.

Therefore:
$$BC = \tfrac{1}{2} BD = \tfrac{1}{2} AB$$

Also $ ABC = 60°$

Now $ AB = 1$

Thus, $ BC = \tfrac{1}{2}$

By Pythagoras' Theorem:
$$AC^2 = AB^2 - BC^2$$
$$= 1^2 - \tfrac{1}{2}^2$$
$$= \tfrac{3}{4}$$

Thus $ AC = \sqrt{3}/2$

It follows that:

$$\sin 30° = BC = \tfrac{1}{2} = \cos 60°$$
$$\sin 60° = AC = \sqrt{3}/2 = \cos 30°$$
$$\tan 30° = \sin 30°/\cos 30° = 1/\sqrt{3} = \cot 60°$$
$$\tan 60° = \sin 60°/\cos 60° = \sqrt{3} = \cot 30°$$
$$\operatorname{cosec} 30° = 1/\sin 30° = 2 = \sec 60°$$
$$\operatorname{cosec} 60° = 1/\sin 60° = 2/\sqrt{3} = \sec 30°$$

The trigonometrical ratios of 45° may be found as follows.

PLANE TRIGONOMETRY

Let the radius of the circular arc depicted in fig. 1·11 be of any unit length. Rotate OX about O to OA so that the angle AOB is 45°. The triangle OAB is, therefore, isosceles, and AB is equal to OB. By Pythagoras' Theorem:

$$OA^2 = AB^2 + OB^2$$
$$= 2AB^2$$

Thus: $AB^2 = OA^2/2$

and $AB = 1/\sqrt{2} = OB$

Therefore: $\sin 45° = \cos 45° = 1/\sqrt{2}$

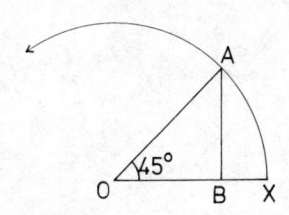

$\tan 45° = \cot 45° = 1$

$\sec 45° = \operatorname{cosec} 45° = \sqrt{2}$

Fig. 1·11

Exercises on Chapter 1

1. Find by scale drawing:
   (i) tan 26°
   (ii) cos 62°
   (iii) sec 48°

2. Find by scale drawing the acute angle whose:
   (i) cotangent is 0·7
   (ii) secant is 0·4
   (iii) sine is 0·6

3. If the sine of an acute angle is 0·4 find, without using tables, the remaining five trigonometrical ratios of the same angle.

4. Explain why cos 56° equals sin 34°.

5. Prove: (i) $\sin^2 A + \cos^2 A = 1$
   (ii) $\operatorname{cosec}^2 a - 1 = \cot^2 a$

6. Show that tan 100° is equal to negative tan 80°.

7. Prove: $\cos 30° = \sqrt{3}/2$.

8. Explain what is meant by the versine of an angle.

9. What is the principal advantage of a versine table compared with a cosine table?

10. If the haversine of an angle is 0·6 find the angle by scale drawing.

11. Derive the trigonometrical ratios of 0°, 90°, 180° and 270°.

12. Prove: $\sin 45° \operatorname{cosec} 45° \sec 30° \sin 60° = 1$.

13. Prove graphically that sin (180° − A) is equal to sin A given that A is an acute angle.

14. Demonstrate graphically that the secant of 38° is approximately equal to the tangent of 38°.

15. Find angle θ given that: $\operatorname{vers} \theta = \tfrac{3}{4} \cos \theta$.

CHAPTER 2

CIRCULAR MEASURE

1. The Radian

The length of the circumference of any circle is a constant number of times the length of the diameter of the same circle. This constant number is denoted by the Greek letter Pi or π. It is an incommensurable quantity; but, to four places of decimals, it is:

$$\pi = 3 \cdot 1416$$

$$= 22/7 \text{ approximately}$$

It follows that the length of the circumference of a circle is 2π times the length of the radius of the same circle. Thus, if the radius is fitted around the circumference, as shown in fig. 2·1, it will be found that 2π, or 6·28 .. radii will equal the length of the circumference.

Since the angle subtended at the centre of a circle by the circumference is 360°, it is evident that the angle subtended by an arc whose length is equal to the radius of the circle is $360/2\pi$, or $360/6 \cdot 28 ..$ degrees. This angle is 57°·3 or 3438′ approximately. It is the unit of circular measure known as the radian, the symbol for which is a small letter c.

Circular measure provides a method of measuring or denoting angles which simplifies the solutions of certain navigational problems. Thus, instead of saying that the angle at the centre of a circle subtended by the circumference is 360°, we may say that it is 2π radians.

Thus:

360° is equivalent to 2π radians

180° is equivalent to π^c

and 57°·3 is equivalent to 1^c approximately.

To find the number of radians corresponding to a given number of degrees, the latter is divided by 57·3. To find the number of radians in a given number of minutes of arc the latter is divided by 3438.

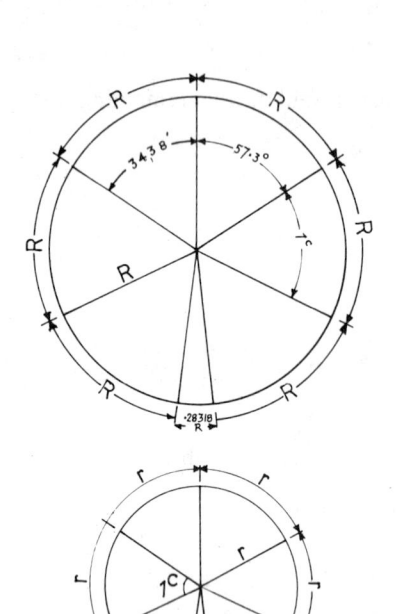

Fig. 2·1

CIRCULAR MEASURE

For example:

$$360° = 360/57 \cdot 3 = 6 \cdot 28 \ldots^c$$
$$90° = 90/57 \cdot 3 = 1 \cdot 57 \ldots$$
$$121° = 121/57 \cdot 3 = 2 \cdot 11 \ldots^c$$
$$5° \, 43' \cdot 8 = 343' \cdot 8 = 343 \cdot 8/3438 = 0 \cdot 1^c$$

If the length of the arc of a circle of given radius is known it is an easy matter to find the number of radians contained at the centre of the circle subtended by that arc. A circle of radius 10 inches has a circumference of $2\pi \cdot 10$ inches or $62 \cdot 83''$. The circular measure of the angle at the centre subtended by the circumference is $6 \cdot 28 \ldots^c$. An arc which is one-quarter of the circumference in length, that is to say $15'' \cdot 707$, subtends an angle which is equal to one-quarter of $6 \cdot 28 \ldots$, which is $1 \cdot 57 \ldots^c$. This may be found by dividing the length of the arc by that of the radius. Similarly, an arc of $23''$ in a circle of radius $10''$ subtends an angle of $23/10$, or $2 \cdot 3^c$.

Thus:

Circular Measure = Arc length/Radius

or

Arc length = Circular Measure × Radius

Example 2·1—A circle has a radius of $20 \cdot 0$ miles. What length of its perimeter subtends an angle of $2 \cdot 3^c$ at its centre?

$$\text{Arc length} = \text{Circular Measure} \times \text{Radius}$$
$$= 2 \cdot 3 \times 20 \cdot 0$$
$$= 46 \cdot 0$$

Answer—Length = $46 \cdot 0$ miles.

Example 2·2—Find the radius of a circle if an arc of $3 \cdot 4''$ subtends an angle of $1 \cdot 60$ radians at its centre.

$$\text{Radius} = \text{Arc length}/\text{Circular Measure}$$
$$= 3 \cdot 4/1 \cdot 60$$
$$= 2 \cdot 12 \ldots$$

Answer—Radius = $2 \cdot 12 \ldots$ inches.

The following examples illustrate some of the practical applications of circular measure to navigation.

Example 2·3—A headland is kept abeam at a distance of $5 \cdot 0$ miles. Find the distance travelled by the ship in changing the bearing of the headland $95°$.

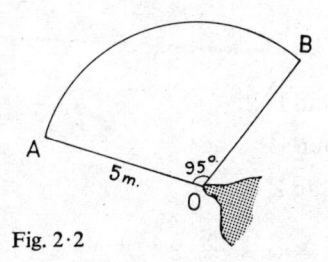

$$\text{Radius} = 5 \cdot 0 \text{ miles}$$
$$AOB = 95°$$
$$= 95/57 \cdot 3 \text{ radians}$$
$$\text{Arc } AB = \text{Radius} \times AOB^c$$
$$= 5 \cdot 0 \, (95/57 \cdot 3)$$
$$= 8 \cdot 3$$

Fig. 2·2

Answer—Distance travelled = $8 \cdot 3$ miles.

Example 2·4—The horizontal angle between the extremities of a small circular island having a diameter of 1·1 miles is 5° 30'. Find the distance off.

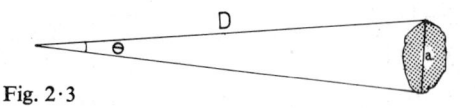

$\theta = 5° 30' = 330/3438^c$

$a = 1·1$ miles

$D = a/\theta = 1·1 (3438/330) = 11·5$

Fig. 2·3

Answer—Distance off = 11·5 miles.

2. Trigonometrical Ratios of Small Angles

Circular measure is particularly useful when dealing with very small angles because, as fig. 2·4 shows, the lengths of the sine, tangent and arc, of a small angle, are very nearly equal to one another. The smaller the angle the more nearly so is this. In practice the degree of smallness of the angle involved determines the accuracy required in the final result. In general, for navigational purposes, angles of less than about 6° may be considered to be small.

Referring to fig. 2·4 it may readily be seen that the arc *BD* is very nearly equal to each of the lines *CD* and *BE*.

Now:

$$BD/OB = BOD^c$$

$$BE/OB = \tan BOD$$

$$CD/OB = \sin BOD$$

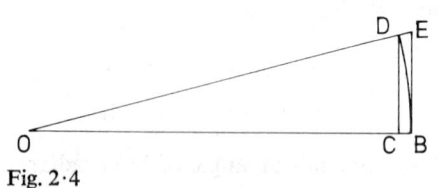

If arc *BD* and lines *BE* and *CD* are considered to be equal in length; then, for practical purposes:

$$BD/OB = BE/OB = CD/OB$$

Fig. 2·4

Therefore: $\theta^c = \tan \theta = \sin \theta$

This means that when the circular measure of a small angle is required, the sine or tangent of the angle may be used without introducing material error. Conversely, when the sine or the tangent of a small angle is required, the angle in radians may be used.

For small values of θ the graphs of sine θ and tangent θ, against angle θ, are almost coincident straight lines. It follows that the sine or tangent of a small angle is proportional to the angle itself. That is to say:

$$\sin \theta \propto \theta$$

$$\tan \theta \propto \theta$$

It follows that:

$$\sin 1°/1° = \sin \theta/\theta$$

and,

$$\tan 1°/1° = \tan \theta/\theta$$

In other words:

$\sin \theta \simeq \theta° \sin 1°$ and $\tan \theta \simeq \theta^c \tan 1°$

or, $\sin \theta \simeq \theta' \sin 1'$ and $\tan \theta \simeq \theta' \tan 1'$

or, $\sin \theta \simeq \theta'' \sin 1''$ and $\tan \theta \simeq \theta'' \tan 1''$

The following examples should be studied carefully.

CIRCULAR MEASURE

Example 2·5—The sine of 1° to four decimal places is 0·0175. Find the sine of 4° and verify from trigonometrical tables that your answer is correct to three decimal places.

$$\sin 1° = 0·0175 \text{ to 4 decimal places}$$

Thus:
$$\sin 4° = 4 \times 0·0175$$
$$= 0·070 \text{ to 3 decimal places}$$

From tables:
$$\sin 4° = 0·0698 \text{ to 4 decimal places}$$

Example 2·6—Find without using tables:

(i) $\qquad\qquad\qquad\qquad\sin 0° \ 30'$

(ii) $\qquad\qquad\qquad\qquad\cot 0° \ 20'$

(i) $\qquad\qquad\sin \theta = \theta^c$ when θ is small

Therefore:
$$\sin 30' \simeq 30/3438 = 0·0088 \text{ approximately}$$

(ii)
$$\cot 20' = 1/\tan 20'$$
$$\tan 20' \simeq 20/3438$$

Therefore:
$$\cot 20' = 3438/20$$
$$= 171·9 \text{ approximately}$$

Answers—$\sin 30' = 0·0088$
$\qquad\qquad\cot 20' = 171·9$.

Exercises on Chapter 2

1. Find the values of the following angles in radians: (i) 137°, (ii) 59° 51′, (iii) 37′ 52″.
2. Given the following arc lengths and corresponding radii, find the angles in radians:
   (i) Arc length 32″, radius 15″.
   (ii) Arc length 7·2 miles, radius 3·73 miles.
   (iii) Arc length 120·0 feet, radius 0·073 nautical miles (*Note:* 1 nautical mile = 6080 feet).
3. Find the length of an arc subtended by an angle of $2·731^c$ in a circle of radius (i) 17·2 cm., (ii) 100·0 yards.
4. Prove that: $a = r\theta$ where a is the length of an arc of a circle of radius r, and θ is the angle subtended by the arc in circular measure.
5. Show that if θ is a small angle:
$$\sin \theta \simeq \tan \theta \simeq \theta^c$$
6. A nautical mile on a spherical Earth is defined as the length of an arc of a great circle of the Earth, the extremities of the arc subtending an angle of 1′ at the Earth's centre. What is the Earth's diameter in nautical miles?
7. A ship is conned around a point of land at a constant distance of 3·5 miles. Find the distance between the instants when the point bore respectively N. 20° E. and S. 50° E.
8. One angle of a plane triangle is $\pi/4^c$, another is $3\pi/8^c$. Find the third angle in degrees and radians.

14   THE ELEMENTS OF NAVIGATION AND NAUTICAL ASTRONOMY

9. A tower 200·0 feet high subtends a vertical angle of 1° 40′. Find the distance off without using tables.
10. Show that for a small angle θ: $\sin \theta \simeq \theta' \sin 1'$.
11. Given $\sin 1° = 0·0174524$, find approximate values of: (i) $\sin 5°$, (ii) $\operatorname{cosec} 2°$, (iii) $\tan 30'$, (iv) $\cot 89°$.
12. Given $\tan 1' = 0·0002909$, find approximate values of: (i) $\tan 1°$, (ii) $\cot 40'$. Comment upon the degree of accuracy of the answers given to questions 11 and 12.

CHAPTER 3

THE TRAVERSE TABLE AND THE SOLUTION OF PLANE RIGHT-ANGLED TRIANGLES

1. Introduction

Perhaps the most useful of all nautical tables is the Traverse Table. This table, simple in construction, is nothing more than an orderly collection of solutions of plane right-angled triangles. Although the traverse table may be used for solving any plane right-angled triangle, its principal uses in the hands of a navigator are to find:

(*a*) the direction of one given terrestrial position from another,

(*b*) the distance between two given positions on the Earth's surface,

(*c*) a ship's position after she has travelled from a given position in a given direction for a given distance.

These purely navigational problems are investigated in Part 2: in this chapter we shall examine the principles of the traverse table.

2. Plane Right-angled Triangles and their Traverse Table Solutions

The hypotenuse of any right-angled plane triangle lies opposite to the right angle, and the two non-90° angles are complementary angles.

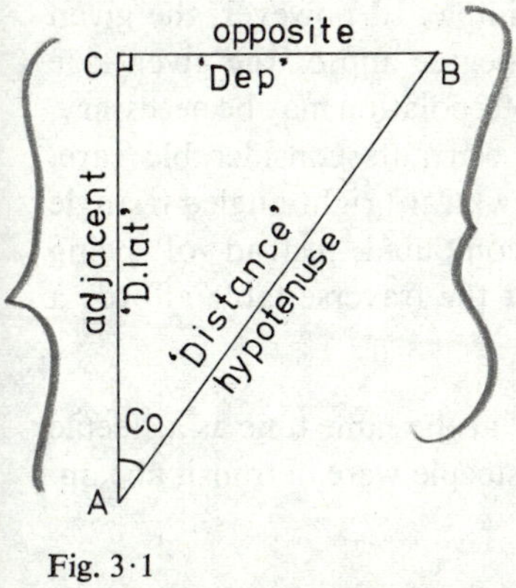

Consider the plane right-angled triangle illustrated in fig. 3·1. The lengths of the sides *AC* and *BC* are functions of the angle *A* (and angle *B*), and of the length of the hypotenuse *AB*.

Values of the sides *AC* and *BC* are tabulated in the traverse table against values of angle *A* (and *B*) and of the hypotenuse *AB*. The tabulated lengths of *AB* are usually given at intervals of one unit from 0 to 600 units.

The values of the three sides of every triangle that can be solved directly by means of the traverse table are tabulated in three vertical columns labelled Hypotenuse (Distance); Opposite (Departure); and Adjacent (D. Lat.). A complete page is given for each whole degree of angle *A* (and *B*), which is designated Course Angle.

Fig. 3·1

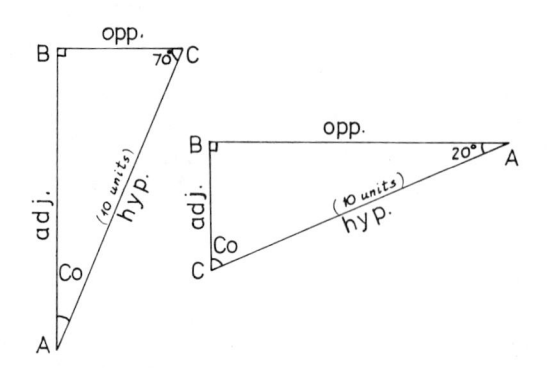

To solve a 20° right-angled triangle having a hypotenuse of 10 units, the traverse table is entered at the page corresponding to a Course Angle of 20°. Abreast of 10 in the Hypotenuse (Distance) column will be found the values of the sides adjacent and opposite to the angle 20° in the triangle to be solved. In this case, the side opposite is 3·4 units and the side adjacent to the 20° angle is 9·4 units.

Fig. 3·2   To solve a 70° right-angled triangle having a hypotenuse of length 10 units, the table may be entered at angle 20°—which is the complement of 70°—and the value of the side adjacent to the angle 70° will be found in the column labelled Opposite; and that of the side opposite to the angle 70° will be found in the column labelled Adjacent. The reason for this will be clear from an examination of fig. 3·2.

It is not necessary to extend the traverse table beyond 45°. To solve any right-angled triangle the table may be entered with the smaller of the two non-90° angles, one of which must be smaller than or equal to 45°. The values of the lengths of the opposite and adjacent sides may then be lifted from the table remembering that the larger of the two sides forming the right angle is that which faces the larger of the two non-90° angles.

To facilitate the use of the traverse table angles between 45° and 90° are printed at the bottoms of the pages, such that the sum of the angles at the top and bottom of the page is 90°. Thus, at the top of the page which is labelled 25°, the angle 65°—the complement of 25°—will be found at the bottom. The columns labelled Opposite and Adjacent, respectively, at the top of the page are labelled Adjacent and Opposite, respectively, at the bottom of the page.

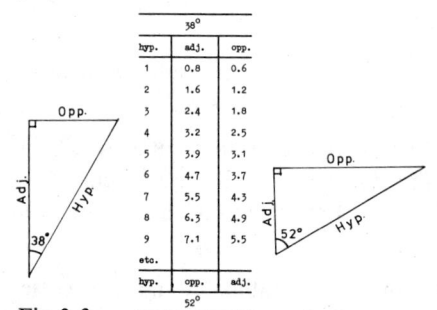

The layout of the traverse table is illustrated in fig. 3·3.

The solution of the straightforward right-angled triangle problem, in which the given angle and side have integral values, is simple. If, however, the given angle is not an integral degree and/or the given side is a fractional quantity, interpolation may be necessary, and this is tedious and warrants considerable care. It is for this reason that awkward right-angled triangle problems are usually computed instead of being solved by inspection. It should be borne in mind, however, that the traverse table affords a ready check on even complex right-angled triangle solutions.

Fig. 3·3

Example 3·1—From a vessel heading 110° a lighthouse bore 050° at the same time as a steeple bore 072°. After travelling for 8·0 miles the lighthouse and the steeple were in transit abeam. Find the distance between the lighthouse and the steeple.

Hint—Draw a diagram and then plan a solution.

THE TRAVERSE TABLE

In fig. 3·4 A and B denote the positions of the vessel at each of the times of observation. C denotes the steeple and D the lighthouse.

Plan—1. Using angle DAB and side AB in triangle DAB, find the side BD.
2. Using angle CAB and side AB in triangle ACB, find the side BC.
3. Subtract BC from BD to give the required distance.

From traverse table:
$$BD = 13·86 \text{ miles}$$
$$BC = 6·25 \text{ miles}$$
$$CD = BD - BC = 7·61 \text{ miles}$$

Answer—Distance = 7·6 miles (to the nearest 0·1 of a mile).

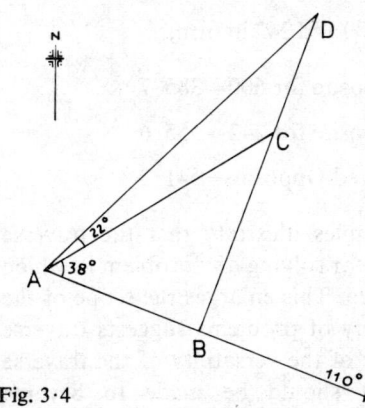

Note—To find the sides BD and BC the traverse table was entered with 80, not 8, in the column labelled Adjacent. The decimal point was then shifted one place to the left to give the result to the nearest second place of decimals.

It should be appreciated that the accuracy of the result of any computation can never be greater than that of the data used in the computation. If, for example, the distance given in Example 3·1 had been 8 miles *to the nearest mile*, BD and BC *to the nearest mile* would have been 14 and 6 respectively, and the required distance would have been 8 miles *to the nearest mile*.

Fig. 3·4

Example 3·2—The shadow of a vertical flag staff is 25 feet long at a time when the Sun's altitude is 46° 15′. Find the length of the staff.

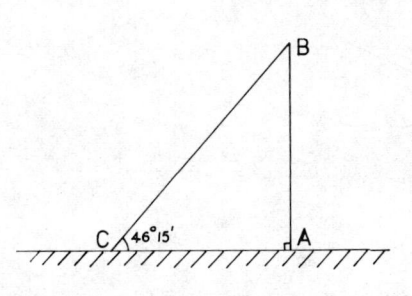

In fig. 3·5 AB is the required length.
$$AB/AC = \tan BCA$$

Therefore:
$$AB = AC \cdot \tan BCA$$
$$= 25 \times \tan 46° 15'$$

From traverse table, interpolating between angles 46° and 47°, with 25 in the column labelled Adjacent (D. Lat.), AB, from the column labelled Opposite (Departure), is 26 feet to the nearest unit.

Fig. 3·5

Answer—Length = 26 feet.

Although the Hypotenuse (Distance) column in the traverse table extends only to 600· units, this in no way limits the use of the table. When the value of the hypotenuse of a triangle exceeds 600, the other two sides may be solved by entering the table with any fraction of the hypotenuse, such as half or quarter, and then multiplying the tabulated values by the reciprocal of the fraction. An alternative method is to find the values corresponding to the hypotenuse value of 600, and to add to these the values corresponding to a hypotenuse equal to the excess of the given hypotenuse over 600.

To add the lengths of the adjacent and opposite sides in a 40° right-angled triangle having a hypotenuse of 842·0, the two methods are used with reference to fig. 3·6 (*a*) and (*b*), respectively.

Method 1—Enter traverse table at angle 40°, and with hypotenuse equal to half the given value, that is to say, with 421·0, we have:

Tabulated adjacent = 322·5. This multiplied by 2 gives 645·0.

Tabulated opposite = 270·6. This multiplied by 2 gives 541·2.

Method 2—Enter traverse table with angle 40°, and with 600 and 242 in turn.

Adjacent for 600 = 459·6	Opposite for 600 = 385·7
Adjacent for 242 = 185·4	Opposite for 242 = 155·6
Required Adjacent = 645·0	Required Opposite = 541·3

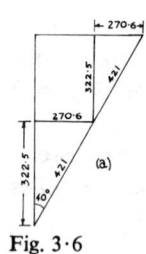

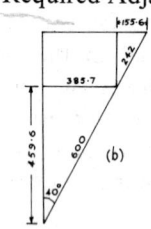

Fig. 3·6

The following examples illustrate that the traverse table may be employed for solving any problem in which a simple ratio is involved. This enlarges the scope of the table, and a wide variety of problems suggests traverse table solutions. In view of the versatility of the traverse table every endeavour should be made to become efficient in its use.

Example 3·3—Find, by means of the traverse table: (i) cot 50°, (ii) hav 32°.

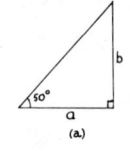

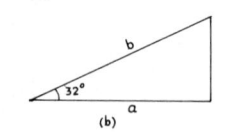

Fig. 3·7

(i) Referring to fig. 3·7 (*a*)

$$\cot 50° = a/b$$

$$= a/100 \text{ (Note the convenient denominator.)}$$

Enter traverse table at angle 50°. The value of *a* is found in the column labelled Adjacent abreast of 100 in the column labelled Opposite.

$$\cot 50° = 83·9/100$$

$$= 0·839$$

THE TRAVERSE TABLE

(ii) Referring to fig. 3·7 (*b*)

$$\text{hav } 32° = \tfrac{1}{2}(1 - \cos 32°) = \tfrac{1}{2}(1 - a/b)$$

Let $b = 100$

Enter traverse table at angle 32°. The value of *a* will be found abreast of 100 in the column labelled Hypotenuse (Distance).

Thus:
$$\begin{aligned}\text{hav } 32° &= \tfrac{1}{2}(1 - 84\cdot8/100) \\ &= \tfrac{1}{2}(1 - 0\cdot848) \\ &= \tfrac{1}{2}\cdot 0\cdot152 \\ &= 0\cdot076\end{aligned}$$

Answers—cot 50° = 0·842; hav 32° = 0·076.

Example 3·4—Find the area of a triangle *ABC* given that $AB = 14\cdot0$ feet, $AC = 20\cdot0$ feet, and angle $BAC = 58°$.

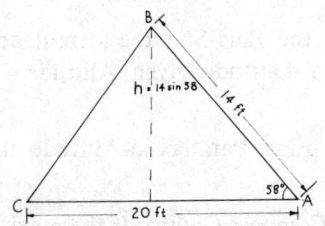

Area of triangle = ½ base × height
$$= \tfrac{1}{2}\cdot 20\cdot0\cdot 14\cdot0\cdot \sin 58°$$

From traverse table:
$$\sin 58° = 0\cdot848$$

Therefore:
$$\begin{aligned}\text{Area} &= 10\cdot0\cdot 14\cdot0\cdot 0\cdot848 \\ &= 118\cdot7 \text{ sq. ft.}\end{aligned}$$

Fig. 3·8   *Answer*—Area = 118·7 sq. ft.

Example 3·5—Given that 38 statute miles is equal to 33 nautical miles, find the number of nautical miles in 187 statute miles.

The problem is equivalent to finding the angle in a triangle such that the ratio of any two sides containing the angle is 38:33.

From the traverse table it may readily be seen that the ratio between Hypotenuse and Adjacent is 38:33 when the angle is 30°. The page of the traverse table for 30° is, therefore, a conversion table for converting nautical into statute miles or *vice versa*. Refer to fig. 3·9.

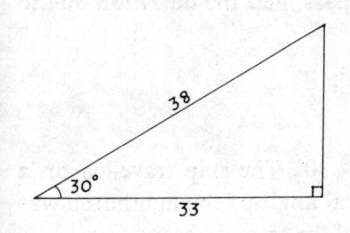

Fig. 3·9

Answer—187 statute miles = 162 nautical miles.

Exercises on Chapter 3

1. Explain clearly the construction of the traverse table. Why is it unnecessary to extend the traverse table beyond 45°?
2. Explain why the traverse table may be used to find the trigonometrical ratios of any angle. Find, by means of the traverse table, (i) sin 50°, (ii) sec 64°, (iii) tan 36°.

3. Explain how the traverse table may be used as a conversion table. Convert, using the traverse table, 18° Centigrade into degrees Fahrenheit.

4. What pages of the traverse table may be used for converting pounds weight into kilograms, given that 2·2 kg. = 1 lb. wt.

5. Given the relationship: Convergency = D. Long. . sin Lat., find Convergency if latitude is 56° and D. Long. is 16°.

6. If 84 Lire are equivalent to 26 Francs, find (i) the number of Lire in 130 Francs, and (ii) the number of Francs in 524 Lire.

7. Given the relationships: Departure = D. Long. . cos Latitude and: Error in Latitude = Error in Departure . cot Azimuth, find the Error in D. Long. if the Error in the Latitude is 11'·0, the Azimuth 047°, and the Latitude 29°.

8. Find the area of an equilateral triangle given the length of a side as 14·8 miles.

9. Two adjacent sides of a plane triangle have lengths of 10·0 feet and 8·5 feet respectively, and the included angle is 36°. Find the area of the triangle.

10. Given the relationship: D. Long./D.M.P. = Tan course, find course if D. Long. is 457' E. and D.M.P. is 642 N.

11. To find Latitude from an observation of the altitude of the Pole Star the formulae: Latitude = Altitude − c, and: c = p . cos P are used. Find Latitude given Altitude = 42° 10'; p = 61'; and P = 37° 30'.

12. Parallax-in-Altitude = Horizontal Parallax . cos Altitude. Find Parallax-in-Altitude if Altitude is 36° 00' and Horizontal Parallax is 60'·0.

13. Compass Deviation due to Force P is proportional to Sine Compass Course. If the maximum deviation due to P is 10°·0 W. on East by Compass, find the deviation due to P on (i) 030° Compass, (ii) 240° Compass.

14. Deviation due to Force Q is proportional to the Cosine of the Compass Course. If the maximum deviation due to Q is 6°·5 E. on North by Compass, find the deviation due to Q on (i) 320° Compass, (ii) 120° Compass.

15. If $x = B . \sin \theta$, and $x = 10°·0$ when $\theta = 90°$, find x when $\theta = 40°$.

16. If $y = C . \cos \theta$, and $y = 6°·5$ when $\theta = 0°$, find y when $\theta = 60°$.

17. A lighthouse bearing 010° subtended a vertical angle of 1° 30'. The ship travelled on a course of 340° for a distance of 10·0 miles, when the vertical angle of the lighthouse was again 1° 30'. Find the distance off the lighthouse at the time of the second observation.

18. The vertical angle of a cliff was 30°. At a point 100·0 feet nearer to the foot of the cliff the vertical angle was 60°. Find the height of the cliff.

19. The vertical angle of a cliff was 20°. At a point 1,000 feet nearer to the foot of the cliff it was 70°. Find the height of the cliff.

20. Find the course to steer to make good a course of due East in order to counteract the effect of a current setting due South at 4·0 knots given the speed of the ship through the water as 12·0 knots.

21. From a vessel heading due North two beacons were observed to bear 030° and 045° respectively. After having travelled for 10·0 miles the beacons were in transit abeam. Find the distance between the beacons.
22. A vertical rod 10·0 feet in length casts a shadow 4·0 feet long at noon when the Sun is at an equinox. Find the Latitude of the place.
23. Find the area of a regular pentagon the length of the side of which is 5·0 inches.
24. The length of a shadow of a vertical pole of length 10·0 feet is 20·0 feet. Find the Sun's altitude.

CHAPTER 4

COMPOUND ANGLES

A knowledge of the contents of this chapter is necessary if it is required to understand the derivation of certain navigational formulae.

1. Trigonometrical Ratios of the Sum of Two Angles

In fig. 4·1:

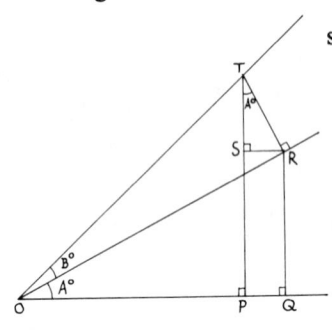

$$\sin(A+B) = \frac{TP}{OT}$$

$$= \frac{RQ + ST}{OT}$$

$$= \frac{RQ}{OT} + \frac{ST}{OT}$$

$$= \frac{RQ \cdot}{\cdot OT} + \frac{ST \cdot}{\cdot OT}$$

$$= \frac{RQ}{OR} \cdot \frac{OR}{OT} + \frac{ST}{TR} \cdot \frac{TR}{OT}$$

Fig. 4·1

That is: $\sin(A+B) = \sin A \cos B + \cos A \sin B$

$$\cos(A+B) = \frac{OP}{OT}$$

$$= \frac{OQ}{OT} - \frac{SR}{OT}$$

$$= \frac{OQ \cdot}{\cdot OT} - \frac{SR \cdot}{\cdot OT}$$

$$= \frac{OQ}{OR} \cdot \frac{OR}{OT} - \frac{SR}{TR} \cdot \frac{TR}{OT}$$

That is: $\cos(A+B) = \cos A \cos B - \sin A \sin B$

Note that if $A = B$, then:

$$\sin(A+B) = \sin 2A = 2 \sin A \cos A$$
$$\cos(A+B) = \cos 2A = \cos^2 A - \sin^2 A$$
$$= 1 - 2\sin^2 A$$
$$= 2\cos^2 A - 1$$

COMPOUND ANGLES

Also note that:
$$\sin A = \sin(A/2 + A/2) = 2 \sin A/2 \cos A/2$$
$$\cos A = \cos(A/2 + A/2) = \cos^2 A/2 - \sin^2 A/2$$
$$= 1 - 2 \sin^2 A/2$$
$$= 2 \cos^2 A/2 - 1$$
$$\text{vers } A = 1 - \cos A$$

But: $\quad 1 = \sin^2 A + \cos^2 A$

Thus: $\quad \text{vers } A = 2 \sin^2 A/2$

And: $\quad \text{hav } A = \sin^2 A/2$

2. Trigonometrical Ratios of the Difference of Two Angles

In fig. 4·2:
$$\sin(A-B) = \frac{QR}{OR}$$
$$= \frac{SP}{OR} - \frac{TR}{OR}$$
$$= \frac{SP \cdot}{\cdot OR} - \frac{TR \cdot}{\cdot OR}$$
$$= \frac{SP}{OS} \cdot \frac{OS}{OR} - \frac{TR}{RS} \cdot \frac{RS}{OR}$$

Fig. 4·2

That is: $\quad \sin(A-B) = \sin A \cos B - \cos A \sin B$

$$\cos(A-B) = \frac{OQ}{OR}$$
$$= \frac{OP}{OR} + \frac{ST}{OR}$$
$$= \frac{OP}{OS} \cdot \frac{OS}{OR} + \frac{ST}{SR} \cdot \frac{SR}{OR}$$

That is: $\quad \cos(A-B) = \cos A \cos B + \sin A \sin B$

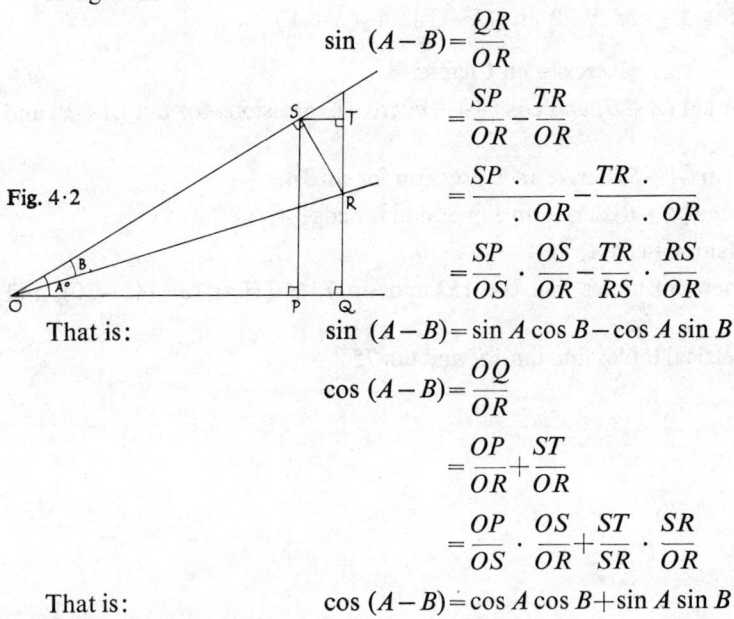

3. Products as Sums and Differences

$$\sin(A+B) = \sin A \cos B + \cos A \sin B$$
$$\sin(A-B) = \sin A \cos B - \cos A \sin B$$

By addition: $\quad \sin(A+B) + \sin(A-B) = 2 \sin A \cos A$

By subtraction: $\quad \sin(A+B) - \sin(A-B) = 2 \cos A \sin B$

$$\cos(A+B) = \cos A \cos B - \sin A \sin B$$
$$\cos(A-B) = \cos A \cos B + \sin A \sin B$$

By addition: $\quad \cos(A+B) + \cos(A-B) = 2 \cos A \cos B$

By subtraction: $\quad \cos(A+B) - \cos(A-B) = -2 \sin A \sin B$

or: $\quad \cos(A-B) - \cos(A+B) = 2 \sin A \sin B$

4. Sums and Differences as Products

Let
$$X = \tfrac{1}{2}(X+Y) + \tfrac{1}{2}(X-Y)$$
and
$$Y = \tfrac{1}{2}(X+Y) - \tfrac{1}{2}(X-Y)$$

Then:
$$\begin{aligned}
\sin X + \sin Y =\ & \sin \tfrac{1}{2}(X+Y)\cos \tfrac{1}{2}(X-Y) \\
& + \cos \tfrac{1}{2}(X+Y)\sin \tfrac{1}{2}(X-Y) \\
& + \sin \tfrac{1}{2}(X+Y)\cos \tfrac{1}{2}(X-Y) \\
& - \cos \tfrac{1}{2}(X+Y)\cos \tfrac{1}{2}(X-Y) \\
=\ & 2\sin \tfrac{1}{2}(X+Y)\cos \tfrac{1}{2}(X-Y)
\end{aligned}$$

Similarly:
$$\sin X - \sin Y = 2\cos \tfrac{1}{2}(X+Y)\sin \tfrac{1}{2}(X-Y)$$
$$\cos X + \cos Y = 2\cos \tfrac{1}{2}(X+Y)\cos \tfrac{1}{2}(X-Y)$$
$$\cos X - \cos Y = -2\sin \tfrac{1}{2}(X+Y)\sin \tfrac{1}{2}(X-Y)$$
$$\cos Y - \cos X = 2\sin \tfrac{1}{2}(X-Y)\sin \tfrac{1}{2}(X+Y)$$

Exercises on Chapter 4

1. Using the expansions for $\sin(A+B)$ and $\cos(A+B)$ derive expressions for $\tan(A+B)$ and $\cot(A+B)$.
2. Using the expansion for $\sin(A+B)$ derive an expression for $\sin 3A$.
3. Using the expansion for $\cos(A+B)$ derive an expression for $\cos 3A$.
4. Prove that: $\sin 2A + 1 = (\sin A + \cos A)^2$.
5. Without using trigonometrical tables find $\cos 15°$ and $\sin 75°$.: (*Hint* $15 = (45-30)$, and $75 = (45+30)$.)
6. Without using trigonometrical tables find $\tan 15°$ and $\tan 75°$.

CHAPTER 5

OBLIQUE-ANGLED TRIANGLES AND THEIR SOLUTIONS

1. Introduction

Any triangle which does not contain a right angle is known as an Oblique-Angled Triangle. Any oblique-angled triangle may be divided to form two right-angled triangles, simply by dropping a perpendicular from any corner of the triangle onto the opposite side or side produced. In this way an oblique-angled triangle may be solved by the methods described in Chapter 3. But such methods are indirect: in this chapter we shall be concerned with the DIRECT methods of solving oblique-angled triangles.

2. The Sine Formula

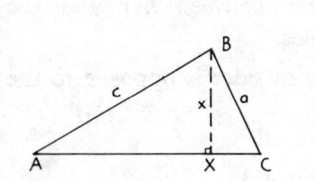

Let the triangle ABC illustrated in fig. 5·1 denote any oblique-angled triangle. From any vertex, say B, drop a perpendicular onto the opposite side or side produced to X. Let the length of this perpendicular be x. In the two right-angled triangles so formed we have:

Fig. 5·1

$$x = c \sin A$$

and $\qquad x = a \sin C$ (Remember that $\sin \theta = \sin (180 - \theta)$)

Therefore: $\quad c \sin A = a \sin C$

By dropping a perpendicular from either of the other two vertices onto the opposite side or side produced it may be shown that, in general:

$$a/\sin A = b/\sin B = c/\sin C$$

This relationship is known as the Sine Formula. Stated in words, it is:

> The ratio between any two sides of a plane triangle is equal to the ratio between the sines of the angles opposite to the respective sides.

When two sides of a plane triangle and an angle opposite to one of the sides are given, or when two angles and a side opposite to one of them are given, the unknown parts of the triangle may be solved by means of the Sine Formula.

Example 5·1—The horizontal angle at a point A between two other points B and C is 30°. The horizontal angle between A and C at point B is 100°. Find the distance between B and C if the distance between A and B is 8·20 miles.

26   THE ELEMENTS OF NAVIGATION AND NAUTICAL ASTRONOMY

Referring to fig. 5·2:

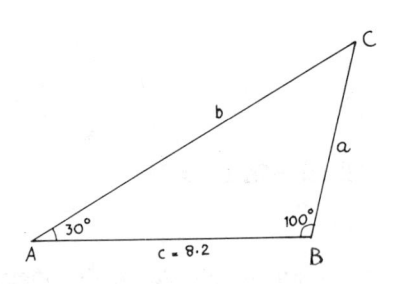

To find a:

$$a/c = \sin A/\sin C$$
$$a = c \sin A \operatorname{cosec} C$$

$$\log c = 0·91381$$
$$\log \sin A = \bar{1}·69897$$
$$\log \operatorname{cosec} C = 0·11575$$
$$\log a = 0·72853$$
$$a = 5·35$$

Fig. 5·2   *Answer*—BC = 5·35 miles.

If two sides and an angle of a plane triangle are given, and the shorter of the two given sides is opposite to the given angle, two values will satisfy each of the unknown parts of the triangle. This is known as the Ambiguous Case. This is illustrated in Example 5·3.

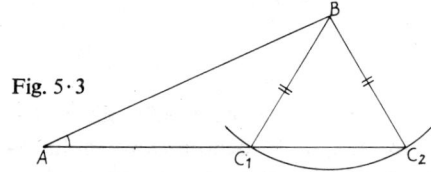

Fig. 5·3

Example 5·2—A vessel heading 340° bears 200° at a distance of 10·50 miles from a lighthouse. Find the distance the vessel travels so that the distance between her and the lighthouse is 8·00 miles.

It is evident from fig. 5·4 that, because the smaller of the given sides is opposite to the given angle, two values satisfy the solution.

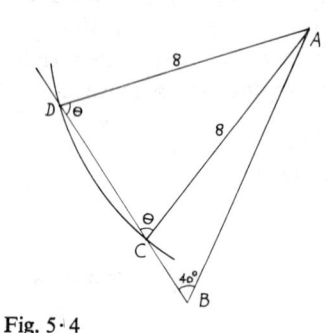

(1) *To find* θ

$$\sin \theta / AB = \sin 40° / AD$$
$$\sin \theta = (10·50 \cdot \sin 40°)/8·00$$
$$\log 10·50 = 1·02119$$
$$\log \sin 40° = \bar{1}·80807$$
$$\log \text{product} = 0·82926$$
$$\log 8·00 = 0·90309$$
$$\log \sin \theta = \bar{1}·92617$$
$$\theta = 57° 31\tfrac{3}{4}'$$

Fig. 5·4

(2) *To find BD*

$$BAD = 180° - (\theta + 40°)$$
$$= 82° 28\tfrac{3}{4}'$$
$$BD/\sin A = AD/\sin 40°$$
$$BD = 8·00 \cdot \sin 82° 28\tfrac{1}{4}' \cdot \operatorname{cosec} 40°$$
$$\log 8·00 = 0·90309$$
$$\log \sin A = \bar{1}·99624$$
$$\log \operatorname{cosec} 40° = 0·19193$$
$$\log BD = 1·09126$$
$$BD = 12·3 \text{ miles}$$

(3) *To find CD*

$$CD/2 = 8 \cdot 00 \ . \ \cos \theta$$
$$CD = 16 \cdot 00 \ . \ \cos 57° 31\tfrac{3}{4}'$$
$$\log 16 \cdot 00 = 1 \cdot 20412$$
$$\log \cos \theta = \bar{1} \cdot 72987$$
$$\overline{\log CD = 0 \cdot 93399}$$
$$\overline{CD = 8 \cdot 60 \text{ miles}}$$

(4) *To find BC*

$$BC = BD - CD$$
$$= 12 \cdot 3 - 8 \cdot 6$$
$$= 3 \cdot 7 \text{ miles}$$

Answer—Distance travelled = $3 \cdot 7$ miles or $12 \cdot 3$ miles.

3. The Cosine Formula

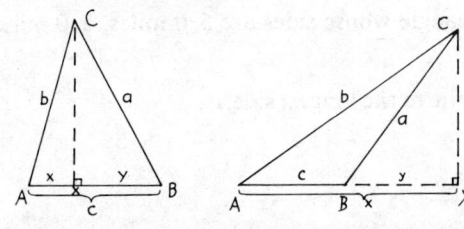

Fig. 5·5

Let *ABC* illustrated in fig. 5·5 be any plane triangle. Drop a perpendicular from any vertex, say *C*, onto the opposite side or side produced.

In the acute-angled triangle *ABC*: $c = x + y$
In the obtuse-angled triangle *ABC*: $c = x - y$
In both triangles: $c = b \cos A + c \cos B$...(I)

(Note that *B* is an obtuse angle and that its cosine is negative.)

By dropping perpendiculars from *A* and *B* onto sides *BC* and *AC* respectively. It may be shown that:

$$b = c \cos A + a \cos C \dotfill \text{(II)}$$
$$a = b \cos C + c \cos B \dotfill \text{(III)}$$

From equations (I), (II) and (III), the plane Cosine Formula is deduced as follows:

Multiply equation (I) by a; equation (II) by b; and equation (III) by c, thus:

$$a^2 = ab \cos C + ac \cos B$$
$$b^2 = bc \cos A + ab \cos C$$
$$c^2 = bc \cos A + ac \cos B$$

By subtracting each of these from the sum of the other two, we have:

$$b^2 + c^2 - a^2 = 2bc \cos A$$

or
$$\cos A = \frac{b^2 + c^2 - a^2}{2bc}$$

or
$$a^2 = b^2 + c^2 - 2bc \cos A$$

Similarly:

$$\cos B = \frac{a^2+c^2-b^2}{2ac}$$

or
$$b^2 = a^2+c^2-2ac \cos B$$

And:
$$\cos C = \frac{a^2+b^2-c^2}{2ab}$$

or
$$c^2 = a^2+b^2-2ab \cos C$$

These formulae are not suitable for use with logarithms, since addition and subtraction do not require logarithms. Their use, therefore, should be confined to solving triangles whose sides are integral numbers which can easily be squared arithmetically, and the whole problem—except perhaps for the extraction of a square root—completed by simple arithmetic.

The plane Cosine Formula may be used for:

1. Finding an angle given the three sides of a triangle.
2. Finding a side of a triangle given the opposite angle and the other two sides.

Example 5·3—Find the largest angle in a plane triangle whose sides are 5·0 miles, 8·0 miles and 4·0 miles, respectively. Refer to fig 5·6.

(Note that the largest angle of a triangle is opposite to the longest side.)

By the Cosine Formula:

$$\cos A = \frac{b^2+c^2-a^2}{2bc}$$

$$= \frac{16+25-64}{40}$$

$$= -\frac{23}{40}$$

$$= -0\cdot 575$$

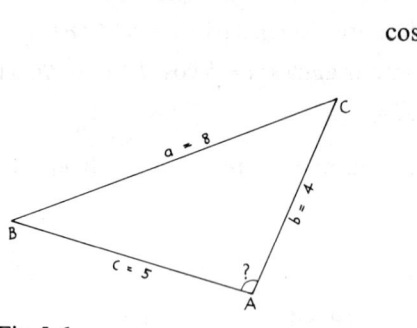

$$A = 180° - 54° \, 54'$$

$$= 125° \, 06'$$

Fig. 5·6

Answer—Largest Angle = 125° 06′.

4. The Haversine Formula

The Haversine Formula, derived from the Cosine Formula, is useful for solving an angle in a plane triangle given the three sides of the triangle. Being suitable for use with logarithms, it is a more useful formula than the Cosine Formula for solving angles.

In any plane triangle: *ABC*:

$$\cos A = \frac{b^2+c^2-a^2}{2bc}$$

OBLIQUE-ANGLED TRIANGLES AND THEIR SOLUTIONS

By multiplying each side by -1, adding $+1$ to each side, and dividing each side by 2, we have:

$$\frac{1-\cos A}{2} = \frac{1-\dfrac{b^2+c^2-a^2}{2bc}}{2}$$

The left-hand side of this expression is equal to the haversine of A, thus:

$$\text{hav } A = \frac{2bc - b^2 - c^2 + a^2}{4bc}$$

$$= \frac{a^2 - (b^2 + c^2 - 2bc)}{4bc}$$

$$= \frac{a^2 - (b-c)^2}{4bc}$$

$$= \frac{(a-b+c)(a+b-c)}{4bc}$$

$$= \frac{(a+b+c-2b)(a+b+c-2c)}{4bc}$$

$$= \frac{(2s-2b)(2s-2c)}{4bc} \quad \text{where } s=(a+b+c)/2$$

$$= \frac{4(s-b)(s-c)}{4bc}$$

or:
$$\text{hav } A = \frac{(s-b)(s-c)}{bc}$$

Similarly:
$$\text{hav } B = \frac{(s-a)(s-c)}{ac}$$

$$\text{hav } C = \frac{(s-a)(s-b)}{ab}$$

Example 5·4—Point Q lies 17·540 miles from Point P. Point R lies 13·600 miles from Point Q and 18·770 miles from Point P. Find the angle between the directions of Q and R from P.

In fig. 5·7:

$$p = 13\cdot600$$
$$q = 18\cdot770$$
$$r = 17\cdot540$$
$$2\overline{)49\cdot910}$$
$$s = 24\cdot955$$

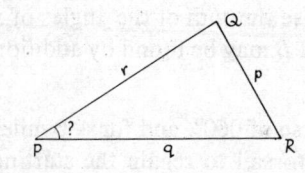

Fig. 5·7

By the Haversine Formula:

$$\text{hav } P = \frac{(s-q)(s-r)}{qr}$$

$$= \frac{6\cdot 185 \times 7\cdot 415}{18\cdot 770 \times 17\cdot 540}$$

$\log 6\cdot 185 = 0\cdot 79134$
$\log 7\cdot 415 = 0\cdot 87011$
$\log \text{num}^r = 1\cdot 66145$
$\log 18\cdot 770 = 1\cdot 27346$
$\log 17\cdot 540 = 1\cdot 24403$
$\log \text{den}^r = 2\cdot 51749$
$\log \text{hav } P = \bar{1}\cdot 14396 \qquad P = 43° 50'$

Answer—Required Angle $= 43° 50'$.

5. The Tangent Formula

We have, by the Sine Formula, for triangle ABC:

$$a/\sin A = b/\sin B = c/\sin C = x$$

Therefore:
$$a = x \sin A$$

and
$$b = x \sin B$$

Therefore:

$$\frac{a+b}{a-b} = \frac{x \sin A + x \sin B}{x \sin A - x \sin B}$$

$$= \frac{\sin A + \sin B}{\sin A - \sin B}$$

$$= \frac{2 \sin \tfrac{1}{2}(A+B) \cos \tfrac{1}{2}(A-B)}{2 \cos \tfrac{1}{2}(A+B) \sin \tfrac{1}{2}(A-B)}$$

$$= \tan \tfrac{1}{2}(A+B) \cot \tfrac{1}{2}(A-B)$$

Or:
$$\frac{a+b}{a-b} = \frac{\tan \tfrac{1}{2}(A+B)}{\tan \tfrac{1}{2}(A-B)}$$

This formula may be used instead of the Cosine or Haversine Formula for finding an angle of a plane triangle given any two sides and the included angle.

The quantity $\tfrac{1}{2}(A+B)$ is equal to the complement of C, because the sum of the angles of a plane triangle is 180°. Thus, if $\tfrac{1}{2}(A-B)$ is known, the angles A and B may be found by addition and subtraction.

Example 5·5—A yacht sails for a distance of 5·0 miles on a course of 060°, and for 4·0 miles on a course of 190°. Find the course to steer and the distance to sail to regain the starting position.

OBLIQUE-ANGLED TRIANGLES AND THEIR SOLUTIONS

In fig. 5·8:
$$C+A=130°$$
$$\tfrac{1}{2}(C+A)=65°$$

Plan:
(i) Find C by the Tangent Formula.
(ii) Find b by the Sine Formula.

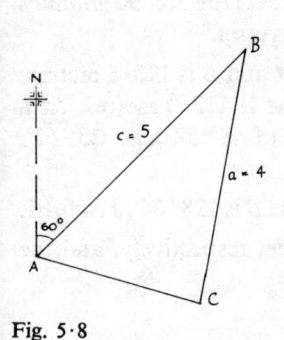

$$\frac{\tan\tfrac{1}{2}(C-A)}{\tan\tfrac{1}{2}(C+A)}=\frac{c-a}{c+a}$$

$$\tan\tfrac{1}{2}(C-A)=\frac{(c-a)}{(c+a)}\cdot\tan\tfrac{1}{2}(C+A)$$

$$=1/9 \cdot \tan 65°$$
$$=1/9 \cdot 2\cdot14451$$
$$=0\cdot23828$$

$$\tfrac{1}{2}(C-A)=13°\ 24'$$
$$\tfrac{1}{2}(C+A)=65°\ 00'$$
$$C=78°\ 24'$$

Course = N. 68° 24' W.

$$b=5\sin 50°\ \text{cosec}\ 78°\ 24'$$
$$\log 5 = 0\cdot69897$$
$$\log \sin 50° = \bar{1}\cdot88425$$
$$\log \text{cosec} = 0\cdot00896$$
$$\log b = 0\cdot59218$$
$$b = 3\cdot91$$

Fig. 5·8

Answer—Course = $291\tfrac{1}{2}°$, Distance = 3·9 miles.

Exercises on Chapter 5

1. From a vessel at anchor a Point A bore 030° and a lighthouse B bore 070°. B lies 190° at a distance of 5·40 miles from A. Find the distance between B and the vessel.

2. The horizontal angle between two vessels A and B is 62°. A is 4·0 miles from an observer and B is 5·0 miles from him. Find the distance between the vessels.

3. In a plane triangle ABC, AB is 3·0 miles; AC is 7·0 miles; and BC is 6·0 miles. Find the three angles of the triangle.

4. In a plane triangle PQR, Q is 97° 00′; PQ is 7·40 miles and QR is 6·70 miles. Find the distance PR.

5. Two vessels A and B are 6·30 miles apart. The horizontal angle between B and a rock C is 20° 32′, and that between A and C at B is 70° 15′. Find the distance BC.

6. Find the required course to steer to counteract the effect of a current which sets 100° at a rate of 4·0 knots, given the course to make good is 025° and the ship's speed through the water is 13 knots.

7. A point of land bears 305° at a distance of 46·0 miles. Find the course to steer and the time taken to reach the point in a current which sets 045° at 3·5 knots, and the speed of the vessel through the water is 15·5 knots.

8. A point of land bore 034°. After travelling for 5·60 miles on a course of 268°, it bore 050°. Find the distance off at the time of the second observation.

9. A point of land bore 066° from a vessel heading 030°. After travelling for 30 minutes the point bore 146° at a distance of 6·0 miles. Find the speed of the vessel.

10. A, B and C, are three buoys in a harbour. The distance between A and B is 125·6 metres; that between B and C is 130·4 metres and that between C and A is 112·0 metres. At a vessel D, A and B, and A and C, subtend, respectively, angles of 48° 58′ and 25° 52′. Find the distances between the vessel and each of the buoys.

11. In a plane triangle ABC, AB is 562 yards, BC is 320 yards, and angle B is 128° 04′. Find AC.

12. In a triangle ABC, the sides AB and BC are 345 feet and 232 feet respectively, and the angle A is 37° 20′. Find the angle B.

CHAPTER 6

SPHERICAL TRIGONOMETRY

1. The Geometry of the Sphere

A Sphere is a three-dimensional shape every point on the surface of which is equidistant from a fixed point known as the Centre of the Sphere. It may be defined as the shape swept out by rotating a circle about any fixed diameter through an angle of 180°. Such a circle is the largest possible circle that may be drawn on the surface of the sphere produced.

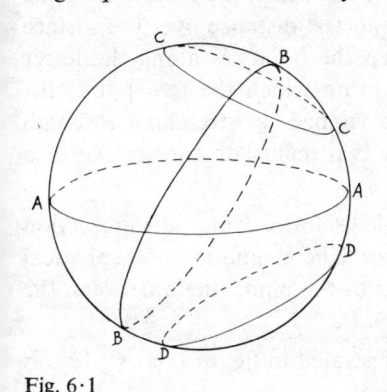

It will be noticed in fig. 6·1 that the centre of the sphere illustrated lies on the planes of the circles AA and BB. Any circle on the surface of a sphere, on whose plane the centre of the sphere lies, is known as a Great Circle of that sphere. AA and BB in fig. 6·1 are examples of great circles. A great circle divides a sphere into two Hemispheres. Any circle on the surface of a sphere which is not a Great Circle is known as a Small Circle. In fig. 6·1 CC and DD are examples of small circles. Notice that the word "small" in this context has no reference to the actual size of a circle: it is used merely to distinguish between great circles on whose plane the centre of the sphere rests from those circles on whose planes the centre does not rest.

Fig. 6·1

An arc of a great circle is measured in angular units: it is a measure of the angle at the centre of the sphere subtended by the two radii which terminate at the extremities of the arc. The measure of a complete great circle is 360°; that of a semi-great circle is 180°; and that of a quadrant or quarter of a great circle is 90°. The measure of an arc of a great circle is known as a Spherical Distance.

Two points on a great circle which are diametrically opposed to one another, that is to say, two points which are separated by a spherical distance of 180°, are known as Antipodal Points, each being an Antipodes of the other.

A point on the surface of a sphere which is 90° from every point on a particular great circle is known as the Pole of that great circle. The diameter of a sphere which connects the two poles of a great circle is known as the Axis of the Great Circle.

Any semi-great circle which connects the poles of a given great circle is referred to as a Secondary to the given great circle which is known, in this case, as the Primary Great Circle. A secondary cuts its primary great circle at an angle of 90°. It follows that the axis of a primary great circle lies in the plane of every secondary.

34 THE ELEMENTS OF NAVIGATION AND NAUTICAL ASTRONOMY

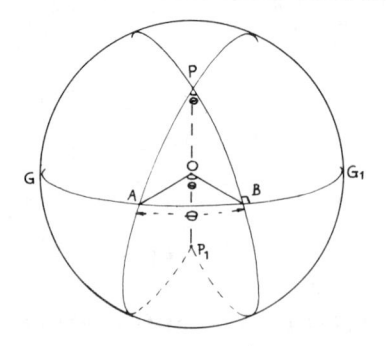

Fig. 6·2

In fig. 6·2, P and P_1 are the poles of a primary great circle GG_1. PA and PB are 90° arcs of the secondaries PAP_1 and PBP_1 respectively. The point G is the antipodes of the point G_1. Note that the arc AB has a spherical distance equal to the angle AOB at the centre of the sphere.

The shortest distance between two points on the surface of a sphere is, of course, a chord joining the two points; but the shortest distance over the surface of the sphere between the points is along the lesser arc of the great circle on which the two points lie. This readily may be verified by stretching a length of cord between any two points on the surface of a model globe.

A Spherical Angle is formed at the intersection of two great circle arcs. The magnitude of a spherical angle is equal to that of the plane angle between the tangents to the great circles at the point of intersection.

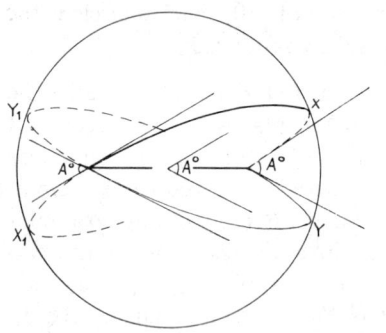

Fig. 6·3

The angle between the two great circles XX_1 and YY_1 illustrated in fig. 6·3, is $A°$. This is equal to the angle between the tangents as indicated in the figure.

A Spherical Triangle is formed on the surface of a sphere by the intersection of three great circle arcs.

Fig. 6·4 depicts a typical spherical triangle formed by great circle arcs XY, YZ and ZX.

The sum of the three angles of any plane triangle is always 180°. The sum of the three angles of any spherical triangle, however, is always greater than 180°, by an amount known as Spherical Excess. It is impossible to construct a plane triangle on the surface of a sphere, but the smaller is a spherical triangle on a given sphere the more nearly is the sum of its three angles equal to 180°. In some navigational problems small spherical triangles are treated as if they are plane, and no material error results.

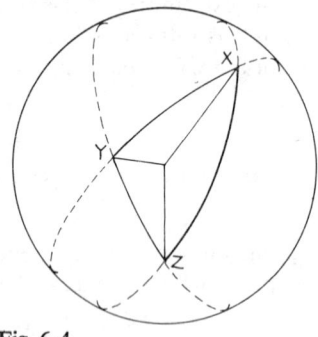

Fig. 6·4

The maximum value of any one angle of a spherical triangle is 180°, or 2 right angles. The sum of the three angles of a spherical triangle must, therefore, be less than 540° or 6 right angles. It follows that a spherical triangle cannot be larger than a hemisphere.

The spherical distance of any side of a spherical triangle never exceeds 180° or 2 right angles. Because a spherical triangle cannot be larger than a hemisphere, it follows that the sum of the three sides cannot be more than 360° or 4 right angles.

It is emphasized that the sides of a spherical triangle are arcs of great circles—arcs of small circles *NEVER* form the sides of a spherical triangle.

2. Spherical Trigonometry

The main use of spherical trigonometry is in the solving of spherical triangles. If any three parts of a spherical triangle are known, any of the remaining parts may be solved direct by one of the three so-called Fundamental Formulae of spherical trigonometry. These are the Sine Formula, the Cosine Formula and the Four Parts Formula. In addition to the fundamental formulae there are numerous *derived* formulae. The spherical trigonometrical formulae used in the practice of navigation and nautical astronomy will be considered in Chapter 7.

In Chapter 10 we shall see that the Earth's shape is not that of a perfect sphere. Nevertheless the spherical formulae used by navigators, despite the fact that they give accurate results only for perfectly spherical surfaces, yield results which are generally sufficiently accurate for all navigational problems.

Exercises on Chapter 6

1. Define: Sphere, Great Circle, Poles of a Great Circle, Axis of a Great Circle.

2. Define: Secondary, Primary Great Circle.

3. What are antipodal points?

4. Show that the magnitude of a spherical angle is equal to that of the plane angle between the tangents at the intersection points of the two great circle arcs which form the spherical angle.

5. Prove that the sum of the three angles of a spherical triangle cannot exceed 540°, and that the sum of the three sides cannot exceed 360°.

CHAPTER 7

THE STEREOGRAPHIC PROJECTION AND THE GRAPHICAL SOLUTIONS OF SPHERICAL TRIANGLES

1. Introduction

The representation on a plane surface of points and lines on an object as they appear in the eye, is known as a Perspective or Geometrical Projection. The plane surface is called the Plane of the Projection, and the position of the eye as the Point of Projection. A straight line extending from the Point of Projection to any point on the object is called a Line of Projection.

There are many methods of projecting a spherical surface onto a plane surface. All maps and charts, for example, are projections of the whole or part of the spherical Earth's surface, and some of these are perspective projections. Non-perspective projections, such as the Mercator Projection (see Chapter 12), are known as Conventional Projections.

2. The Stereographic Projection

A very useful method of projecting a spherical surface onto a plane surface is the stereographic projection, which is a perspective projection. In the stereographic projection a great circle of the sphere to be projected is assumed to lie in the plane of the projection. This great circle is known as the Primitive, and the point of the projection is one of the poles of the primitive great circle.

On a stereographic projection of a sphere all projected arcs of circles, great and small, are straight lines or arcs of circles. This property makes it possible to construct a stereographic projection geometrically by means of straightedge and drawing compasses. Other interesting properties of the stereographic projection are that more than a hemisphere can be projected, and that the projection is Orthomorphic, which means that angles at any point on the plane of projection are without distortion.

The stereographic projection is often used for Star Maps, and it has been used as the basis of a variety of instruments designed for solving spherical triangles, especially in nautical astronomy. Although we shall indicate how to construct a stereographic projection and how to measure angles and spherical distances, our main intention here is to assist students in visualizing the relative positions of points and arcs on a spherical surface, especially that of the celestial sphere, thereby leading them to an understanding and appreciation of certain navigational and nautical astronomical problems.

3. The Principles of the Stereographic Projection

Fig. 7·1 serves to illustrate the principle of the stereographic projection.

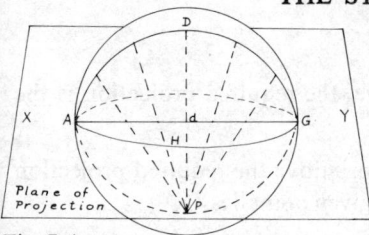

THE STEREOGRAPHIC PROJECTION

Fig. 7·1

The great circle *AHG* illustrated in fig. 7·1 lies in the plane of the projection. It is, therefore, the primitive, and it is upon the plane of the primitive that the sphere's surface is to be projected.

Consider a great circle *ADGP* which stands perpendicular to the plane of the projection. If lines of projection are drawn from the point of projection to several points on the great circle, the points at which these lines cut the plane of the projection lie on the straight line *AdG*. This line, therefore, is the projection of the great circle arc *ADG*, and the point *d*, which is the projection of *D*, lies at the centre of the primitive.

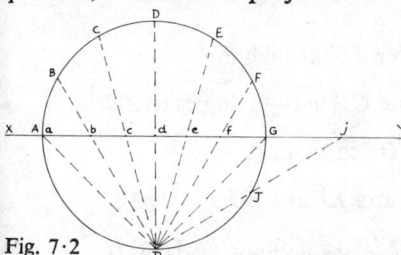

Fig. 7·2

In fig. 7·2, *P* is the point of projection and the straight line *AdGY* lies in the plane of the projection. Points *A*, *B*, *C*, *D*, *E*, *F*, *G* and *J*, are points on a great circle which is a secondary to the primitive. These points are 30° apart and they are projected at points *a* to *j* respectively. It will be noticed that the projected lengths of these equal arcs are not uniform. The projected arc *Ab* is longer than that of *CD*, and the projection of arc *GJ*, which is *Gj*, is even longer than the projection *ab* of arc *AB*. At the edge of the primitive there is no distortion of arcs; within the primitive distortion is such that equal arcs of the sphere are projected as increasingly smaller lines as the centre of the projection is approached; and that outside the primitive the distortion is such that equal arcs of the sphere are projected as increasingly longer lines as distance from the primitive increases.

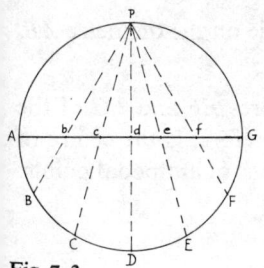

All great circles which cut the primitive at right angles, that is to say, all secondaries to the primitive, are projected as straight lines which pass through the centre of the projection.

Imagine the sphere illustrated in fig. 7·2 to be rotated about the diameter *AdG* through an angle of 90°. The points *a*, *b*, *c*, etc., remain stationary because they lie on the axis of rotation. The points *B*, *C*, *D*, etc., however, move to new positions on the plane of the projection. The projection viewed from directly above now appears as in fig. 7·3.

Fig. 7·3

Fig. 7·3 serves to demonstrate that when constructing a stereographic projection all projecting is made from a point on the circumference of the primitive. It is to be realized, however, that the completed projection is a view of the sphere's surface from the pole of the primitive.

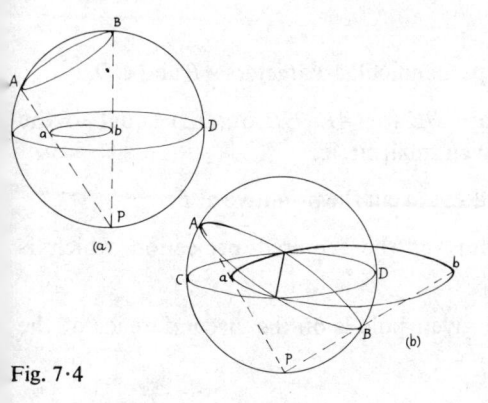

Fig. 7·4 illustrates that all circles on the sphere are projected as circles or straight lines. Let the point *P*, in fig. 7·4, be the point of projection, and *CD* the primitive. In fig. 7·4(*a*) *AB* denotes any small circle on the sphere, and in fig. 7·4(*b*) *AB* denotes any great circle on the sphere. In both cases the circle is projected as a circle *ab*.

Fig. 7·4

4. To Project a Great Circle about a Given Point as Pole

Case 1—If the given point is at the centre of the primitive, the required projection is the primitive itself.

Case 2—If the given point is on the circumference of the primitive the required projection is a diameter lying perpendicular to the diameter on which the given point lies.

Referring to fig. 7·5, let the circle *ACBD* be the primitive and the point *P* the pole of the required projection.

Case 3—If the given point is within the primitive, as illustrated in fig. 7·5. The procedure in this case is as follows:

(i) Draw diameter *AB* through *P*.

(ii) Draw diameter *CD* at right angles to *AB*.

(iii) Draw chord *DE* through *P*.

(iv) Describe 90° arcs *EF* and *EG*.

(v) Draw chord *FD* to cut diameter *AB* at *I*.

(vi) Draw chord *DG* (produced) to cut diameter *AB* (produced) at *H*.

(vii) Bisect *IH* at *O*, which is the centre of the required projection, radius *OI* or *OH*.

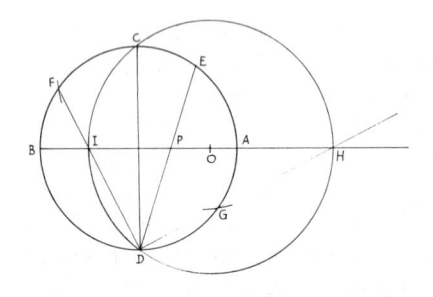

Fig. 7·5

In practice it is customary to find the centre *O*, which is known to lie on the diameter *AB*, or *AB* produced, by trial and error.

The construction described above is demonstrated thus. Because arcs *EF* and *EG* of the primitive are each 90°, the projections *IP* and *PH* are each 90°. *P*, therefore, is the centre of the great circle which passes through *I* and *H*, and *I* and *H* are projections of antipodal points.

5. To Project a Small Circle about a Given Point as Pole

Case 1—If the given pole is at the centre of the projection.

Referring to fig. 7·6:

(i) Draw any two perpendicular diameters *AB* and *CD*.

(ii) Mark off an arc *BE* (or *AE*, *CE* or *DE*) equal to the radius of the given small circle.

(iii) Draw the chord *EA* to cut the primitive at *F*.

(iv) *PF* is the radius of the required projection which is centred at *P*.

Case 2—If the given pole is on the circumference of the primitive.

Fig. 7·6

THE STEREOGRAPHIC PROJECTION

Referring to fig. 7·7:

(i) Let *B* be the given pole.
(ii) Draw the perpendicular diameters *AB* and *CD*.
(iii) Mark off arcs *BE* and *BF* each equal to the radius of the given small circle.
(iv) Draw *EC* to cut *AB* at *G*.
(v) Draw *CF* and produce to cut *AB* produced at *H*.
(vi) Bisect *GH* at *O* which is the centre of the required projection of radius *OG* or *OH*.

Case 3—If the given pole is within the primitive but not at the centre.

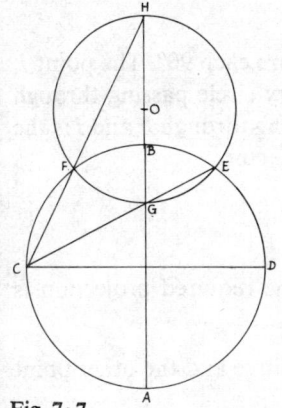

Fig. 7·7

Referring to fig. 7·8:

(i) Let *ACBD* be the primitive and *P* the given pole.
(ii) Draw diameter *AB* through *P* the given pole.
(iii) Draw diameter *CD* perpendicular to *AB*.
(iv) Draw the chord *DP* and produce to *E*.
(v) Centre at *E* and describe arcs *EF* and *EG*, each equal to the radius of the given small circle.
(vi) Draw *DF* to cut diameter *AB* at *I*, and *DG* produced to cut *AB* produced at *H*.

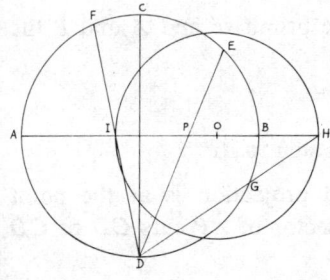

Fig. 7·8

(vii) Bisect *IH* at *O*, which is the required centre, the radius of the required projection being *OI* or *OH*. (It is instructive to compare this construction with that given in fig. 7·5.)

6. To Find the Locus of Centres of All Great Circles which Pass Through a Given Point

Referring to fig. 7·9:

(i) Let the circle *ACBD* be the primitive and *P* the given point.
(ii) Draw diameter *AB* through *P*.
(iii) Draw diameter *CD* perpendicular to *AB*.
(iv) Draw the chord *CE* through *P*.
(v) Centre at *E* and radius 90°, describe arc *EF*.
(vi) Centre at *F* and radius 90°, describe arc *FG*.
(vii) Draw chord *CF* to cut *AB* at *H*.
(viii) Draw *CG* and produce to cut *AB* produced at *I*.
(ix) Bisect *PI* at *O*.

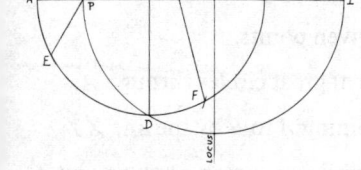

Fig. 7·9

(x) Draw a perpendicular to *AB* through *O*, which is the required locus.

The proof of this construction is as follows:

Because *EF* and *FG* are each 90°, it follows that arcs *PH* and *HI* are each 90°. The point *I*, therefore, is 180° from *P*, so that *P* and *I* are antipodal points. Every circle passing through *P* and *I* must be a great circle. The locus of centres of all circles passing through *P* and *I* is the perpendicular bisector of *PI*, so that the required locus must be this bisector.

7. To Project a Great Circle through Two Given Points

Case 1—If one of the points is at the centre of the primitive, the required projection is the diameter drawn through the two points.

Case 2—If one of the points lies on the circumference of the primitive and the other point is not at the centre or the circumference of the primitive.

Referring to fig. 7·10:

(i) Let the circle *ACBD* be the primitive and *A* and *P* the two given points.

(ii) Draw diameter *AB*.

(iii) Draw diameter *CD* perpendicular to *AB*.

(iv) The centre of the required projection is at the point where the perpendicular bisector of *AP* cuts *CD* or *CD* produced.

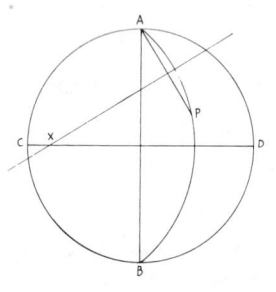

Fig. 7·10

This is so because:

1. All great circles which pass through *A* must pass through the antipodes of *A*, that is to say, through *B*.

2. The locus of centres of all great circles which pass through *A* and *B* must lie on the diameter *CD* or *CD* produced.

3. The centres of all circles which pass through *A* and *P* must lie on the perpendicular bisector of the chord *AP*.

Therefore, the centre of the great circle which passes through *A* and *P*, must lie at point *X* which is the point at which the perpendicular bisector of *AP* cuts *CD*.

Case 3—If neither of the given points lies at the centre or the circumference of the primitive.

Referring to fig. 7·11:

(i) Let *A* and *B* be the two given points.

(ii) Find the locus of centres of great circles through *A*.

(iii) Bisect the straight line joining *A* to *B* by the line *XY*.

(iv) The centre of the required projection is the point *O* at which the line *XY* cuts the locus of centres of great circles through *A*.

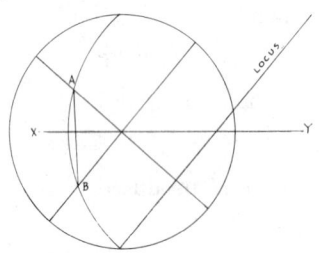

Fig. 7·11

THE STEREOGRAPHIC PROJECTION

8. To Measure a Given Arc of a Projected Great Circle

Case 1—If the given arc is part of the primitive, the required measure is the angle at the centre of the primitive contained between the extremities of the given arc.

Case 2—If the arc is part of a great circle which is projected as a straight line.

Referring to fig. 7·12:

(i) Let *AB* be the projected arc.

(ii) Draw a diameter through *A* and *B*.

(iii) Draw a diameter perpendicular to *AB*.

(iv) Draw chords *CE* and *CF* through *A* and *B* respectively.

(v) Arc *EF* is the required measure of the arc *AB*.

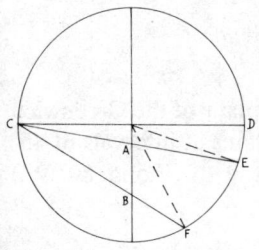

Fig. 7·12

Case 3—If the given arc is part of a great circle which is inclined to the primitive.

Referring to fig. 7·13:

(i) Let *AB* be the given arc.

(ii) Find the pole *P* of the projected circle on which arc *AB* lies.

(iii) Draw *PA* and *PB* and produce each to cut the primitive at *C* and *D* respectively.

(iv) Arc *CD* is a measure of the arc *AB*.

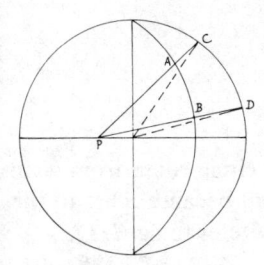

Fig. 7·13

9. To Measure a Projection of a Spherical Angle

The angle between two great circle arcs is equivalent to the plane angle between their radii or the tangents of the projected great circles at the point of intersection.

In fig. 7·14 the angle between the projected great circles *AB* and *CD* is equal to the plane angle *POR* or the plane angle *SOQ*.

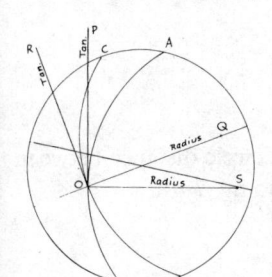

Fig. 7·14

10. Examples

The following examples serve to show how the stereographic projection may be used to solve spherical triangles by construction. The student is advised to read and understand Chapter 4 before considering Examples 1 to 4 inclusive, and to read and understand Parts 4 and 5 before considering Examples 5 to 10 inclusive.

Example 7·1—Construct a stereographic projection of the Earth's northern hemisphere. Project meridians at intervals of 30° from the Greenwich meridian, and the parallels of Latitude of 30° and 60° North. Project the positions of London (Lat. 51° N., Long. 0°); New York (Lat. 41° N., Long. 74° W.); Moscow (Lat 56° N., Long. 38° E.); and Tokyo (Lat. 36° N., Long. 139° E.). Refer to fig. 7·15.

42   THE ELEMENTS OF NAVIGATION AND NAUTICAL ASTRONOMY

Fig. 7·15

Example 7·2—Construct a stereographic projection of the Earth on the plane of the Greenwich meridian. Project the parallels of 30° and 60° N. and S., and the meridians at intervals of 45° from the meridian of Greenwich. Project the positions of Panama (Lat. 9° N., Long. 80° W.) and Cape Horn (Lat. 55° S., Long. 66° W.). Refer to fig. 7·16.

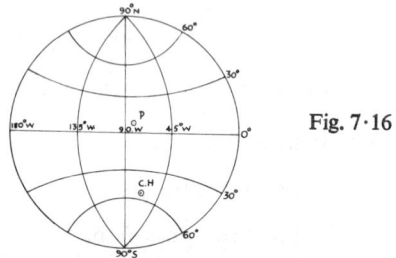

Fig. 7·16

Example 7·3—Find, by scale drawing, the initial course of the great circle route from *A* in Lat. 20° N., Long. 100° W., to *B* in Lat. 60° N., Long. 40° W. (Note that because *B* lies to the eastwards of *A*, it is convenient to project on the plane of *A*'s meridian.) Refer to fig. 7·17.

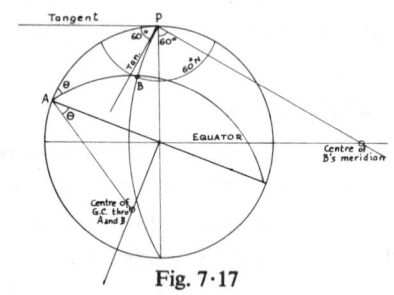

Fig. 7·17

Answer—Initial Course = 033°.

Example 7·4—Find, by means of a stereographic projection, the great circle distance between *X* in Lat. 30° S., Long. 90° E., and *Y* in Lat. 40° N., Long. 40° E. Refer to fig. 7·18.

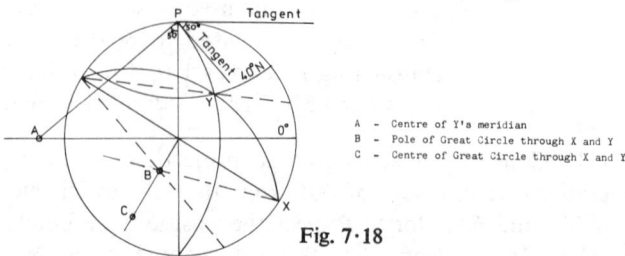

A = Centre of Y's meridian
B = Pole of Great Circle through X and Y
C = Centre of Great Circle through X and Y

Fig. 7·18

Answer—Great Circle Distance = 85° or 5,100 miles.

THE STEREOGRAPHIC PROJECTION

Example 7·5—Construct a stereographic projection of the celestial sphere on the plane of the celestial equator. Project celestial meridians at intervals of 45° from the meridian of the First Point of Aries. Project parallels of declination 30° and 60° N., and the positions of Capella (dec. 46° N., S.H.A. 283°), and Arcturus (dec. 19$\frac{1}{2}$° N., S.H.A. 147°) Refer to fig. 7·19.

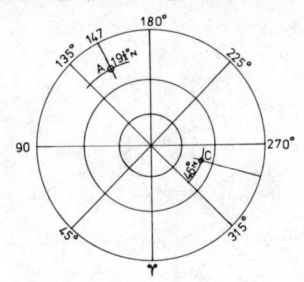

Fig. 7·19

Example 7·6—Construct a stereographic projection of the celestial sphere on the plane of the celestial horizon of an observer in Lat. 50° N. Project the celestial equator and hour circles at intervals from 3 hr. from the Observer's lower celestial meridian. (*Note that*—Latitude of Observer = Altitude of Celestial Pole.) Refer to fig. 7·20.

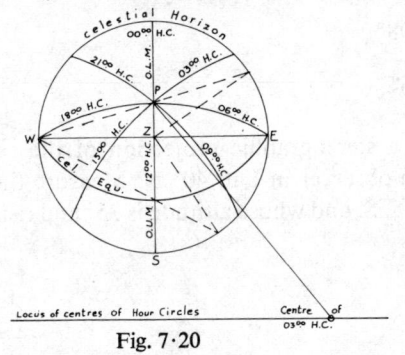

Fig. 7·20

Example 7·7—Construct a stereographic projection of the celestial sphere on the plane of the celestial horizon of an observer in Lat. 30° S. Project the celestial equator and parallels of declination of 30° N., 30° S. and 60° S. Refer to fig. 7·21.

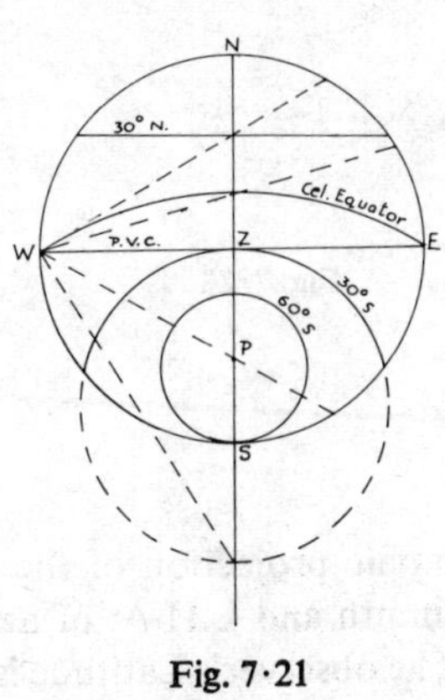

Fig. 7·21

44 THE ELEMENTS OF NAVIGATION AND NAUTICAL ASTRONOMY

Example 7·8—Construct a stereographic projection of the celestial sphere on the plane of the celestial meridian of an observer in Lat. 30° N. Project the position of a star whose declination is 40° N., and whose altitude is 50° and decreasing. Measure the azimuth and the Local Hour Angle of the star. Refer to fig. 7·22.

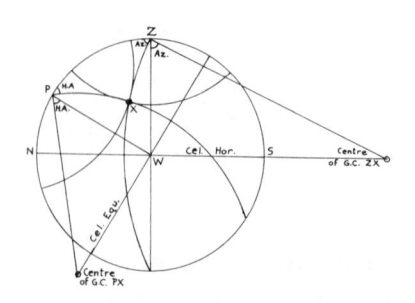

Fig. 7·22

Answer—Azimuth = 298°

L.H.A. = 46°.

Example 7·9—Construct a stereographic projection of the celestial sphere on the plane of the celestial horizon of an observer in Lat. 40° N. Measure the L.H.A. and the azimuth of a star whose declination is 35° S. and whose altitude is 35° and rising. Refer to fig. 7·23.

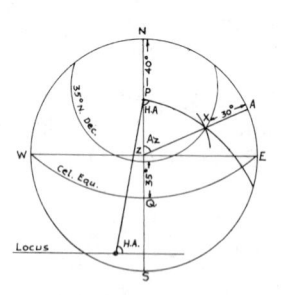

Fig. 7·23

Answer—Azimuth = 065°

L.H.A. = 80°.

Example 7·10—Construct a stereographic projection of the celestial sphere on the plane of the celestial equator. Measure the azimuth and L.H.A. of a star whose declination is 20° N. and whose altitude is 30° and setting. The observer's Latitude is 50° N. Refer to fig. 7·24.

THE STEREOGRAPHIC PROJECTION

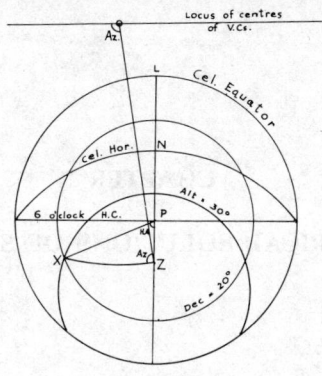

Fig. 7·24

Answer—Azimuth = 266°

L.H.A. = 67°.

11. Figure Drawing

Although in the practice of navigation and nautical astronomy solutions by scale drawing are not, in general, sufficiently accurate, freehand sketches often help to clarify problems. The student is recommended to illustrate his navigational and astronomical problems by freehand sketches, and to continue to do so until he is confident of solving his problems without the assistance of diagrams.

CHAPTER 8

THE TRIGONOMETRICAL SOLUTIONS OF SPHERICAL TRIANGLES

1. The Spherical Sine Formula

In any spherical triangle ABC:

$$\sin a/\sin A = \sin b/\sin B = \sin c/\sin C$$

Let ABC in fig. 8·1 be any spherical triangle on a sphere whose centre is at O. Drop a perpendicular from A onto the plane BOC at P. Drop a perpendicular from P onto the radii OC and OB at Y and X respectively. Join A to Y and A to X.

Because AY and AX are in the planes of the arcs AC and AB respectively, therefore:

Plane angle AYP = Spherical angle ACB
Plane angle AXP = Spherical angle ABC

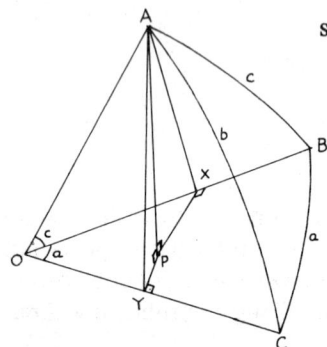

Fig. 8·1

Therefore:

$$\sin b/\sin B = (AY/AO)/(AP/AX) = (AY \cdot AX)/(AO \cdot AP) \quad \ldots \ldots \quad \text{(I)}$$
$$\sin c/\sin C = (AX/AO)/(AP/AY) = (AX \cdot AY)/(AO \cdot AP) \quad \ldots \ldots \quad \text{(II)}$$

From (I) and (II):

$$\sin b/\sin B = \sin c/\sin C$$

By dropping a perpendicular from B or C onto the opposite plane, it may similarly be shown that:

$$\sin b/\sin B = \sin a/\sin A$$

Therefore:

$$\sin a/\sin A = \sin b/\sin B = \sin c/\sin C$$

Example 8·1—In the spherical triangle PZX, $P = 30° 00'$, $PX = 100° 00'$, and $ZX = 40° 00'$. Find Z.

In fig. 8·2:

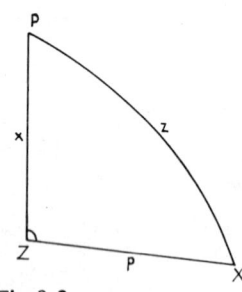

$\sin Z/\sin z = \sin P/\sin p$
$\sin Z = \sin P \sin z \operatorname{cosec} p$
$\log \sin P = \bar{1} \cdot 69897$
$\log \sin z = \bar{1} \cdot 99335$
$\log \operatorname{cosec} p = 0 \cdot 19193$
$\log \sin Z = \bar{1} \cdot 88425 \qquad Z = 180° - 50° 00' = 130° 00'$

Fig. 8·2   *Answer*—$Z = 130° 00'$

THE TRIGONOMETRICAL SOLUTIONS OF SPHERICAL TRIANGLES

It will be remembered that the sine of an angle between 90° and 180° has a positive value. Care, therefore, must be taken when using the Spherical Sine Formula to ensure that the correct quadrant is designated for the computed angle. In other words, it must not be thought that the computed angle is always less than 90°.

2. The Spherical Cosine Formula

In any spherical triangle ABC:

$$\cos A = \frac{\cos a - \cos b \cos c}{\sin b \sin c}$$

or:

$$\cos a = \cos A \sin b \sin c + \cos b \cos c$$

Let ABC in fig. 8·3 be any spherical triangle on the sphere whose centre lies at O. At A draw tangents to the arcs AB and AC. These tangents lie in the planes of their respective arcs. Thus, the first must meet OB produced at D, and the second must meet OC produced at E. Join D to E.

Because AD and AE are tangents, the plane angle DAE is equal to the spherical angle BAC, and the angles OAE and OAD are right angles.

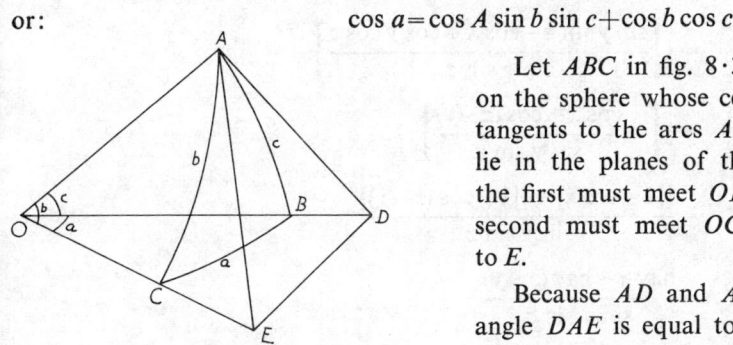

Fig. 8·3

By the Plane Cosine Formula:

$$DE^2 = OD^2 + OE^2 - 2\,OD\,OE \cos a \quad \ldots \ldots \ldots \ldots \text{(I)}$$
$$DE^2 = AD^2 + AE^2 - 2\,AD\,AE \cos A \quad \ldots \ldots \ldots \ldots \text{(II)}$$

Subtract (II) from (I):

$$0 = (OD^2 + OE^2 - 2\,OD\,OE \cos a - AD^2 - AE^2 + 2\,AD\,AE \cos A$$
$$= (OD^2 - AD^2) + (OE^2 - AE^2) - 2\,OD\,OE \cos a + 2\,AD\,AE \cos A$$
$$= 2\,OA^2 - 2\,OD\,OE \cos a + 2\,AD\,AE \cos A$$

Therefore:

$$\cos A = \frac{OD\,OE \cos a - OA^2}{AD\,AE}$$

By dividing throughout by $OD\,OE$, we get:

$$\cos A = \frac{\cos a - \cos b \cos c}{\sin b \sin c}$$

or:

$$\cos a = \cos A \sin b \sin c + \cos b \cos c$$

The Spherical Cosine Formula suffers from two disadvantages:

1. It is not convenient for logarithmic computations.
2. The cosines of angles in the second quadrant are negative so that great care must be taken in handling signs.

A formula similar to the Spherical Cosine Formula, but which does not suffer from the disadvantages of the Spherical Cosine Formula, is the Spherical Haversine Formula, which is easily derived from the former.

3. The Spherical Haversine Formula

In any spherical triangle ZYX:

$$\text{hav } X = \frac{\text{hav } x - \text{hav } (y \sim z)}{\sin y \sin z}$$

or:
$$\text{hav } x = \text{hav } X \sin y \sin z + \text{hav } (y \sim z)$$

Proof:
$$\text{hav } X = \tfrac{1}{2}(1 - \cos X)$$
$$= \tfrac{1}{2}\left[1 - \left(\frac{\cos x - \cos y \cos z}{\sin y \sin z}\right)\right]$$
$$= \tfrac{1}{2}\left[\frac{\sin y \sin z - \cos x + \cos y \cos z}{\sin y \sin z}\right]$$
$$= \tfrac{1}{2}\left[-\frac{\cos x + \cos(z \sim y)}{\sin y \sin z}\right]$$
$$= \frac{\tfrac{1}{2}[1 - \cos x] - \tfrac{1}{2}[1 - \cos(z \sim y)]}{\sin y \sin z}$$

That is:
$$\text{hav } X = \frac{\text{hav } x - \text{hav }(z \sim y)}{\sin y \sin z}$$

or:
$$\text{hav } x = \text{hav } X \sin y \sin z + \text{hav }(z \sim y)$$

By doubling each side we have the Spherical Versine Formula, viz:

$$\text{vers } x = \text{vers } X \sin y \sin z + \text{vers }(z \sim y)$$

The principal advantage of the haversine or versine formula is that all the trigonometrical functions used for solving triangles by its means are positive.

Example 8·2—In the spherical triangle ABC, $a = 50°\ 00'$, $b = 60°\ 00'$, $c = 100°\ 00'$. Find A.

Referring to fig. 8·4:

$a = 50°\ 00'$
$b = 60°\ 00'$
$c = 100°\ 00'$
$(c - b) = 40°\ 00'$

hav A = hav a − hav $(c-b)$ cosec b cosec c

nat hav $a = 0 \cdot 17861$
nat hav $(c-b) = \underline{0 \cdot 11698}$
nat hav $\theta = \underline{0 \cdot 06163}$
log hav $\theta = \overline{2} \cdot 78978$
log cosec $b = 0 \cdot 06247$
log cosec $c = \underline{0 \cdot 00665}$
log hav $A = \overline{2} \cdot 85890$ $\qquad A = 31°\ 11'$

Answer—$A = 31°\ 11'$.

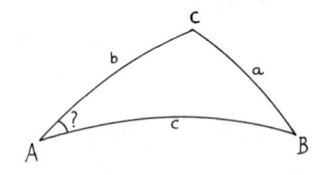

Fig. 8·4

THE TRIGONOMETRICAL SOLUTIONS OF SPHERICAL TRIANGLES   49

Example 8·3—In the spherical triangle XYZ, $X = 40°\,00'$, $z = 30°\,00'$, $y = 80°\,00'$. Find x.

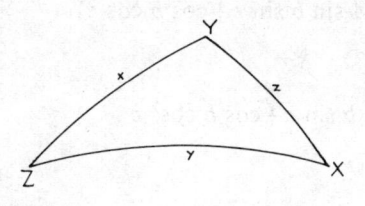

Fig. 8·5

Referring to fig. 8·5:

$$X = 40°\,00'$$
$$y = 80°\,00'$$
$$z = 30°\,00'$$
$$(y \sim z) = 50°\,00'$$

$$\text{hav } x = \text{hav } X \sin y \sin z + \text{hav}(y \sim z)$$

$$\log \text{hav } X = \bar{1}\cdot 06810$$
$$\log \sin y = \bar{1}\cdot 99335$$
$$\log \sin z = \bar{1}\cdot 69897$$
$$\log \text{hav } \theta = \bar{2}\cdot 76042$$
$$\text{nat hav } \theta = 0\cdot 05760$$
$$\text{nat hav}(y \sim z) = 0\cdot 17861$$
$$\text{nat hav } x = 0\cdot 23621 \qquad x = 58°\,09'$$

Answer—$x = 58°\,09'$.

4. The Four Parts Formula

In any spherical triangle if three of any four adjacent parts are known the unknown of the four parts may be found direct by means of the Four Parts Formula.

Referring to the spherical triangle ABC in fig. 8·6:

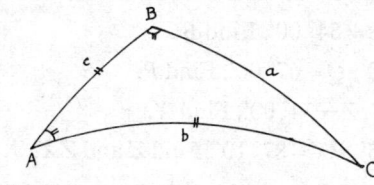

Fig. 8·6

B is the Outer Angle (O.A.)
c is the Inner Side (I.S.)
A is the Inner Angle (I.A.)
b is the Outer Side (O.S.)

The Four Parts Formula relating to these parts is:

$$\cos c \text{ (I.S.) } \cos A \text{ (I.A.)} = \sin c \text{ (I.S.) } \cot b \text{ (O.S.)} - \sin A \text{ (I.A.) } \cot B \text{ (O.A.)}$$

Proof:

By the Spherical Cosine Formula:

$$\cos b = \cos B \sin c \sin a + \cos c \cos a \quad \ldots \ldots \ldots \quad \text{(I)}$$
$$\cos a = \cos A \sin b \sin c + \cos b \cos c \quad \ldots \ldots \ldots \quad \text{(II)}$$

By the Spherical Sine Formula:

$$\sin a = (\sin A \sin b)/\sin B \quad \ldots \ldots \ldots \ldots \ldots \quad \text{(III)}$$

Substitute (II) for $\cos a$ in (I), and (III) for $\sin a$ in (I).

Thus:

$$\cos b = \cos B \sin c \,[(\sin A \sin b)/\sin B] + \cos c\,(\cos A \sin b \sin c + \cos b \cos c)$$

That is:

$$\cos b = \cot B \sin c \sin A \sin b + \cos c \cos A \sin b \sin c + \cos b \cos^2 c$$

From which:

$$\cos b - \cos b \cos^2 c = \cot B \sin c \sin A \sin b + \cos c \cos A \sin b \sin c$$

That is:

$$\cos b\,(1 - \cos^2 c) = \sin b \sin c\,(\sin A \cot B + \cos c \cos A)$$

$$\frac{\cos b \sin^2 c}{\sin b \sin c} = \sin A \cot B + \cos c \cos A$$

That is:

$$\cot b \sin c = \sin A \cot B + \cos c \cos A$$

or:

$$\cos c \cos A = \sin c \cot b - \sin A \cot B$$

Exercises on Chapter 8

1. In the spherical triangle ABC: $B = 75° \, 00'$, $C = 55° \, 00'$, $AC = 67° \, 00'$. Find AB.
2. In the spherical triangle XYZ: $XZ = 105° \, 00'$, $XY = 95° \, 00'$, $XZ = 54° \, 00'$. Find X.
3. In the spherical triangle PQR: $PQ = 65° \, 10'$, $PR = 106° \, 23'$, $Q = 43° \, 10'$. Find RQ.
4. In the spherical triangle DEF: $DE = 30° \, 00'$, $DF = 60° \, 00'$, $EF = 50° \, 00'$. Find E.
5. In the spherical triangle ABC: $A = 35° \, 00'$, $AB = 65° \, 00'$, $B = 54° \, 00'$. Find BC.
6. In the spherical triangle PQR: $PQ = 100° \, 00'$, $RQ = 54° \, 00'$, $Q = 67° \, 00'$. Find P.
7. In the spherical triangle XYZ: $XY = 65° \, 00'$, $ZX = 78° \, 00'$, $Z = 34° \, 00'$. Find X.
8. In the spherical triangle PZX: $P = 03$ h. 24 m., $PZ = 54° \, 55'$, $PX = 87° \, 10'$. Find Z and ZX.

CHAPTER 9

NAPIER'S RULES

1. Napier's Rules for Solving Right-angled Spherical Triangles

Any spherical triangle, right-angled or otherwise, may be solved using one or more of the formulae described in the preceding chapter. If, however, a spherical triangle contains a right angle, a shorter and simpler solution than that in which a formula for oblique-angled triangles is used, is made possible by Napier's Rules.

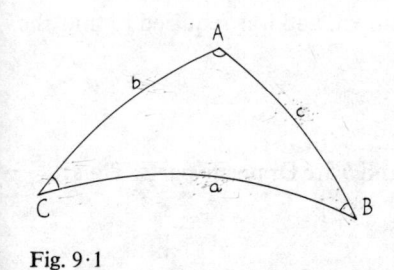

In the spherical triangle illustrated in fig. 9·1 suppose that A, a and c are known, and that it is required to find C.

By the Spherical Sine Formula:

$$\sin C = \sin c \sin A \operatorname{cosec} a \quad \ldots \ldots \quad (I)$$

Suppose that A, b and c in triangle ABC are known, and that it is required to find a:

Fig. 9·1

By the Spherical Cosine Formula:

$$\cos a = \cos A \sin b \sin c + \cos b \cos c \quad \ldots \ldots \ldots \quad (II)$$

Suppose that A, B and c are known and that it is required to find a:

By the Four Parts Formula:

$$\cos c \cos B = \sin c \cot a - \sin B \cot A$$

or:

$$\cot a = (\cos c \cos B + \sin B \cot A)/\sin c \quad \ldots \ldots \quad (III)$$

Now suppose that the angle A in the triangle ABC is 90°. Then, because $\cos 90° = 0$ and $\sin 90° = 1$, the three formulae (I), (II) and (III) reduce, respectively, to:

$$\sin C = \sin c \operatorname{cosec} a \quad \ldots \ldots \ldots \ldots \quad (IV)$$

$$\cos a = \cos b \cos c \quad \ldots \ldots \ldots \ldots \quad (V)$$

$$\cot a = \cot c \cos B \quad \ldots \ldots \ldots \ldots \quad (VI)$$

It is possible to derive ten simple formulae which, collectively, provide the means for solving every possible case of a right-angled spherical triangle. Instead of deducing from these ten formulae so many distinct rules for the solution of the various cases, the whole, by the assistance of an ingenious contrivance, may be comprehended in two remarkably simple rules. These rules, named after their illustrious inventor, are known as Napier's Rules For Circular Parts.

The parts of a given right-angled spherical triangle, not including the right angle, are written in order—either clockwise or anti-clockwise—in the five sectors of a cartwheel as illustrated in fig. 9·2.

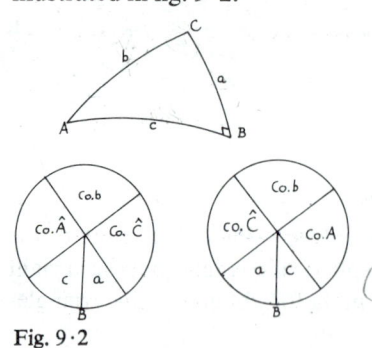

Referring to fig. 9·2 it will be noticed that the two angles and the side opposite to the right angle are prefixed with the letters "co" which denotes complement.

Of any three parts in the cartwheel one must be a Middle part, and the other two must be either Opposite or Adjacent parts.

Napier's Rules are:

s*I*ne m*I*ddle part = product of the c*O*sines of the *O*pposites

s*I*ne m*I*ddle part = product of the t*A*ngents of the *A*djacents

Fig. 9·2

If, in the triangle ABC in fig. 9·2, the sides b and c are known, and it is required to find the remaining three unknown parts, the rules are:

(a) *To find C*

Of the three parts C, b and c, c is the Middle part, and C and b are Opposite parts. Thus:

$$\sin c = \cos \text{co-}C \cos \text{co-}b$$

That is: $\sin c = \sin C \sin b$

and: $\sin C = \sin c \; \text{cosec} \; b$ . . . . (I)

(b) *To find a*

Of the three parts a, b and c, b is the Middle part and a and c are Opposite parts.

$$\sin \text{co-}b = \cos a \cos c$$

That is: $\cos b = \cos a \cos c$

and: $\cos a = \cos b \sec c$ . . . . (II)

(c) *To find A*

Of the three parts A, b and c, A is the Middle part and b and c are Adjacent parts.

$$\sin \text{co-}A = \tan c \tan \text{co-}b$$

That is: $\cos A = \tan c \cot b$ . . . . (III)

It is imprudent, in circumstances when it can be avoided, to solve a right-angled triangle using a part which has previously been calculated, and which may, therefore, be in error. Any error in such a part used to solve another part will cause unnecessary error in that part. When solving the three unknown parts of a right-angled spherical triangle it is advisable, therefore, to derive the three formulae before commencing the calculations. By so doing,

NAPIER'S RULES

not only is the time spent in entering tables reduced, but the possibility of blundering in the calculation is also reduced.

Before commencing to solve a right-angled spherical triangle it is advisable to ascertain whether or not the value of any unknown part is greater or less than 90°. This is easily done by means of the device now to be described.

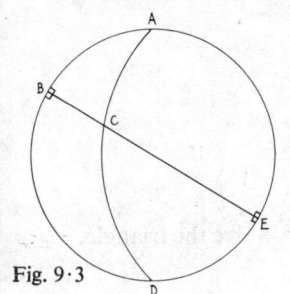

Fig. 9·3 illustrates each of the four cases of right-angled spherical triangles.

In the triangle *ABC ALL* parts, except the right angle, are *LESS* than 90°.

In the triangle *BCD* only *D* and *BC* are less than 90°

In the triangle *CDE* only *C* and *DE* are less than 90°

In the triangle *ACE* only *AC* is less than 90°.

Fig. 9·3

By constructing such a simple figure the relative values of the unknown parts are readily seen.

Example 9·1—In the spherical triangle *PQR*: $P=45°$, $r=60°$, $Q=90°$. Find the remaining parts.

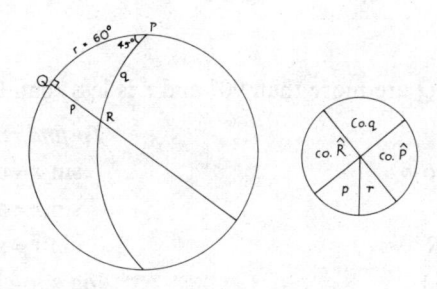

Fig. 9·4

Referring to fig. 9·4:

All parts are *LESS* than 90°

To find p:

$\sin r = \tan p \tan \text{co-}P$

$\sin r = \tan p \cot P$

$\tan p = \sin r \tan P$

$r = 60°$ log sin $= \bar{1}\cdot 93753$

$P = 45°$ log tan $= 0\cdot 00000$

log tan $= \bar{1}\cdot 93753$

$p = 40° 54'$

To find R:

$\sin \text{co-}R = \cos r \cos \text{co-}P$

$\cos R = \cos r \sin P$

$\cos R = \cos r \sin P$

log cos $= \bar{1}\cdot 69897$

log sin $= \bar{1}\cdot 84949$

log cos $= \bar{1}\cdot 54846$

$R = 69° 17'$

To find q:

$\sin \text{co-}P = \tan r \tan \text{co-}q$

$\cos P = \tan r \cot q$

$\cot q = \cot r \cos P$

$\log \cot = \bar{1} \cdot 76144$

$\log \cos = \bar{1} \cdot 84949$

$\log \cot = \bar{1} \cdot 61093$

$q = 67° \ 47'$

Answer—$p = 40° \ 54'$, $R = 69° \ 17'$, $q = 67° \ 47'$.

Example 9·2—In the spherical triangle PQR: $P = 90°$, $R = 45°$, $p = 110°$. Solve the triangle.

Fig. 9·5

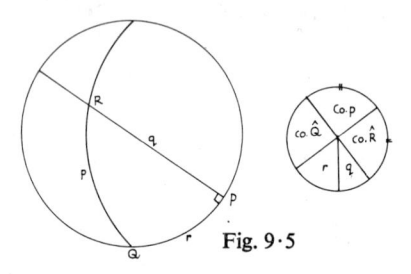

Referring to fig. 9·5:

q and Q are more than 90° and r is less than 90°

To find q:

$\sin \text{co-}R = \tan q \tan \text{co-}p$

$\cos R = \tan q \cot p$

$\tan q = \tan p \cos R$

$p = 110°$ $\log \tan = 0 \cdot 43893(-)$

$R = 45°$ $\log \cos = \bar{1} \cdot 84949(+)$

$\log \tan q = 0 \cdot 28842(-)$

$q = 117° \ 14'$

To find r:

$\sin r = \cos \text{co-}p \cos \text{co-}R$

$\sin r = \sin p \sin R$

$\sin r = \sin p \sin R$

$\log \sin = \bar{1} \cdot 97299(+)$

$\log \sin = \bar{1} \cdot 84949(+)$

$\log \sin r = \bar{1} \cdot 82248(+)$

$r = 41° \ 38'$

To find Q:

$\sin \text{co-}p = \tan \text{co-}Q \tan \text{co-}R$

$\cos p = \cot Q \cot R$

$\cot Q = \cos p \tan R$

$\log \cos = \bar{1} \cdot 53405(-)$

$\log \tan = 0 \cdot 00000(+)$

$\log \cot = \bar{1} \cdot 53405(-)$

$Q = 108° \ 53'$

Answer—$q = 117° \ 14'$, $r = 41° \ 38'$, $Q = 108° \ 53'$.

NAPIER'S RULES 55

2. Napier's Rules for Solving Quadrantal Spherical Triangles

A Quadrantal Spherical Triangle is one in which one of the sides has a value of 90°. Quandrantal triangles may be solved by a modification of Napier's Rules for Right-angled Triangles. The modifying rule is:

"In a quadrantal triangle, if both Adjacents or both Opposites are both sides or both angles, change the final sign."

This modifying rule is derived from the fact that a quadrantal triangle may be solved by first solving a "related" right-angled spherical triangle. The following example serves to show this.

Example 9·3—In the quadrantal triangle ABC: $BC = 90° \ 00'$, $BAC = 60° \ 00'$, $ABC = 30° \ 00'$. Find ACB.

Referring to fig. 9·6:

To find C use the right-angled spherical triangle ABD in which:

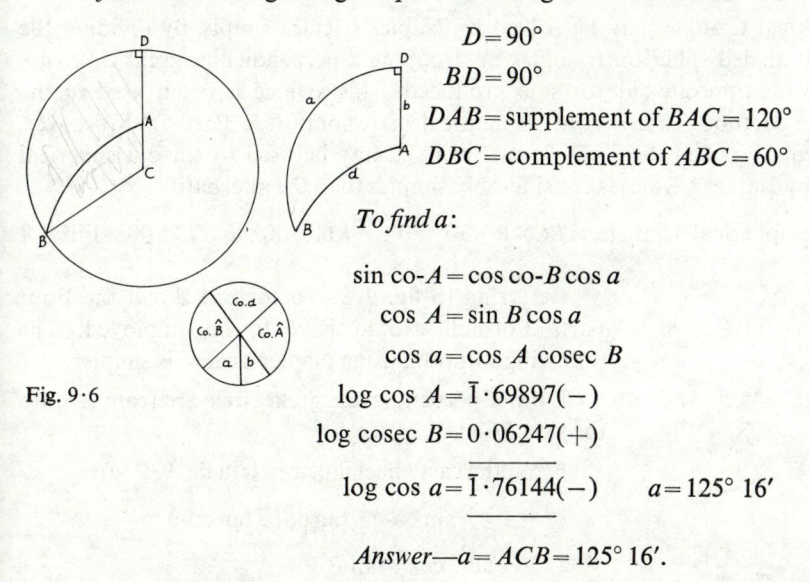

$$D = 90°$$
$$BD = 90°$$
$$DAB = \text{supplement of } BAC = 120°$$
$$DBC = \text{complement of } ABC = 60°$$

To find a:

$$\sin \text{co-}A = \cos \text{co-}B \cos a$$
$$\cos A = \sin B \cos a$$
$$\cos a = \cos A \ \text{cosec} \ B$$
$$\log \cos A = \bar{1} \cdot 69897 (-)$$
$$\log \text{cosec} \ B = 0 \cdot 06247 (+)$$
$$\log \cos a = \bar{1} \cdot 76144 (-) \qquad a = 125° \ 16'$$

Answer—$a = ACB = 125° \ 16'$.

Fig. 9·6

To solve a quadrantal triangle using the modifying rule, the procedure is as follows:

The parts of the quadrantal triangle are written in order in the sectors of the cartwheel as in fig. 9·7.

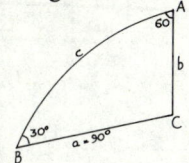

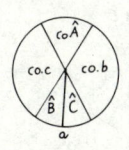

Fig. 9·7

Notice that in fig. 9·7 the angle opposite to the 90°-side and the other two sides are prefixed with the "co" to denote complement.

To find C:

$$\sin \text{co-}A = \cos B \cos C$$

Note that Both the Opposites (B and C) are Angles in this case. Therefore the final sign will have to be changed as shown below.

$$\cos A = \cos B \cos C$$
$$\cos C = \cos A \sec B$$
$$\log \cos A = \bar{1} \cdot 69897 (+)$$
$$\log \sec B = 0 \cdot 06247 (+)$$
$$\log \cos C = \bar{1} \cdot 76144 (+) \quad \text{This sign becomes } (-)$$

Therefore:
$$C = 180° - 54° \, 44'$$
$$= 125° \, 16'$$

Answer—$C = 125° \, 16'$.

3. The Solution of Oblique Spherical Triangles by Napier's Rules

Any oblique spherical triangle may be solved by Napier's Rules simply by dividing the triangle into two right-angled spherical triangles by dropping a perpendicular great circle arc from any vertex onto the opposite side or side produced. This artifice is often used in the construction of Short Method Tables used in nautical astronomy (see Part 5). Moreover, in many cases, particularly when the Four Parts Formula may be used to solve a spherical triangle, the solution by Napier's Rules is considerably simpler than the alternative.

Example 9·4—In the spherical triangle ABC, $B = 30° \, 00'$; $c = 60° \, 00'$, $b = 70° \, 00'$. Find A using Napier's Rules.

Referring to fig. 9·8: To solve A direct the Four Parts Formula would have to be employed. The following solution, using Napier's rules, is simpler.

Drop a perpendicular great circle arc from A onto the side BC at X.

From the cartwheel illustrated in fig. 9·9 (*a*):

$$\sin \text{co-}c = \tan \text{co-}B \tan \text{co-}\theta$$
$$\cot \theta = \cos c \tan B$$

$$\log \cos c = \bar{1} \cdot 69897$$
$$\log \tan B = \bar{1} \cdot 76144$$
$$\log \cot \theta = \bar{1} \cdot 46041 \quad \theta = 73° \, 54'$$

$$\sin x = \sin B \sin c$$
$$\log \sin B = \bar{1} \cdot 69897$$
$$\log \sin c = \bar{1} \cdot 93753$$
$$\log \sin x = \bar{1} \cdot 63650 \quad x = 25° \, 39'$$

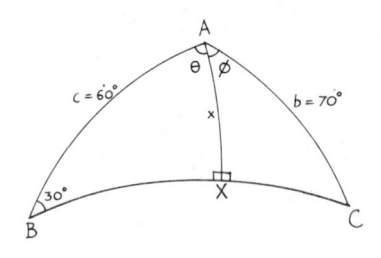

Fig. 9·8

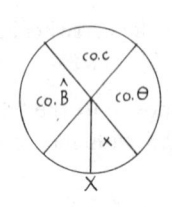

Fig. 9·9 (*a*)

From the cartwheel illustrated in fig. 9·9(b):

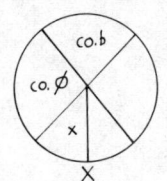

$$\sin \text{co-}\varphi = \tan \text{co-}b \tan x$$
$$\cos \varphi = \cot b \tan x$$
$$\log \cot b = \bar{1}\cdot 56107$$
$$\log \tan x = \bar{1}\cdot 68158$$
$$\log \cos \varphi = \bar{1}\cdot 24265 \qquad \varphi = 79° 56'$$

$$A = \theta + \varphi$$
$$= 73° 54' + 79° 56'$$
$$= 153° 50'$$

Answer—$A = 153° 50'$

Exercises on Chapter 9

1. In the spherical triangle ABC: $A = 40° 00'$, $B = 90° 00'$, $C = 65° 00'$. Find the three sides.
2. In the spherical triangle EFG, $F = 90° 00'$, $E = 28° 45'$, $EF = 75° 15'$. Find the unknown parts.
3. In the spherical triangle XYZ: $Z = 90° 00'$, $ZY = 40° 30'$, $Y = 102° 30'$. Find the unknown parts.
4. In the spherical triangle PQR: $R = 90° 00'$, $P = 115° 35'$, $PQ = 98° 40'$. Solve the triangle.
5. In the spherical triangle ZYX: $Z = 90° 00'$, $X = 140° 00'$, $ZX = 123° 00'$. Solve the triangle.
6. In the spherical triangle ABC: $AB = 90° 00'$, $C = 65° 00'$, $B = 150° 00'$. Solve the triangle.
7. In the spherical triangle XYZ: $XY = 90° 00'$, $YZ = 55° 00'$, $ZX = 70° 00'$. Find the unknown parts.
8. In the spherical triangle PZX: $PX = 50° 00'$, $ZX = 30° 00'$, $P = 35° 00'$. Explain why two values may be assigned to angle Z. Compute these values.
9. In the spherical triangle ABC: $AB = 75° 00'$, $BC = 60° 00'$, $A = 50° 00'$. Find B and C.
10. In the spherical triangle XYZ: $XZ = 100° 00'$, $X = 30° 00'$, $Z = 40° 00'$. Find YZ.

PART 2

THE SAILINGS

The "Sailings" embrace the several methods used to find the course to steer and the distance to travel in going from one position on the Earth's surface to another. The fact that vessels travel over a spherical surface made for difficulties in connection with sailing problems which were not overcome until the advent of the Mercator Chart in the sixteenth century. In this Part we shall first consider the shape and size of the Earth; the methods of defining position on the Earth's surface; the nature of the tracks traced out by vessels moving over the sea; and the principles of the Mercator Chart. Following this a discussion on the several methods of computing courses and distances will be presented.

CHAPTER 10

THE SHAPE AND SIZE OF THE EARTH

1. The Earth

The Earth's shape is not quite spherical. In many navigational problems, however, the Earth is considered to be a perfect sphere—an assumption which leads to no appreciable error.

It is believed that the notable Pythagoras, of right-angled triangle fame, taught that the Earth "is a ball suspended in space". It was not, however, until about three centuries after the time of Pythagoras, that the first recorded attempt at measuring the Earth was made by the Greek philosopher Eratosthenes. Eratosthenes noticed that at noon on the longest day of the year, at Syene in the upper Nile valley, the buildings cast no shadows. At Alexandria, situated to the north of Syene, the buildings did cast shadows at noon on the longest day of the year. Eratosthenes accounted for this by arguing that the Earth must be spherical, and that the parallel rays of the very remote Sun cast shadows of different lengths at the two places.

Fig. 10·1 illustrates the method used by Eratosthenes for determining the circumference of the Earth. By finding the angle θ by measurement, and estimating the distance between *A* and *B*, the Earth's circumference may be deduced from the relationship:

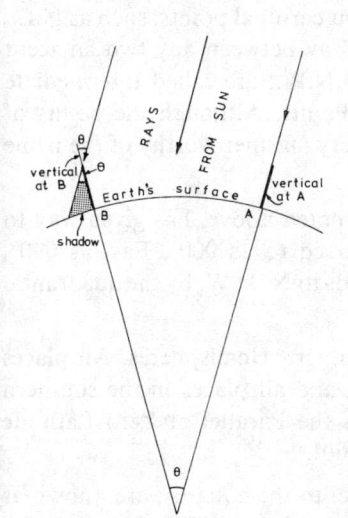

Fig. 10·1

Arc *AB* : θ° : : Circumference : 360°

From which:

$$\text{Circumference} = \frac{360 \cdot AB}{\theta}.$$

The Earth rotates about a fixed diameter known as the Earth's polar axis. The rate of the Earth's rotation is relatively slow: it spins once in the same time as it takes for the hour hand of a clock to make two circuits of the dial.

The direction towards which points on the Earth's surface are carried around the Earth's polar axis is known as East. The direction opposite to East is called West.

The extremities of the Earth's axis are known as the Earth's Poles. These two points are the poles of a great circle which lies in the plane of the Earth's rotation. This great circle is known as the Equator.

The Earth's pole at which the Earth's rotation is clockwise when viewed from above it is known as the North Pole. The other is called the South Pole.

An observer facing the direction of the North Pole from any point on the Earth's surface, would be looking in a direction which is 90° to the left of East. This direction is called North. The direction opposite to North is called South.

The four directions North, South, East and West, which are abbreviated to N., S., E. and W., respectively, are known as the Cardinal Points of the Compass. All other horizontal directions may be referred to the two adjacent cardinal points. Thus we may signify that a lighthouse bears N. 28° E., meaning that the horizontal angle between the direction of North and that of the lighthouse is 28°.

The complete "compass", that is to say, the horizontal circle through the cardinal points, is divided into 32 points. The term Point of the Compass sometimes refers to a direction and sometimes to the 32nd part of the compass, which is an arc of $11\frac{1}{4}°$.

The directions which lie midway between any two adjacent cardinal points, such as N.E., S.W., are called Half Cardinal Points. Those which lie midway between any two adjacent cardinal and half cardinal points, such as E.N.E., W.S.W., N.N.W., are called Intermediate or Three-Letter Points. The remaining points are called By-Points. Although the points of the compass are not used so extensively as in days gone by, every mariner worthy of the name should be able to Box the Compass.

The quadrantal system of denoting horizontal directions, noted above, has given way to the superior Three Figure Notation, in which North is referred to as 000°, East as 090°, South as 180°, West as 270°, and so on to 359° which corresponds to N. 1° W. by the quadrantal notation.

The equator divides the Earth into the Northern and Southern Hemispheres. All places in the northern hemisphere are said to have North Latitude, and all places in the southern hemisphere, South Latitude. The equator may be defined as the Parallel of Zero Latitude because every point on the equator has a Latitude of 00° 00′ 00″.

Small circles on the surface of the Earth, which are parallel to the equator, are known as Parallels of Latitude. All points on a particular parallel of Latitude have the same Latitude. The Latitude of a place on the Earth's surface, assuming the Earth to be perfectly spherical, is the angle at the Earth's centre, measured in the plane of a secondary to the equator, from the plane of the equator to the place.

Secondary great circles to the equator are called Meridians. Thus, the Latitude of a place is defined as the arc of a meridian intercepted between the equator and the place. Strictly speaking, meridians are semi-great circles which terminate at the Earth's Poles. In other words a secondary to the equator forms two antipodal meridians.

2. Describing a Terrestrial Position

Navigators employ one of two general methods of describing a position on the Earth's surface. The more common method is to state the parallel of Latitude and the meridian on which the position to be described rests. The parallel of Latitude is denoted by stating the Latitude of the place, and the meridian is denoted by stating an angle called Longitude. Whereas the datum parallel from which Latitude is measured is the equator, the datum meridian from which Longitude is measured is the meridian of Greenwich. This meridian, the Prime Meridian, is generally called the Greenwich Meridian.

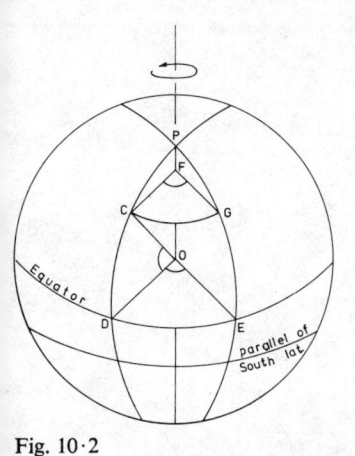

In fig. 10·2, the angle DOC is the Latitude of the point C (and of every other point on C's parallel of Latitude). If the meridian on which the point G lies is the Greenwich Meridian, then the West Longitude of C is given by the angle EOD.

But: Angle EOD = Angle GFC
= Angle EPD
= Arc ED

Notice in fig. 10·2 that the Latitude of the pole is 90°.

The Longitude of a place is the smaller angle at either pole, or the lesser arc of the equator, contained between the Greenwich Meridian and the meridian on which the place lies. Every point on the Greenwich Meridian has a Longitude of 00° 00′ 00″.

Fig. 10·2

The Greenwich and the Antipodal, or 180th meridian, divides the Earth into the Eastern and Western Hemispheres. All places which lie to the east of the Greenwich Meridian and to the west of the 180th meridian, are said to have East Longitude. All places which lie to the west of the Greenwich Meridian and to the east of the 180th meridian have West Longitude.

By examining a world map it may be verified that Cardiff is in Latitude 51° 30′ N. Longitude 03° 10′ W., and that Capetown is in Latitude 34° 00′ S. Longitude 18° 30′ E.

The Difference of Latitude—abbreviated to D. Lat.—between two places is the arc of any meridian contained between the parallels of Latitude of the two places. If the two places have Latitudes of the same name, the D. Lat. is found by subtracting the smaller from the greater Latitude. If the two places have Latitudes of different names, the D. Lat. is found by adding the two Latitudes. D. Lat. is sometimes named North or South, according as the ship is moving northerly or southerly, respectively.

The Difference of Longitude—abbreviated to D. Long.—between two places is the smaller angle at either pole, or the lesser arc of the equator, contained between the meridians of the two places. If the two places have Longitudes of the same name the D. Long. is found by subtracting the smaller from the greater Longitude. When the two places have Longitudes of different names, and the Greenwich Meridian lies within the arc of D. Long., the D. Long. is found by adding the Longitudes. When, however, the 180th meridian lies within the arc of D. Long. between two places, the D. Long. is found by adding the Longitudes and subtracting the sum from 360° 00′ 00″. D. Long. is named East or West, according as the ship moves easterly or westerly, respectively.

Example 10·1—Find the D. Lat. and D. Long. between the following pairs of positions:

(*a*) From   Lat. 20° 38′ N.   Long. 96° 54′ W.
      To   Lat. 15° 22′ N.   Long. 35° 34′ W.

(*b*) From   Lat. 15° 10′ S.   Long. 36° 06′ E.
      To   Lat. 06° 08′ N.   Long. 06° 55′ E.

(c) From   Lat. 20° 33′ S.   Long. 04° 00′ W.
    To   Lat. 15° 36′ S.   Long. 05° 38′ E.

(d) From   Lat. 54° 45′ N.   Long. 176° 25′ E.
    To   Lat. 00° 55′ S.   Long. 164° 52′ W.

(a) From   Lat. 20° 38′ N.   Long. 96° 54′ W.
    To   Lat. 15° 22′ N.   Long. 35° 34′ W.

    D. Lat. 05° 16′ S.   D. Long. 61° 20′ E.

(b) From   Lat. 15° 10′ S.   Long. 36° 06′ E.
    To   Lat. 06° 08′ N.   Long. 06° 55′ E.

    D. Lat. 21° 18′ N.   D. Long. 29° 11′ W.

(c) From   Lat. 20° 33′ S.   Long. 04° 00′ W.
    To   Lat. 15° 36′ S.   Long. 05° 38′ E.

    D. Lat. 04° 57′ N.   D. Long. 09° 38′ W.

(d) From   Lat. 54° 45′ N.   Long. 176° 25′ E.
    To   Lat. 00° 55′ S.   Long. 164° 52′ W.

    D. Lat. 55° 40′ S.   D. Long. 18° 43′ E.

In the alternative method of describing a terrestrial position the direction of the position and its distance from some known reference point are stated. The reference point is usually a prominent headland, a lighthouse, or an important landmark. The direction is given by stating a Bearing. The Bearing of an object indicates its compass direction. Thus we may say that a ship is in a position with Cape Hatteras bearing 265° at a distance of 16 miles. This means that the ship lies 085°—which is the opposite direction to the bearing of the Cape—16 miles from Cape Hatteras.

3. The True Shape of the Earth

Thus far the shape of the Earth has been considered to be a perfect sphere. For certain problems in navigation, notably in connection with the mariner's chart and the nautical unit of distance, it is necessary for us to consider the Earth's true shape. The actual shape of the Earth is that of an oblate Spheroid of Revolution. An oblate spheroid is the shape that would be swept out by rotating an ellipse about its minor diameter.

The Ellipticity of the terrestrial spheroid, that is to say, the ratio between the difference of the lengths of the equatorial and polar radii, and the length of the equatorial radius, is approximately 1/300. This very small fraction indicates that the Earth is almost a perfect sphere. The Earth's principal radii are:

$$\text{Equatorial radius} = 6{,}378{,}249 \text{ metres}$$
$$\text{Polar radius} = 6{,}356{,}515 \text{ metres}$$

THE SHAPE AND SIZE OF THE EARTH

The true shape of the Earth affects our earlier definition of Latitude, so that it is necessary to examine this closely.

The Vertical at any place is the direction perpendicular to the horizontal plane which touches the Earth's surface at the place. The angle contained between the vertical at a place and the plane of the equator is known as the Geographical Latitude of the place. It is the Geographical Latitude that is measured in astronomical observations for Latitude. For this reason it is often called True or Astronomical Latitude. When the term Latitude is used without qualification, it is understood to mean Geographical Latitude.

The angle at the Earth's centre contained between the equator and any place on the Earth's surface and measured in the plane of the meridian of a place, is called the Geocentric Latitude of the place. Except for places on the equator or at either pole, the Geocentric Latitude of a place is always smaller numerically than the Geographical Latitude of the place. For this reason Geocentric Latitude is sometimes called Reduced Latitude.

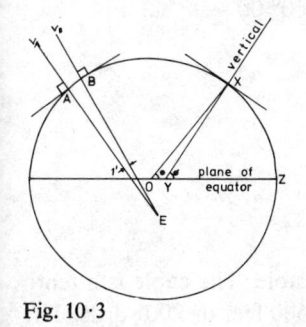

In fig. 10·3:

Geographical Latitude of $X = ZYX = \varphi$

Geocentric Latitude of $X = ZOX = \theta$

The maximum difference between the Geographical Latitude of a place and its Geocentric Latitude occurs when the Latitude of the place is 45°. The Geographical Latitude and Geocentric Latitude of any point on the equator is 00° 00′. The Geographical Latitude and the Geocentric Latitude of either pole is 90° 00′.

Fig. 10·3

The difference between the lengths of the equatorial and polar radii is 21,734 metres. This is equivalent to 11·6 nautical miles.

4. The Nautical Mile

The important feature of the navigational unit of distance called the Nautical Mile is that it is related to a meridianal spherical distance of one minute of arc. A nautical mile is the length of an arc of a meridian the Geographical Latitudes of the end points of which differ by 1′ of arc. In fig. 10·3, if the angle between BE and AE, which are the verticals at B and A, respectively, is exactly 1′, the arc-length AB is one nautical mile. It is for this reason that a nautical mile is sometimes defined as the length of an arc of a meridian between two points whose verticals are inclined to one another at an angle of 1′. Thus, if angle AEB is 1′, the arc AB is one nautical mile. The point E is at the centre of curvature of the small piece of the meridian contained between A and B. Because of the oblateness of the Earth, the radius of curvature of the meridian increases as the Latitude increases. As the radius of curvature increases, the arc-length corresponding to an angle of 1′ at the centre of curvature also increases. For this reason, the length of a nautical mile increases as the Latitude increases.

Length of nautical mile in Lat. 0° = 1842·787 metres or 6046 feet

Length of nautical mile in Lat. 90° = 1861·656 metres or 6108 feet

The average length of the nautical mile is 6077 feet, or 1852·221 metres. This corresponds to the length in Latitude 45°. The figure 6077 is rounded off to 6080 and this latter figure is taken as the number of feet in the Standard Nautical Mile. The Standard Nautical Mile in metres is taken as 1852.

The length of the actual nautical mile in any Latitude φ, is given by the formula:

$$\text{Length in Metres} = 1852 - 19 \cos 2\varphi$$
$$\text{or Length in Feet} = 6077 - 31 \cos 2\varphi$$

From this formula it may be verified that the standard nautical mile of 6080 feet or 1852 metres may be used without introducing error only in Latitude 49° approximately. In all other Latitudes, by using a distance-measuring instrument calibrated in standard nautical miles, an error proportional to the distance results. This error is, of course, greatest for any given distance when the Latitude is zero or near 90°. The lengths of a minute of a meridian in Latitudes 0° and 90° are, respectively, 0·995 and 1·005 nautical miles.

Example 10·2—Find the length of the nautical mile in feet in Lat. 60° 00′.

$$\begin{aligned}\text{Length} &= 6077 - 31 \cos (2 \times 60)° \\ &= 6077 - 31 \cos 120° \\ &= 6077 + 15 \cdot 5 \\ &= 6092 \cdot 5\end{aligned}$$

Answer—Length = 6092·5 feet.

A commonly used sub-multiple unit of the nautical mile is the cable. The cable is a tenth of a nautical mile. In practice it is usual to reckon a cable as being 600 feet or 200 yards. The nautical unit of speed is the Nautical Mile per Hour. This is called the Knot.

5. Reduction of the Geographical Latitude

We have noted above that the difference between the lengths of the equatorial and polar radii of the Earth is 11·6 nautical miles. The reduction R of the Geographical Latitude may, therefore, be found from the following formula:

$$R = 11 \cdot 6 \sin 2\varphi$$

where φ is Geographical Latitude.

Example 10·3—Find the Geocentric Latitude of a place whose Geographical Latitude is 30° 00′.

$$\begin{aligned}\text{Reduction} &= 11 \cdot 6 \cdot \sin (2 \times 30)° \\ &= 11 \cdot 6 \cdot \sin 60° \\ &= 10 \cdot 0\end{aligned}$$

$$\begin{aligned}\text{Geocentric Latitude} &= \text{Geographical Latitude} - \text{Reduction} \\ &= 30° \ 00′ - 10′ \\ &= 29° \ 50′\end{aligned}$$

Answer—Geocentric Latitude = 29° 50′.

THE SHAPE AND SIZE OF THE EARTH

6. The Geographical Mile

The length of a minute of arc of the equator is called a Geographical Mile. It is of interest to note that the equator is the only true great circle on the Earth. Meridians, because of the Earth's oblate shape, are semi-ellipses.

The geographical mile is 6087 feet or 1855·4 metres. This distance is used for computing the distance along a parallel of Latitude for purposes of surveying and large-scale mapping.

Exercises on Chapter 10

1. Define Great Circle; Small Circle. Give examples of terrestrial great and small circles.
2. Describe the method used by Eratosthenes for measuring the circumference of the Earth.
3. State four proofs of the Earth's rotundity.
4. Describe the true shape of the Earth. Explain why, in most navigational problems, the Earth may be assumed to be a perfect sphere.
5. What is a Meridian? In what direction would a ship be sailing were she steered along a meridian?
6. Define Bearing. Explain why the bearing of every point on the Earth is 180° from the Earth's North Pole.
7. Define Statute Mile; Geographical mile; Cable.
8. Explain clearly the meaning of Reduction of the Latitude.
9. Describe two systems of defining terrestrial positions.
10. Explain clearly the derivation of the Standard Nautical Mile.
11. Define Prime Meridian; Eastern Hemisphere; D. Lat.; D. Long.
12. What is the antipodal position of Lat. 20° S. Long. 15° W.?
13. A ship sailed due North for two days at 10 knots along the Prime Meridian. Find her final Latitude if her Departure position was in Lat. 54° 00′ S.
14. A ship sailed due West along the equator for 18 hours at 16 knots. Find her final position if her Departure position was in Long. 10° 30′ W.
15. What is the D. Lat. and D. Long. between the following pairs of positions:
    (a) From   Lat. 10° 43′ S. Long. 05° 56′ W.
        To   Lat. 06° 34′ S. Long. 18° 05′ E.
    (b) From   Lat. 34° 18′ N. Long. 177° 08′ E.
        To   Lat. 22° 52′ N. Long. 06° 18′ W.
16. What is meant by the term Ellipticity as it applies to the Earth?
17. Explain how the radius of curvature of the meridians changes with Latitude. How does this change affect the seaman's unit of distance?
18. Describe the error that results by using a patent log calibrated in Standard Nautical Miles when sailing in very low or very high Latitudes.
19. Define Geocentric Latitude; Geographical Latitude. What is the Geocentric Latitude of a place whose True Latitude is 60° 00′ N.?
20. Calculate the length in feet of the nautical mile in Lat. 42° S.

CHAPTER 11

THE RHUMB LINE

1. Introduction

Ships are steered from place to place, when out of sight of land, by means of a magnetic or gyro compass which indicates a fixed horizontal direction irrespective of the movements of the ship. Compass Points are marked on the outer edge of the compass card; and radial lines, extending from the centre of the card to the several points, are known as Rhumbs. When a ship's head is steadied in a certain compass direction the fore-and-aft line of the ship lies in the vertical plane of a rhumb, and it is easy to visualise the path and track of the ship as extensions of the rhumb. For this reason a line of constant course is known as a Rhumb Line.

A rhumb line is usually defined as a line on the Earth's surface which cuts every meridian at the same constant angle. The most convenient path to travel along is a rhumb line path which connects the places of departure and destination. This is so because, in travelling along a rhumb line, the course of the vessel remains constant.

In fig. 11·1, *AB* represents a typical rhumb line. Notice the constant angle which it makes with the meridians it crosses.

Special cases of rhumb lines are the equator, parallels of Latitude, and meridians. The equator and all other parallels of Latitude are rhumb lines because the course of a vessel travelling along a parallel is constantly 090° or 270°.

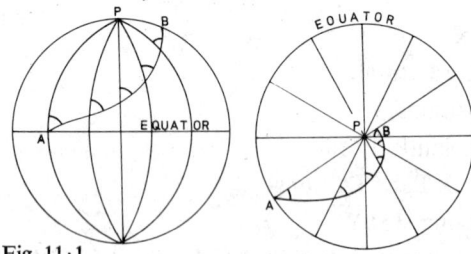

Fig. 11·1

Meridians are rhumb lines of constant course 000° or 180°.

The art of sailing obliquely across the meridians is known as Loxodromics, from the Greek words "loxos" and "dromos" meaning *oblique* and *running* respectively. For this reason all rhumb lines, other than parallels of Latitude and meridians, are sometimes called Loxodromic Curves. When a vessel sails along a Loxodromic Curve her track is an equi-angular spiral which constantly approaches the Earth's Pole. This follows because the meridians close together, or converge, as the Latitude increases. Theoretically a Loxodromic Curve continually gets closer to but never reaches the Earth's Pole.

2. The Sailings

The Sailings comprise the various methods of finding the course and distance from one place on the Earth's surface to another. When the distance travelled by a vessel is relatively small it is usual, when practicable, to travel along the rhumb line connecting the points of departure and destination. For long distances, however, it is often advantageous to travel along the great circle arc connecting the points of departure and destination. In the latter case the distance to travel is less than the rhumb line distance.

There are four methods of Rhumb Line Sailing. Two of these will be dealt with in this chapter.

THE RHUMB LINE

3. Parallel Sailing

When a vessel travels in any direction except due North or due South, she moves some distance towards due East or due West. This distance is known as Departure. Departure may be represented by an arc of a parallel of Latitude cut off between the meridians of the points between which the vessel travels.

If a vessel travels along the equator the D. Long. between the places left and arrived at is numerically equal to the Departure, the D. Long. being given in minutes of arc and the departure in nautical miles. This, of course, assumes the Earth to be a perfect sphere, in which case a nautical mile would be the length of any arc of the Earth's surface, the extremities of the arc subtending an angle of one minute at the Earth's centre.

When a vessel travels along any parallel of Latitude other than the equator; that is to say, when her course is due East or due West, the Departure measured in miles between the points left and arrived at is always numerically less than the D. Long. in minutes of arc between the two points. This is due to the Convergency of the meridians.

The relationship between D. Long., Departure, and Latitude, is given in the Parallel Sailing Formula, in which the Earth is assumed to be a perfect sphere.

4. The Parallel Sailing Formula

In fig. 11·2:

AB = Departure in miles

CD = D. Long. in equatorial minutes of arc, these being equivalent to nautical miles on a spherical Earth.

Now:

$$\frac{\text{Departure}}{\text{D. Long.}} = \frac{AB}{CD}$$

$$= \frac{AE}{CO} \text{ (arcs of concentric circles subtended by the same angle are proportional to their radii)}$$

$$= \frac{AE}{AO} \text{ } (CO = AO\text{: radii of the same sphere})$$

$$= \cos \text{Lat. } A \text{ (or } B\text{)}$$

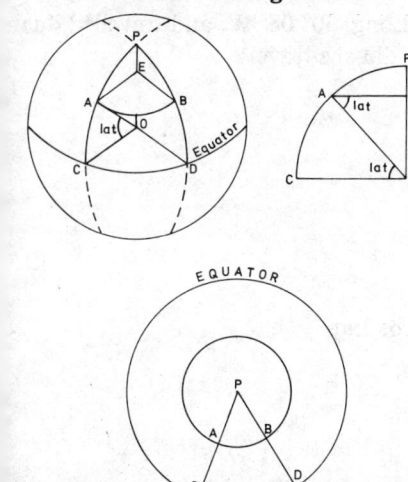

Fig. 11·2

Therefore:

$$\frac{\text{Departure}}{\text{D. Long.}} = \cos \text{Lat.} \quad \ldots \ldots \ldots \ldots \quad (1)$$

$$\frac{\text{D. Long.}}{\text{Departure}} = \sec \text{Lat.} \quad \ldots \ldots \ldots \ldots \quad (2)$$

$$\text{Departure} = \text{D. Long.} \cdot \cos \text{Lat.} \quad \ldots \ldots \ldots \quad (3)$$

$$\text{D. Long.} = \text{Departure} \cdot \sec \text{Lat.} \quad \ldots \ldots \ldots \quad (4)$$

These four relationships are variations of the Parallel Sailing Formula.

THE ELEMENTS OF NAVIGATION AND NAUTICAL ASTRONOMY

The traverse table is almost invariably used for solving problems in which the Parallel Sailing Formula is involved. The three columns of the table are labelled with supplementary headings: the top of the Distance column is labelled D. Long., and the top of the D. Lat. column is labelled Dep. In a traverse table which extends only to 45°, the bottoms of the Distance and Departure columns are labelled D. Long. and Dep. respectively.

Example 11·1—A vessel travelled 100 miles due East along the parallel of 50° 30′ N. If the Longitude of the point she left was 03° 50′ W., find her final Longitude.

$$\text{D. Long.} = \text{dep . sec Lat.}$$
$$= 100 \text{ . sec } 50° 30′$$

From traverse tables:

$$\text{D. Long.} = 158′$$
$$= 02° 38′ \text{ E.}$$
$$\text{Long. left} = 03° 50′ \text{ W.}$$
$$\overline{\text{Final Long.} = 01° 12′ \text{ W.}}$$

Answer—Final Longitude = 01° 12′ W.

Example 11·2—A vessel left a position in Lat. 39° 00′ S. Long. 30° 08′ W. and travelled due East until her Longitude was 25° 22′ W. How many miles did she travel?

$$\text{Long. from} = 30° 08′ \text{ W.}$$
$$\text{Long. to} = 25° 22′ \text{ W.}$$
$$\overline{\text{D. Long.} = 04° 46′ \text{ E.}}$$
$$= 286′ \text{ E.}$$

$$\text{Dep.} = \text{D. Long. . cos Lat.}$$
$$= 286 \text{ . cos } 39°$$

From Traverse Table:

$$\text{Dep.} = 222·3 \text{ miles}$$

Answer—Distance = 222·3 miles.

Example 11·3—What is the Latitude where the D. Long. is numerically equal to three times the Departure?

$$\cos \text{Lat.} = \frac{\text{Departure}}{\text{D. Long.}}$$
$$= \frac{\text{Dep.}}{3 \times \text{Dep.}}$$
$$= \frac{1}{3}$$
$$\text{Lat.} = 70° 32′ \text{ N. or S.}$$

Answer—Latitude = 70° 32′ N. or S.

THE RHUMB LINE

5. Plane Sailing

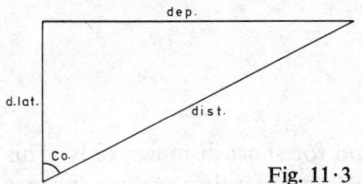

When a vessel travels along a rhumb line, the acute angle which her fore-and-aft line makes with the meridians she crosses is known as the Course Angle. When a vessel travels along any rhumb line except a meridian or parallel of Latitude; the Distance steamed, the difference of Latitude, and the Departure between the first and final positions, may be regarded as forming the sides of a plane right-angled triangle with the Course Angle opposite to the side representing the Departure. This plane right-angled triangle is called the Plane Sailing Triangle. It must be borne in mind that this triangle does *NOT* represent a triangle on the Earth's surface: it is simply an artifice which shows the relationship between Rhumb Line Course, Distance, D. Lat. and Departure. If the Distance and Course Angle are known the D. Lat. and Departure may be found by solving a Plane Sailing Triangle. The formulae used in solving plane sailing triangles are called the Plane Sailing Formulae as indicated in fig. 11·3.

Fig. 11·3

They are,

$$\text{Departure} = \text{Distance} \cdot \sin \text{Course} \quad\quad\quad\quad (1)$$

$$\text{D. Lat.} = \text{Distance} \cdot \cos \text{Course} \quad\quad\quad\quad (2)$$

By dividing (1) by (2) we get:

$$\frac{\text{Dep.}}{\text{D. Lat.}} = \tan \text{Course} \quad\quad\quad\quad (3)$$

It should be noted that the Plane Sailing Formulae are true irrespective of the magnitude of the distance. The Plane Sailing Formulae, and their general applicability, will now be proved.

6. Proof of Plane Sailing Formulae

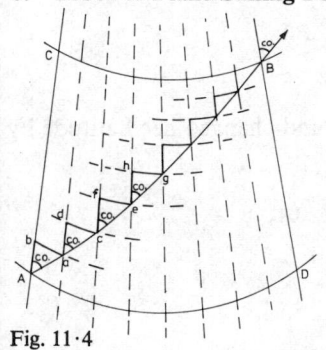

Fig. 11·4 represents a portion of the Earth's surface showing parts of two meridians and two parallels of Latitude. The rhumb line between A and B is drawn. Let the rhumb line Course Angle be denoted by θ.

Imagine the distance AB to be divided into a sufficiently large number of small pieces, so that the triangles Aab, acd, cef, etc., may be considered to be plane. Strictly speaking the pieces Aa, ac, ce, etc., should be infinitely small. On this assumption the following proof holds good.

Fig. 11·4

Ab, ad, cf, etc., are pieces of D. Lat.

ab, cd, ef, etc., are pieces of Departure.

Then:

$$ab + cd + ef + \text{etc.} = Aa \cdot \sin \theta + ac \cdot \sin \theta + ce \cdot \sin \theta + \text{etc.}$$
$$= \sin \theta \, (Aa + ac + ce + \text{etc.})$$

But, $\quad Aa + ac + ce + \text{etc.} = \text{Distance } AB$

And, $\quad ab + cd + ef + \text{etc.} = \text{Departure between } A \text{ and } B$

72 THE ELEMENTS OF NAVIGATION AND NAUTICAL ASTRONOMY

Therefore:
$$\text{Departure} = \text{Distance} \cdot \cos \theta$$

Similarly it may be proved that:
$$\text{D. Lat.} = \text{Distance} \cdot \sin \theta$$

It is often thought that the rules of plane sailing hold good for short distances only. This is entirely wrong: the D. Lat. or Departure between any two places, regardless of their distance apart, may be calculated using Plane Sailing.

Example 11·4—A vessel travelled for 245·0 miles on a Course of 062° S. Find the D. Lat. and Departure.

Course Angle = 62° Distance = 245·0 miles

From Traverse Tables:

Departure = 216·3′ E.

D. Lat. = 115·0′ N.

Answer—Departure = 216·3 miles East.

D. Lat. = 115·0 miles North.

Example 11·5—A vessel left position Lat. 41° 44′ S. on a course of 158°. Find the distance she had travelled on reaching the parallel of 48° 16′ S.

Lat. from = 41° 44′ S.   Course Angle = 22°
Lat. to = 48° 16′ S.   D. Lat. = 392′ S.

D. Lat. = 06° 32′ S.

From Traverse Tables:

Distance = 423 miles

Answer—Distance = 423 miles.

Example 11·6—A vessel travelled for a distance of 346·0 miles and changed her Latitude by 4° 00′. If the course had been in the S.E. quadrant, find it.

D. Lat. = 240·0′ S.

Distance = 346·0 miles

From Traverse Tables:

Course Angle = 46°

Course S. 46° E. = 134°

Answer—Course = 134°.

Example 11·7—A vessel travelled for a distance of 928·0 miles on a course of 306°. Find the D. Lat. and Departure.

Distance = 928·0 miles

= (600+328) miles

Course Angle = 54°

From Traverse Tables:

$$\text{D. Lat.} = 352\cdot7 + 192\cdot8$$
$$= 545\cdot5' \text{ N.}$$
$$\text{Departure} = 485\cdot4 + 265\cdot4$$
$$= 750\cdot8' \text{ W.}$$

Answer—D. Lat. = 545·5′ N.
Departure = 750·8′ W.

7. Traverse Sailing

When a vessel in travelling from one place to another has to make several courses, the irregular track she makes is known as a Traverse. This name is derived from the circumstance that a sailing vessel, in making a passage, would have to cross and recross the desired path several times because of the direction and/or change in direction of the wind during the passage. The problem of finding the course and distance the vessel would have made had it been possible for her to sail directly from the Departure position to the destination; that is to say, the Course and Distance Made Good, is known as Resolving a Traverse. This method of zig-zag sailing was, therefore, known as Traverse Sailing.

When making a traverse, the several legs of the track may be considered to be the hypotenuses of plane sailing triangles. The D. Lat. and Departure of each of these triangles may be solved by means of Plane Sailing. By summing the D. Lats. and Departures of the several plane sailing triangles, the D. Lat. and Departure between the points left and arrived at may be found.

The record of the courses and distances sailed on each leg of a traverse was known, in by-gone days, as the Ship's Reckoning. By means of the reckoning, and a knowledge of the initial Latitude, the Latitude of the vessel at any time could be found without recourse to observations. A Latitude so found was known as a Dead Reckoning, or D.R. Latitude.

In the modern practice of navigation, a D.R. position is one that has been worked up from the last Observed Position, making no allowance for current and/or leeway. The name Observed Position is given to any position obtained from observations of celestial or terrestrial objects, or from any electronic navigation instrument such as radar, Decca Navigator or Radio Direction Finder.

When observations are not possible, and a navigator wishes to know his vessel's position, he applies to the last Observed Position, courses and distances travelled through the water since the time of the Observed Position, and so finds his D.R. Position. To the D.R. Position he applies an estimated allowance for current, leeway, and any other disturbing factor which has influenced the vessel's movement. The position so found is referred to as an Estimated Position (E.P.). An Estimated Position is the most reliable position obtainable when direct observations are not available.

The traverse table lends itself admirably to the solution of traverse sailing. This is the reason, in fact, why the traverse table is so-named.

8. The Departure Position

When it is necessary to venture into the open sea it is essential, before the land is lost to sight, that the position of the vessel be found from terrestrial observations, in order to obtain

74   THE ELEMENTS OF NAVIGATION AND NAUTICAL ASTRONOMY

a reliable Observed Position from which the course may be set. Such a position is known as a Departure Position. It is customary to describe a Departure Position as a bearing and distance from some conspicuous land- or sea-mark.

9. Current

The movement of the surface layers of the sea, due to meteorological causes, is known as current. The direction towards which the water in a current is moving is known as the **Set**, and the speed is known as the Rate of the current. The distance which a vessel is set in any given interval of time is called the Drift of the current.

If current is the only external factor influencing the movement of a vessel, the set and drift is equivalent to the course and distance from a D.R. Position to a corresponding Observed Position.

Example 11·8—A vessel takes her Departure from a position with Cape Sable in Lat. 43° 25′ N. Long. 65° 30′ W., bearing 062° distance 12·0 miles. Course was set to 210°, log zero. When the log registered 14 the course was altered to 300° and when it registered 32 the course was altered to 223°. Find the vessel's D.R. Latitude when the log registered 49. Find also the course and distance the vessel made good.

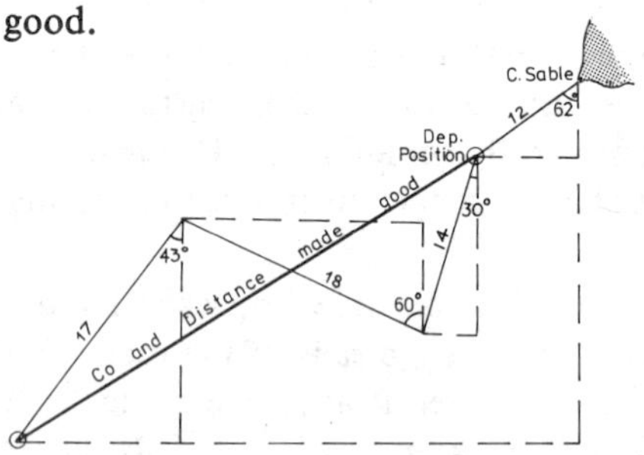

Fig. 11·5

In fig. 11·5:

Departure Bearing is N. 62° E.

Departure Course is S. 62° W.

	D. Lat.		Dep.	
	N.	S.	E.	W.
Dep. course S. 62° W. distance 12 miles	—	5·6	—	10·6
1st course S. 30° W. distance 14 miles	—	12·1	—	7·0
2nd course N. 60° W. distance 18 miles	9·0	—	—	15·6
3rd course S. 43° W. distance 17 miles	—	12·4	—	11·6
		30·1′ S.		
		9·0′ N.		
	D. Lat. (for Lat.)	21·1′ S.		34·2 = **Dep.**
		5·6		
	D. Lat. (for Co.)	15·5′ S.		

From Traverse Tables:

Course and Distance Made Good = 246° × 38 miles

Lat. Cape Sable = 43° 25′ N.

D. Lat. = 21·1′ S.

D.R. Lat. ship = 43° 03·9′ N.

Answers—Course Made Good = 246°

Distance Made Good = 38 miles

D.R. Latitude = 43° 04′ N.

It is to be noted at this stage that it is not possible to find the final Longitude by using Plane or Traverse Sailing. When sailing obliquely across meridians, the Latitude changes constantly, and difficulty arises in finding the correct Latitude to use for converting Departure into D. Long. and *vice versa*. To find the position (Latitude *AND* Longitude) of a vessel after she has travelled on a given course for a given distance such that her Latitude *AND* Longitude change, is the problem of Mercator Sailing or Middle Latitude Sailing. These problems will be discussed in Chapter 13. The next chapter deals with the principle and the construction of the Mercator Chart, this leading to a discussion on the problems of Mercator and Middle Latitude Sailing.

Exercises on Chapter 11

1. Describe the properties of a rhumb line.
2. Define: Departure. What is the relationship between Departure and D. Long?
3. Explain clearly why the traverse table may be used for converting Departure into D. Long.
4. Construct a traverse table for distance 652 miles and course angles at 10°-intervals from 000°.
5. Devise a graphical method, suitable for all Latitudes, for converting Departures into D. Long. for distances up to 100 miles.
6. What is meant by Plane Sailing?
7. Define: D.R. Position; Estimated Position; Observed Position.
8. Prove that: D. Lat. = Distance . cos Course, for all distances on a spherical Earth.
9. A vessel travels 200 miles due East in Latitude 40° 30′ N. Find her change in Longitude.
10. How many miles must a vessel travel along the parallel of Latitude 56° 00′ S. in order to change her Longitude 10° 00′?
11. A vessel travels 250 miles due West and changes her Longitude by 8° 00′. Find the Latitude of the parallel along which she travelled.
12. Find the length of the parallel of Latitude 35°.
13. At what speed is a point in Latitude 60° 00′ carried around the Earth's axis?
14. A vessel travels for 12 hours at a speed of 10·0 knots due East along the parallel of Latitude 50° 10′ S. Her Longitude changes 03° 24′. Find the set and rate of the current.

76   THE ELEMENTS OF NAVIGATION AND NAUTICAL ASTRONOMY

15. A vessel leaves a position in Latitude 40° 30′ N. Long. 16° 00′ W., and makes good the following courses and distances:
    (i) due East 300 miles   (iii) due West 300 miles
    (ii) due North 300 miles   (iv) due South 300 miles.
    Find her final position.

16. A vessel leaves a position in Lat. 30° 00′ S. Long. 178° 05′ E. and travels 200 miles due East. Find her final Longitude.

17. A vessel left a position in Lat. 20° 00′ S. Long. 18° 00′ E. and travelled due South for 120 miles, when her position was found to be in Lat. 22° 00′ S. Long. 18° 12′ E. Find the set and drift of the current.

18. Two vessels are 50 miles apart in Latitude 35° 00′ N. They both travel due South until they are 55 miles apart. What is their present Latitude and how far has each vessel travelled?

19. A vessel on a course of 305° changed her Latitude by 4° 25′. Find the Departure and distance.

20. Find the departure and the change in Latitude after having travelled on a course of 163° for a distance of 312 miles.

21. Find the distance and change in Latitude after having made a Departure of 218 miles on a course of 218°.

22. A vessel steamed between South and East, and in so doing made a Departure of 59 miles and changed her Latitude by 81 miles. Find the distance and course made good.

23. A vessel took her Departure off the South West coast of Ireland with the Fastnets bearing 037° distant 12·0 miles. The log was set to zero and the following courses were steered:
    205° until the log registered 82
    208° until the log registered 196
    220° until the log registered 326
    Find the course and distance made good and the D.R. Latitude of the vessel when the log registered 326.

24. A vessel left a position off Callao in Lat. 14° 50′ S. Long. 76° 55′ W., and made the following courses and distances:
    195° for 45 miles
    165° for 160 miles
    170° for 82 miles
    The current was estimated to have set 290° for 27 miles. Find the course and distance made good and the vessel's present estimated Latitude.

25. A vessel left a position off the Cape of Good Hope with the Cape bearing 090° distance 20 miles. She travelled for 265 miles on a course of 330° and for 345 miles on a course of 324°. During the interval the current was estimated to have set 020° for 25 miles. Find the estimated course and distance made good by the vessel, and her estimated Latitude at the end of the interval.

26. How many miles are there in one degree of D. Long. in Lat. 42°?

27. A vessel took her Departure from a position in the mouth of the River Plate in Lat. 34° 49′ S. Long. 54° 50′ W., and sailed along the parallel to a position off the Cape of Good Hope in Lat. 34° 49′ S. Long. 20° 00′ E. Find the distance travelled.

CHAPTER 12

THE MERCATOR CHART

1. The Navigator's Chart

The two main requirements of a navigational chart are that:

(1) Rhumb lines should be projected as straight lines so that course lines may be laid down easily.

(2) Angles, such as Course and Bearing Angles, should be projected without distortion.

The Mercator Chart fulfils these requirements and, for this reason, nearly all navigational charts for use at sea are of this type.

The principle on which the Mercator Chart is constructed was first used in the sixteenth century by a German cartographer named Gerhard Kaufman, the Latinised version of whose name (which in English means merchant) is Mercator. There seems to be doubt that Kaufman understood the exact mathematical principle of the chart which bears his name, and credit is given to the famous Elizabethan scholar Edward Wright for discovering the mathematical principle of the Mercator Chart. Wright published a description, and also a table for facilitating the construction of Mercator Charts, in an important book first published in the closing decade of the sixteenth century.

When the surface of a sphere is projected onto a plane surface there is bound to be distortion. The amount and type of distortion depends upon the method of projecting the spherical surface onto the plane surface. The Mercator Chart is based on a projection which is not a perspective projection. The Mercator projection is a mathematical projection described by cartographers as a Conventional or Non-Perspective Projection. In the Mercator projection, as in all other conventionals, distances representing the spacing of parallels and meridians must be calculated, in order to project. It is not possible to construct an accurate Mercator Chart geometrically, like, for example, a stereographic projection.

In order that angles on the projection are not distorted the exaggeration of the representation of any small area of the sphere's surface must be equal in the North/South to that in the East/West direction. A projection in which angles are not distorted is known as an Orthomorphic projection.

2. Features of a Mercator Chart

The characteristic features of a Mercator Chart are:

(1) All meridians are projected as equidistantly spaced parallel straight lines.

(2) All parallels of Latitude are projected as parallel straight lines perpendicular to the projected meridians.

78   THE ELEMENTS OF NAVIGATION AND NAUTICAL ASTRONOMY

(3) All rhumb lines are projected as straight lines.
(4) All arcs of great circles, with the exceptions of arcs of the equator or any meridian, are projected as curves which are concave to the projected equator.

Because meridians are projected as parallel straight lines, whereas on the globe they converge towards the poles, it follows that the exaggeration of arcs of parallels of Latitude increases polewards. In order for the map to be orthomorphic the distances between successive parallels of Latitude must also increase polewards in the same ratio.

3. The Defects of a Mercator Chart

Although the Mercator Chart satisfies the principal needs of the navigator it does have defects. The principal defects of the Mercator Chart are:

(1) Every Latitude has a different scale of distance.
(2) Great circle arcs, except those of the equator or meridians, are projected as curves. This makes for difficulty in the practice of Great Circle Sailing.

The variation in the Latitude scale causes areas to be exaggerated proportional to their Latitudes. It will be noticed that on a Mercator map of the world, Greenland appears larger than the continent of South America, and yet the range of Latitude of Greenland is no more than about a quarter of that of South America.

4. Distortion of the Mercator Projection

The degree of exaggeration of lengths along parallels and meridians will now be examined. In doing so it will be convenient to think of the Earth reduced in size to a model globe from which the chart is to be projected. Let the radius of the globe be R.

Exaggeration of the Parallels of Latitude

On the Globe:

$$\text{Length of the Equator} = 2\pi R$$

$$\text{Length of the Pole} = 0 \text{ (Pole is a point)}$$

Therefore:

$$\text{Length of any Parallel of Latitude } \theta = 2\pi R \cos \theta$$

On the chart:

$$\text{Length of Equator} = 2\pi R$$

$$\text{Length of any Parallel} = 2\pi R$$

Now:

$$\text{Exaggeration} = \frac{\text{Length on Chart}}{\text{Length on Globe}}$$

Therefore:

$$\text{Exaggeration} = \frac{2\pi R}{2\pi R \cos \theta}$$

$$= \frac{1}{\cos \text{Lat.}}$$

$$= \sec \text{Lat.}$$

A parallel of Latitude, therefore, is projected with exaggeration which is proportional to the secant of the Latitude of the parallel.

Now the trigonometrical ratio of the secant, changes from unity when the angle is 0°, to infinity when the angle is 90°. It is impossible to represent a line that has been exaggerated to an infinite extent. Therefore, the poles of the Earth, whose Latitudes are 90°, cannot be represented on a Mercator Chart. Not only is it impossible to project the poles, but it is impracticable to project areas of very high Latitude. But this does not concern surface mariners, the vessels of whom trade in more temperate climes than those of very high Latitudes. The parallel of Latitude 60° is exaggerated two-fold because the secant of 60° is 2: the parallel of Latitude $70\frac{1}{2}°$ is exaggerated three-fold, because the secant of $70\frac{1}{2}°$ is 3, and so on.

Exaggeration of the Meridians

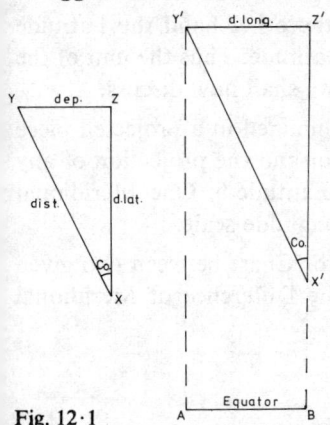

Consider the rhumb line, illustrated in fig. 12·1, which cuts the meridians at an angle θ. A part XY of this rhumb line, if sufficiently small, may be regarded as forming the hypotenuse of a right-angled plane triangle containing the Course Angle θ. Again, if the length of this hypotenuse is sufficiently small the side opposite the course angle in the right-angled triangle may be considered to be equal to the Departure between the end points of the rhumb line XY. The side coinciding with the meridian through X then represents the D. Lat. between the end points.

If the point X is projected at point X^1 on a Mercator Chart; then, because the chart is orthomorphic, the angle θ is represented without distortion. This applies equally to the angles at Y and Z, so that these points are projected at Y^1 and Z^1 respectively, such that the triangle $X^1Y^1Z^1$ on the chart is similar to the triangle XYZ on the globe. This is strictly true only when the triangle XYZ is infinitely small. For this reason the term orthomorphism, when applied to a map projection, has a special meaning: *shape is preserved only for infinitely small areas.*

Because meridians are projected on a Mercator Chart as parallel straight lines, the Departure between X and Y is represented by Y^1Z^1. This length also represents the D. Long. between X and Y.

$$\text{Exaggeration of Arc } XZ \text{ of Meridian} = \frac{\text{Length on Chart}}{\text{Length on Globe}}$$

$$= \frac{X^1Z^1}{XZ}$$

The triangles XYZ and $X^1Y^1Z^1$, are similar, so that the ratio between corresponding sides is constant.

Therefore:
$$\frac{X^1Z^1}{XZ} = \frac{Y^1Z^1}{YZ}$$

That is:
$$\text{Exaggeration} = \frac{\text{D. Long.}}{\text{Dep.}}$$

Fig. 12·1

By the Parallel Sailing Formula:

$$\frac{\text{D. Long.}}{\text{Dep.}} = \sec \text{Lat.}$$

Therefore:

$$\text{Exaggeration of Arc } XZ \text{ of Meridian} = \sec \text{Lat.}$$

The exaggeration of the projection of an arc of a meridian is proportional to the secant of the Latitude. This is to the same extent as the exaggeration of the projection of an arc of a parallel of Latitude. The Mercator projection is, therefore, orthomorphic.

5. Meridional Parts

The Longitude scale on a Mercator Chart is constant. On the other hand the Latitude scale is variable: it increases proportionally to the secant of the Latitude. Thus the unit of the Longitude scale is a convenient unit for certain purposes which we shall now discuss.

The number of minute-of-arc units of the Longitude scale contained in a projected piece of a meridian on a Mercator Chart between the projected equator and the projection of any given parallel of Latitude θ, is called the Meridional Parts for Latitude θ. One Meridional Part (m.pt.), therefore, is equivalent to a minute of arc of the Longitude scale.

The length of any piece of a projected meridian on a Mercator Chart between two given projected parallels of Latitude, expressed in m.pts., is called the Difference of Meridional Parts (D.M.P.) between the Latitudes of the two parallels.

Meridional Parts are useful in two applications:

(1) in constructing Mercator Charts

(2) in Mercator Sailing.

Fig. 12·2 illustrates a part of a Mercator Chart with the rhumb line connecting projected positions X^1 and Y^1.

The number of units of the constant Longitude scale contained in arcs $X^1 Z^1$ and $Z^1 Y^1$ are the D.M.P. and D. Long. respectively between X^1 and Y^1.

Provided that the correct part of the Latitude scale is used, the number of units of the variable Latitude scale contained in arcs $X^1 Z^1$ and $Z^1 Y^1$ are D. Lat. and Departure, respectively, between X^1 and Y^1.

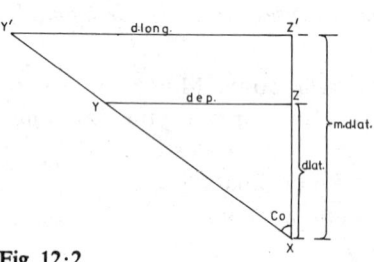

Fig. 12·2

It follows that:

$$\frac{\text{D.M.P.}}{\text{D. Lat.}} = \frac{\text{D. Long.}}{\text{Dep.}}$$

By the Parallel Sailing Formula:

$$\frac{\text{D. Long.}}{\text{Dep.}} = \cos \lambda$$

THE MERCATOR CHART

where λ is an angle known as the Middle Latitude (see Chapter 13).
Therefore:

$$\frac{\text{D.M.P.}}{\text{D. Lat.}} = \sec \lambda$$

This relationship is known as the Mercator Principle.

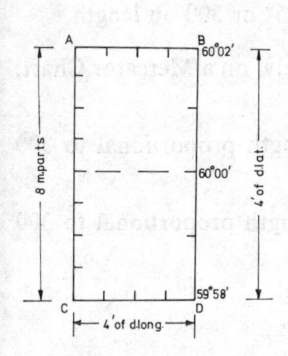

Fig. 12·3 illustrates a part of a Mercator Chart. The D. Lat. between A and C is 4′. If the Middle Latitude is taken as 60° 00′, the D.M.P. between A and C, by the Mercator Principle, is:

$$\begin{aligned}\text{D.M.P.} &= \text{D. Lat. } \sec 60° 00' \\ &= 4 \sec 60° 00' \\ &= 4 \times 2 \\ &= 8\end{aligned}$$

Fig. 12·3

Therefore, the piece of the meridian contained between A and C has been doubly magnified.

The D.M.P. between two Latitudes may be measured direct from a Mercator Chart simply by finding the number of minutes of the Longitude scale in the piece of any meridian contained between the parallels of the two Latitudes.

Before a Mercator Chart can be constructed, a table of meridional parts must be available. We shall now discuss how Edward Wright devised his table of m.pts. from which the first mathematically correct chart was constructed.

After explaining the Mercator principle of orthomorphism and expressing it in mathematical terms, viz. D.M.P./D. Lat = sec Middle Latitude, Wright divided the part of a meridian between the equator and a given parallel of Latitude into a number of equal pieces. He then multiplied the number of Latitude minutes in each piece by the secant of the *MEAN* Latitude of each piece. This gave a value for the Meridional Parts in each piece. By adding the m.pts. of the pieces together, a value for the m.pts. for the given Latitude was obtained. When computing m.pts. in this way the degree of accuracy of the results depends upon the number of pieces into which the part of the meridian between the equator and the given parallel is divided. The greater the number of pieces the more accurate is the result. The inaccuracy of a result arising through not taking a sufficiently large number of pieces, is due to the Mean Latitude not being equivalent to the Middle Latitude of the piece. In order to compute the exact number of m.pts. for any given Latitude it is required to take an infinite number of pieces. The precise computation, therefore, requires the use of the integral calculus. In the calculus notation the number of m.pts. (M) in any Latitude φ is given by:

$$M = \int_{\theta=0}^{\theta=\varphi} \sec \theta \; . \; d\theta$$

Edward Wright computed his table of m.pts. before the integral calculus had been invented. Wright's table, which was based on the assumption that the Earth is a perfect sphere, was computed by dividing the meridian into 1′ arc lengths.

82 THE ELEMENTS OF NAVIGATION AND NAUTICAL ASTRONOMY

The following example illustrates the principle of Wright's method of computing meridional parts.

Example 12·1—Compute the approximate Meridional Parts for Latitude 30° 00′ assuming the Earth to be a perfect sphere.

Method—Divide the arc of the meridian contained between the equator and the parallel of Latitude of 30° into (say) six equal pieces. Each piece is, therefore, 5° or 300′ in length.

The piece between Lat. 0° and Lat. 5° is represented, approximately, on a Mercator Chart, by a length proportional to 300 sec 2½°.

The piece between Lat. 5° and Lat. 10° is represented by a length proportional to 300 sec 7½°.

The piece between Lat. 10° and Lat. 15° is represented by a length proportional to 300 sec 12½° and so on.

Thus:

$$\text{M.pts. for Lat. } 30° = 300 \sec 2\tfrac{1}{2}° + 300 \sec 7\tfrac{1}{2}° +$$
$$300 \sec 12\tfrac{1}{2}° + 300 \sec 17\tfrac{1}{2}° +$$
$$300 \sec 22\tfrac{1}{2}° + 300 \sec 27\tfrac{1}{2}°$$
$$= 300\,(\sec 2\tfrac{1}{2}° + \sec 7\tfrac{1}{2}° + \sec 12\tfrac{1}{2}° +$$
$$\sec 17\tfrac{1}{2}° + \sec 22\tfrac{1}{2}° + \sec 27\tfrac{1}{2}°)$$

$$\sec\ 2\tfrac{1}{2}° = 1\cdot00095$$
$$\sec\ 7\tfrac{1}{2}° = 1\cdot00863$$
$$\sec 12\tfrac{1}{2}° = 1\cdot02428$$
$$\sec 17\tfrac{1}{2}° = 1\cdot04853$$
$$\sec 22\tfrac{1}{2}° = 1\cdot08239$$
$$\sec 27\tfrac{1}{2}° = 1\cdot12738$$

$$\text{sum} = 6\cdot29216$$
$$\times 300$$
$$\overline{1887\cdot64800}$$

Answer—M.pts. Lat. 30° = 1887·65 approximately.

Note—Had the arc of the meridian been divided into a number of parts greater than six, the result would have been more accurate than that obtained above.

6. Meridional Parts for the Terrestrial Spheroid

The m.pts. table given in nautical tables such as Norie's and Burton's, are computed for a terrestrial spheroid having an ellipticity of about 1/300. Admiralty charts on the Mercator projection are constructed using these tables.

THE MERCATOR CHART

7. Constructing Mercator Charts

The Graticule, or network of projected parallels and meridians, of a Mercator Chart is drawn to a convenient scale using a straightedge. The first thing to do is to choose a suitable scale of Longitude. Given the range of Longitude of the proposed chart, the East/West extent of the chart may then be found. A straight line of this length is then drawn across the lower part of the sheet on which the graticule is to be constructed. This line is the projection of one of the limiting parallels of Latitude of the area to be portrayed. Straight lines, to represent the projected meridians, are then erected perpendicularly from this projected parallel. The range of Latitude is divided into a number of equal arcs. The D.M.P. between the limiting Latitudes of each of these arcs is found using the m.pts. table, and the spacing of the projected parallels is then computed. The following example illustrates this method.

Example 12·2—Construct a Mercator Chart between the limits of Latitudes 10° and 50° N., and between the meridians of 90° and 150° W. Project parallels and meridians every 10°.

$$\text{Range of Longitude} = 60°$$

Let the scale of Longitude be 1 unit to represent 10° or 600′ of Longitude.

$$\text{Width of chart} = \frac{60}{10} = 6 \text{ units}$$

m.pts. Lat. 10° = 599·01
m.pts. Lat. 20° = 1217·14

$$\text{D.M.P.} = 618\cdot13 \text{ represented by } \frac{618\cdot13}{600} = 1\cdot030 \text{ units}$$

m.pts. Lat. 20° = 1217·14
m.pts. Lat. 30° = 1876·67

$$\text{D.M.P.} = 659\cdot53 \text{ represented by } \frac{659\cdot53}{600} = 1\cdot099 \text{ units}$$

m.pts. Lat. 30° = 1876·67
m.pts. Lat. 40° = 2607·64

$$\text{D.M.P. } 730\cdot97 \text{ represented by } \frac{730\cdot97}{600} = 1\cdot218 \text{ units}$$

m.pts. Lat. 40° = 2607·64
m.pts. Lat. 50° = 3456·53

$$\text{D.M.P.} = 848\cdot89 \text{ represented by } \frac{848\cdot89}{600} = 1\cdot415 \text{ units}$$

Range of Latitude 40° will be represented by:

$$1\cdot030 + 1\cdot099 + 1\cdot218 + 1\cdot415 = 4\cdot762 \text{ units}$$

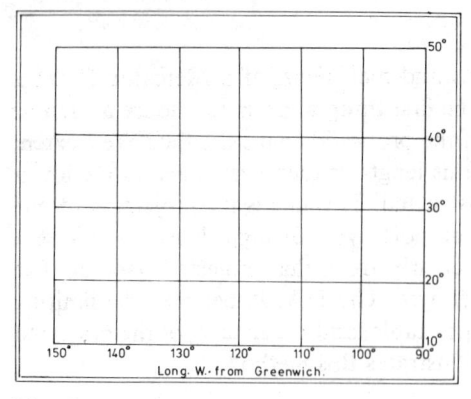

Fig. 12·4

Fig. 12·4 illustrates the required graticule. An alternative method of constructing a Mercator Chart, which is suitable only when the range of Latitude is small, is shown in fig. 12·5 which illustrates Example 12·3. In this method, after having chosen a Longitude scale and projected one of the limiting parallels from which the projected meridians are erected, an angle equal to the Middle Latitude of the limiting Latitudes is constructed as shown in fig. 12·5. The hypotenuse of the triangle, by the Mercator Principle, is proportional in length to the D.M.P. between the two limiting parallels.

Example 12·3—Construct a Mercator Chart for the area contained between parallels of 52° and 55° N. and the meridians 06° and 10° W. Insert parallels and meridians at one-degree intervals.

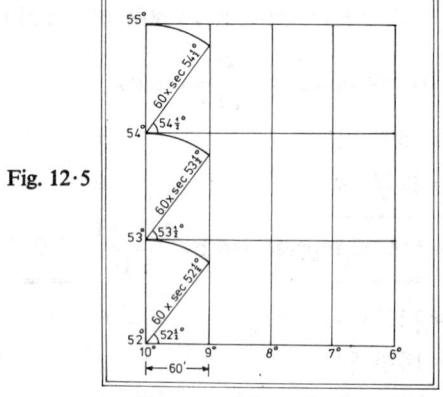

Fig. 12·5

Range of Longitude = 4°

Let the Scale of Longitude be 1 unit to 1°.

Width of chart = 4 units

The construction is illustrated in fig. 12·5 which is the required projection.

Exercises on Chapter 12

1. What are the main requirements of a navigational chart?
2. Describe the features of a Mercator Chart. State the advantages and disadvantages of a Mercator Chart to a navigator.
3. Explain carefully the mathematical principle of the Mercator projection.
4. What is the meaning of the term Orthomorphism as it applies to a map projection?
5. Define: Meridian Parts for Lat. θ; Difference of Meridional Parts.
6. Explain how Edward Wright constructed his table of meridional parts.
7. Explain carefully how the graticule of a Mercator Chart of a small area, such as a harbour or estuary, may be constructed.

THE MERCATOR CHART

8. Describe how a small circle of diameter 600 miles lying on the equator appears on a Mercator Chart.

9. How many units of the Longitude scale of a Mercator Chart are contained in a part of a meridian between the parallels of 53° 20′ N. and 53° 40′ N.?

10. If the scale of Longitude on a Mercator Chart is 1 inch to 1°, find the scale of Latitude in Latitude 65° 15′ N.

11. If 1′ of Longitude on a Mercator Chart is represented by 2·52 in., what length represents 1′ of Latitude in Lat. 42° 10′ S.?

12. If 1′ of Latitude on a Mercator Chart is represented by 2·30 in. in Lat. 32° 20′ N., find the scale of Longitude.

13. If 1′ of Latitude on a Mercator Chart is represented by 3·25 in. in Latitude 47° 30′, find the scale of Latitude in Lat. 40° 00′.

14. Construct a Mercator Chart for the area between the limits of Latitude 50° 00′ N. and 58° N., and between the meridians of 4° 00′ W. and 20° W., using a Longitude scale of 1 in. to 1°. Insert parallels and meridians at one-degree intervals.

CHAPTER 13

MERCATOR SAILING AND MIDDLE LATITUDE SAILING

1. Introduction

When a vessel has travelled a given distance along a meridian her Departure is zero and her change in Latitude, in minutes of arc, is numerically equal to the number of miles she has sailed.

When a vessel has travelled a given distance along a parallel of Latitude her change in Latitude is zero and her Departure in miles is equal to the distance she has sailed. If the D. Long. corresponding to an unknown distance sailed along a parallel is known the distance may be found by means of the Parallel Sailing Formula.

In both cases, of travelling along a meridian and along a parallel of Latitude, the sailing problems of finding position or course and distance present no difficulty.

It now remains to examine the general sailing problems in which a vessel changes both Latitude *AND* Longitude by travelling on an oblique rhumb line path.

There are two methods of solving the general sailing problem:

(1) by Mercator Sailing

(2) by Middle Latitude Sailing.

2. Mercator Sailing

If a rhumb line path is drawn between two places A_1 and B_1 on a Mercator Chart, the plane right-angled triangle having the rhumb line path as its hypotenuse and containing the Course Angle, may conveniently be called the Chart Triangle. The two sides which meet to form the right angle in the Chart Triangle, when measured on the constant scale of Longitude, give the D. Long. and D.M.P. respectively between A_1 and B_1.

The Chart Triangle is geometrically similar to the Plane Sailing Triangle corresponding to the rhumb line path AB which is projected onto the Mercator Chart as A_1B_1.

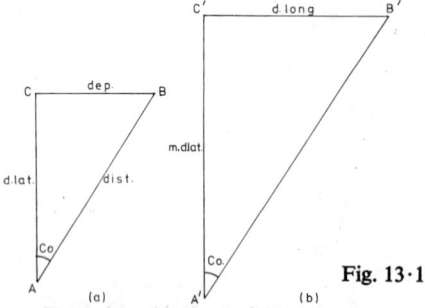

Fig. 13·1 (*a*) and (*b*) illustrate corresponding Plane Sailing and Chart Triangles.

The two general rhumb line sailing problems are:

(1) Finding the rhumb line course and distance from one given position to another.

(2) Finding the position of arrival after having travelled on a given rhumb line course for a given distance from a given position.

Fig. 13·1

To solve the first of these problems, using Mercator Sailing, the procedure is as follows:

(1) Find D. Lat. and D. Long.

(2) Using m.pts. table find D.M.P.

MERCATOR SAILING AND MIDDLE LATITUDE SAILING

(3) In the Chart Triangle, using D. Long. and D.M.P. find rhumb line Course Angle.
(4) In the Plane Sailing Triangle, using Course Angle and D. Lat., find the rhumb line Distance.

An example will make this clear.

Example 13·1—Find the rhumb line Course and Distance using Mercator Sailing, from *A* in Lat. 49° 50′ N. Long. 05° 30′ W. to *B* in Lat. 37° 50′ N. Long. 25° 40′ W.

Lat. A = 49° 50′ N.	m.pts. = 3441·05	Long. A = 05° 30′ W.
Lat. B = 37° 50′ N.	m.pts. = 2441·23	Long. B = 25° 40′ W.
D. Lat. = 12° 00′ S.	D.M.P. = 999·82	D. Long. = 20° 10′ W.
= 720′ S.		= 1,210′ W.

Referring to fig. 13·2:

In the Chart Triangle:

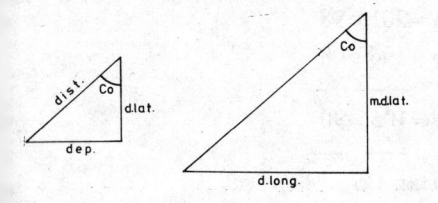

$$\tan Co = \frac{D. Long.}{D.M.P.}$$

$\log$ D. Long. = 3·08279
$\log$ D.M.P. = 2·99991

$\log \tan Co$ = 0·08288

Co = S. 50° 26′ W.

Fig. 13·2

In the Plane Sailing Triangle:

Distance = D. Lat. sec Co
$\log$ D. Lat. = 2·85733
$\log \sec Co$ = 0·19588

$\log$ Distance = 3·05321

Distance = 1130 miles

Answers—Course = 230½°
Distance = 1130 miles.

It should be noted that the answers are found by solving two right-angled plane triangles. The traverse table, therefore, may be used to check the calculated answers.

The second general rhumb line sailing problem is solved thus:

(1) In the Plane Sailing Triangle, using Course Angle and Distance find D. Lat.
(2) Find the final Latitude and thence the D.M.P.
(3) In the Chart Triangle, using D.M.P. and Course Angle, find the D. Long.
(4) Find the final Longitude.

88   THE ELEMENTS OF NAVIGATION AND NAUTICAL ASTRONOMY

The following example illustrates the method of solution.

Example 13·2—A vessel leaves a position in Lat. 32° 00′ S. Long. 116° 05′ E. and sails for a distance of 1243 miles on a course of 322°. Find her D.R. position after making this run.

Referring to fig. 13·3:

In the Plane Sailing Triangle:

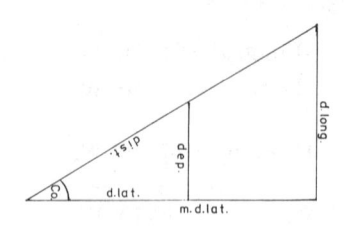

Fig. 13·3

D. Lat. = distance cos Co
log Distance = 3·09447
log cos Co = $\bar{1}$·89050
───────
log D. Lat. = 2·98497
───────

D. Lat. = 966·0′ N.
Lat. from = 16° 06′ N.
Lat. to = 32° 00′ S.
───────
= 15° 54′ S.

m.pts. Lat. from = 2015·98
m.pts. Lat. to = 960·08
───────
D.M.P. = 1055·90
───────

In the Chart Triangle:

D. Long. = D.M.P. tan Co
log D.M.P. = 3·02362
log tan Co = $\bar{1}$·89281
───────
log D. Long. = 2·91643
───────

D. Long. = 825′ W.
= 13° 45′ W.
Long. from = 116° 05′ E.
───────
Long. to = 102° 20′ E.
───────

Answer—Lat. to = 15° 54′ S.
Long. to = 102° 20′ E.

3. Rhumb Line Sailing when Course Angle is Large

When the Course Angle is large—more than about 60°—the change in the secant of the course angle is large for a small change in the angle. Examination of the secant table will reveal this.

It will be noticed that when finding the rhumb line Distance in example 13·1, the Course was found using the tangent table, and then, for finding the Distance, the secant table was

MERCATOR SAILING AND MIDDLE LATITUDE SAILING

used. Now sec θ is equivalent to tan θ . cosec θ. Therefore, to facilitate finding the rhumb line Distance in problems in which the Course Angle is large, instead of working as we have done in Example 13·1, it is better to use the formula:

$$\text{Distance} = \text{D. Lat.} \cdot \tan \text{Co} \cdot \text{cosec Co}$$

The tangent of the course angle is found from the formula:

$$\tan \text{Co} = \frac{\text{D. Long.}}{\text{D.M.P.}}$$

and the cosecant of the course angle may be lifted from the tables without difficulty. When an angle is large, the change in its cosecant as the angle increases, is small.

The following example illustrates a case in which the course angle is large.

Example 13·3—A vessel sails from *A* in Lat. 40° 00′ S. Long. 149° 00′ E. to *B* in Lat. 37° 00′ S. Long. 173° 00′ E. Find the Course and Distance using Mercator Sailing.

Lat. *A* = 40° 00′ S.	m.pts. Lat. *A* = 2607·6	Long. *A* = 149° 00′ E.
Lat. *B* = 37° 00′ S.	m.pts. Lat. *B* = 2378·5	Long. *B* = 173° 00′ E.
D. Lat. = 03° 00′ N.	D.M.P. = 229·1	D. Long = 24° 00′ E.
= 180′ N.		= 1,440′ E.

Referring to fig. 13·4:

In the Chart Triangle:

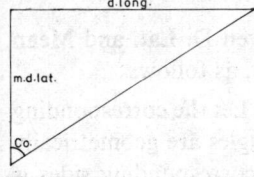

Fig. 13·4

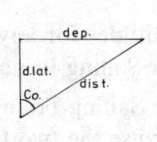

$$\tan \text{Co} = \frac{\text{D. Long.}}{\text{D.M.P.}}$$

log D. Long. = 3·15836
log D.M.P. = 2·36003

log tan Co = 0·79833

Co = N. 81° E.

In the Plane Sailing Triangle:

$$\text{distance} = \text{D. Lat.} \cdot \tan \text{Co} \cdot \text{cosec Co}$$

log D. Lat. = 2·25527
log tan Co = 0·79833
log cosec Co = 0·00542

log Distance = 3·05902

Distance = 1146 miles

Answers—Course 081°
Distance 1146 miles.

4. Middle Latitude Sailing

Fig. 13·5 illustrates a portion of the Earth's surface with the rhumb line connecting the points A and B.

The D. Lat. between A and B, in fig. 13·5, is equal to the arc of the meridian AC or BD. The Departure between A and B, however, is greater than arc BC and less than arc AD. The actual Departure between A and B may be represented by the arc EF. The Latitude of the parallel on which this arc lies is referred to as the Middle Latitude.

Considering the Earth to be a perfect sphere the Middle Latitude is always greater than the average or Mean Latitude between any two points in the same hemisphere on the Earth. This is not always the case, however, on a spheroidal Earth.

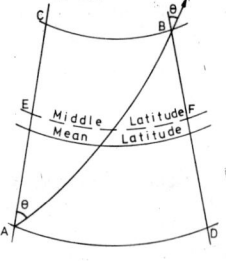

Fig. 13·5

Middle Latitude may be defined as the angle the cosine of which is equal to the ratio between Departure and D. Long. Thus:

$$\text{Cos Middle Latitude} = \frac{\text{Departure}}{\text{D. Long.}}$$

This relationship is known as the Middle Latitude Sailing Formula.

The magnitude of the difference of the Middle and Mean Latitudes between two places depends upon the D. Lat. and the Mean Latitude of the two places. A table giving differences between Middle and Mean Latitudes for all convenient values of D. Lat. and Mean Latitude is given in collections of nautical tables such as Norie's and Burton's.

5. Mean to Middle Latitude Correction Table

The difference between the Mean and Middle Latitudes for any given D. Lat. and Mean Latitude may be calculated by means of the Mercator Sailing Principle, as follows:

The triangle ABC in fig. 13·6 represents any Plane Sailing Triangle. Let the corresponding Mercator Chart Triangle be represented by XYZ. Because the two triangles are geometrically similar the ratios between any corresponding sides is a constant amount.

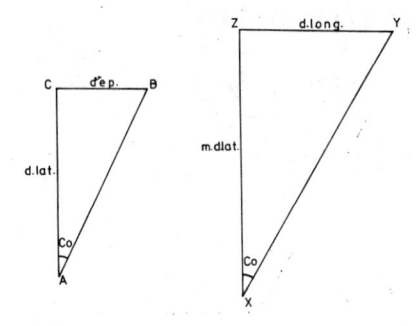

Therefore: $\dfrac{\text{Dep.}}{\text{D. Lat.}} = \dfrac{\text{D. Long.}}{\text{D.M.P.}}$

or: $\dfrac{\text{Dep.}}{\text{D. Long.}} = \dfrac{\text{D. Lat.}}{\text{D.M.P.}}$

But $\dfrac{\text{Dep.}}{\text{D. Long.}} = \cos \text{Mid. Lat.}$

Therefore: $\dfrac{\text{D. Lat.}}{\text{D.M.P.}} = \cos \text{Mid. Lat.}$

Fig. 13·6

The difference between the Middle and Mean Latitudes for any given range of D. Lat. may, therefore, be computed using the Mercator Principle. The following example makes this clear.

Example 13·4—Find the correction to apply to the Mean Latitude 50° 00' if the D. Lat. is 14° 00'.

MERCATOR SAILING AND MIDDLE LATITUDE SAILING

Limiting parallels are 57° 00′ and 43° 00′

$$\text{m.pts. Lat. } 57° = 4{,}162 \cdot 97$$
$$\text{m.pts. Lat. } 43° = 2{,}847 \cdot 13$$
$$\text{D.M.P.} = 1{,}315 \cdot 84$$
$$\text{D. Lat.} = 14 \times 60$$
$$= 840'$$

$$\cos \text{Mid. Lat.} = \frac{\text{D. Lat.}}{\text{D.M.P.}}$$

$$\log \text{D. Lat.} = 2 \cdot 92428$$
$$\log \text{D.M.P.} = 3 \cdot 11921$$
$$\log \cos \text{Mid. Lat.} = \bar{1} \cdot 80507$$

Mid. Lat. = 50° 20′
Mean Lat. = 50° 00′
Correction = 20′ to Add to Mean Lat.

From Table (Norie's or Burton's)

Correction = 20′ to Add to Mean Lat.

The following examples serve to show that the general sailing problems (Examples 13·1, 13·2 and 13·3), solved by Mercator Sailing, may be solved using Middle Latitude Sailing: the results being the same in both cases.

Example 13·5—Find the course and distance using Middle Latitude Sailing, from *A* in Lat. 49° 50′ N. Long. 05° 30′ W. to *B* in Lat. 37° 50′ N. Long. 25° 40′ W.

Lat. *A* = 49° 50′ N.	Lat. *A* = 49° 50′ N.	Long. *A* = 05° 30′ W.
Lat. *B* = 37° 50′ N.	Lat. *B* = 37° 50′ N.	Long. *B* = 25° 40′ W.
sum = 87° 40′	D. Lat. = 12° 00′ S.	D. Long. = 20° 10′ W.
Mean Latitude = 43° 50′		
Correction = +6′	= 720′ S.	= 1,210′ W.
Mid. Latitude = 43° 56′		

$$\tan \text{Co} = \frac{\text{Dep.}}{\text{D. Lat.}}$$

Dep. = D. Long. . cos Mid. Lat.

Referring to figure 13·7:

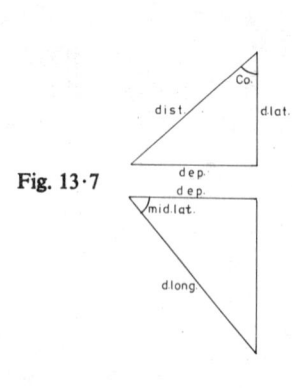

$$\log \text{D. Long.} = 3 \cdot 08279$$
$$\log \cos \text{Mid. Lat.} = \bar{1} \cdot 85742$$
$$\log \text{Dep.} = 2 \cdot 94021$$
$$\log \text{D. Lat.} = 2 \cdot 85733$$
$$\log \tan \text{Co} = 0 \cdot 08288$$
$$\text{Co} = \text{S. } 50° \; 26' \; \text{W.}$$

Distance = D. Lat. . sec Co
$$\log \text{D. Lat.} = 2 \cdot 85733$$
$$\log \sec \text{Co} = 0 \cdot 19588$$
$$\log \text{Distance} = 3 \cdot 05321$$
$$\text{Distance} = 1130 \text{ miles}$$

Fig. 13·7

Answers—Course = $230\tfrac{1}{2}°$

Distance = 1130 miles.

Example 13·6—A vessel left a position in Lat. 32° 00′ S. Long. 116° 05′ E. and travelled for a distance of 1234 miles on a rhumb line course of 322°. Find her D.R. position at the end of this run.

Referring to fig. 13·8:

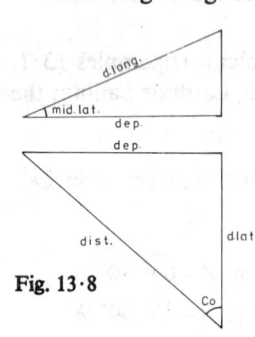

D. Lat. = Distance . cos Co
$$\log \text{Distance} = 3 \cdot 08447$$
$$\log \cos \text{Co} = \bar{1} \cdot 89050$$
$$\log \text{D. Lat.} = 2 \cdot 98479$$
$$\text{D. Lat.} = 966'$$
$$= 16° \; 06' \; \text{N.}$$

Lat. from = 32° 00′ S.

Lat. to = 15° 54′

Sum = 47° 54′

Mean Lat. = 23° 57′

Correction = −8′

Mid. Lat. = 23° 49′

Fig. 13·8

MERCATOR SAILING AND MIDDLE LATITUDE SAILING

$$\text{D. Long.} = \text{Dep. sec Mid. Lat.}$$
$$= \text{D. Lat. tan Co . sec Mid. Lat.}$$

log D. Lat. = 2·98497
log tan Co = $\bar{1}$·89281
log sec Mid. Lat. = 0·03865
log D. Long. = 2·91643
D. Long. = 825′ W.
     = 13° 45′ W.
Long. from = 116° 05′ E.
Long. to = 102° 20′ E.

Answer—Final Position: Lat. = 15° 54′ S.
Long. = 102° 20′ E.

Example 13·7—A vessel left a position in Lat. 40° 00′ S. Long. 149° 00′ E. to a position in Lat. 37° 00′ S. Long. 173° 00′ E. Find by Middle Latitude Sailing, the course and distance travelled.

Lat. from = 40° 00′ S.	Lat. from = 40° 00′ S.	Long. from = 149° 00′ E.
Lat. to = 37° 00′ S.	Lat. to = 37° 00′ S.	Long. to = 173° 00′ E.
Sum = 77° 00′	D. Lat. = 3° 00′ N.	D. Long. = 24° 00′ E.
Mean Lat. = 38° 30′ S.	= 180′ N.	= 1,440′ E.
Correction = −17′		
Mid. Lat. = 38° 13′ S.		

Referring to fig. 13·9:

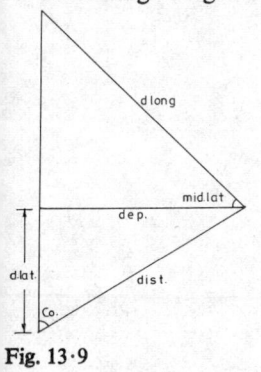

Fig. 13·9

$$\text{Dep.} = \text{D. Long. . cos Mid. Lat.}$$
$$\tan \text{Co} = \frac{\text{Dep.}}{\text{D. Lat.}}$$
$$= \frac{\text{D. Long. . cos Mid. Lat.}}{\text{D. Lat.}}$$

log D. Long. = 3·15836
log Cos M.L. = $\bar{1}$·89524
     = 3·05360
log D. Lat. = 2·25527
log tan Co = 0·79833
Co = N. 80° 57′ E.

$$\text{Distance} = \text{D. Lat.} \cdot \sec \text{Co}$$
$$= \text{D. Lat.} \cdot \tan \text{Co} \cdot \text{cosec Co}$$

log D. Lat. = 2·25527
log tan Co = 0·79833
log cosec Co = 0·00543
log Distance = 3·05903
Distance = 1146 miles

Answers—Course = 081°
Distance = 1146 miles.

6. Crossing the Equator

When it is necessary to find the rhumb line Course and Distance between two places which lie on different sides of the equator, it is more convenient to use Mercator Sailing than Middle Latitude Sailing. In this case the D.M.P. is found by adding together the m.pts. for the two Latitudes.

Example 13·8—Find, by Mercator Sailing, the course and distance from A in Lat. 10° 00′ S. Long. 90° 00′ W. to B in Lat. 08° 30′ N. Long. 60° 00′ W.

Lat. A = 10° 00′ S.	m.pts. Lat. A = 599·0	Long. A = 90° 00′ W.
Lat. B = 08° 30′ N.	m.pts. Lat. B = 508·4	Long. B = 60° 00′ W.
D. Lat. = 18° 30′ N.	D.M.P. = 1107·4	D. Long. = 30° 00′ E.
= 1,110′ N.		= 1,800′ E.

Referring to fig. 13·10:

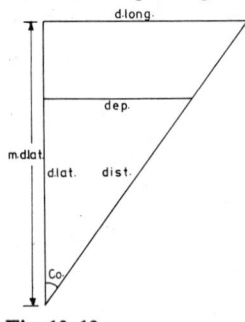

Fig. 13·10

$$\tan \text{Co} = \frac{\text{D. Long.}}{\text{D.M.P.}}$$

log D. Long. = 3·25527
log. DM.P. = 3·04430
log tan Co = 0·21097
Co = N. 58° 24′ E.

Distance = D. Lat. . sec Co
log D. Lat. = 3·04532
log sec Co = 0·28068
log Distance = 3·32600
Distance = 2118 miles

Answers—Course = 058½°
Distance = 2118 miles.

MERCATOR SAILING AND MIDDLE LATITUDE SAILING

7. The Day's Run

In merchant vessels it is customary to determine the ship's position as accurately as possible at each noon. The distance travelled over the ground between successive noons is known as the Day's Run. This distance divided by the Steaming Time gives the average speed of the vessel for the day. When coasting, the Day's Run is measured direct from the chart, and the difference between the measured distance and the distance recorded by the patent log is usually regarded as Favourable or Adverse Current.

When navigating out of sight of land, the distance made good is found by calculation. Measuring the distance accurately from a chart is not possible in this case, because the ocean chart used has too small a scale of distance.

If, during the day, the vessel has made one course only, the distance made good is calculated by Mercator or Middle Latitude Sailing. If, however, the vessel has made more than one course, the distance on each leg is estimated as accurately as possible, and the several distances are then summed to give the total run for the day.

In days of sail, when a vessel was forced to make a zig-zag course towards her destination, it was usual to ascertain the direct course and distance made good for the day. This has no value to a navigator on a power-driven vessel: he is interested only in the actual distance his vessel has steamed over the ground.

The vessel's D.R. Noon Position, reckoned from the observed position of the previous noon, is compared with the Observed Noon Position, to give the set and drift of the current experienced during the day.

An estimation of the distance travelled by a vessel may be found by means of a knowledge of the propeller revolutions. The engineer officer, at each noon, records the reading of the Counter, which registers the number of revolutions of the propeller. From successive noon recordings, the revolutions made by the propeller during the day may be found. If the Pitch of the propeller is known the distance which should have been covered may be found.

$$\text{Engine Distance} = \frac{\text{Revs.} \times \text{Pitch}}{6080} \text{ miles}$$

Because of several factors including:

(a) current
(b) hull resistance
(c) faulty propeller

the Engine Distance is not generally the same as the Vessel's Distance as found from observations. The Engine Distance is usually greater than the Vessel's Distance by an amount called Slip.

Slip is usually calculated as a percentage of the Engine Distance from the formula:

$$\text{Slip (\%)} = \frac{\text{Engine Distance} - \text{Vessel's Distance}}{\text{Engine Distance}} \times 100$$

If the slip is estimated and the Engine Distance is known, the Vessel's Distance may readily be computed.

96   THE ELEMENTS OF NAVIGATION AND NAUTICAL ASTRONOMY

Example 13·9—If the Engine Distance is found to be 240·0 miles, and the slip is estimated to be 4%, find the Vessel's Distance.

$$\text{Engine Distance} = 240 \cdot 0 \text{ miles}$$
$$\text{Slip} = 4\%$$

Now,
$$\text{Slip} = \frac{\text{Engine Dist.} - \text{Vessel's Dist.}}{\text{Engine Dist.}} \times 100$$

Therefore,
$$4 = \frac{240 \cdot 0 - \text{Vessel's Dist.}}{240 \cdot 0} \times 100$$

and
$$\text{Vessel's Distance} = \frac{240 \cdot 0 - 96 \cdot 0}{10}$$
$$= 230 \cdot 4 \text{ miles}$$

Answer—Vessel's Distance = 230·4 miles.

8. The Day's Work

The process of finding a vessel's course and distance made good between successive noon positions, finding the average speed, and finding the set and rate of the current experienced during the day, is known as the Day's Work. An example of a typical Day's Work is as follows.

Example 13·10—At noon on 2nd January, Malin Head (Lat. 55° 22′ N. Long. 07° 24′ W.) was observed to bear 170° Distance 10·0 miles. The log was set to zero and the course was set to 330°. At the following noon the log registered 302·0 and the observed position was Lat. 59° 50·0′ N. Long. 11° 54·0′ W. The steaming time was 24 hrs. 00 mins. Find the Day's Run, the average speed and the set and drift of the current experienced during the day.

$$\text{Departure Brg.} = 170°$$
$$\text{Departure Co and Distance} = 350° \text{ by } 10 \cdot 0 \text{ miles}$$

D. Lat. = 10′ N.	Dep. = 1·7′ W.	D. Long. = 3·0′ W.
Lat. Point = 55° 22′ N.		Long. Point = 07° 24′ W.
D. Lat. = 10′ N.		D. Long. = 3′ W.
Lat. Ship = 55° 32′ N.		Long. Ship = 07° 27′ W.

$$\text{Co} = 330° \qquad \text{Distance} = 302 \text{ miles}$$

From Traverse Table: D. Lat. =   261·5′ N.
                                    = 4°   21·5′ N.
                    Lat. ship 2nd = 55°   32·0′ N.
                    D.R. Lat. 3rd = 59°   53·5′ N.

m.pts. Lat. from = 4004·8
m.pts. Lat. to = 4494·1
D.M.P. = 489·3

MERCATOR SAILING AND MIDDLE LATITUDE SAILING

Referring to fig. 13·11:

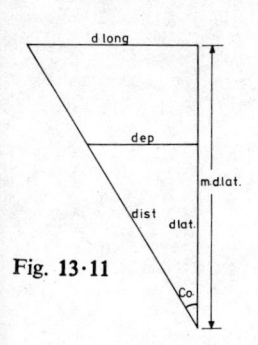

$$\text{D. Long.} = \text{D.M.P.} \tan Co$$

$$\log \text{D.M.P.} = 2\cdot 68958$$
$$\log \tan Co = \bar{1}\cdot 76144$$

$$\log \text{D. Long.} = 2\cdot 45102$$

$$\text{D. Long.} = 282\cdot 5' \text{ W.}$$
$$= 04° 42\cdot 5' \text{ W.}$$

Long. ship 2nd = 07° 27·0' W.

D.R. Long. 3rd = 12° 09·5' W.

Fig. 13·11

D.R. Lat. 3rd = 59° 53·5' N.	D.R. Long. = 12° 09·5' W.
Obs. Lat. 3rd = 59° 50·0' N.	Obs. Long. = 11° 54·0' W.
D. Lat. = 3·5' S.	D. Long. = 15·5' E.

$$\text{Dep.} = 7\cdot 8' \text{ E.}$$

From Traverse Table:

$$\text{Set} = \text{S. } 66° \text{ E.} \quad \text{Drift} = 8\cdot 6 \text{ miles}$$

Obs. Lat. 2nd = 55° 32' N.	m.pts. = 4004·8	Obs. Long. = 07° 27' W.
Obs. Lat. 3rd = 59° 50' N.	m.pts. = 4487·2	Obs. Long. = 11° 54' W.
D. Lat. = 4° 18' N.	D.M.P. = 482·4	D. Long. = 4° 27' W.
= 258' N.		= 267' W.

$$\tan Co = \frac{\text{D. Long.}}{\text{D.M.P.}}$$

$$\log \text{D. Long.} = 2\cdot 42651$$
$$\log \text{D.M.P.} = 2\cdot 68341$$

$$\log \tan Co = \bar{1}\cdot 74310$$

$$Co = \text{N. } 28° 58' \text{ W.}$$

$$\text{Distance} = \text{D. Lat.} \sec \text{Co}$$

$$\log \text{D. Lat.} = 2 \cdot 41162$$
$$\log \sec \text{Co} = 0 \cdot 05804$$

$$\log \text{Distance} = 2 \cdot 46966$$

$$\text{Distance} = 295 \text{ miles}$$

$$\text{Average Speed} = \frac{\text{Distance}}{\text{Stmg. Time}}$$

$$\log \text{Distance} = 2 \cdot 46966$$
$$\log \text{Stmg. Time} = 1 \cdot 38021$$

$$\log \text{Speed} = 1 \cdot 08945$$

$$\text{Average Speed} = 12 \cdot 29 \text{ knots}$$

Answers—Course Made Good = 331°
Speed Made Good = 12·29 knots
Set = 114°
Drift = 8·6 miles.

Exercises on Chapter 13

1. Compare the methods of Mercator and Middle Latitude Sailing.
2. Show how the difference between the Mean and Middle Latitudes may be found using the Mercator Principle.
3. Verify that the correction to the Mean to find the Middle Latitude is −46′ when the Mean Latitude is 19° and the D. Lat. is 10°.
4. Explain how the Engine Distance as found from Counter readings may be used to estimate the ship's position.
5. If the slip is estimated to be 2·5% and the Engine Distance is 346 miles, find the Vessel's Distance.
6. Find the Course and Distance from a position off the North Andaman, in Lat. 13° 05′ N. Long. 92° 10′ E. to a position off Trincomalee in Lat. 08° 20′ N. Long. 82° 00′ E.
7. A vessel travelled 820 miles on a course of 060°, from Lat. 40° 00′ N. Long. 40° 00′ W. Find her present position.
8. A vessel left a position off the Lizard in Lat. 49° 50′ N. Long. 05° 10′ W. and travelled for a distance of 800 miles on a Course of 230°. What was the bearing and distance of Fayal in Lat. 38° 32′ N. Long. 28° 40′ W. after making this run?

MERCATOR SAILING AND MIDDLE LATITUDE SAILING

9. A vessel sailed from a position in Lat. 40° 00′ N. Long. 60° 00′ W. and arrived in a position Lat. 42° 00′ N. Long. 20° 00′ W. Find the Course and Distance.

10. A vessel left a position in Lat. 00° 50′ S. Long. 91° 00′ W. (off Galapagos Islands) and travelled on a course of 136° until her Latitude was 12° 30′ S. What distance did she travel and what is her present Longitude assuming that the current set 316° throughout.

11. Find the present Latitude and the distance travelled since leaving a position in Lat. 26° 00′ S. Long. 109° 30′ W., if the course had been 040° and the present Longitude is 90° 30′ W.

12. Find the course and distance from a position in Lat. 05° 30′ S. Long. 32° 06′ W. to a position in Lat. 08° 14′ N. Long. 25° 00′ W.

13. A vessel left a position in Lat. 38° 40′ N. Long. 09° 00′ E. and travelled a distance of 120 miles on a course of 130°. What was the bearing and distance of Cape Bon in Lat. 37° 06′ N. Long. 11° 05′ E. at the end of the run?

14. A vessel left a position in Lat. 42° 40′ N. Long. 170° 20′ E. and travelled 705 miles on a course of 112°. Find her position at the end of this run.

15. At noon, Tuskar Rock in Lat. 52° 12′ N. Long. 06° 12′ W. bore 042° at a distance of 12·0 miles. The log was set to zero and the course was set to 200°. At 1430 hours, log 25, the course was altered to 242°. At 2000 hours, log 79, the course was again altered to 260°. Find the Estimated Position of the vessel for midnight when the log registered 117, allowing for a current the average set and rate of which was estimated to be 100°, 1·2 knots.

16. A vessel found her position to be Lat. 31° 10′ N. Long. 72° 22′ W. She travelled 120 miles on a course of 340°, and then for 32 miles on a course of 320°, when her position was found to be Lat. 35° 15′ N. Long. 73° 25′ W. Find the set and rate of the current experienced.

17. A vessel took Departure with Cape St. Mary in Lat. 25° 39′ S. Long. 45° 06′ E., bearing 328° distance 12·0 miles. Course was set to 142°. What was the ship's E.P. when the log registered 498, assuming that the current had set 090° for 28 miles during the interval.

18. A vessel found that she was in Lat. 42° 06′ S. Long. 50° 15′ W. at noon. The log was set to zero and the course was set to 042°. At the following noon, when the log registered 345, the position was found to be in Lat. 37° 54′ S. Long. 45° 12′ W. Find the set and rate of the current.

CHAPTER 14

GREAT CIRCLE SAILING

1. Introduction

The advantage of rhumb line sailing is that the Course Angle is constant. To travel along a rhumb line path, the navigator simply joins the points of departure and destination with a straight line on the Mercator Chart. He then measures the Course Angle: this being the inclination of the parallel rulers with the projected meridians. The Course is set and, if it is maintained during the voyage, the vessel will fetch up at her desired destination. The principal disadvantage of employing rhumb line sailing is that the path does not coincide with the shortest route between the points of departure and destination.

The shortest route over the Earth's surface between two terrestrial points is along the shorter arc of the great circle on which the two points lie. Now the angle which a great circle track makes with the meridians constantly changes, except in those special cases when the great circle track is also a rhumb line track. Thus, if it is desired to travel along the shortest route from one place to another, the great circle path must be followed. In this event the course must be constantly changed as the voyage proceeds. For long ocean voyages the great circle path should be followed, when it is safe and practicable to do so, in the interests of economy.

The difference between the distances along the rhumb line and great circle arc between any two places depends upon the:

(1) distance between the places

(2) D. Long. between the places

(3) Latitudes of the places.

A considerable difference results when the D. Long. between the two places is great and the Latitudes of the places are high. If the D. Long. is small the rhumb line path, which lies almost due North or South, almost coincides with the great circle path. If the Latitudes of the two places are low the rhumb line path lies almost along the equator in which case it approximates to the great circle path.

In order to measure courses along a great circle route it is necessary to plot the route on a Mercator Chart. Because of the distortion of the Mercator Chart, the great circle track between any two places not on the equator or on the same meridian, is projected as a curved line concave to the equator. The rhumb line track appears as a straight line, and the two tracks together give the false impression that the rhumb line distance is shorter than that of the great circle. On a terrestrial globe, however, it is readily seen that great circle tracks are shorter than corresponding rhumb line tracks.

Notice in figs. 14·1 and 14·2 that the rhumb line makes a constant angle θ with every meridian it crosses, whereas the direction of the great circle changes constantly.

GREAT CIRCLE SAILING

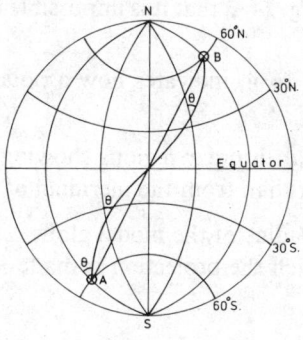

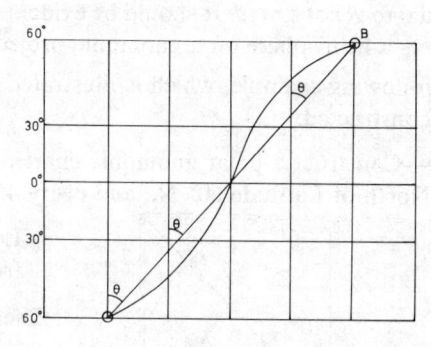

Fig. 14·1

Fig. 14·2

On the globe, as depicted in fig. 14·1, the rhumb line appears as a curved line. The great circle arc, in contrast, appears as a straight line when viewed in its plane.

It will be noticed in fig. 14·1 that the rhumb line *AB* and the great circle arc *AB* intersect at the equator. This is the case only when the points *A* and *B* are antipodal to one another.

On a Mercator Chart, as depicted in fig. 14·2, the rhumb line *AB* appears as a straight line, whereas the great circle arc between *A* and *B* appears as a curved line.

Great sailing routes may be used to great advantage for many ocean passages, and for some coastal passages too. A good example of a suitable great circle coastal track is the route along the eastern seaboard of North America between Long Island and Florida. To ascertain which is the better route to choose when a choice presents itself, a Gnomonic Chart will be found to be of great use. A gnomonic chart is constructed on the Gnomonic Projection. Its most important feature is that great circle arcs are projected as straight lines.

2. The Gnomonic Chart

In the gnomonic projection, the sphere's surface is projected outwards from the sphere's centre onto a plane which is tangential to the sphere. The tangential point may be at the North or South Pole of the globe, in which case the resulting projection is called a Polar Gnomonic; or it may be at a point on the equator of the globe, in which case the projection is called a Transverse or Equatorial Gnomonic. If the tangential point is at a position other than the pole or a point on the equator (and this is generally the case with gnomonic charts used on board ship), the projection is called an Oblique Gnomonic.

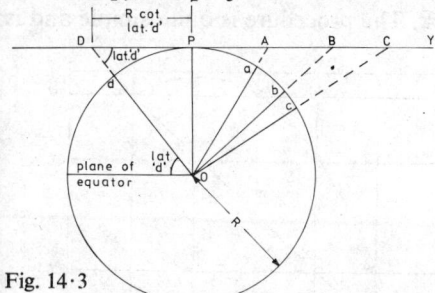

Fig. 14·3

On a polar gnomonic projection all meridians appear as straight lines which intersect, like the spokes of a bicycle wheel, at the projection of the Earth's Pole. Parallels of Latitude appear as concentric circles centred at the projection of the pole. The radius of any parallel is proportional to the cotangent of the Latitude of the parallel. A polar gnomonic projection is easily constructed as illustrated in figs. 14·3 and 14·4.

In fig. 14·3, the plane of a polar gnomonic projection is tangential to the globe at *P*. Points *a*, *b*, *c* and *d*, are projected from the centre of the globe onto the plane of the projection at *A*, *B*, *C* and *D*, respectively. Note that the radius of the parallel of Latitude on which *d*

102   THE ELEMENTS OF NAVIGATION AND NAUTICAL ASTRONOMY

lies is equal to $R \cot \text{Lat}. d$. It should be evident from fig. 14·3 that it is impossible to represent as much as a hemisphere on a gnomonic projection.

The following example, which is illustrated by fig. 14·4, indicates how a polar gnomonic chart is constructed.

Example—Construct a polar gnomonic chart of the North polar regions showing every tenth parallel North of Latitude 60° N., and every 45th meridian from the meridian of Greenwich.

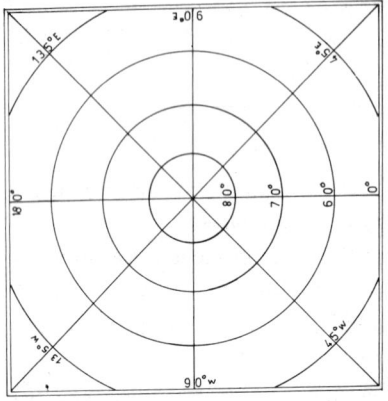

Let the radius of the model globe
from which the projection is made = 1 unit

Then:

Radius of projected parallel of Lat. 60° = 1 . cot 60°
= 0·577 units

Radius of projected parallel of Lat. 70° = 1 . cot 70°
= 0·364 units

Radius of projected parallel of Lat. 80° = 1 . cot 80°
= 0·176 units

Fig. 14·4

Fig. 14·4 illustrates the required projection.

A gnomonic chart having the tangential point at any position other than the Earth's Pole, is more difficult to construct than the polar gnomonic. In such a case the parallels of Latitude, except the equator, are projected as hyperbolae.

A gnomonic chart is not suitable for measuring distances and courses. Position, however, may be lifted with ease.

3. Practical Great Circle Sailing

A gnomonic chart has great value in the practice of Great Circle Sailing as an auxiliary to a Mercator Chart. When it is desired to lay down a great circle track on a Mercator Chart, the track is first laid down on a gnomonic chart which is merely a straight line connecting the points of departure and destination. Positions of several points on this line are then lifted, and these are transferred to the Mercator Chart. A fair curve is then drawn through these plotted points: this being the required great circle track. The procedure is quite simple and is illustrated in figs. 14·5 and 14·6.

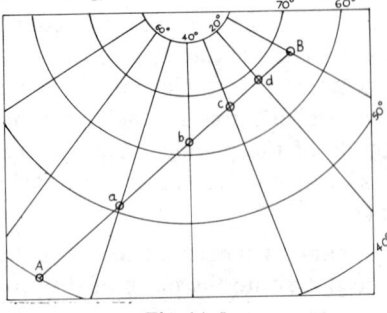

Fig. 14·5

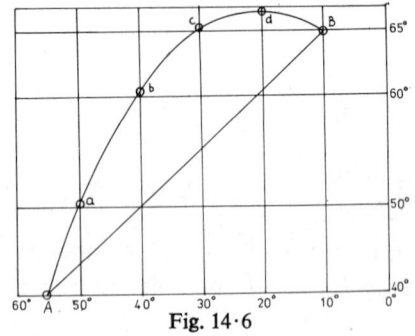

Fig. 14·6

GREAT CIRCLE SAILING

It will be noticed in figs. 14·5 and 14·6 that the nearest approach to the pole of the great circle track AB is at position d. At this point the track cuts the meridian at an angle of 90°. To the eastwards of the point d, the course at any point on the track is South-Easterly, and to the westwards of point d the course is North-Easterly. At the point of highest Latitude on a great circle track the course is due East or due West.

The point on a great circle track at which the course changes from Northerly to Southerly, or from Southerly to Northerly, is known as a Vertex of the great circle. The course at the vertex is due East or due West. Every great circle has two vertices, one in the northern hemisphere and the other in the southern hemisphere. The vertices of a great circle are antipodal points.

Fig. 14·7 illustrates two views of a great circle which crosses the equator at the points X and Y. The vertices of this great circle are at V_1 and V_1.

Because the meridian of the vertex crosses the great circle track and the equator at an angle of 90°, the triangle XV_1Q is isosceles, and the arc XV is, therefore, equal to the arc XQ.

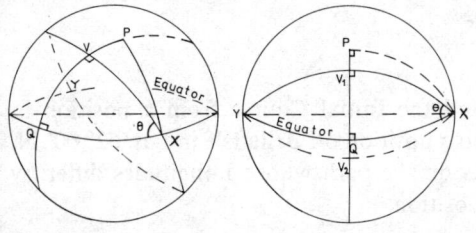

Fig. 14·7

But: $\quad\quad$ arc $XV_1 =$ arc $V_1Y = 90°$

Therefore: $\quad\quad$ arc $XQ = 90°$

Also, $\quad\quad$ arc $VQ =$ angle V_1XQ

It follows that the Latitude of either vertex is equal to the angle at which the great circle track crosses the equator. This angle is equal to the complement of the Course Angle at the equator.

$\quad\quad$ Latitude of Vertex = Complement of Course Angle at Equator

It is also evident from fig. 14·7 that the Longitude of the vertex differs by 90° from the Longitude of either of the two points where the great circle track crosses the equator.

To steer along a great circle would render it necessary for the course to be continually altered. This is impracticable, if not impossible. In practice, when employing Great Circle Sailing, the course is altered frequently, and the ship, therefore, is steered along a series of short rhumb line tracks which, collectively, approximate to the great circle track. This method of sailing is sometimes called Approximate Great Circle Sailing.

In fig. 14·8, which illustrates part of a Mercator Chart, the straight lines AB, BC, CD, etc., represent rhumb line tracks which approximate to the great circle track from A to H. Where the curve of the great circle track is most pronounced, as at arc CG, the course is altered more frequently than at other parts of the track.

The general practice in great circle sailing is to find, from a gnomonic chart or otherwise, the new initial course every time the ship's position is found, and to alter heading if necessary, to this course.

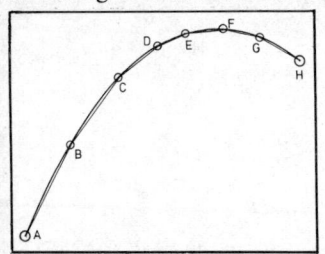

Fig. 14·8

In the event of a gnomonic chart not being available the positions of points on the great circle track, which are needed for plotting the track on the Mercator Chart, may be found by

104 THE ELEMENTS OF NAVIGATION AND NAUTICAL ASTRONOMY

means of Azimuth or *A B C* tables. Short Method navigation tables may also be employed for this purpose; and, in the last resort, the positions may be computed using spherical trigonometry. The trigonometrical method of solution, although providing useful practice at computation, is seldom used at sea. The method is:

(1) Find the Great Circle Distance using the Haversine Formula for a side.

(2) Find the Initial Course using the Haversine Formula for an angle.

(3) Find the position of the vertex by means of Napier's Rules.

(4) Calculate the Latitudes of points whose Longitudes differ by regular amounts from the Longitude of the vertex.

The following example illustrates the method.

Example 14·2—Find the Great Circle Distance and the Initial Course from a position in Lat. 51° 10′ N. Long. 10° 00′ W. (off S.W. Ireland) to a position off Belle Isle in Lat. 52° 00′ N. Long. 55° 00′ W. Calculate the Latitudes of points on the path whose Longitudes differ by multiples of 10° from the Longitude of the initial position.

In Fig. 14·9:

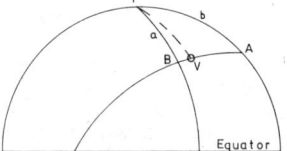

Fig. 14·9

A represents the Initial Position

B represents the Destination

P represents the Earth's North Pole

V represents the Northern Vertex

Given:

$$a = \text{co. Lat. } B = 38° \ 00'$$
$$b = \text{co. Lat. } A = 38° \ 50'$$
$$\hat{P} = \text{D. Long. } AB = 45° \ 00'$$

To find p, the Great Circle Distance:

$$\text{hav } p = \text{hav } P \sin a \sin b + \text{hav } (a \sim b)$$

$$\log \text{hav } P = \bar{1} \cdot 16568$$
$$\log \sin a = \bar{1} \cdot 78934$$
$$\log \sin b = \bar{1} \cdot 79731$$

$$\log \text{hav } \theta = \bar{2} \cdot 75233$$

$$\text{nat hav } \theta = 0 \cdot 05654$$
$$\text{nat hav}(a \sim b) = 0 \cdot 00005$$

$$\text{nat hav } p = 0 \cdot 05659 \qquad p = 27° \ 30'$$
$$\text{Distance} = 1650 \text{ miles}$$

GREAT CIRCLE SAILING

To find A, the Initial Course:

$$\text{hav } A = \{\text{hav } a - \text{hav } (b \sim p)\} \text{ cosec } b \text{ cosec } p$$

nat hav $a = 0\cdot 10599$

nat hav $(b \quad p) = 0\cdot 00975$

———————

nat hav $\theta = 0\cdot 09624$

log hav $\theta = \bar{2}\cdot 98336$
log cosec $b = 0\cdot 20269$
log cosec $p = 0\cdot 33559$

———————

log hav $A = \bar{1}\cdot 52164 \qquad A = 70° 24\cdot 8'$

——————— Initial Course $= 289\tfrac{1}{2}°$

To find x, the co-Lat of the vertex:

(refer to fig. 14·10)

$\sin x = \cos \text{co } b \cos \text{co } A$

$\sin x = \sin b \sin A$

log sin $b = \bar{1}\cdot 79731$
log sin $A = \bar{1}\cdot 97412$

———————

log sin $x = \bar{1}\cdot 77143 \qquad x = 36° 13'$

Fig. 14·10 $\qquad\qquad\qquad\qquad\qquad$ Lat. $V = 53° 47'$

To find X, the D. Long. between A and V:

(refer to fig. 14·11)

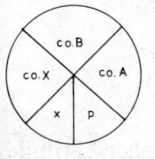

$\sin \text{co}.b = \tan \text{co}.X \tan \text{co}.A$

$\cos b = \cot X \cot A$

$\cot X = \cos b \tan A$

log cos $b = \bar{1}\cdot 89152$

Fig. 14·11 $\quad$ log tan $A = 0\cdot 44883$

———————

log cot $X = 0\cdot 34035 \qquad X = 24° 33'$

$\qquad\qquad\qquad\qquad\qquad$ Long. $V = 34° 33'$ W.

To find y_1, y_2, y_3, y_4, the co-Lats. of a, b, c, and d

(refer to fig. 14·12)

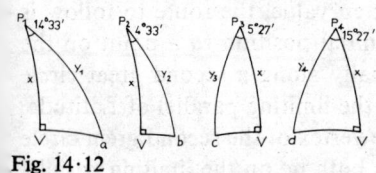

$\sin \text{co } P = \tan x \tan \text{co } y$

$\cos P = \tan x \cot y$

$\cot y = \cos P \cot x$

Fig. 14·12

Note—x has a constant value in the following computations.

$$\log \cos P_1 = \bar{1}\cdot 98584 \qquad\qquad \log \cos P_2 = \bar{1}\cdot 99863$$
$$\log \cot x = 0\cdot 13529 \qquad\qquad \log \cot x = 0\cdot 13529$$

$$\log \cot y_1 = 0\cdot 12113 \qquad\qquad \log \cot y_2 = 0\cdot 13392$$

$$y_1 = 37°\ 07' \qquad\qquad\qquad y_2 = 36°\ 18'$$
$$\text{Lat. } a = 52°\ 53'\ \text{N.} \qquad\qquad \text{Lat. } b = 53°\ 42'\ \text{N.}$$

$$\log \cos P_3 = \bar{1}\cdot 99803 \qquad\qquad \log \cos P_4 = \bar{1}\cdot 98402$$
$$\log \cot x = 0\cdot 13529 \qquad\qquad \log \cot x = 0\cdot 13529$$

$$\log \cot y_3 = 0\cdot 13332 \qquad\qquad \log \cot y_4 = 0\cdot 11931$$

$$y_3 = 36°\ 20' \qquad\qquad\qquad y_4 = 37°\ 14'$$
$$\text{Lat. } c = 53°\ 40'\ \text{N.} \qquad\qquad \text{Lat. } d = 52°\ 46'\ \text{N.}$$

Answers—Initial Course = $289\tfrac{1}{2}°$

Distance = 1650 miles.

Positions of Points:

(*a*) Lat. 52° 53′ N. Long. 20° 00′ W.

(*b*) Lat. 53° 42′ N. Long. 30° 00′ W.

(*c*) Lat. 53° 40′ N. Long. 40° 00′ W.

(*d*) Lat. 52° 46′ N. Long. 50° 00′ W.

4. Composite Great Circle Sailing

It will be noticed in fig. 14·6 that every point on a great circle path is in a higher Latitude than the point on the rhumb line path which lies on the same meridian. The great circle path always carries a vessel into Latitudes higher than does the rhumb line path between the same two places. Many great circle paths climb into very high Latitudes and are, accordingly, unsuitable for navigation. Other great circle arcs are unsuitable for marine use because they cross land.

If it is desired to travel along the shortest route between two places, such that the Latitude of the ship during the voyage is never greater than some given value, the route to follow is a Composite route, involving a great circle path from the initial position to a point on the limiting parallel; thence along the limiting parallel; and finally along a second great circle path to the destination. The middle part of the route, along the limiting parallel of Latitude, extends between the vertex of the first great circle path to the vertex of the second great circle path. In other words the vertices of the two great circle arcs both lie on the limiting parallel of Latitude.

GREAT CIRCLE SAILING

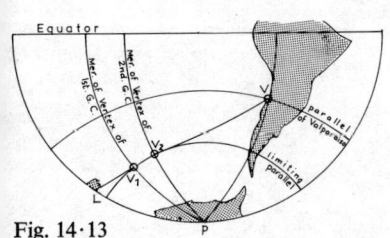

Fig. 14·13 illustrates the composite great circle path between Port Lyttelton and Valparaiso. The limiting parallel of Latitude for this particular route is taken to be Lat. 50° 00′ S.

The tedious task of computing the initial and final courses, the distance, and positions of points along the track, resolves itself into solving two right-angled spherical triangles and a Parallel Sailing problem. In practice, a gnomonic chart would normally be used to facilitate Composite Great Circle Sailing.

Fig. 14·13

The following example, which is illustrated by fig. 14·14, serves to show how a Composite Creat Circle Sailing problem is computed.

Example 14·3—Find the distance and the initial course for the composite track from a position off the Cape in Lat. 40° 00′ S. Long. 20° 00′ E., to a position off Tasmania in Lat. 12° 00′ S. Long. 145° 00′ E. The limiting Latitude is to be 48° 00′ S.

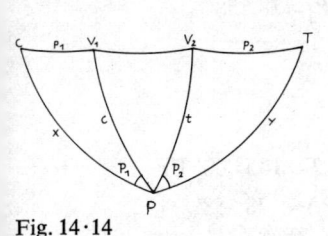

Given:
$x = 50° 00′$
$y = 78° 00′$
$c = 42° 00′$
$t = 42° 00′$

Fig. 14·14

Find p_1, P_1 and C for Distance, D. Long., and Initial Course, in the triangle PCV_1.

Find p_2, P_2, in triangle PTV_2, for Distance and D. Long.

In Triangle PCV_1:

$\sin \text{co } x = \cos p \cos c$
$\cos p = \cos x \sec c$
$\cos P = \cot x \tan c$

log cos $x = \bar{1} \cdot 80807$
log sec $c = 0 \cdot 12893$

log cos $P_1 = \bar{1} \cdot 93700$

$P_1 = 30° 07′$

Distance = 1807 miles

log cot $x = \bar{1} \cdot 92381$
log tan $c = \bar{1} \cdot 95444$

log cos $P_1 = \bar{1} \cdot 87825$

$P_1 = 40° 56′$

$\sin c = \sin C \cdot \sin x$
$\sin C = \text{cosec } x \cdot \sin c$

log cosec $x = 0 \cdot 11575$
log sin $c = \bar{1} \cdot 82551$

log sin $C = \bar{1} \cdot 94126$

$C = 60° 52'$

Initial Course $= 119°$

In Triangle PTV_2.

$$\sin co\ y = \cos p \cos t$$
$$\cos p = \cos y \sec t$$
$$\cos P = \cot y \tan t$$

$\log \cos y = \bar{1}\cdot 31788$	$\log \cot y = \bar{1}\cdot 32748$
$\log \sec t = 0\cdot 12893$	$\log \tan t = \bar{1}\cdot 95444$
$\log \cos p_2 = \bar{1}\cdot 44681$	$\log \cos P_2 = \bar{1}\cdot 28192$
$p_2 = 73° 45'$	$P_2 = 78° 58'$

Distance $= 4425$ miles

Long. $P = 20° 00'$ E.	Long. $T = 145° 00'$ E.
D. Long. $(P_1) = 40° 56'$	D. Long. $(P_2) = 78° 58'$
Long. $V_1 = 60° 56'$ E.	Long. $V_2 = 66° 02'$ E.

D. Long. $V_1 V_2 = 5° 6'$
$\phantom{\text{D. Long. }V_1 V_2} = 366'$

From Traverse Table:

Dep. $= 245$ miles

Total Distance $= p_1 + \text{Dep.} + p_2$
$\phantom{\text{Total Distance}} = 1807 + 245 + 4425$
$\phantom{\text{Total Distance}} = 6477$ miles

Answers—Initial Course $= 119°$
Distance $= 6477$ miles.

Exercises on Chapter 14

1. Compare Great Circle Sailing with Rhumb Line Sailing.
2. In what circumstances does Great Circle Sailing provide the most economical route?
3. State the two disadvantages of Great Circle Sailing.

GREAT CIRCLE SAILING

4. Describe the properties of a gnomonic projection.
5. Define Polar Gnomonic; Transverse Gnomonic; and Oblique Gnomonic Projections.
6. Describe how a polar gnomonic projection is constructed.
7. What is the radius of parallel of Latitude 65° on a polar gnomonic projection constructed from a model globe of radius 16·500 inches?
8. Explain carefully how a gnomonic chart is used to facilitate Great Circle Sailing.
9. In the absence of a gnomonic chart how would you lay down a great circle path on a Mercator Chart?
10. Find out how ABC and Azimuth tables are used to facilitate Great Circle Sailing.
11. What is meant by the vertex of a great circle?
12. The northern vertex of a great circle is in Lat. 36° 25′ N. Long. 126° 18′ E. What is the position of the southern vertex?
13. In what circumstances do the vertex of a great circle route lie (a) within and (b) outside the route?
14. The Initial Course is 072°. The Final Course is 168°. Where does the vertex of this Great Circle route lie relative to the final position?
15. What is a Composite Great Circle route?
16. Prove that the shortest route between two places when the ship is not permitted to sail polewards of a given Latitude is a Composite Great Circle route.
17. Describe the appearance of a Composite Great Circle route on (a) a Mercator Chart and (b) a polar gnomonic chart.
18. A great circle cuts the equator in Long. 40° 00′ W. at an angle of 29° with the meridian. What is the position of the northern vertex?
19. A vessel on a great circle track crosses the equator on a course of 060° in Long. 170° E. What is the position of the southern vertex of the great circle along which the ship is sailing?
20. A vessel on a great circle track has a course of 090° in Lat. 40° 10′ N. Long. 160° 00′ E. Where does this great circle cross the equator, and what angle does it make on crossing the equator?
21. Find the great circle Distance and the Initial Course from a position off the Lizard in Lat. 49° 50′ N. Long. 05° 15′ W. to a position off the Bermudas in Lat. 32° 29′ N. Long. 64° 00′ W. Find the positions of the points along the route which differ by 10° of Longitude from the Longitude of the vertex.
22. Find the great circle Distance and the Initial Course from a position off Newfoundland in Lat. 46° 20′ N. Long. 53° 00′ W., to a position off San Salvador in Lat. 24° 10′ N. Long. 74° 15′ W.
23. Find the Initial Course and the Distance along the great circle route from a position off Gibraltar in Lat. 36° 00′ N. Long. 06° 00′ W. to a position in the Mona Passage in Lat. 19° 00′ N. Long. 68° 00′ W. Find the positions of points along the route whose Longitudes differ by multiples of 10° from 10° W.

24. Find the Initial Course and the Great Circle Distance from a position off New Zealand in Lat. 38° 00′ S. Long. 179° 00′ E. to a position in the Gulf of Panama in Lat. 06° 00′ N. Long. 79° 00′ W.

25. Find the Great Circle Distance and the Initial Course from a position off the Cape of Good Hope in Lat. 39° 00′ S. Long. 20° 00′ W. to a position in the Bass Strait in Lat. 40° 00′ S. Long. 142° 00′ E.

26. Find the Initial Course and the Distance along the composite great circle route from A in Lat. 42° 00′ S. Long. 175° 00′ E. to B in Lat. 56° 00′ S. Long. 70° 00′ W., so that the maximum Latitude is 56° 00′ S.

27. Find the Initial Course and the Distance along the composite great circle route from a position off the S.W. coast of Ireland in Lat. 51° 20′ N. Long. 10° 00′ W. to a position in the Belle Isle Strait in Lat. 52° 00′ N. Long. 55° 00′ W. so that the limiting parallel is 53° 00′ N. Calculate the positions of points on the route which differ by multiples of 10° of Longitude from Longitude 20° 00′ W.

28. What is the least distance from the North Pole of an aircraft flying on a great circle route from Tokyo in Lat. 35° 40′ N. Long. 139° 46′ E. to Bergen in Lat. 60° 24′ N. Long. 05° 19′ E?

PART 3

INTRODUCTION TO CHARTWORK

Part 3, which comprises Chapters 15 to 24 inclusive, deals essentially with Coastal Navigation, in which the Chart and Compass are the principal instruments used by navigators.

Following a general discussion on charts and their use, consideration is given to the several methods of establishing a vessel's position—or "fixing" as the navigator describes the process—from observations of charted land- or sea-marks.

In many sea areas, especially those of North-West Europe, the range of the tide is of great importance when navigating coastwise. In such areas the mariner, especially when his vessel is close inshore, pays particular attention to the depth of water under his vessel's keel: the possibility of running aground because of insufficient water in tidal regions, should always be in the mind of the prudent navigator. A description of the tide and of its importance in navigation is, therefore, included in this part of the book.

CHAPTER 15

INTRODUCTION TO CHARTWORK

1. Coastal Navigation

Coastal Navigation is that branch of the main subject which includes the various methods of ascertaining the position of a vessel from observations of lighthouses, beacons and other conspicuous land- or sea-marks. Many of the methods used are based on geometrical principles. It is desirable, therefore, for a navigator to have a good working knowledge of the fundamentals of plane geometry so that he is able to understand the principles of the methods he may use.

2. Charts

A chart is a map used by a navigator. Depicted on a chart is all the essential data that may assist him in navigating his vessel. Charts may be classified into four main groups according to the specific purpose they serve:

  (i) *Ocean Charts*—Ocean charts are small-scale charts used mainly for deep sea navigation. For this reason coastal detail is not depicted. A small-scale chart is one on which a relatively long distance of the Earth's surface is represented by a relatively short distance on the chart. A large-scale chart, on the other hand, is one on which a relatively short distance of the Earth is represented by a relatively long distance on the chart.

  (ii) *Coastal Charts*—Coastal charts are large-scale charts used when navigating close inshore. Included in the considerable information given on a coastal chart are: the nature of the shore and coastline; positions and characteristics of lights, radio beacons, towers, and other prominent features observable from the offing, which may aid the coastal navigator; depths of water; current and tidal information; and positions of rocks, shoals, buoys and other floating and fixed sea-marks.

  (iii) *Plan Charts*—Plans are very large-scale charts on which are depicted detailed information of small areas such as harbours or estuaries.

  (iv) *Miscellaneous Charts*—In this group are included all those charts that are not included in the other groups. These are: gnomonic charts for facilitating great circle sailing; charts of the magnetic elements of the Earth, lattice charts for use with hyperbolic navigation systems, route charts, and wind and current charts.

3. The Natural Scale of a Chart

The natural scale of a chart is the ratio, expressed as a fraction, between a unit of length on the chart and the number of such units of the Earth's surface which it represents. If, say, 1 cm. on a chart represents 1 km. on the Earth, the natural scale of the chart is 1/100000. If 1 cm. on the chart represents one nautical mile the natural scale is 1/185200, because there are 1852 metres in 1 nautical mile.

Charts having a natural scale of more than 1/50000 are usually constructed on the gnomonic projection. Most plan charts are of this type. The area to be represented, in this case being relatively small, the tangential plane almost coincides with the spherical Earth's surface over the limits of the area charted. It should be noted that on plan charts parallels of Latitude and meridians are not projected as straight lines, and that projected meridians do not cross projected parallels at 90°. However, lack of parallelism of projected meridians and parallels is hardly detectable on most plans because of the smallness of the area represented.

Because of the advantages offered to the navigator by the Mercator projection all ocean and coastal charts employ this projection.

All vessels should be adequately equipped with up-to-date charts and Sailing Directions. Sailing Directions are supplied by the Hydrographic Department of the Navy to cover the world. These, familiarly known as Pilots, contain valuable information of assistance to navigators. The Pilots are complementary to the charts, and the relevant Pilot should be consulted before and during a voyage as necessary.

It is the duty of the Master of a vessel to ensure that his vessel is properly supplied with charts and Sailing Directions and other publications necessary for the safe navigation of his vessel. Compiling the requirements for a voyage, and seeing to it that the necessary charts and Pilots are on board before the commencement of the voyage, is the duty of the Navigating Officer.

The charts published by the Hydrographic Department of the Navy are among the best procurable. The Department issues more than 4000 charts covering the whole world. The *Catalogue of Hydrographic Charts*, which is obtainable at small charge, contains detailed information of all the charts and sailing directions published by the Hydrographic Department. This information includes: sizes and scale of charts; limits of Latitudes and Longitudes; and dates of publication. The *Catalogue* also contains index maps from which a navigator may readily see what charts are necessary for a given voyage.

4. Description of a Chart

The margin of a chart contains much useful information. In the bottom right-hand corner will be found the Catalogue Number of the chart. The dimensions of the plate from which the chart is printed and particulars of the plate or reproduction method are also given in the bottom right-hand corner of the chart.

In the bottom left-hand corner of the chart will be found the inscription "Small Corrections". Small corrections are made by hand from information promulgated by the Hydrographic Department in small booklets published weekly, monthly, quarterly and annually. These booklets, entitled *Notices to Mariners*, contain details of changes to charted information. The Notices are numbered and, when a chart has been corrected from information given in a particular Notice, the number of the Notice is recorded on the chart. The year of the Notice is given in heavy type and the number of the Notice is given in light type, thus:

Small Corrections: **1970.** 16,87,96. **1971.** 40 87,

Temporary and Preliminary information should be inserted in pencil on the chart of largest scale, and recorded thus:

230/1970 (T)   363/1975 (P)

these meaning, respectively, that temporary information was placed on the chart on the 230th day of 1970, and preliminary information was placed on the chart on the 363rd day of 1975.

Any necessary large correction is made to a chart before it is issued. A large correction often necessitates a new edition of the chart. The date on which a large correction was made, or a new edition issued, is inserted in the middle of the lower margin of the chart.

The day on which the chart was printed is recorded by giving the day number of the year and the last two figures of the year number in the top right-hand corner of the chart. A chart printed on the 4th of February 1975, for example, would be marked 35·75.

The Title of a chart, usually found clear of the sea area, contains information which should be studied carefully before using the chart. Details given in the title include:

(i) Name of the area depicted.

(ii) Details of the survey.

(iii) Units (metres or fathoms) of depths.

(iv) Level below which charted depths are given.

(v) Natural Scale.

(vi) A note stating that bearings are True and given from seaward.

(vii) Certain abbreviations used.

A particular chart is described by giving its Title and Catalogue Number; the date of the edition; the date of printing, and the date of the latest small correction.

To distinguish a good chart from an indifferent one, the navigator may be assisted by the following:

(i) The date of the survey of the area depicted. In general, the more recent the survey the more reliable is the chart.

(ii) Amount of detail charted. Lack of detail often indicates an incomplete survey. If the coastline, for example, is shown as a dotted instead of a full line, it is probable that the survey was incomplete.

(iii) The spacing of the charted depths. This affords a good indication of the value of a chart. Not only the number of soundings given, but their spacing and arrangement, provide good tests of the reliability of the chart. Charted depths of a well-surveyed area usually appear systematically in parallel or radial straight lines and there are no large areas where no depths are given. The soundings on a chart compiled from scanty survey material are marked here and there more or less haphazardly.

On metric charts published by the Hydrographic Department shore areas lying between the High Water line and the 5-metre isobath are tinted blue.

5. Chart Abbreviations and Symbols

All abbreviations and symbols employed on charts of the Hydrographic Department are given on Chart 5011. This chart should be studied carefully and the information it gives should be familiar to all coastal navigators.

The description of the nature of the sea bed is denoted by the initial letter of the deposit or material. In general, capital letters are used for nouns and small letters for descriptive words. Thus: yG indicates yellow gravel; and rS indicates red sand.

Symbols for lighthouses depend upon the candle power of the light displayed. The usual symbols are:

The less elaborate is the symbol the smaller is the candle power of the light.

For purposes of identification navigational lights have a variety of characteristics. A light which gives a continuous steady light is described as a Fixed Light which is denoted on the chart by the abbreviation F. A Flashing Light, which is denoted by Fl., is one in which the length of the eclipse period between the flashes is greater than the length of the flashes. An Occulting Light, denoted by Occ., is one in which the length of the eclipse is less than that of the flash. Lights which exhibit flashes of the same length as the eclipses are known as Isophase Lights. Some lighthouses exhibit lights consisting of a combination of flashes of different lengths as letters of the Morse Code. These are called Morse Code Lights.

A light that shows different colours on the same bearing or over the same arc of the horizon is called an Alternating Light, and is denoted on a chart by the abbreviation Alt. A light that shows different colours on different arcs is called a Sector Light. These are indicated on the chart by dotted arcs, as shown in fig. 15·1.

The radii of the arcs of sector lights are not to be mistaken for the ranges of visibility. The light illustrated in fig. 15·1 shows Green from 330° to 000°; White from 270° to 330°; and Red from 240° to 270°. Remember that all bearings given on a chart are true and are given *FROM* seawards.

A light that flashes at a rate of more than 60 per minute is called a Quick Flashing Light, the abbreviation for which is Qk.Fl.

Following the abbreviated description of the type of light the Period of the light is given. This is the interval in seconds for a complete sequence of the characteristic of the light.

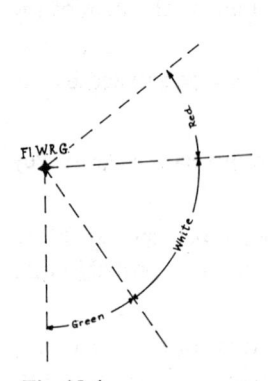

Fig. 15·1

The height of the centre of the lens of a lighthouse above the level of Mean High Water at Spring Tide (M.H.W.S.) is given on large-scale charts, and so also is the range of visibility in miles. The charted visibility of a light is given as an integral number of miles, and it represents the distance the light is visible in clear weather assuming the eye of the observer to be located 15 feet above the level of the sea. This distance is known as the Charted Range.

All lights are to be taken as being white unless otherwise indicated in which case R denotes Red and G denotes Green.

The abbreviated description of a light, red in colour, which exhibits a group of three flashes at intervals of 30 seconds, and which has a height above M.H.W.S. of 105 feet and a charted range of 15 miles is:

Gp Fl (3) R 30 secs 105 ft 15 M

INTRODUCTION TO CHARTWORK

The charted information of this light on small-scale charts is less complete. On an ocean chart, for example, the abbreviated description might be:

$$\text{Gp Fl R}$$

In the symbol for a Light Vessel, which is ⛴, the small circle denotes the exact position of the light vessel.

Buoys are depicted thus: △ for conical; ⌷ for can-shaped; and ◯ for spherical buoys. The colours and patterns of buoys are indicated by the following abbreviations:

B	Black
R	Red
Y	Yellow
G	Green
BWHB	Black and White Horizontal Bands
RWHS	Red and White Horizontal Stripes
RWCheq	Red and White Chequered
BWVS	Black and White Vertical Stripes

All charted heights on the land area of a chart are given in feet or metres above the level of M.H.W.S. All depths, on the other hand, are given below a level which is approximately at the level of Mean Low Water Springs (M.L.W.S.). The actual level is known as Chart Datum (C.D.). Drying heights on banks and offshore zones are given in feet or metres above the level of Chart Datum. The figures which indicate these heights are underlined

Tidal Stream and Current information is depicted thus:

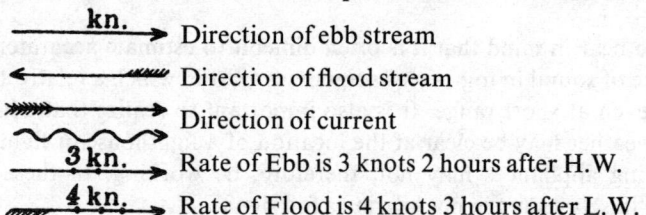

⎯⎯kn.⟶ Direction of ebb stream

⟵⎯⎯⎯ Direction of flood stream

⟶⟶⟶ Direction of current

⎯⎯3 kn.⟶ Rate of Ebb is 3 knots 2 hours after H.W.

⎯⎯4 kn.⟶ Rate of Flood is 4 knots 3 hours after L.W.

On some coastal charts tables of tidal stream information are given for certain reference positions on the chart which are denoted thus:

◇A◇ ◇B◇ ◇C◇

The commonly used symbols for Wrecks, Rocks and other obstructions are:

✤ Rock with less than 6 ft. of water above it at Chart Datum.

✢ Rock awash at Chart Datum.

◌ In general denoted a danger area.

Wₖ④ Wreck, exact depth over it indicated.

⊞ Submerged wreck non-dangerous to surface navigation.

✣ Dangerous wreck—depth unknown.

⊥   Wreck with superstructure visible at L.W.

Wk   Wreck visible at L.W. (large-scale charts).

(o o o)   Wreck invisible at L.W. (large-scale charts).

Foul   Non-dangerous except for anchoring or trawling.

   Swept by wire drag to depth indicated.

The sea-bed, especially in coastal waters, is often obstructed by submarine telegraph and power cables and with pipelines. It is important that vessels do not anchor in the vicinity of these. The location of cables and pipelines is indicated on charts published by the Hydrographic Department by a wavy line and a line composed of dots and dashes, respectively. These lines are usually given in magenta to render them conspicuous in artificial light.

A variety of fog signalling devices are in current use. These include the Diaphone, denoted on charts by the abbreviation Fog Dia., which produces a very powerful air signal of low pitch, the sound ending with a distinctive grunt caused by the abrupt lowering of the pitch of the sound. The Nautophone, denoted by Fog Nauto., produces a note of high pitch, whereas the Reed, denoted by Fog Reed, gives a high pitch of power less than that of the nautophone. Other fog signalling devices include the Fog-Whistle, -Horn, -Gun, -Gong, -Siren, as well as the modern Tyfon.

Fog Detector Lights are sometimes incorporated at lighthouses and light vessels. A Fog Detector Light is a concentrated light of high frequency—and hence bluish in colour—which sweeps across the horizon, and which may be reflected by droplets of water in the atmosphere. The reflected energy triggers off a device which automatically sets the fog signalling apparatus in operation.

It is important to bear in mind that it is often difficult to estimate accurately the direction and range of a source of sound in fog. Areas sometimes exist in which a relatively strong sound signal is not heard even at short range. It is also important to realize that, although a vessel may be in fog, the weather may be clear at the location of a lighthouse or light vessel nearby, and the fog signalling appliances may not, therefore, be working. In these circumstances a navigator may feel falsely secure in not hearing a fog signal.

Symbols associated with radio and radar stations and beacons are coloured magenta:

⊙   denotes a coast radio station

⊙ Racon   denotes a Radar Responder Beacon

⊙ Ramark   denotes a Radar Beacon which transmits continuously and which provides information by means of which the navigator on a vessel in the vicinity may identify the beacon

⊙ Consul Bn   denotes a Consol Station

⊙ R°D.F.   denotes a Radio D.F. Station.

6. Hints when using Charts

1. A chart in use at any time should be properly corrected from *Notices to Mariners* up to the time of its being used.

INTRODUCTION TO CHARTWORK

2. It is important to investigate any Cautionary Notices which are sometimes displayed in not-too-conspicuous positions on charts. These Notices may include:
   (i) Prohibited and Dangerous Areas.
   (ii) Areas of Abnormal Variation.
   (iii) Exceptional Tidal Streams.
   (iv) Ice Warnings.
3. Always use the chart of largest scale available. The reasons for this include:
   (i) More detail is shown than on charts of smaller scale.
   (ii) Any plotting errors are reduced to a minimum.
   (iii) It is more up-to-date than charts of smaller scale. The plate of a large-scale chart is always corrected before those of smaller scale of the same area.
   (iv) Errors of distortion have the least effect.
4. When transferring a position from one chart to another use bearing and distance from a charted point common on both charts. Remember that the navigator is not so much interested in Latitude and Longitude as he is in his position relative to the land.
5. Ascertain the vessel's position as soon as possible after transferring from one chart to another.
6. Never have more than one chart on the table at any one time: the scales may be wrongly used.
7. A soft well-sharpened pencil should be used for all chart work: never harder than an HB and never softer than a B.
8. All courses should be checked and re-checked before setting.
9. Write information such as times of bearings, log readings, and so on, on that part of the chart which has been used: never clutter up the path ahead.
10. Use the meridian lines in conjunction with the parallel rulers for measuring course- and bearing-angles, and for laying down course-lines.
11. Keep the chart storage space as dry as possible: a damp atmosphere often leads to distortion of charts.
12. Remember that a chart is an invaluable part of a vessel's equipment. Treat it with the respect that it deserves.

Exercises on Chapter 15

1. Distinguish between a Plan Chart, a Coastal Chart and an Ocean Chart.
2. Ask to see the chart outfit of your vessel. What miscellaneous charts are carried on board?
3. Distinguish between a small-scale chart and a large-scale chart.
4. What is meant by the Natural Scale of a chart?
5. What is the Natural Scale of a chart on which 1 inch represents (*a*) 10 nautical miles, (*b*) 1 cable?
6. The Natural Scale of a chart is 1/50000. What is its linear scale in cms. per nautical mile?
7. Describe fully the information to be found in the margin of a chart.

8. Describe the information found in the Title of a Chart.
9. What are Cautionary Notices?
10. Enumerate the features of a reliable chart.
11. Define: Flashing Light, Occulting Light, Group Flashing Light, Group Occulting Light.
12. Distinguish between an Alternating Light and a Sector Light.
13. Describe how the following sector light is depicted on a chart: Light shows Green from 040° to 160°, thence White from 160° to 350°. (*N.B.* Remember that bearings are true and that they are always given from seawards.)
14. What tidal stream information may be given on a chart?
15. Describe the tables of tidal stream information that may be given on some coastal charts.
16. Describe the symbols used for depicting wrecks on charts.
17. Describe the *Chart Catalogue* published by the Hydrographic Department. Explain how you would draw up a list of the charts necessary for a given voyage.
18. What information is given in Sailing Directions? (Ask to examine a copy of the Sailing Directions, and study the contents.)
19. What are *Notices to Mariners*? Explain how charts are kept up-to-date.
20. Describe the information pertaining to Radio and Radar stations that may be depicted on charts.
21. Describe the symbols for the various forms of fog-signalling devices.
22. What is a Fog Detector Light?

CHAPTER 16

CORRECTING THE COURSE

1. The Three Norths

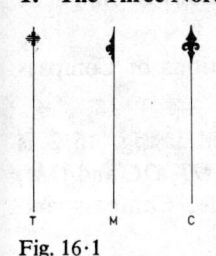

Fig. 16·1

(i) *True North*—This is the horizontal direction along the plane of any meridian towards the northern extremity of the Earth's axis of rotation.

(ii) *Magnetic North*—This is the horizontal direction indicated by the North-Seeking, or Red, end of a compass needle when it is under the influence of the Earth's magnetism and no other disturbing force.

(iii) *Compass North*—This is the horizontal direction indicated by the North-Seeking, or Red, end of the compass needle.

2. The Earth's Magnetism

The Earth has magnetic properties such that lines of magnetic force emanate from a position in South Victoria Land and follow approximate great circle paths to a position in Hudson Bay. These two positions are known, respectively, as the North and South Magnetic Poles.

The North Magnetic Pole of the Earth is said to have Blue Polarity, and the South Magnetic Pole Red Polarity.

Every artificial magnet, such as a compass needle, has similar properties to those of the Earth-Magnet. The fundamental law of magnetism is that a force acts between any two magnetic poles. If the two poles are both Red or both Blue, the force is one of repulsion. If they have opposite polarities, the force is one of attraction. The Red end of a compass needle is attracted by the Earth's Blue pole. It is for this reason that the Red end of the compass needle is the North-Seeking end.

The magnetic force which endeavours to hold a compass needle in the plane of the magnetic meridian is very meagre: for this reason the compass needle is very easily disturbed. The chief disturbing force is that which results from the magnetism possessed by the vessel on board which the compass is housed. Cargo in the vessel may also have a disturbing influence. The vessel's magnetism is acquired by induction from the Earth-Magnet. The effect of the vessel's magnetism at the position of the compass usually results in the axis of the compass needle being forced out of the plane of the magnetic meridian. Thus, the directions of the three norths, defined in paragraph 1, are usually different from one another.

It will readily be appreciated that the courses and bearings laid down on a chart are related to True North, whereas courses steered by a helmsman, and bearings observed by a navigator, are related to Compass North. It is, therefore, necessary to convert courses and bearings from one system to another.

Conversion of courses and bearings from True to Compass, and *vice versa*, is a navigational duty which every responsible navigator will perform with the utmost care and with subsequent check.

3. The Three Courses

(i) *True Course*—This is the direction of a vessel's head relative to the direction of True North. It is the horizontal angle between the direction of True North and the vessel's fore-and-aft line ahead.

(ii) *Magnetic Course*—This is the horizontal angle between the directions of Magnetic North and the fore-and-aft line of the vessel in the forward direction.

(iii) *Compass Course*—This is the horizontal angle between the directions of Compass North and the fore-and-aft line of the vessel ahead.

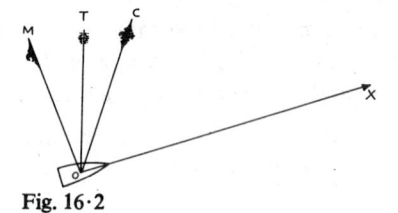

Suppose that the vessel illustrated in fig. 16·2 is heading in the direction OX; and that OT, OC and OM, are the respective directions of True, Compass and Magnetic Norths, Then:

TOX = True Course
COX = Compass Course
MOX = Magnetic Course

Fig. 16·2

4. The Three Bearings

(i) *True Bearing*—This is the horizontal angle between the directions of True North and "bearing" of an object. It is, in other words, the angle between the directions of True North and the object. Notice that the term "bearing" may be used in the sense of it being an angle or a direction.

(ii) *Magnetic Bearing*—This is the angle, in the horizontal plane, between the directions of Magnetic North and an object.

(iii) *Compass Bearing*—This is the horizontal angle between the directions of Compass North and the object.

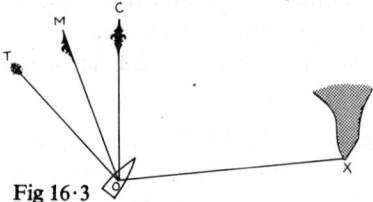

In fig. 16·3, O denotes the position of a vessel, X that of a point of land, and OT, OM and OC, denote, respectively, the directions of True North, Magnetic North and Compass North.

TOX = True Bearing of X
MOX = Magnetic Bearing of X
COX = Compass Bearing of X

Fig 16·3

5. Compass Error

Compass Error is the horizontal angle between the directions of True North and Compass North. When Compass North lies to the right of True North, Compass Error is named East. When Compass North lies to the left of True North, Compass Error is named West. The operation of converting True Courses and Bearings into their corresponding Compass Courses and Bearings requires the application of Compass Error. The process is simple provided that the conventional system of naming Compass Error, given above, is memorized. A rough diagram also assists, as indicated in the following examples.

CORRECTING THE COURSE 123

Example 16·1—Find the Compass Course if the True Course is 070° and the Compass Error is 5° E.

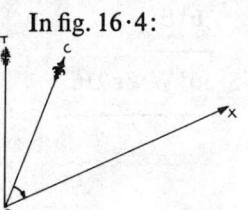

In fig. 16·4:

Fig. 16·4

OT = True North
OC = Compass North
TOC = Compass Error
OX is direction of vessel's heading
TOX = True Course   = 070°
TOC = Compass Error = 5° E.
                         ———
COX = Compass Course = 065°

Example 16·2—Find the True Bearing of a lighthouse if its Compass Bearing is 140° and the Compass Error is 10° W.

In fig. 16·5:

OT_N = True North
OC_N = Compass North
COT = Compass Error   = 10° W.
COX = Compass Bearing = 140°
TOX = True Bearing   = 130°

Fig. 16·5

6. Variation

Variation is the horizontal angle between the directions of True and Magnetic Norths. It is defined as the angle between the True and Magnetic Meridians at any place.

When Magnetic North lies to the right of True North, the Variation is named East. When Magnetic North lies to the left of True North, Variation is named West.

The Variation at any place may be found by inspection from a Variation Chart on which is drawn lines connecting places where the variation is the same. These lines are known as Isogonic lines or Isogonals: a Variation Chart is, therefore, sometimes known as an Isogonic Chart. The angle of variation is sometimes printed on the compass roses of navigational charts.

Variation at any place is not constant: it undergoes a gradual change with time. The navigator should be careful to apply any necessary secular correction to variation found from a chart for a date different from that of the year the chart was published.

Variation applied to a True Course or Bearing gives a corresponding Magnetic Course or Bearing, and *vice versa*.

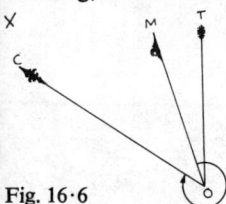

Example 16·3—Find the Magnetic Course if the True Course is 320° and the Variation is 8° W.

In fig. 16·6:

TOX = True Course   = 320°
TOM = Variation   = 8° W.
MOX = Magnetic Course = 328°

Fig. 16·6

Example 16·4—Find the True Bearing of a star if its Magnetic Bearing is S. 30° W. and the Variation is 6° E.

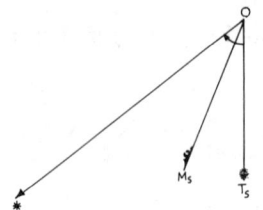

In fig. 16·7:

$MO\star$ = Magnetic Bearing = S. 30° W.
TOM = Variation = 6° E.

$TO\star$ = True Bearing = S. 36° W. or 216°

Fig. 16·7

7. Deviation

Deviation is the horizontal angle between the directions of Magnetic and Compass Norths. It is due to the magnetism of the vessel and/or her cargo. Deviation for an uncompensated compass normally changes with the vessel's course.

When Compass North lies to the right of Magnetic North, Deviation is named East: when Compass North lies to the left of Magnetic North, it is named West.

To convert a Compass Course or Bearing into a corresponding Magnetic Course or Bearing, or *vice versa*, Deviation must be applied.

Example 16·5—Find the Compass Course to steer a Magnetic Course of 030° if the Deviation for the ship's heading is known to be 10° E.

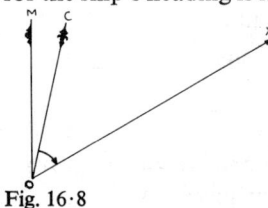

In fig. 16·8:

MOX = Magnetic Course = 030°
MOC = Deviation = 10° E.

COX = Compass Course = 020°

Fig. 16·8

Example 16·6—The Magnetic Bearing of a lighthouse is required. The Compass Bearing is 330° and the Deviation for the heading of the vessel, when the bearing was observed, is 5° W.

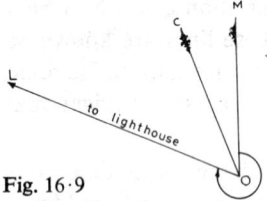

In fig. 16·9:

COL = Compass Bearing = 330°
MOC = Deviation = 5° W.

MOL = Magnetic Bearing = 325°

Fig. 16·9

From the foregoing remarks it will be seen that the Compass Error is a combination of Variation and Deviation.

Example 16·7—Given Variation = 15° W., and Deviation = 5° E. Find the Compass Error.

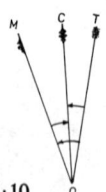

In fig. 16·10:

TOM = Variation = 15° W.
MOC = Deviation = 5° E.

TOC = Compass Error = 10° W.

Fig. 16·10

Example 16·8—Given Compass Error = 4° E., Variation = 15° E. Find the Deviation for the corresponding heading of the vessel.

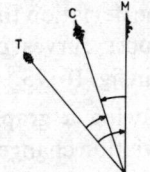

CORRECTING THE COURSE

In fig. 16·11:

TOC = Compass Error = 4° E.
TOM = Variation = 15° E.

MOC = Deviation = 11° W.

Fig. 16·11

Example 16·9—Given Compass Error = 20° W., Variation 15° W. Find the Deviation for the corresponding heading of the vessel.

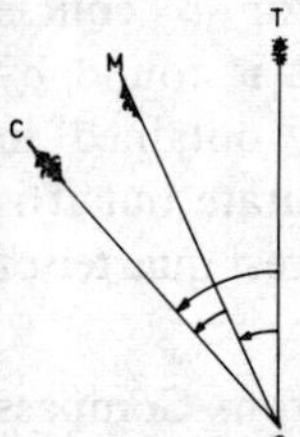

In fig. 16·12:

TOC = Compass Error = 20° W.
TOM = Variation = 15° W.

MOC = Deviation = 5° W.

Fig. 16·12

8. The Deviation Card or Table

An attempt is made to neutralize the magnetism of the vessel at the position of the magnetic compass, but complete success at doing so is not normally to be expected, and any uncompensated magnetism due to the vessel results in residual deviations. Before leaving harbour, and before making a landfall, whenever residual deviations for anticipated courses are not known, it is wise to "swing" the vessel and to draw up a table of residual deviations. This is done by steadying the vessel on successive courses, and comparing the True and Compass Bearings of a fixed shore object or a celestial body, for each course, in order to find the deviation for that course. This is found by combining the Compass Errors (found from the observations) with the Variation. It is customary to find the deviations at 10°-intervals around the compass.

It should be borne in mind that the deviations shown on a Deviation Card apply only to the time and place at which they were determined. Deviation changes with geographical position and it may be affected by the condition of loading of the vessel, and by the cargo she has on board.

The deviations should be verified by observation frequently and particularly as soon as possible after each alteration of course. It is usual, when on a steady course, to check the deviation at least once each watch, and to record the details of the observation in a Deviation Book. On many vessels the Deviation Book serves as a working Deviation Card.

A Deviation Card may give deviations against Compass Course, Magnetic Course, or both Compass and Magnetic Courses. A portion of a Deviation Card of the latter type is illustrated below:

Compass Course	Deviation	Magnetic Course
000°	2° E.	002°
010°	$3\frac{3}{4}$° E.	014°
020°	$4\frac{3}{4}$° E.	025°
030°	$5\frac{1}{4}$° E.	035°
040°	$5\frac{1}{4}$° E.	045°
050°	$4\frac{1}{2}$° E.	054°
060°	$3\frac{1}{4}$° E.	063°
070°	$\frac{3}{4}$° E.	071°
080°	$1\frac{2}{3}$° W.	078°

126   THE ELEMENTS OF NAVIGATION AND NAUTICAL ASTRONOMY

If a graph of deviations against course is drawn it is an easy matter to find the deviation for any Compass of Magnetic Course. The Deviation Card given above produces curves of deviation which are illustrated in fig. 16·13.

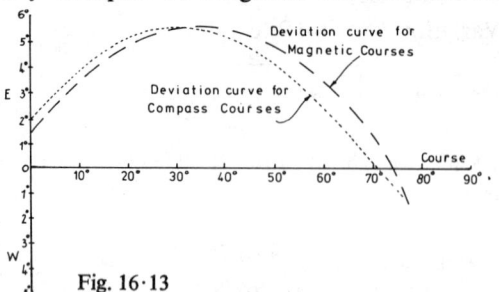

In practice, instead of drawing a graph it is usual to assume that the deviation changes linearly between successive tabulated values given on the Deviation Card. In this circumstance the deviation for a course lying between tabulated values is found by simple proportion. The result obtained in this method is not strictly accurate but it is usually sufficiently accurate for practical purposes in which deviations to the nearest quarter of a degree are good enough.

Fig. 16·13

The task of converting courses and bearings from True to Compass or from Compass to True is mastered only after considerable practice. It is recommended that novices draw diagrams to assist them in solving these very important problems. If the principles are learnt thoroughly there is no need to employ any of a profusion of mnemonical rules which, instead of helping students, often make confusion doubly sure.

Example 16·10—Find the Compass Course to steer a True Course of 336° if the variation is 10° W. Use the attached Deviation Card.

Compass Course	Deviation		
000°	4° E.	True Course =	336°
350°	6° E.	Variation =	10° W.
340°	9° E.	Magnetic Course =	346°
330°	13° E.	Deviation =	11° E.
320°	10° E.	Compass Course =	335°
310°	6° E.		
300°	5° E.		

Explanation of how the deviation was found:

Two adjacent compass courses were chosen such that when the deviations were applied to them two magnetic courses were found. It was arranged that these lay one on each side of the magnetic course of the vessel, viz. 346°. The change in the deviation between one of the two magnetic headings found and the vessel's magnetic course was found by simple proportion. This change was applied to the chosen magnetic heading and the deviation for the vessel's magnetic course, therefore, found. The computation is as follows:

Compass Course = 340°	Compass Course = 330°
Deviation = 9° E.	Deviation = 13° E.
Magnetic Course = 349°	Magnetic Course = 343°

For a change of 6° in Magnetic Course, Deviation changes 4°.
For a change of 1° in Magnetic Course, Deviation changes 4/6°.
For a change of 3° in Magnetic Course, Deviation changes 4/6 × 3°.

$$\text{That is to say:} \quad 2°$$
$$\text{Deviation on } 349° \text{ (M)} = 9° \text{ E.}$$

$$\text{Therefore: Deviation on } 346° \text{ (M)} = 11° \text{ E.}$$

9. Leeway

When the wind blows in any direction except right ahead or right astern, the path of a vessel under way is orientated to leeward of the direction of the vessel's fore-and-aft line.

Leeway, which is illustrated in fig. 16·14, is defined as the angle between the direction of the fore-and-aft of a vessel and the direction she makes through the water.

The angle of leeway may be ascertained by facing aft at the stern of a vessel and estimating the angle between the fore-and-aft line and the direction of the wake. The amount of leeway depends upon the strength of the wind; the direction it blows relative to the vessel; and the areas of the profile of the vessel above and below the waterline.

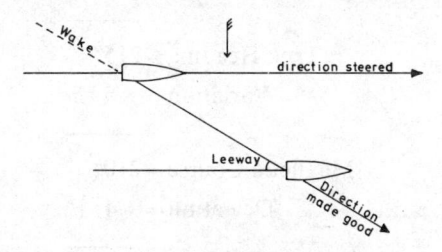

Fig. 16·14

In days of sail a semicircle was engraved on the taffrail of the vessel. The diametrical line of the semicircle was athwartships and the arc was marked in points and quarter points of the compass. At the centre of the semicircle was fitted a small swivel. Whenever the log was hove for measuring the speed of the vessel, before taking in the line it was slipped into the swivel, and the leeway read from the engraved semicircle.

Note that although a vessel proceeds along the direction she makes through the water, the vessel's head lies in the direction in which she is being steered, and it is this direction that governs the amount of compass deviation. Note also that the direction in which the vessel travels through the water is in the opposite direction to that of her wake, and that it is along this line that distance through the water is measured. A streamed log lies in the wake and it registers, therefore, the distance travelled through the water.

The two examples which follow should be studied carefully. Note particularly that in Example 16·11, in which it is required to find the Compass Course, the leeway is applied to windward; in Example 16·12, in which it is required to find the True Course, and in which the Compass Course is given, the leeway is applied to leeward. This is always the case: for finding Compass Course apply leeway to windward: but if given the Compass Course, apply leeway to leeward.

128   THE ELEMENTS OF NAVIGATION AND NAUTICAL ASTRONOMY

Example 16·11—Find the Compass Course to steer a True Course of 225°. Leeway due to a South wind is 10°; Variation is 5° E.; and deviation is obtainable from the attached Deviation Card.

Compass Course	Deviation
180°	4° W.
190°	1° W.
200°	2° E.
210°	6° E.
220°	9° E.
230°	10° E.
240°	3° E.
250°	2° W.
260°	6° W.

(C) = 200°   (C) = 210°
Deviation = 2° E.   Deviation = 6° E.
(M) = 202°   (M) = 216°

4/14 × 8° = 2°

Required Deviation = 4° E.

True Course = 225°
Leeway = 10°
———
True Heading = 215°
Variation = 5° E.
———
Magnetic Course = 210°
Deviation = 4° E.
———
Compass Course = 206°

Example 16·12—The Compass Course is 220°. Find the True Course made good given that the Variation is 5° W., and the leeway due to a South wind is 10°, and that the deviation is obtainable from the Card given in Example 16·11.

Compass Course = 220°
Deviation = 9° E.
———
Magnetic Course = 229°
Variation = 5° W.
———
True Course through water = 224°
Leeway = 10°
———
True Course made good = 234°

CORRECTING THE COURSE

Exercises on Chapter 16

1. Define: True North, Magnetic North, Compass North.
2. Define: True Bearing, Magnetic Bearing, Compass Bearing.
3. Define: True Course, Magnetic Course, Compass Course.
4. Define: Variation, Deviation. State the conventional rules for naming these angles.
5. What is Compass Error? Explain why Compass Error normally changes with the heading of a vessel.
6. What is meant by Residual Deviations? What is a Card of Deviations?
7. Explain how deviation for a given Magnetic Course is found from a table of deviations for Compass Courses.
8. Describe Leeway and its compensation when finding distance made good over the ground.
9. Fill in the blank spaces in the following table:

Compass Course	Variation	Deviation	Compass Error	True Course
318°	5° W.	2° E.	—	—
178°	5° E.	—	—	183°
185°	10° E.	—	14° W.	—
—	4° W.	—	6° W.	082°
—	—	4° W.	10° E.	257°
000°	7° W.	2° W.	—	—
272°	—	5° W.	6° W.	—
114°	—	4° W.	10° E.	—

10. Whilst heading 200° (C) the Compass Bearing of a lighthouse was 304°. The True Bearing at the time was 292°. Find the deviation for the vessel's heading given that the variation is 16° W.

11. A point of land was abeam to port. The vessel's heading was 124° (C). The True Bearing of the point was 040°. Find the deviation for the vessel's heading given that the variation is 6° E.

12. Find the Compass Course to steer in order to counteract the effect of a southerly gale given that the True Course to make good is 070°; the variation is 8° E., and the leeway is 10°. Use the Deviation Card given on page 130.

13. The course of a vessel is 320° (C). Variation = 5° E. Find the True Course made good if the wind is N.E. and the leeway is 5°. Use the Deviation Card given on page 125.

14. Convert the following True Courses into Compass Courses. Variation = 10° W. Use the Deviation Card given on page 125. (a) 120°, (b) 200°, (c) 176°, (d) 275°, (e) 265°.

15. Heading 157° (C) the following Compass Bearings were observed. Lighthouse 317°, Steeple 242°, Peak 170°. Variation = 10° E. Find, using the Deviation Card given below, the corresponding True Bearings.

Compass Course	Deviation	Compass Course	Deviation
000°	4° E.	180°	10° W.
010°	7° E.	190°	13° W.
020°	12° E.	200°	15° W.
030°	14° E.	210°	17° W.
040°	16° E.	220°	18° W.
050°	17° E.	230°	19° W.
060°	18° E.	240°	19° W.
070°	18° E.	250°	19° W.
080°	17° E.	260°	16° W.
090°	16° E.	270°	13° W.
100°	14° E.	280°	9° W.
110°	12° E.	290°	5° W.
120°	10° E.	300°	2° W.
130°	7° E.	310°	1° E.
140°	3° E.	320°	4° E.
150°	1° W.	330°	6° E.
160°	5° W.	340°	9° E.
170°	7° W.	350°	7° E.

CHAPTER 17

THE POSITION LINE

1. Fixing by Cross Bearings

If the Compass Bearing of a lighthouse or other conspicuous shore mark is observed, and the Compass Error for the particular heading of the vessel applied, the True Bearing of the observed object is determined. This enables a navigator to project or plot a straight line on his chart and to say, with certainty, that his vessel's position may be fixed on that straight line. Such a line is called a Position Line.

A position line obtained from a bearing of a fixed terrestrial mark is drawn from the charted position of the mark in a direction opposite to that of the bearing of the mark. If, for example, a mark bears due North, the position line is drawn in a direction due South of the charted position of the mark. Again, if a mark bears, say, 310°, the position line is drawn 130° from the charted position of the mark.

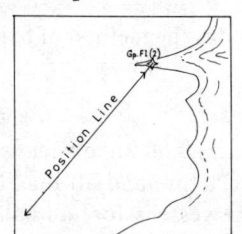

Fig. 17·1 illustrates a position line obtained from a bearing of a lighthouse. It is customary to indicate a position line by means of single arrowheads as illustrated in fig. 17·1.

Fig. 17·1

It is impossible to ascertain a vessel's position from a single position line. Before a fix is possible two pieces of information are required. The most common and perhaps the most reliable, method of fixing is by simultaneous observations of the bearings of each of two conspicuous and suitably-placed shore marks, thus obtaining two position lines. This method of fixing is illustrated in fig. 17·2, in which it is evident that the fix is at the intersection of the two position lines.

In practice, when possible, it is usual to observe three bearings, thus to obtain three position lines. The third bearing is referred to as a Check Bearing. This method of fixing a vessel is known as Fixing by Cross Bearings.

It should be borne in mind that a position found from cross bearings is accurate only if the following conditions apply:

1. The bearings are accurately observed.
2. The Compass Error is known and properly applied.
3. The position lines are laid down accurately.
4. The charted positions of the observed marks are correct.

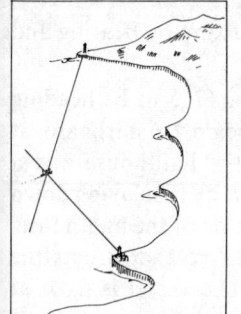

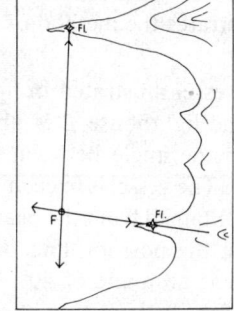

Fig. 17·2

131

2. The Cocked Hat

When three position lines obtained from three simultaneous bearings are laid down on a chart they do not usually intersect at a common point. The probable reason for this is that one or more of the conditions stated above are not met. The three position lines usually form a small triangle which is known as a Cocked Hat.

When, on laying down three position lines they intersect to form a cocked hat, the navigator is given evidence of error of some sort. If, after careful checking, the cocked hat remains, the vessel's position is usually fixed at the apex of the triangle nearest to danger. This is illustrated in fig. 17·3.

Fig. 17·3

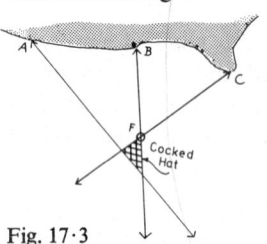

If three position lines obtained, respectively, from bearings of *A*, *B* and *C*, are laid down on the chart as illustrated in fig. 17·3, the vessel's position would be reckoned to be at *F*.

3. Transits

Very frequently, when coasting, two charted shore marks are observed to lie in the same direction from the vessel. Such marks are said to be in Transit, and their bearing is said to be a Transit Bearing. A transit bearing affords an excellent means of finding or checking the Compass Error. The True Bearing of marks in transit may be found from the chart, and the difference between this and the Compass Bearing is the Compass Error for the particular heading of the vessel.

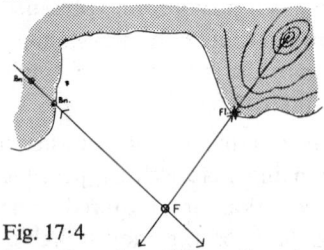

Fig. 17·4

Fig. 17·4 illustrates the method of fixing by means of transit bearings.

4. Relative Bearings

A Relative Bearing of an object is the object's bearing relative to the vessel's fore-and-aft line. It is the angle between the direction of the vessel's fore-and-aft line and that of the observed mark. The relative bearing of an object which lies in a direction 40° on the starboard bow is stated to be Green 040°. The relative bearing of an object which lies in a direction 120° on the port bow is stated to be Red 120°, and so on.

5. Fixing by Bearing and Angle

If it is desired to fix a vessel by cross bearings, and it is found that one of the marks to be observed cannot be seen at the compass position—it being obscured by a mast or funnel—a fix may be obtained from a single bearing and a horizontal angle measured by means of a sextant.

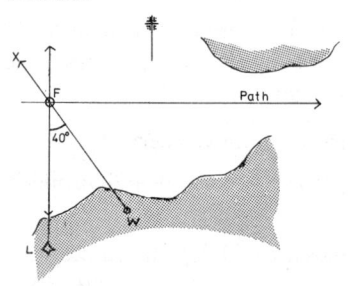

Fig 17·5

Fig. 17·5 illustrates the method of fixing by bearing and angle.

Suppose the vessel illustrated in fig. 17·5 to be heading 090°, and that the lighthouse *L* is abeam to starboard at the same time as the angle between the lighthouse and a windmill *W* is 40°. The vessel is fixed at *F* by first laying down the position line obtained from the bearing of the lighthouse, and then drawing the position line *WX* so that it cuts the first position line at an angle of 40°. The vessel is fixed at the intersection of the two position lines.

6. Fixing by Bearing and Sounding

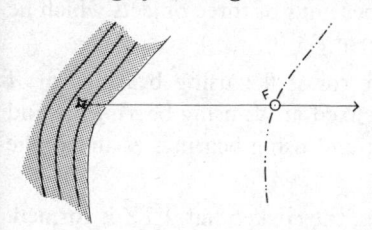

A sounding often assists in fixing a vessel. A sounding is, in a sense, a vertical position line.

Suppose the beacon, illustrated in fig. 17·6, was observed to bear 270° at the same time as a corrected sounding was 20 fathoms. The vessel is fixed at the position where the position line obtained from the bearing of the beacon cuts the 20-fathom isobath; provided, of course, that the position line cuts the sounding line in one place only.

Fig. 17·6

7. Fixing by Sector Light

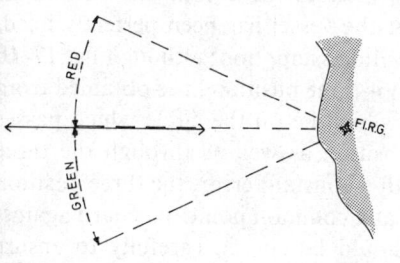

A useful method of obtaining a position line is by noting when the colour of a sector light changes. This is illustrated in fig. 17·7.

Referring to fig. 17·7: if, from a vessel on a northerly course the colour of the light is seen to change from Green to Red, the vessel may be fixed on the dotted line on the chart which indicates the bearing of the lighthouse at the change of the colour of the light.

Fig. 17·7

8. Choosing Marks for Fixing

When it is necessary to observe bearings for the purpose of fixing a vessel, it is better, when a choice is available, to observe near objects instead of more remote marks.

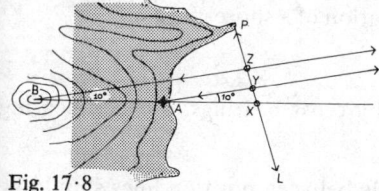

Referring to fig. 17·8, suppose that a vessel is at X on the position line PL. If a bearing of lighthouse A is observed and the position line obtained from the observation is laid down incorrectly to the extent of, say, 10°, the vessel will be erroneously fixed at Y. If a bearing of the peak B is observed and the resulting position line laid down incorrectly to the same extent, the vessel will be erroneously fixed at Z. Now, although the error in laying down is the same in both cases, the fix obtained from the observation of the nearer object A is closer to the vessel's true position than the fix obtained from the more remote peak B.

Fig. 17·8

It is interesting to note that an error of 1° in laying down a position line displaces the position line one mile for every 60 miles between the observer and the observed mark. For small angles the displacement is roughly proportional to the angular error. This means that an error of 10° produces a displacement of the position line to the extent of about 10 miles for each 60 miles of distance; or 5 miles for each 30 miles of distance, and so on.

The most reliable fix by cross bearings applies when the bearings of two near objects are about 90° apart.

9. Angle of Cut

The angle contained between two position lines is known as the Angle of Cut. The angle of cut should be 90°, or as near to 90° as possible, to ensure a reliable fix, the angle of cut should not be less than about 30°.

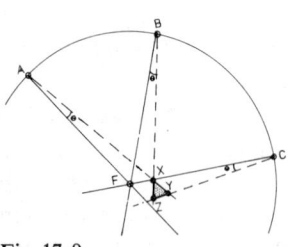

Fig. 17·9 illustrates the effect of a constant error in laying down three position lines from bearings of three objects which lie at the same distance from a vessel at *F*.

Suppose that the constant error is θ°: using bearings of *A* and *B*, the vessel is erroneously fixed at *X*; using bearings *A* and *C* she is erroneously fixed at *Y*; and using bearings *B* and *C* she is erroneously fixed at *Z*.

Fig. 17·9

Using the three position lines, the cocked hat *XYZ* is formed. It should be noticed that the vessel's true position, in this case, is outside the cocked hat.

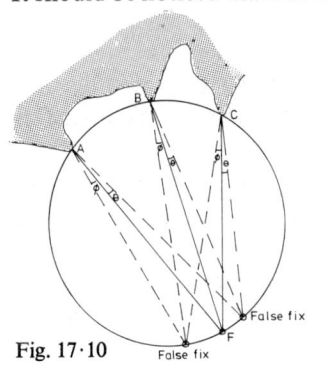

If three position lines intersect at a common point it is reasonable to assume that the vessel has been perfectly fixed. This, in many cases, is a valid assumption; although fig. 17·10 serves to demonstrate that if three position lines obtained from three bearings of objects which lie on the circle which passes through the observer's position as well as through the three objects, are laid down with a constant error, the three position lines will always intersect at a common point. To guard against this possibility objects should be chosen carefully to ensure that the observer does not lie on the circle through the three observed objects.

Fig. 17·10

Exercises on Chapter 17

1. Describe how a position line is obtained from an observation of a shore mark.
2. Describe clearly how a vessel is fixed by cross bearings.
3. What is a check bearing? Explain why, when fixing by cross bearings, a check bearing should be observed when it is possible to do so.
4. What is meant by Angle of Cut? Explain why the angle between position lines should be as near to 90° as possible.
5. Define: Relative Bearing, Transit Bearing.
6. What is a Cocked Hat? Explain how a vessel's position is ascertained when a cocked hat is formed.
7. Explain clearly how a vessel may be fixed from a single bearing and angle.
8. Explain how a vessel may be fixed by a bearing and a sounding.
9. Why should relatively near objects be chosen in preference to more remote objects when fixing by cross bearings?
10. Explain how a sector light may be used for fixing.
11. What precautions should be taken when choosing objects for fixing by cross bearings?
12. "When a cocked hat is formed the vessel's true position may lie outside the triangle." Investigate this statement.

CHAPTER 18

THE TRANSFERRED POSITION LINE

1. Introduction

As mentioned in Chapter 17, a vessel cannot be fixed from a single observation. When only a solitary terrestrial object is visible a bearing of that object gives a position line. This single position line has but little value at the time of the observation but, because it may be transferred, it has potential value.

2. Transferring a Position Line

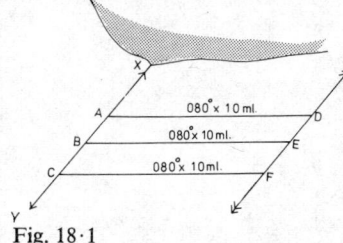

Suppose that the point X in fig. 18·1 represents a conspicuous shore mark the bearing of which gives the position line XY. The vessel may be fixed somewhere on this line; but, in the absence of additional observational information, it is impossible for a navigator to say where, precisely, his vessel is located on the line. Suppose that immediately after the time at which the observation was made the vessel made good a distance of 10 miles on a course of 080°. Had the vessel been at position A at the time of the observation; then, after having made the run, she would have been at D. Had the vessel, however, been at B at the time of the observation, she would have been at E after having made the run. Similarly, had she been at C when the observation was made, she would have arrived at F after having made the run. It is evident that a straight line may be drawn through the points D, E and F, and that this line is parallel to the position line XY.

Fig. 18·1

It is clear that, regardless of where the vessel may have been assumed to have been on the position line XY, a straight line drawn from the assumed position in a direction 080° at a distance of 10 miles from it, is bound to terminate somewhere on the straight line on which D, E and F, are located. This line is called a Transferred Position Line. To identify it from the original position line it is marked with double arrowheads as indicated in fig. 18·1.

A transferred position line is plotted in the following way: Mark off from any point on the first position line a line to represent the course and distance made good from the time of the observation to that at which the transferred position line is required. Through the end of this line draw a line parallel to the original position line. This is the transferred position line.

The vessel may make more than one course during the interval between the times of an original position line and the transferred position line; and, in many cases, allowance will have to be made for current and/or leeway. But, so long as the courses and current and leeway effects are known with accuracy and that the lines which represent them are laid down carefully, a transferred position line is no less valuable than a position line obtained from observation.

The following example illustrates the use of a transferred position line.

Example 18·1—A navigator on a vessel proceeding up Channel observed Eddystone Lighthouse bearing 020° at 0800 hr. The vessel was travelling at 10 knots on a course of 040°.

At 0830 hr. the course was altered to 080°. The current was estimated to be setting 120° at a rate of 3·0 knots. Plot the transferred position line for 0930 hr.

Referring to fig. 18·2: the vessel's position at 0930 hr. is somewhere on the line *XY*.

3. The Running Fix

When it is desired to find a vessel's position at a time when simultaneous observations for cross bearings are not possible, a fix may be obtained as follows:

Two bearings of a single shore mark are observed—the interval of time between the instants at which the bearings are observed being such that the bearing of the mark changes appreciably —more than about 30°—during the interval. If the course and distance made good over the ground in the interval are known, the vessel may be fixed on the second position line at the point where the transferred original position line cuts it.

This method of fixing is known as Fixing by Open Bearings or as the Running Fix.

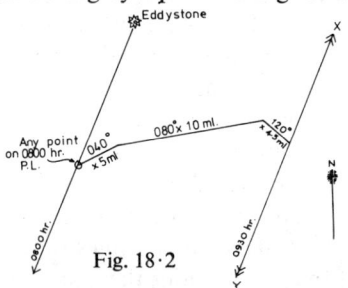

Fig. 18·2

Example 18·2—At 2200 hr. Klein Curaçao Lighthouse bore 305°. The vessel's speed was 12 knots and her course 275°. The current was estimated to be setting at the rate of 2·5 knots in a direction of 230°. Find the distance off the lighthouse when it was abeam at 2300 hr.

From fig. 18·3:

Answer—(By scale drawing) Beam Distance = 9·8 ml.

It is not essential, as in Example 18·2, for the second position line to have been obtained from an observation of the object used for obtaining the first position line. It may happen that the bearing of that object did not change sufficiently to give a good angle of cut during the interval during which it was within range of visibility. In this case a second object may be observed, and a position line obtained from it, when crossed with the original position line transferred, enables the navigator to fix his vessel.

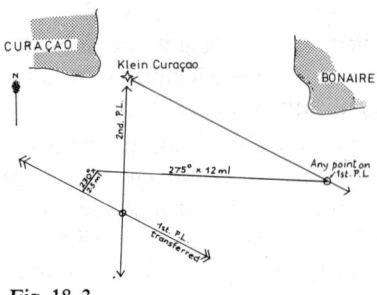

Fig. 18·3

Example 18·3—Point *A* bore 265° at 0800 hr. The vessel travelled for 18·0 ml. on a course of 355°, during which time the current set 340° for 5·0 ml. At the end of the run Point *B*, located 30·0 ml. due North of Point *A*, bore 324°. Find the distance from Point *B* at the time of the second observation.

Referring to fig. 18·4:

Answer—(By scale drawing) Distance = 7·8 ml.

THE TRANSFERRED POSITION LINE

It must be borne in mind that to obtain an accurate position using the Running Fix method, the course and distance made good over the ground during the interval between the times at which the first and second observations were made must be known with certainty. It follows that, because these effects cannot be known exactly, the running fix method should not be used when a cross bearing fix is possible. Of course, when prominent shore marks are not available, the navigator has no alternative but to use a running fix, but such should not be relied upon implicitly, especially when knowledge of currents or tidal streams is uncertain, or when the weather is rough and the effect of wind is uncertain.

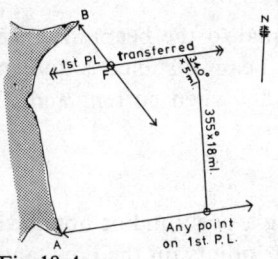

Fig. 18·4

4. Additional Use of a Position Line

Suppose a vessel is travelling off the coastline illustrated in fig. 18·5, and that she is bound for the Port X. Suppose that the approach course to the port is 024°—a direction that can be measured from the chart before the vessel arrives off the port. If the time at which the conspicuous object O is bearing 024° is noted, and if the time taken for the vessel to make good a distance equal to PQ is calculated; then, after the interval of time that elapses from the time of the observation if the vessel's course is altered to 024°, she will make the Port X irrespective of her position on the transferred position line illustrated in fig. 18·5.

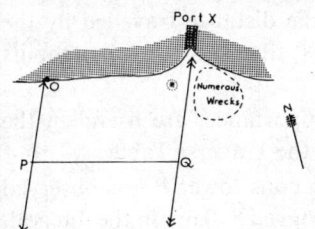

Fig. 18·5

This wrinkle is particularly useful in cases when the weather might become thick, or in cases in which coast and harbour are not well lighted or marked.

5. Doubling the Angle on the Bow

If the relative bearing of a shore mark is observed, provided that it is less than 45°, the distance run between the time of the observation and that at which the relative bearing has doubled, is equal to the distance off the mark at the time of the second observation. This applies only when the vessel has not been affected by current or wind during the interval.

Fig. 18·6 illustrates the principle of the method described above—a method known as Doubling the Angle on the Bow.

Referring to fig. 18·6:

    If $CBX = 2 CAB$

    Then: $ACB = CAB$ and the triangle ABC is isosceles

    Therefore: $AB = BC$

    Or: Distance run = Distance off at Second Observation

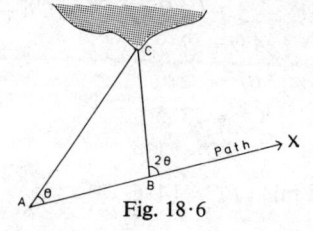

Fig. 18·6

If the angle θ in fig. 18·6 is equal to 45° the distance run is equal to the beam distance when the relative bearing has doubled. This method of finding the beam distance is known as the Four Point Bearing Problem. It does not give accurate results when current and/or leeway influence the progress of the vessel.

6. The Four Point Bearing Problem with Leeway and Current

The beam distance off a shore mark may be found by transferring a position line obtained from an observation of the shore mark made at the time it was four points on the bow, and crossing it with the position line obtained from a beam bearing observation. It may also be computed using plane trigonometry. The traverse table is used to facilitate this problem: a diagram also assists in the solution.

In the trigonometrical method of solving the four point bearing problem with current and leeway it is necessary to find four quantities, these being the unknown sides of two right-angled triangles. One of these triangles has for its hypotenuse the distance travelled by the vessel in the interval between the four point and beam bearings; and the other has for its hypotenuse the drift of the current in the interval.

The following examples, although only of slight practical importance, are useful in the understanding of the problem as well as affording practice at using the Traverse Table.

Example 18·4—At 0200 hr. from a vessel heading 080°, a conspicuous tower P was observed to bear 45° on the port bow. At 0300 hr. it bore abeam. The vessel logged 8·0 ml. in the interval. The leeway due to a North wind was 10°, and the current was setting 140° at 4·0 knots. Find the distance off the tower at 0300 hr.

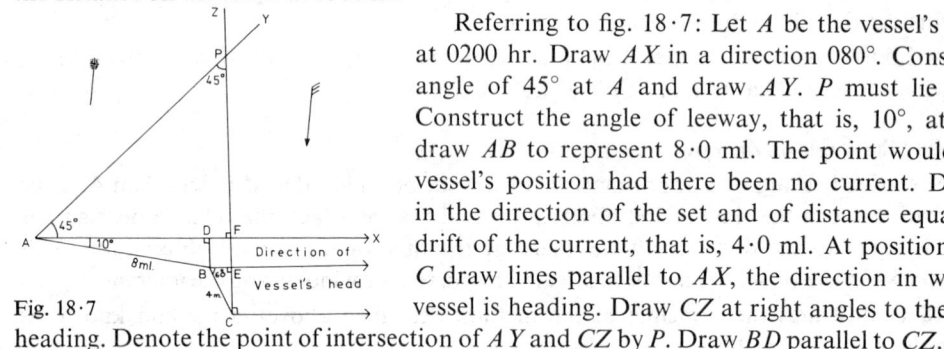

Referring to fig. 18·7: Let A be the vessel's position at 0200 hr. Draw AX in a direction 080°. Construct an angle of 45° at A and draw AY. P must lie on AY. Construct the angle of leeway, that is, 10°, at A, and draw AB to represent 8·0 ml. The point would be the vessel's position had there been no current. Draw BC in the direction of the set and of distance equal to the drift of the current, that is, 4·0 ml. At positions B and C draw lines parallel to AX, the direction in which the vessel is heading. Draw CZ at right angles to the vessel's heading. Denote the point of intersection of AY and CZ by P. Draw BD parallel to CZ.

Fig. 18·7

Solution—Using Traverse Table: find AD, BD, BE and CE.

$$\text{Beam Distance} = PC$$
$$= PF + FC$$
$$= AF + FC$$
$$= (AD + DF) + (CE + EF)$$
$$= (AD + BE) + (CE + BD)$$
$$AD = 7·9$$
$$BE = 2·0$$
$$CE = 3·5$$
$$BD = 1·4$$

Answer—Beam Distance = 14·8 ml. $PC = 14·8$

Example 18·5—From a vessel heading 170° Point P is observed to be 45° on the starboard bow. After travelling for 30 minutes at 12·0 knots, during which time the current set 040° at the rate of 4·0 knots, and the leeway due to a West wind was 10°, the point was abeam. Find the beam distance.

Referring to fig. 18·8:

Find by Traverse Table: *AB, BC, CD* and *DE*.

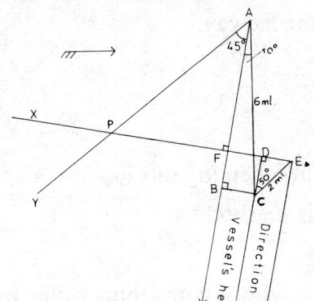

$$\text{Beam Distance} = PE$$
$$= PF + EF$$
$$= (AB - CD) + (BC + DE)$$
$$AB = 5\cdot9$$
$$CD = \underline{1\cdot3}$$
$$4\cdot6$$
$$BC = 1\cdot0$$
$$DE = \underline{1\cdot5}$$
$$PE = \underline{7\cdot1}$$

Fig. 18·8

Answer—Beam Distance = 7·1 ml.

7. Special Angles

The Four Point Bearing method of finding the beam distance, regardless of whether wind and/or current affect the progress of the vessel, suffers from the disadvantage that the required distance is not known until the vessel is actually abeam, and therefore at the position of greatest danger relative to the observed point. If it is desired to know the beam distance before the vessel is at its nearest approach to the point, resort may be made to the use of Special Angles. These are two relative bearings such that the distance run on a steady course between the times of observations is equal to the beam distance.

In fig. 18·9, θ and φ are Special Angles.

It may readily be proved that the cotangents of Special Angles differ by unity.

Referring to fig. 18·9:

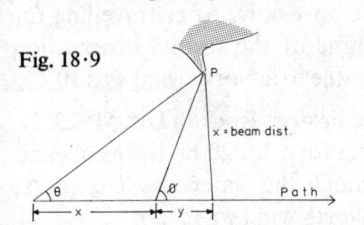

Fig. 18·9

$$\text{Cot } \theta = (x+y)/x$$
$$\text{Cot } \varphi = y/x$$
$$\text{Cot } \theta - \text{cot } \varphi = (x+y)/x - y/x$$
$$= (x+y-y)/x$$
$$= 1$$

Thus, if θ is, say, equal to 25°, φ must be 34¾°. The angles 25° and 34¾° are, therefore, special angles because their cotangents differ by unity.

$$\cot 25° = 2\cdot14451$$
$$\cot 34\tfrac{3}{4}° = \underline{1\cdot14451}$$
$$\cot 25° - \cot 34\tfrac{3}{4}° = \underline{1\cdot00000}$$

140   THE ELEMENTS OF NAVIGATION AND NAUTICAL ASTRONOMY

The most useful pair of special angles are $26\frac{1}{2}°$ and 45°. This follows because the distance run by the vessel between the times of the observations is equal to the distance to run to bring the point abeam, as well as being equal to the beam distance.

When a shore mark is fine on the bow its relative bearing changes comparatively slowly. Therefore, when using special angles, the relative bearing at the first observation should not be too fine on the bow.

It cannot be too strongly emphasized that the beam distance obtained from this method is accurate only when there is no current and the vessel is not making leeway.

Exercises on Chapter 18

1. Explain clearly how a position line is transferred.
2. Explain how a single position line may be used to make harbour when fog sets in.
3. What are the advantages and disadvantages of the Running Fix method?
4. Explain the principle of Doubling the Angle on the Bow.
5. What are Special Angles? Prove that special angles are those whose cotangents differ by unity.
6. At 1000 hr. Start Point in Lat. 50° 13′ N. Long. 03° 38′ W. bore 020°. After travelling for one hour at 12·0 knots on a course of 080°, in a current setting 290° at 2·0 knots, Start Point bore 342°. Find by scale drawing the distance off the point at 1100 hr.
7. Sombrero Light in Lat. 18° 36′ N. Long. 63° 28′ W. bore 258°. After travelling for a distance of 6·0 ml. on a course of 185°, and 5·0 ml. on 210°, it bore 346°. Find the distance off and the Latitude and Longitude of the vessel at the time of the second observation.
8. Ascension Island in Lat. 07° 57′ S. Long. 14° 21′ W. bore 260°. After travelling for 20 ml. on a course of 185° it bore 320°. Find the distance off at each of the times of observation.
9. At 0700 hr. Flamborough Head in Lat. 54° 07′ N. Long. 00° 05′ W. bore 320° (C). The vessel's course was 350° (C). Variation = 13° W. Deviation 3° E. At 0800 hr. the light bore 255° (C). Find the vessel's position at the time of the first observation given that the speed through the water was 8·0 knots and the current was setting 180° at a rate of 2·5 knots. Find also the distance off the light at 0800 hr.
10. Heading 082° at 12·0 knots Point *A* bore 4 points on the port bow. After travelling for 45 minutes it was abeam. Find the distance off at the time of the second observation given that the current was 135° at 4·0 knots, and the leeway due to a North wind was 10°.
11. Heading 086° (C). Variation 5° 0′ E., Deviation 6° 0′ E., the Fastnet Rock in Lat. 51° 23′ N. Long. 09° 36′ W. bore 45° on the port bow. After travelling for 1 hr. 20 m. it was abeam. Find the beam distance given that the vessel's speed through the water was 7·0 knots; the current was 200° at 4·0 knots; and the leeway due to a North wind was 5°.
12. Heading 172° at 16·0 knots a lighthouse bore 45° on the starboard bow. 30 minutes later it was abeam. Find the beam distance if the current has been setting 100° at 4·0 knots and the leeway due to an East wind has been 10°.
13. A point of land bore 22° on the port bow. 25 minutes later it bore 44° on the port bow. Find the distance off the point when it is abeam given that the vessel's speed is 12·0 knots and that there is no current or wind.

THE TRANSFERRED POSITION LINE

14. Heading 222° (C). Compass Error 10° E. Speed 10·0 knots through the water. A point of land bore 45° on the starboard bow. The point was abeam 45 minutes later. Find the beam distance assuming that the current has been setting 265° at 4·0 knots and the leeway due to a North wind had been 5°.

15. Heading 330° at 15·0 knots a light vessel bore 015°. 40 minutes later it was abeam. Find the distance off at the time of the second observation given that the current was 000° at 6·0 knots and the leeway due to an East wind had been 10°.

CHAPTER 19

POSITION LINE BY VERTICAL ANGLE: DISTANCE OF THE HORIZON

1. Distance off by Vertical Angle

If the vertical angle subtended by an object of known height above sea level is measured by means of a sextant; then, provided that the base of the object is within the visible horizon of the observer, the distance off may be found trigonometrically, on the assumption that the vertical height of the object and the horizontal distance off form two adjacent sides of a right-angled plane triangle. This is illustrated in fig. 19·1.

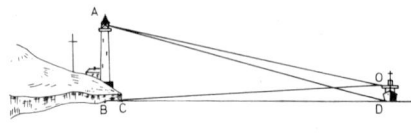

Fig. 19·1

If the distance BD in fig. 19·1 is large in comparison with distances BC and OD, the angle AOC is very nearly equal to the angle ADB, and it may be used instead of ADB for finding BD—without introducing material error.

If the Vertical Sextant Angle (V.S.A.) of a lighthouse or other vertical object is θ, and the height of the top of the vertical object above sea level is h feet, the distance off in nautical miles is given by the formula:

$$\text{Distance off} = (h \cot θ)/6080$$

If h is given in metres, the corresponding formula is:

$$\text{Distance off} = (h \cot θ)/1852$$

It is not necessary ever to compute a problem in which the distance off an object of known angular height is to be found: most collections of Nautical Tables include a table designed to give the answers by inspection.

It must be remembered that the charted height of a lighthouse is the vertical distance from the level of M.H.W.S. to that of the centre of the lens of the light. Great care, therefore, is necessary when measuring the vertical angle of a lighthouse. The telescope of highest magnification available should be used, and an allowance should be made for the state of the tide, if an accurate and reliable result is expected. Quite often H.W. level is distinctly visible as a dark tide mark on rocky shores, or it may be identified by a line of stranded flotsam. If this is so the angle between the direction of the centre of the lens and that of the H.W. mark vertically below it should be measured. It is useful to note that if no allowance is made for the state of the tide, the calculated distance off will be less than the actual distance.

If the distance off a charted mark is known, a circular position line may be determined. With distance off as radius centred at the charted position of the mark, the resulting arc of the circle is a line somewhere on which the vessel's position may be fixed. Such a line is a position line; or, more strictly in this case, a position circle. It is useful to remember that when

POSITION LINE BY VERTICAL ANGLE: DISTANCE OF THE HORIZON

it is necessary to transfer a position circle, the simplest method is to transfer the centre and to draw the transferred position circle centred at the transferred centre.

The distance off determined from a V.S.A. observation may be found by means of a neat little formula which has the advantage over that given above in that no trigonometrical tables are required in its use. This formula is derived with reference to fig. 19·2.

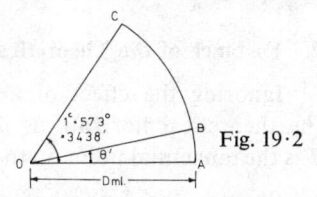

Fig. 19·2

A knowledge of circular measure and the use of radians is essential for a clear understanding of the formula now to be derived. The reader is referred to Chapter 2.

From fig. 19·2:

$$\frac{\text{Arc } AC}{\text{Radius } OA} = 1 \text{ (because } COA = 1 \text{ radian)}$$

$$\frac{\text{Arc } AB}{\text{Chord } AB} = 1 \text{ very nearly (because } \theta \text{ is a small angle)}$$

Thus:
$$\frac{\text{Arc } AC}{\text{Radius } OA} = \frac{\text{Arc } AB}{\text{Chord } AB}$$

and:
$$\frac{\text{Arc } AC}{\text{Arc } AB} = \frac{\text{Radius } AO}{\text{Chord } AB}$$

That is:
$$\frac{COA}{AOB} = \frac{D}{h}$$

or:
$$\frac{3438}{\theta'} = \frac{D}{h}$$

Thus: $\quad D \text{ (in ml.)} = \frac{h \cdot 3438}{\theta' \cdot 6080} \quad (h \text{ in ft.}) \quad \ldots \ldots \ldots \ldots \ldots$ (I)

Or: $\quad D \text{ (in ml.)} = \frac{h \cdot 3438}{\theta' \cdot 1852} \quad (h \text{ in metres}) \quad \ldots \ldots \ldots \ldots$ (II)

Formulae I and II reduce, respectively, to:

$$D \text{ (in ml.)} = 0 \cdot 565 \, h/\theta$$

and
$$D \text{ (in ml.)} = 1 \cdot 856 \, h/\theta$$

The formula giving distance off in terms of height of object and Vertical Angle may be transposed to give V.A. in terms of D and h. The angle θ may be required when it is desired to set the sextant with the angle θ corresponding to the least safe distance to pass a lighthouse. The angle, when used in this way, is called a Vertical Danger Angle (V.D.A.).

The use of a V.D.A. facilitates the process of rounding a headland when the nearest approach to a lighthouse located on the headland is to be not less than a given distance on account of the existence of some off-lying danger.

By transposition:

$$\theta' = \frac{h}{D} \cdot 0{\cdot}565 \quad (h \text{ in feet})$$

or:

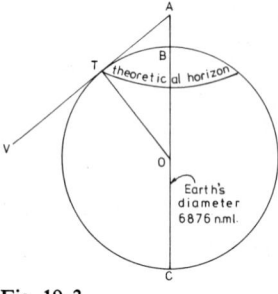

$$\theta' = \frac{h}{D} \cdot 1{\cdot}856 \quad (h \text{ in metres})$$

Fig. 19·3

2. Distance of the Theoretical Horizon

Ignoring the effect of atmospheric refraction, the range of the theoretical horizon, as illustrated in fig. 19·3, is BT where T is the tangential point on the line AV.

Because the height of an observer's eye is usually small compared with the distance of the visible horizon, AT is so nearly equal to BT that no appreciable error is introduced by assuming that AT is equal to BT.

The line AT in fig. 19·4 is a tangent to the circle, and the line ABC, which cuts the circle at any points B and C, is a secant. It may be proved that for any tangent AT and any secant AC:

$$AT^2 = AB \cdot AC$$

Fig. 19·4

From this proposition, and referring back to fig. 19·3:

$$AT^2 = AB \cdot AC$$

Therefore: $BT^2 = AB \cdot AC$

and $BT = \sqrt{AB \cdot AC}$

But AB is so very small compared with BC that AC may be treated for all practical purposes as being equal to the Earth's diameter.

Therefore:
$$BT = \sqrt{AB \cdot BC}$$
$$= \sqrt{\text{Ht. of Observer's eye} \cdot \text{Earth's Diameter}}$$
$$= \sqrt{\frac{\text{Ht. in feet} \cdot 6876}{6080}}$$
$$\sqrt{1{\cdot}13 \cdot h}$$
$$= 1{\cdot}06\sqrt{h}$$

Thus:

Distance of Theoretical Horizon $(D) = 1{\cdot}06\sqrt{h}$ where D is in nautical miles and h, the height of the observer's eye above sea level, is in feet.

If h is given in metres instead of feet, the corresponding formula is:

$$D \text{ (in nautical miles)} = 1{\cdot}93 \sqrt{h} \text{ (in metres)}$$

POSITION LINE BY VERTICAL ANGLE: DISTANCE OF THE HORIZON

The effect of normal atmospheric refraction is for an observer's actual, or Visible, horizon to lie at a greater distance than that of his theoretical horizon.

Making allowance for the effect of normal refraction, the distance of the Visible Horizon is about one-twelfth of the distance of the theoretical horizon beyond the theoretical horizon. Thus the distance of the visible horizon is given by the formula:

$$\text{Distance (in ml.)} = 1\cdot 15 \sqrt{h} \text{ (in feet)}$$

or:

$$\text{Distance (in ml.)} = 2\cdot 09 \sqrt{h} \text{ (in metres)}$$

This formula is useful for finding the extreme distance at which a light of known height is just visible.

When a vessel is approaching a lighthouse at night the distance off the light at which it heaves into view is known as the Rising Range. When travelling away from a light, the range at which it disappears is known as the Dipping Range. The dipping or rising range depends upon the heights of the light and of the observer's eye above sea level. It also depends upon the state of the tide.

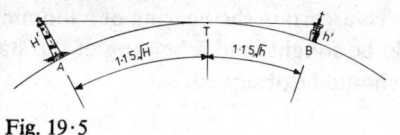

Fig. 19·5 illustrates how the extreme range of a light is found.

The extreme range of the light illustrated in fig. 19·5 may be found by adding the distance AT to that of BT.

Fig. 19·5

$$\text{Extreme range} = 1\cdot 15 \sqrt{h} + 1\cdot 15 \sqrt{H}$$
$$= 1\cdot 15 (\sqrt{h} + \sqrt{H}) \quad (h \text{ in ft.})$$

Example 19·1—Find the extreme range of a light whose height above sea level is 100 feet, if the observer's eye is 49 ft. above sea level.

$$\text{Distance} = 1\cdot 15 (\sqrt{49} + \sqrt{100})$$
$$= 1\cdot 15 \times 17$$
$$= 19\cdot 55 \text{ ml.}$$

Answer—Distance = 19·55 miles.

Sometimes, as noted in Chapter 15, the range of a light is given on the chart. The charted range is calculated on the assumption that the eye of the observed is located at a height of 15 feet above sea level. This height was the average for navigators aboard sailing vessels, and it seems that the charted ranges were given for their benefit. On modern vessels the heights are usually much greater than this, so that the extreme range of a light is usually more than its charted range.

The charted range of a light is always given as a whole number of miles less than the actual range for an observer whose eye is 15 ft. above sea level.

If the height of a light is not given on the chart the extreme range may be found from the charted range, if this is given, as follows:

Example 19·2—The charted range of a light is 12 ml. Find the dipping range to an observer whose height of eye is 49 ft.

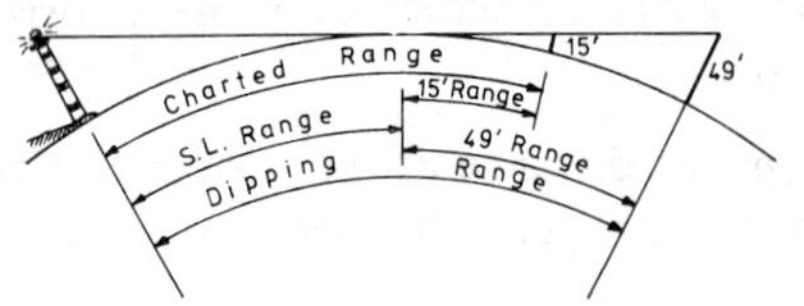

Referring to fig. 19·6:

Charted Range = 12·00 ml.
Range for 15 ft. = 4·45
                   ―――
S.L. Range = 7·55
Range for 49 ft. = 8·05
                   ―――
Dipping Range = 15·60 ml.

Fig. 19·6

Answer—Dipping Range = 15·60 miles.

Great care should be exercised when using a distance off found from an observation of a dipping or rising light. Atmospheric conditions sometimes exist which cause lights to be seen at sometimes a greater, and sometimes a less, range than the theoretical rising or dipping range. When these conditions exist atmospheric refraction is said to be Abnormal. When abnormal refraction is suspected a position obtained by this method should be treated as suspect.

Quite frequently when approaching it a light flashed at a lighthouse is seen by reflection from clouds, or even from the atmosphere itself, before the direct rays of the light are visible. When this is so the light is said to be looming or to loom. To ascertain the bearing of a looming light a star which has the same bearing as the light should be sought, and a bearing of the star instead of that of a point on the horizon vertically below, should be observed.

Exercises on Chapter 19

1. Explain in detail how a Vertical Sextant Angle of a lighthouse should be observed. How is a position line obtained from a V.S.A. observation, and how is such a position line transferred?
2. Explain the derivation of the formula: $D = 1·06 \sqrt{h}$, where D is the range of the theoretical horizon in nautical miles and h is the height of the observer's eye in feet.
3. What is the effect of atmospheric refraction on the range of the horizon? What is the formula used for finding the range of the visible horizon?
4. What is meant by abnormal refraction?
5. What is meant by looming?
6. Find the distance off a lighthouse of height 120 ft. above sea level, if the V.S.A. is 2° 15'·0. Explain why the observer's own height of eye does not enter the problem?
7. Explain how the dipping range of a light may be found.
8. Find the extreme range of a light of height 140 ft. above sea level to an observer whose height of eye is 60 ft.
9. Find the rising range of a light whose charted height is 10 ml. to an observer whose height of eye is 40 ft.
10. Explain the use of a Vertical Danger Angle.
11. Find the V.D.A. when it is necessary to pass not less than 2·5 ml. from a lighthouse of height 320 ft.
12. Prove that if AX is a tangent to a circle at T, and AY is a secant to the same circle cutting it at B and C respectively, $AT^2 = AB \cdot AC$.

CHAPTER 20

POSITION LINE BY HORIZONTAL ANGLE

1. Geometrical Principles

A locus line in geometry is a line every point on which conforms to a given set of conditions. The locus line of points at which the angle between the directions of the two extremities of a straight line is a constant, for example, is the arc of a circle of which the straight line is a chord.

Let AB in fig. 20·1 be any straight line. If a circle is drawn such that AB is a chord of the circle: then, at every point on the arc of the major segment the angle between the directions of A and B from the point is θ. Similarly at any point on the arc of the minor segment the angle between the directions of A and B is φ.

The values of θ and φ depend on the position of the chord with respect to that of the centre of the circle. This is illustrated in fig. 20·2.

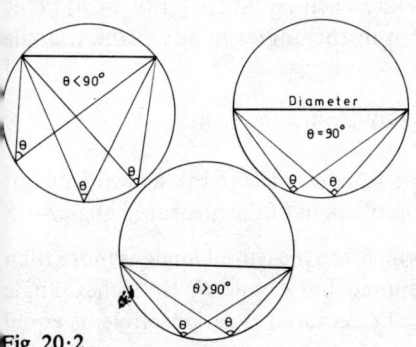

Fig. 20·1

It will be seen in fig. 20·2 that the nearer is the chord to the centre of the circle the greater is the value of θ and the smaller is the value of φ. If the angle between the extremities of the chord is less than 90° the point at which the angle is measured is on the arc of the major segment. If the angle is greater than 90° the point lies on the arc of the minor segment. If the angle is exactly 90° the point lies on the arc of a semicircle, and the chord, in this case, is a diameter.

Fig. 20·2

Another important geometrical proposition is that the angle at the centre of a circle standing on a chord is double the angle standing on the same chord at the circumference. This is illustrated in fig. 20·3.

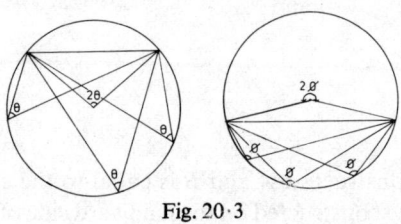

Fig. 20·3

2. Application to Fixing

Suppose the horizontal angle between two shore marks is measured. From this information a navigator knows that his vessel lies on the arc of a circle on which the two observed marks lie. Such an arc, when drawn on a chart, is a position line or, more precisely a Position Circle. The process of plotting a position circle obtained from an horizontal angle observation is as follows:

The charted positions of the observed marks are connected with a straight line. This is a chord of the required position circle. If the measured angle is less than 90° the position line is an arc of a major segment of the circle. If the measured angle is greater than 90° it is an arc of a minor segment, and if the measured angle is 90°, the position line is a semicircle.

When the position line is an arc of a major segment the centre of the required circle and the vessel lie on the same side of the chord. When the position line is an arc of a minor segment the centre of the circle and the vessel lie on opposite sides of the chord. When the measured angle is exactly 90° the centre of the position circle lies at the mid-point of the straight line joining the charted positions of the two observed marks.

The centre of the position circle lies at the apex of an isosceles triangle which has for its base the chord of the circle, or line joining the charted positions of the observed marks. The equal angles of this triangle are each $\frac{1}{2}(180° - 2\theta)$ where θ is the measured angle. This may be verified from fig. 20·4.

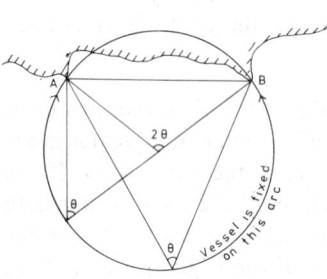

Referring to fig. 20·4, which illustrates the case in which the measured angle is less than 90°, if the measured angle is θ, and A and B represent the observed marks, the angle at the centre of the position circle is equal to 2θ. Hence the angles BAO and ABO are each equal to $\frac{1}{2}(180° - 2\theta)$. This follows because the sum of the angles in any plane triangle is equal to 180°.

Now: $\qquad \frac{1}{2}(180° - 2\theta) = (90° - \theta)$

Fig. 20·4

Therefore: the angle to construct on the seaward side of the line joining the charted positions of the marks is the complement of the measured angle.

Referring to fig. 20·5, which illustrates the case in which the measured angle is more than 90°, if the measured angle is φ and the two marks denoted by A and B, the reflex angle at O, the centre of the required position circle, is equal to 2φ.

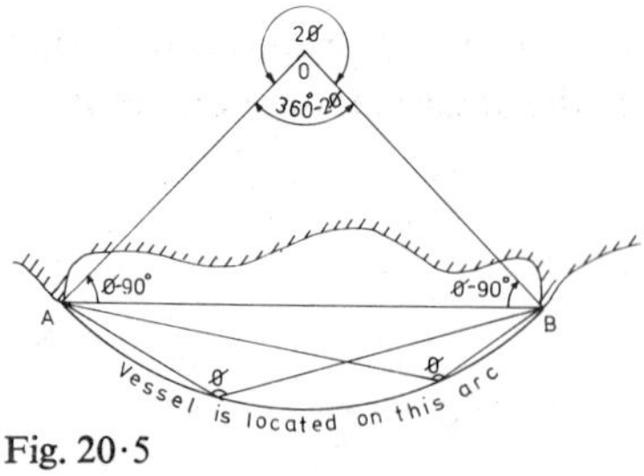

Therefore: $\qquad AOB = 360° - 2\varphi$

and
$$OAB = OBA = \frac{1}{2}(180° - [360° - 2\varphi])$$
$$= \frac{1}{2}(180° - 360° + 2\varphi)$$
$$= \frac{1}{2}(2\varphi - 180°)$$
$$= \varphi - 90°$$

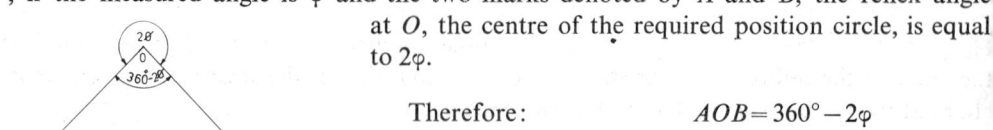

Fig. 20·5

Therefore: the angle to construct at A and B is equal to the excess of the measured angle over 90°, and this angle is to be constructed on the landward side of the chord AB.

POSITION LINE BY HORIZONTAL ANGLE

3. The Horizontal Danger Angle

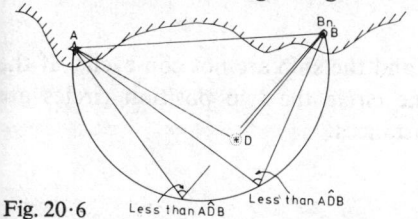

Fig. 20·6

Referring to fig. 20·6, suppose a danger D lies offshore in the vicinity of two well-marked objects A and B. A navigator in passing D has only to ensure that the horizontal angle between the objects A and B does not exceed the angle $A\hat{D}B$, which is measured in advance from the chart. If such be the case, the vessel will pass clear and outside of the danger. The angle at the danger between the two shore objects is referred to as a Horizontal Danger Angle (H.D.A.). The angle to set on the sextant should be smaller than the H.D.A. to ensure a safe clearance.

4. Fixing by Horizontal Angles

If two horizontal angles are measured and two intersecting position circles plotted, the vessel is fixed at one of the intersection points of the two circles. If three marks are observed from which two circles are obtained, one of the intersections lies at one of the observed marks, so that there is no ambiguity in fixing. This method of fixing a vessel has great value and is sometimes preferable to every other available method. The main reasons for this are:

(i) A reliable fix may be obtained even when the compass error is not known. The method, therefore, is extremely useful for fixing a vessel at anchor or for rapid continuous fixing when navigating narrow and tortuous channels when course is frequently being altered, and residual deviations may be large and uncertain.

(ii) When the distance between the vessel and the observed marks is great the necessity of drawing long lines on the chart when fixing by cross bearings may introduce error. Fixing by horizontal angles, in these circumstances, is preferable.

(iii) The method is particularly valuable at times when the vessel is rolling and the compass, as a consequence, unsteady. In such cases compass bearings, especially when observed with a magnetic compass, are unreliable.

Although for convenience and accuracy the sextant is usually employed for measuring horizontal angles, the angle may be measured using the Standard Compass—provided that the card is steady. Fixing a vessel using angles measured with a compass, allows the navigator to check the compass error for the particular heading of the vessel at the time of the observation. Having fixed the vessel it then remains to find the true bearing of one of the observed marks and to compare this with its corresponding compass bearing: the difference being the compass error.

The required horizontal angles may be found from relative bearings. For example, it may be observed that one mark is dead ahead at the same time as another is abeam. The horizontal angle between the marks is, therefore, 90°, and the vessel may be fixed on the semicircle of which the line joining the marks is a diameter.

Reliability of the Horizontal Angle Fix

There are two main factors to consider in relation to the reliability of a fix obtained from horizontal angles. These are:

(i) The Angle of Cut: this should be as near to 90° as possible for a reliable fix. In some instances it is advisable to draw three circles, the third being the circle which passes through the two outer marks. Example 20·5 illustrates a typical case in which the three circles should be drawn.

(ii) Care necessary to ensure that the three marks and the ship are not con-cyclic. If the three marks and the observer lie on the same circle the two position circles are coincident and a fix is impossible in these circumstances.

6. Method of Recording a Fix by Horizontal Angles

If the horizontal angle between points A and B is, say, θ; and that between B and C is φ the observation is recorded thus: $A \ \theta \ B \ \varphi \ C$ in which B lies to the right of A, and C lies to the right of B.

7. Examples of Fixes by Horizontal Angles

The following examples indicate the usefulness of the method of fixing by horizontal angles

Example 20·1—The horizontal sextant angle between point A and steeple B was 70° at the same time as the steeple was in transit with peak C. Fix the vessel.

Referring to fig. 20·7:

Vessel is fixed at F

N.B.—The keen and expert navigator is always on the lookout for suitable transit marks when coasting.

Example 20·2—Light vessel A 60° Buoy B 100° Point C Fix the vessel.

Referring to fig. 20·8:

Vessel is fixed at the intersection of the two position circles

Example 20·3—Peak A bore 010° (C) at the same time as Point B bore 080° (C) and Chimney C bore 130° (C). Fix the vessel and ascertain the compass error for the vessel's heading at the time of the observation

Referring to fig. 20·9:

Vessel is fixed at F

True Bearing of A (from chart) = 340°

Compass Bearing of A = 010°

Compass Error = 30° W.

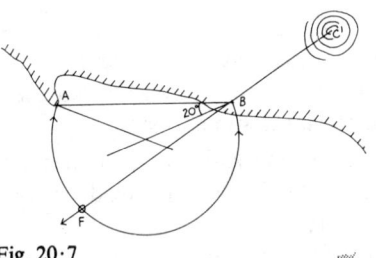

Fig. 20·7

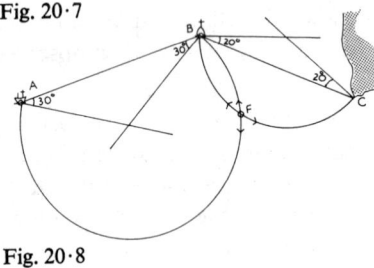

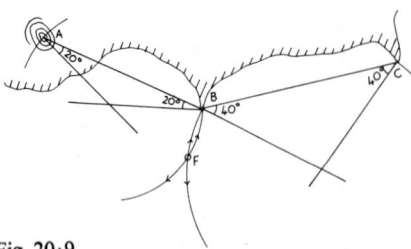

Fig. 20·8

Fig. 20·9

Example 20·4—Buoy A was right ahead, Light vessel B was on the starboard beam, a Buoy C was 4 points abaft the starboard beam. Fix the vessel.

POSITION LINE BY HORIZONTAL ANGLE

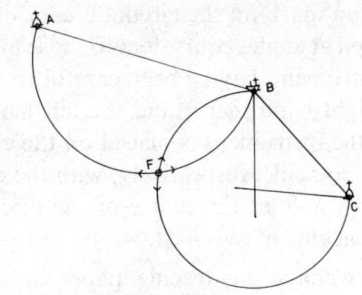

Fig. 20·10

Referring to fig. 20·10:

$$A\ 90°\ B\ 45°\ C$$

Vessel is fixed at F

Example 20·5—Point A 20° Point B 30° Point C. Fix the vessel.

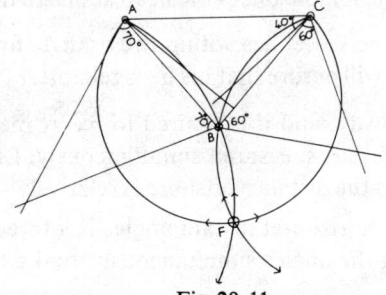

Fig. 20·11

Referring to fig. 20·11:

Vessel is fixed at F

N.B.—The circles through A and B, and through B and C, cut at a very small angle, thus giving an unreliable fix. By drawing the circle through A and C, the horizontal angle between which is $(20+30)°$ that is, 50°, the vessel is reliably fixed at F.

8. Use of Tracing Paper for Fixing by Horizontal Angles

To obviate the need for cluttering the chart with geometrical construction lines, as we have done in the preceding examples, the measured angles may be drawn from a common point on a large sheet of transparent paper. If the paper is then placed on the chart and moved about until the arms of the angles coincide with the charted positions of the observed marks, the vessel may be fixed without drawing any lines on the chart. The fix is located immediately under the point of intersection of the three lines, and may be transferred to the chart with the aid of the point of a dividers.

9. The Station Pointer

A very useful instrument, the principle of which is based on the geometry given above, is the Station Pointer. The station pointer consists of a transparent disc graduated in degrees from 0° to 180° to the right and left. Three radial arms are centred at the centre of the disc; the middle one being fixed and the outer ones moveable. The bevelled edge of the fixed arm coincides with the zero graduation mark on the circumference of the disc. The two moveable arms are capable of being clamped at angles equivalent to those measured by means of a sextant or compass. After setting the instrument, having been careful to set the right-hand arm to the angle between the centre and right-hand mark, and the left-hand arm to the angle between the centre and left-hand mark; the instrument is placed on the chart and adjusted so that the bevelled edges of the three arms coincide, respectively, with the charted positions of the three marks. The vessel is then fixed: a hole at the centre of the disc, having a diameter equal to that of a pencil, facilitating the marking of the chart.

Great caution is necessary when using tracing paper or station pointer for fixing by horizontal angles. Neither device provides an indication of angle of cut, and hence the degree of reliability of the resulting fix. The following rules should be observed when choosing marks for fixing by horizontal angles:

(i) Choose marks which lie more or less on the same straight line; or:

(ii) Choose marks such that the middle one is the nearest one to the observer; or:

(iii) Choose marks such that the vessel lies within the triangle formed by the three marks. By so doing the navigator will ensure that his fix is reliable.

When a vessel is making headway, and it is desired to fix by means of horizontal sextant angles, it is essential that the angles are measured simultaneously. Especially is this important when the vessel's speed is great and the distance offshore is relatively short.

Navigating tricky channels by horizontal sextant angles is a three-man job. Two observers should be employed in measuring the angles simultaneously, and a third should attend to the plotting of the angles on the chart.

Exercises on Chapter 20

1. Explain the principles of obtaining a position line from an observation of a horizontal angle between two charted marks.

2. What precautions are necessary when fixing by horizontal angles?

3. What are the advantages and disadvantages of fixing by horizontal angles compared with fixing by cross bearings?

4. Explain how the compass error may be checked using horizontal compass bearings.

5. What rules should a navigator observe when choosing marks for fixing by horizontal angles?

6. Describe a station pointer and explain how it is used.

7. Prove that, if three shore marks and a vessel lie on the same circle, a fix by horizontal angles is impossible. Hence show the danger of using a station pointer carelessly.

8. Lighthouse Y bears 100° (T) distance 3·0 ml. from a steeple X. A conspicuous chimney Z lies 065° (T) distance 6·0 ml. from the lighthouse. A vessel to the southwards observes X to bear 310° (C) at the same time as Z bears 025° (C) and Y bears 325° (C). Find the distance of the vessel from Y and the compass error for the heading of the vessel at the time of the observation.

POSITION LINE BY HORIZONTAL ANGLE

9. *B* is 086° (T) distance 6·0 ml. from *A*. *C* is 095° (T) distance 5·0 ml. from *B*. A vessel to the northwards makes the following observations:

    A 48° *B* 67° *C*

    Find the distance between the vessel and *B* at the time of the observations.

10. A buoy *A* lies 330° (T) distance 5·0 ml. from a buoy *B*. Buoy *C* lies 265° (T) distance 4·5 ml. from *B*. A vessel at anchor observes the H.S.A.'s between *A* and *B*, and between *B* and *C*, to be 110° and 105° respectively. Find the bearing and distance of the nearest buoy.

11. Point *A*, which lies to the westward of a vessel, was abeam to port at the same time as Point *B* was dead astern. The horizontal angle between *B* and *C* was 10°. *B* is due South (T) of *A* at a distance of 5·0 ml., and *C* is 200° (T) distance 7·0 ml. from *A*. Find the distance of the vessel from *B*.

12. *A* is 5·0 ml. 330° (T) from lighthouse *B*, the height of which is 150 ft. above sea level. The V.S.A. of *B* is 30′ 00″. Find the distance of the vessel from *A* if the H.S.A. between *A* and *B* is 50°.

CHAPTER 21

THE THREE BEARING PROBLEM

1. Principles

If three bearings of a fixed shore mark are observed and the distances run through the water between the times of taking the first and second, and the second and third observations, are known; then, if the observer's vessel has maintained a steady course and speed in the interval, the True Course made good, that is to say, the course made over the ground, may readily be ascertained.

If, instead of the distances travelled, the intervals that elapse between the instants of the observations are known, the same result may be obtained. This follows because distances travelled by a vessel making a uniform speed are proportional to the times taken to make these distances. This principle, simple and useful as it is, seems not to be employed as much as it should be.

The principle is based on the properties of similar triangles. Similar triangles are those which are equi-angular.

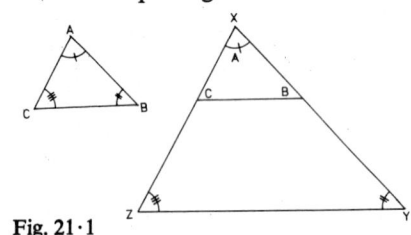

Fig. 21·1

The two triangles ABC and XYZ illustrated in fig. 21·1 are similar triangles. An important property of similar triangles is that the ratio between corresponding sides is constant. Referring to the triangles illustrated in fig. 21·1, we have:

$$AB : XY :: AC : XZ :: BC : YZ$$

or: $AB/XY = AC/XZ = BC/YZ$

Notice that the triangle ABC fits exactly into the corner of triangle XYZ, and that when so fitted the side BC of the triangle ABC is parallel to the side YZ of the triangle XYZ. So that:

$$XB/XY = BC/YZ = XC/XZ$$

2. Practice

The following example illustrates how the course made good may be found using this principle.

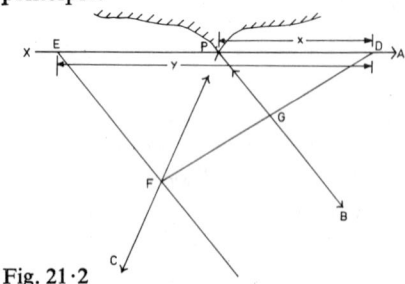

Fig. 21·2

Suppose that the Point P illustrated in fig. 21·2 is observed on three successive instants and that the three position lines PA, PB and PC, are obtained. If the distance travelled between the times of observing the first and second bearings is x ml., and the distance travelled between the times of taking the second and third bearings is y ml., the course made good between the times of the first and third bearings may be found as follows:

154

THE THREE BEARING PROBLEM

Referring to fig. 21·2, produce AP to X. Mark off any distance PD from P towards A to represent x ml. From D mark off on the same scale a distance DE to represent y ml. From E draw a line parallel to the middle position line PB to cut the third position line PC at F. Join D to F. The direction of DF is that of the course made good between the times of the first and third observations.

It must be appreciated that only the direction is found from this information. The line DF does NOT represent the track that the vessel has made good during the interval between the times of taking the first and third bearings. To find the actual track made good additional information is necessary.

The scale of units along the line AX, which is the ratio line, is chosen as convenient. The larger the scale the more accurate will be the result. This follows because the larger is the scale the longer will be the line DF, and the longer is a line the more accurately may its direction be measured.

Fig. 21·3 illustrates the triangles relating to the problem demonstrated above. By comparing triangles DPG and DEF it will be seen that they are similar. Therefore:

$$DP : DE :: DG : DF$$

But: $$DP : DE :: x : y$$

Thus: $$DG : DF :: x : y$$

It follows that because D and E lie on the first and third position lines, respectively, DF must lie in the direction of the course made good.

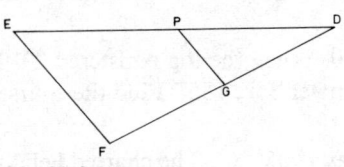

Fig. 21·3

If the vessel is fixed on any of the three position lines or, indeed, at any other position: by a vertical sextant angle; a sounding; a position line obtained from a bearing of a second mark; or, by any other means, the actual track made good over the ground may be found: this, simply, by drawing a line parallel to DF through the fix.

If the actual track of a vessel is known, the effect of any current that may have been acting may be found by comparing the course and distance made good, found from the chart, with the course and distance travelled through the water.

3. Examples

The following examples, which illustrate the three bearing problem, should be studied closely.

Example 21·1—Light vessel A was observed to bear 140° at 1000 hr.; 090° at 1015 hr.; and 040° at 1035 hr. Find the course made good.

Referring to fig. 21·4:
$$AX : AY :: 15 : 20$$
Course direction = XZ
$$= 185° \text{ (By measurement)}$$

Answer—Course = 185°.

156   THE ELEMENTS OF NAVIGATION AND NAUTICAL ASTRONOMY

Example 21·2—A vessel is heading 250° and logging 15·0 knots. Point *A* is observed to bear 310° at 0900 hr.; 355° at 0930 hr.; and 050° at 1030 hr. The distance off the point at 0900 hr. was 13·0 ml. Find the set and rate of the current.

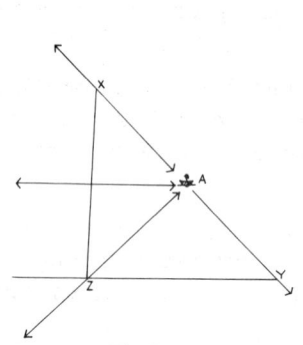

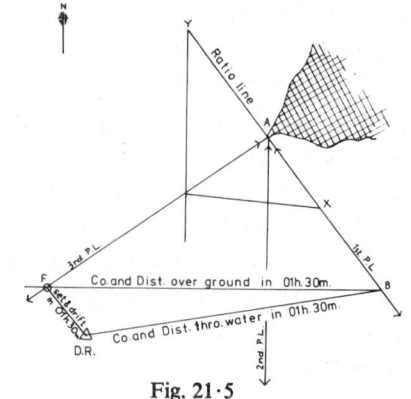

Fig. 21·4   Fig. 21·5

Referring to fig. 21·5:

Set and Rate (by measurement) = 275° × 6·0 knots

Answer—Set = 275°

Rate = 6·0 knots.

Exercises on Chapter 21

1. Explain the principle of the Three Bearing Problem.
2. A lighthouse bore 010° at 1000 hr.; 340° at 1015 hr. and 310° at 1030 hr. Find the course made good.
3. A tower bore 270° at 1000 hr. when the log registered 15·0. When the log registered 20·0 the tower bore 230°, and when the log registered 23·0 the tower bore 185°. Find the course made good.
4. Heading 175°, the V.S.A. of a lighthouse abeam to port was 1° 05′ 00″. The charted height is 200 ft. 25 minutes later the lighthouse bore 065°, and after a further 15 min. it bore 037°. Find the distance off the lighthouse at the time of the third observation.
5. A lighthouse bore 160°. After 15 min. it bore 105° and a sounding fixed the vessel at a distance of 4·0 ml. from the lighthouse. After running for 12 min. the lighthouse bore 085°. Find the distance off the lighthouse at the time of the last observation. The vessel was heading 235° and logging 15·0 knots. Find the set and rate of the current.
6. Heading 322° (C) variation 5° W., deviation 4° E. A lighthouse bore 020° (C). After running 4·0 ml. through the water it bore 038 (C) and after running a further 3·0 miles it bore 063° (C), at which time it was found to be 5·0 ml. off. Find the course made good and the drift of the current the set being known to have been 090°.

CHAPTER 22

THE THREE POSITIONS: CURRENT SAILING

1. The Three Positions

(i) *Observed Position (Obs.) or Fix* (☉)

A position obtained from observations of celestial bodies is known as an Observed Position (Obs.). A position found from observations of land- or sea-marks is known as a Fix. In practice the terms Observed Position and Fix are often used synonymously.

(ii) *Dead Reckoning Position (D.R.)* (△)

A Dead Reckoning or D.R. Position is one that is found from the compass and the log. The course and distance made through the water, as found from log and compass, when applied to the last Observed Position or Fix, gives a D.R. Position.

(iii) *Estimated Position (E.P.)* (+)

As its name implies, an Estimated Position is a position at which a vessel is estimated to be at any given time. It is the most likely, or most probable position of the vessel ascertained without celestial or terrestrial observations. It is found by applying to the D.R. position of the vessel for the time it is required, estimated allowances for the factors which have set the vessel off her course through the water. These factors include current and leeway.

The course and distance from one observed position to another is known as the Course and Distance Made Good. The course and distance from an observed position to an estimated position is known as an Estimated Course and Distance. The course and distance from an observed position to a D.R. position is known as the Course and Distance through the water.

From the above remarks it should be noted that a vessel is seldom at her D.R. position. This follows from the simple reason that there is almost always some factor or factors which influence the motion of a vessel besides the engines and the helmsman.

When observations are unobtainable or unreliable, and it is desired that a vessel's position be plotted on the chart, the D.R. position is first laid down by applying to the last observed position courses and distances made through the water. Allowances are then made for the estimated set and drift of the current; leeway; heave of the sea; bad steering; and alterations of course that may have been made to avoid other vessels, in order to derive an estimated position. A great deal of experience and skill is necessary to estimate accurately the effects of each of these factors responsible for throwing the vessel off her anticipated track. The best navigator may be said to be the one who can give the best estimated position at any time.

2. Examples

The following example serves to illustrate the remarks given in paragraph 1.

Example 22·1—A vessel's observed position is found to be in Lat. $X°$ N., Long. $Y°$ W. The vessel runs on a course of 060° through the water and logs 60 ml. During this time the current was estimated to have set the vessel 160° for 10 ml. The wind, which blew a gale from the N.E., was estimated to have caused the vessel to have drifted 9 ml. Plot the vessel's estimated position. At this time the vessel's position was found to be 070° by 65 ml. from the observed position found earlier. This position was found from shore bearings. Plot the vessel's position and estimate the accuracy of the navigator's estimations of the effects of current and wind combined.

Referring to fig. 22·1, the vessel's observed position, D.R. position, Estimated position, and Fix, are illustrated in the traditional way.

Had the vessel's reckoning, or record of course and distance through the water, been accurately made, the estimation of the combined effect of current and wind was in error to the extent of the distance between the E.P. and the Fix. By measurement, this is about 12 ml.

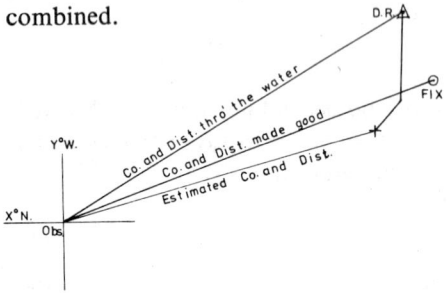

Fig. 22·1

If, during the run, the wind had not affected the vessel's movement, the actual set and drift of the current may be taken as being equivalent to the course and distance from the D.R. position to the Fix. The D.R. position, it should be noted, has value in finding the actual set and drift of a current for any given interval.

3. Current Sailing

When a course line has been laid down on a chart the navigator should estimate the probable effect of current and make allowance for this in finding the required course to steer.

In order to understand the principle of finding the course to steer to counteract the effect of a given current it is necessary to have a knowledge of the principle of the Parallelogram of Velocities.

Suppose that a vessel has to be steered along a path the direction of which is due East. Suppose that a current is known to be setting 045°. Now it should be clear that in order to counteract the effect of this current, the vessel's head must be set in some direction to the southwards of East. The angle between the direction to make good: that is to say, due East, and the direction to steer through the water, depends upon the relative speeds of the vessel and the current. The faster the vessel or the slower the current, the smaller will be the angle, and conversely.

The course to steer through the water may be found by a clumsy trial-and-error method which is illustrated as follows.

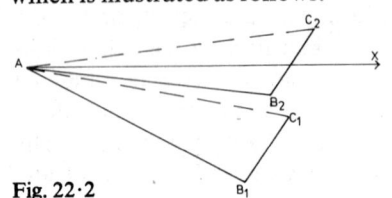

Referring to fig. 22·2, the direction AX is the direction to make good over the ground. Suppose that the vessel is capable of making distance AB in a given time in still water; and suppose that the drift of the current, in the same interval, is equal to BC.

If the vessel's head is set in the direction AB_1, she

Fig. 22·2

THE THREE POSITIONS: CURRENT SAILING

would make good course and distance AC_1 in the time taken to make AB_1 through the water. Because C_1 lies to the southwards of the line AX, the angle which the vessel's head has been directed south of East is, therefore, too great. Had the vessel's head been set in the direction AB_2, she would have made good the course and distance AC_2. Because C_2 lies to the northwards of the line AX, it follows that the angle the vessel's head has made with the direction to make good, is too small. AB must lie in such a direction that the effect of the current will keep the vessel on the line AX. This is illustrated in fig. 22·3.

Fig. 22·3

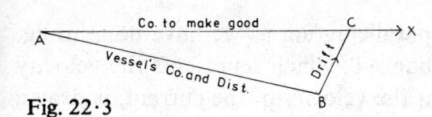

If the set and rate of the current and the vessel's speed remain constant, the vessel will always lie on the line AX, but the vessel's head will always lie in the direction AB.

The direction AB could have been found very simply, compared with the trial-and-error method used above, by the method illustrated in fig. 22·4.

Fig. 22·4

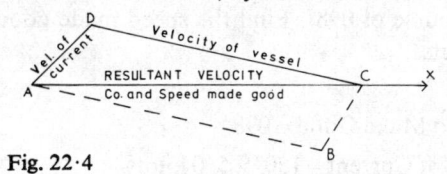

Referring to fig. 22·4; by marking off, from position A, a distance AD to represent the velocity of the current; and, from the end of that line, marking off a distance DC to represent the speed of the vessel, such that this line terminates on the line AX, the direction of the course to steer, that is to say, AB, is immediately ascertained.

It should be remembered that the term Velocity means "Speed in a Given Direction". The lines representing the velocities may be drawn to any convenient scale of knots. Whatever scale is used, the direction AB will always be found to be the same.

The line AC represents the velocity made good. This is the resultant of the velocities, respectively, of the vessel through the water and the current.

When two velocities act simultaneously on a point, the resultant velocity of the point may be found by the principle known as the Parallelogram of Velocities. This is explained thus:

Draw from a common point two lines to represent, respectively, the component velocities. Let these two lines be the adjacent sides of a parallelogram, the diagonal of which, drawn from the same common point, represents the resultant velocity.

There are two, and only two, common problems pertaining to current sailing which are based on the principle of the parallelogram of velocities. These are:

(i) Given a vessel's course and speed through the water, and the set and rate of the current, to find the course and speed made good.

(ii) Given the course to make good, the vessel's speed through the water, and the set and rate of the current, to find the course to steer through the water and the speed made good over the ground.

The following examples should be studied closely.

Example 22·2—A vessel is heading 075° and logging 10·0 knots. The current set 150° at a rate of 5·0 knots. Find the course and speed made good.

160   THE ELEMENTS OF NAVIGATION AND NAUTICAL ASTRONOMY

In fig. 22·5:

AB = Velocity through Water = 075° × 10·0 knots

AD = Velocity of Current = 150° × 5·0 knots

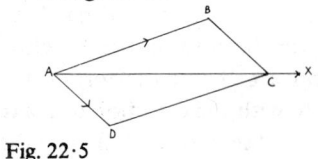

AC = Velocity made good over Ground

By scale drawing—AC = 098° × 12·0 knots

Fig. 22·5

Answer—Course and Speed made good = 098° at 12·0 knots.

N.B.—In practice it is not usual to draw the complete parallelogram as we have done in fig. 22·5. The triangle ABC is sufficient. At the end of the line AB, which represents the velocity of the vessel through the water, the line BC, to represent the velocity of the current, is drawn so as to terminate on the line drawn in the direction to make good.

Example 22·3—Find the course to steer to counteract the effect of a current which sets 150° at the rate of 5·0 knots, in order to make good a course of 098°. Find the speed made good if the speed of the vessel through the water is 10·0 knots.

In fig. 22·6:

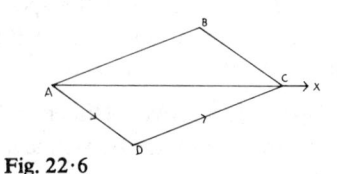

AX = Course to Make Good = 098°

AD = Velocity of Current = 150° × 5·0 knots

DC = Speed through water = 10·0 knots

AC = Course and Speed made good

By scale drawing: AC = 075° × 12 knots

Fig. 22·6

Answer—Course and speed made good = 075° at 12 knots.

In the above example, which should be compared with Example 22·2, it is necessary only to draw the lower half of the parallelogram. From the end of the line AD, which represents the velocity of the current, the line DC is drawn to represent the speed of the vessel. This line terminates on the line which represents the direction to make good.

Example 22·4—A point bears 150° distance 32·0 ml. Find the course to steer, and also the time taken to reach the point, in a current which sets 270° at a rate of 3·0 knots, and the speed of the vessel through the water is 10·0 knots.

In fig. 22·7:

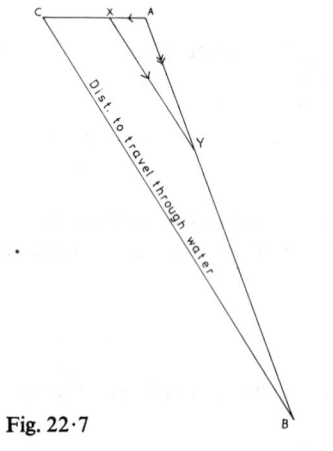

A represents the position of the vessel

B represents the position of the point

AX = velocity of current

XY = speed of vessel in direction to steer

AY = speed of vessel over ground

$$\text{Time Taken} = \frac{\text{Distance made through the water}}{\text{Speed through water}}$$

$$= \frac{\text{Distance made over ground}}{\text{Speed over ground}}$$

Fig. 22·7

THE THREE POSITIONS: CURRENT SAILING

By scale drawing:

$$\text{Time Taken} = \frac{BC}{10} = \frac{40}{10} = 4 \cdot 00 \text{ hr.}$$

Course to steer = 145°

Answer—Time taken = 4 hr. 00 min.
Course = 145°.

Exercises on Chapter 22

1. Distinguish between: Observed Position; D.R. Position; Estimated Position.
2. Explain clearly the principle of the Parallelogram of Velocities, and show how it is used in current sailing.
3. Set course to make good a direction of 170° when the current sets 040° at the rate of 3·0 knots, given that the speed of the vessel through the water is 10·0 knots.
4. Set course to make good 100° if the vessel's speed through the water is 12·0 knots, and the current is estimated to be 040° at 4·0 knots. If the vessel's speed is reduced to 6·0 knots, what adjustment will have to be made to the course steered?
5. A vessel was heading 170° at a speed of 12·0 knots. The current was estimated to be setting 100° at a rate of 3·0 knots. Find the course and speed made good.
6. A point of land bore 295° (C). Variation = 13° W., Deviation = nil on all headings. Set compass course for the point in order to counteract the effect of a current setting 000° at a rate of 5·0 knots, given that the vessel's speed is 12·0 knots.
7. A point of land lies 265° at a distance of 63 ml. A current is estimated to be setting 030° at a rate of 2·5 knots. Find the course to steer to reach the point if the vessel's speed is 12·0 knots. After travelling for 4·00 hr. the point was observed to bear 320° at a distance of 5·0 ml. Find the actual current that has affected the vessel during the interval.
8. A light vessel bears 176° at a distance of 15·0 ml. Find the course to steer to reach the light vessel if the vessel's speed through the water is 10·0 knots. The current is estimated to be setting 285° at a rate of 3·0 knots. Find the time taken to reach the light vessel, and the distance travelled through the water in the interval.
9. The current is estimated to be setting 060° at a rate of 4·0 knots. The vessel's speed is 13·0 knots. The wind is North and leeway is estimated to be 10°. Find the course to steer to make good 282°.
10. Explain why a vessel is seldom at her D.R. position.

CHAPTER 23

POSITION LINE BY RADIO BEARING

1. Introduction

Most large vessels are required by law to be fitted with Medium Frequency Radio Direction Finding equipment, by means of which radio bearings of Radio Beacons and other radio transmitters may be observed. A Radio Beacon transmits a radio signal on a specified frequency. When transmitting, radio energy is broadcast, or thrown out in all directions, from the beacon, and the ray of energy received at a vessel is that which has travelled along the shortest route between the beacon and the vessel. This route is the great circle arc connecting the transmitter and receiver.

The course angle of a ray of radio energy varies along its path, unless the path coincides with a meridian or with the equator. The Direction Finder on board, when used to observe the bearing of a radio beacon, indicates the direction of the ray of energy as it arrives at the vessel. In order to obtain a position line from such an observation, the great circle bearing must be corrected to give the corresponding rhumb-line or Mercatorial Bearing, which may be laid down on the chart as a straight line (see para. 3).

Many shore-based radio stations are equipped with radio finding equipment by means of which the radio bearing of a vessel may be obtained. These stations provide a service whereby the position of a vessel even if not fitted with a direction finder may be found from radio cross bearings observed ashore. The navigator wishing to avail himself of this service contacts the station by radio. On receipt of instructions from the station the vessel transmits her radio call-sign. This signal is received at the station and a bearing of the vessel obtained. The observed bearing is then transmitted to the vessel and a radio position line obtained therefrom.

2. Convergency of the Meridians

It will be remembered that all meridians converge towards the Earth's poles. Moreover, all great circles, except the equator and meridians, cross meridians at ever-changing angles.

The convergency of the meridians of two places A and B is equal to the angle between the tangents lying in the planes of the meridians at the two places. This angle is also equal to that of the change in the direction of the great circle passing through A and B.

In fig. 23·1 the convergency of the meridians at A and B is equal to the angle ACB. This is equal to the angle between AX and BY, the tangents, respectively, to the great circle arc AB at A and B.

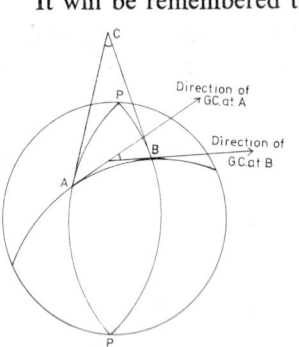

Fig. 23·1

The magnitude of convergency depends upon:

(i) The difference of Longitude between the two places.

(ii) The Middle Latitude of the two places.

POSITION LINE BY RADIO BEARING

If both places lie on the equator, convergency is zero, because the equator does not change its direction, this being due East or due West.

If both places lie on the same meridian, convergency is zero for meridians do not change direction, this being due North or due South.

As Latitude increases convergency between two meridians also increases from zero at the equator to a maximum at the poles where convergency equals D. Long.

$$\text{Convergency} \propto \text{sine Latitude}$$
$$\text{Convergency} \propto \text{D. Long.}$$
$$\text{Convergency (in °s)} = \text{D. Long. (in °s)} \cdot \text{sin Middle Latitude}$$
$$\text{Convergency (in 's)} = \text{D. Long. (in 's)} \cdot \text{sin Middle Latitude}$$

Referring to fig. 23·2:

In Lat. $0°$: convergency $= 0°$
In Lat. $X°$: convergency $= x°$
In Lat. $90°$: convergency $= D° = $ D. Long.

The convergency may conveniently be found from the Traverse Table, as indicated in fig. 23·3.

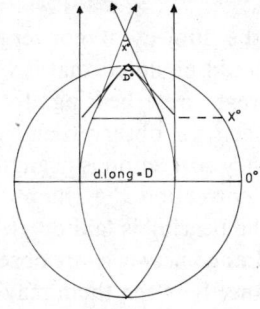

Fig. 23·2

Fig. 23·3

Referring to fig. 23·3:

$$\text{convergency} = \text{D. Long.} \sin \text{Lat.}$$

Enter Traverse Table with Latitude as a Course and D. Long. in the Distance column. Convergency is then lifted from the Departure column.

3. The Half-Convergency Correction

Fig. 23·4 represents a portion of the Earth's northern hemisphere on a Mercator Chart. Let B and A represent a vessel and a radio beacon, respectively. The curved line represents the great circle arc joining A and B.

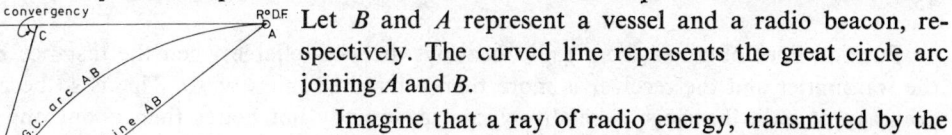

Imagine that a ray of radio energy, transmitted by the beacon A, is received at vessel B. The path of this ray is the curved line AB: its initial direction is along AX and its final direction is along BY—these directions being tangential to the great circle arc AB at A and B, respectively. The change in the direction of the ray between the beacon and the vessel is, therefore, **equal** to the angle XCY, and this is the convergency.

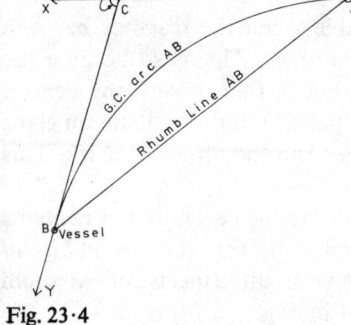

Fig. 23·4

164   THE ELEMENTS OF NAVIGATION AND NAUTICAL ASTRONOMY

The radio bearing of beacon A indicated by the direction finder on board vessel B, is the direction BC. In order to obtain a position line, the direction BA must be found: this being the rhumb-line, or Mercatorial, Bearing.

In triangle ABC the sum of the angles CBA and CAB equals the angle BCX: the exterior angle of the triangle ABC. It is assumed that angle CAB is equal to angle CBA. This is a reasonable assumption because angle XCB is usually very small. Thus, angles CAB and CBA are each equal to half the convergency XCB. To find the rhumb-line bearing of A from B half-convergency is, therefore, applied to the great circle, or observed radio, bearing.

Fig. 23·5 serves to illustrate that the Half-Convergency Correction is to be applied equatorwards to the great circle bearing to obtain the corresponding Mercatorial bearing. This is always the case and applies to the circumstance in which the station takes the bearing of the vessel as well as that in which the vessel takes the bearing of the station.

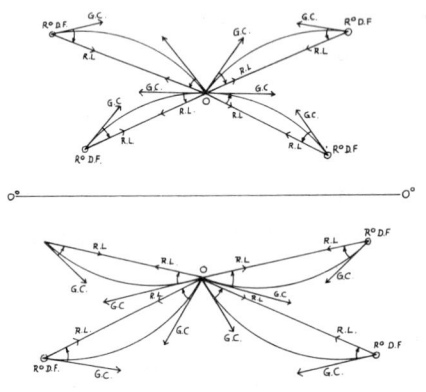

Fig. 23·5

Whilst on the subject of convergency of the meridians, it should be noted that a visual bearing is, in fact, a great circle bearing. In general, the D. Long. between an observer and an observed land- or sea-mark is small, in which case the half-convergency correction is negligible and may, therefore, be ignored. In ignoring the half-convergency correction the rhumb-line and the great circle bearing are considered to be coincident, and the bearing is laid down on the chart as a straight line. In high Latitudes, however, the D. Long. between an observer and an observed mark may be considerable even though the distance between them may be relatively small. In this circumstance it may be necessary to apply the appropriate half-convergency correction to the observed bearing in order to obtain a rhumb-line bearing before plotting on the chart.

The accuracy of a radio bearing depends, in part, on the quality and calibration of the Direction Finder. It also depends upon the structure of the vessel and the location of the D.F. aerial. It is also influenced by the nature of the surface over which the radio energy has travelled in passing from the transmitter to the receiver. These matters are discussed in Chapter 45.

Position lines obtained from radio bearings are not reliable when the distance between the transmitter and the receiver is more than a hundred miles or so. This is so because the accuracy of radio bearings is relatively coarse—usually not better than about one degree. It is interesting to recall than an error of 1° in laying down a position line results in an error of one mile for every 60 miles of distance between the observer and the observed mark. This is illustrated in fig. 23·6.

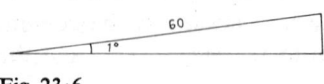

Fig. 23·6

The reader is advised to examine carefully the contents of the two parts of Volume 2 of the *Admiralty List of Radio Signals*, which deal with all aspects of Medium Frequency Radio Direction Finding.

POSITION LINE BY RADIO BEARING

Exercises on Chapter 23

1. Explain carefully how a position line is obtained from a radio bearing.
2. Show that the convergency of the meridians of two places is equal to the D. Long. between the places multiplied by the sine of their Middle Latitude.
3. What is the connection between convergency and departure? Explain how convergency may be found by means of the Traverse Table.
4. Why is the Half-Convergency Correction always applied equatorwards to a great circle bearing in order to find the corresponding rhumb-line bearing?
5. Explain why the correction to apply to a great circle bearing to obtain a corresponding rhumb-line bearing is equal to half the convergency between the meridians of the station and the vessel.
6. A vessel in approximate position Lat. 50° 00′ N. Long. 20° 00′ W., obtains a radio bearing of Land's End Radio Station in Lat. 50° 00′ N. Long. 05° 00′ W. Find the Mercatorial bearing if the radio bearing is 085°.
7. Explain why position lines obtained from radio bearings are not reliable when the distance between the vessel and the beacon is more than a hundred miles or so.
8. What factors influence the accuracy of radio bearings?
9. Explain how the *Admiralty List of Radio Signals*, Volume 2, is used.
10. Explain why it may be necessary to apply a half-convergency correction to a visual bearing taken in high Latitudes.

CHAPTER 24

TIDES

1. The Tide

Observation of the level of the sea surface against a graduated post, or Tide Pole erected vertically on the sea-bed, reveals the vertical oscillation of the sea surface known as the Tide. Notice that, although we talk loosely about the tide coming in or going out, the tide is not a horizontal movement: it is a vertical motion of sea level.

The height of sea level above a given fixed reference level, plotted as a graph against time, produces a Tidal Curve. A typical tidal curve is illustrated in fig. 24·1.

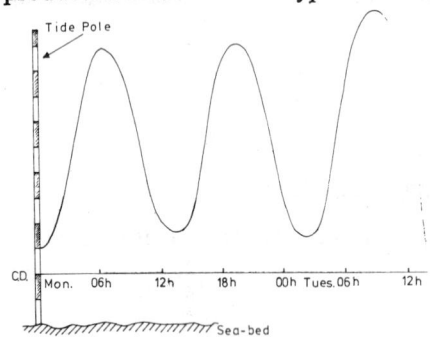

Fig. 24·1

It will be noticed that the curve illustrated in fig. 24·1 approximates to a sine (or cosine) curve. It follows that the rate of rising or falling of sea level is not uniform. The rate is greatest at the mid-time of the instants when the sea level ceases to fall (or rise) and the following occasion when it ceases to rise (or fall). When, at any place, the sea level ceases to rise and before it commences to fall, it is said to be High Water at the place. When the sea level ceases to fall and before it commences to rise it is said to be Low Water at the place.

The interval between the times of any given High Water (H.W.) or Low Water (L.W.), and the succeeding Low Water or High Water, is known as the Duration of the Tide. The interval between the times of successive H.W.s (or L.W.s) is known as the Period of the Tide. The vertical distance between the levels of H.W. (or L.W.) and the following L.W. (or H.W.) is known as the Range of the Tide.

Examination of a tidal curve reveals that the period and range of the tide at the place for which the curve applies are by no means constant. Moreover, the tidal curves of different places vary in period and range.

The level of the sea surface in coastal waters is usually referred to a level known as Chart Datum. This is a level adopted by the hydrographic surveyor and is the level above which charted depths or soundings are given, and this is the reason why it is called Chart Datum (C.D.).

At most places, especially European and North American harbours, the sea level rises and falls twice each day, so that the duration of the tide is about 6 hours and the period of the tide about 12 hours. At such places the tide is said to be Semi-Diurnal, because the period of the tide is about half a day. At other places, notably in the Pacific, the tide is Diurnal, which means that the period of the tide is about 24 hours, so that each day experiences only one H.W. and one L.W.

TIDES

Investigation into the causes of the tide has engaged the attention of numerous philosophers down the ages. So complex are the factors which influence the tide that, even at the present time, the state of understanding of all tidal phenomena has not reached perfection.

Before dealing with the practical problems of the tide as they affect the navigator we shall discuss briefly certain aspects of tidal theory.

2. The Equilibrium Theory of the Tide

The basis of tidal prediction of the times and heights of H.W. and L.W. at given coastal locations is the Equilibrium Theory. According to this theory every particle of water on the Earth is in a state of balance, or equilibrium, under the action of several component tidal forces which act on it.

The tide, according to the Equilibrium Theory, is due to the forces which act between the Earth, Moon, Sun, and water on the Earth; and to the motions of the Moon and Sun relative to the Earth. The Earth's motion on its axis is also considered.

The Equilibrium Theory is due, primarily, to the great English philosopher Sir Isaac Newton, in whose Universal Law of Gravitation it is stated that a force is exerted between every two bodies in the universe, the magnitude of the force being dependent upon the masses of, and distance between, the bodies. The force varies directly as the product of the masses of the bodies and inversely as the square of their distance apart. The law is expressed thus:

$$F \propto \frac{m_1 \cdot m_2}{d^2}$$

where F is the force, m_1 and m_2 the masses of the bodies, and d their distance apart.

3. Effect of Earth's Rotation on Tides

Because the Earth rotates every particle of water on its surface, except particles at the Earth's poles, experiences a centrifugal force which acts perpendicularly to the Earth's axis of rotation. This force varies as the cosine of the Latitude, being maximum at the equator and zero at either pole.

Centrifugal force may be resolved into vertical and horizontal components. The vertical component, which acts vertically upwards, acts against the downward-acting force of gravity. The horizontal component acts towards the equator, so that every particle of water on the Earth tends to move meridianally towards the equator; thus, there is a tendency to cause a piling of water at the equator.

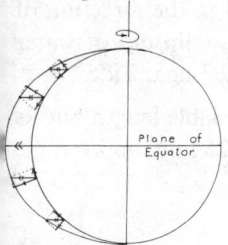

Centrifugal force alone considered, the water level would be at a maximum height above the solid Earth's surface at the equator, and this height would decrease polewards. This is illustrated in fig. 24·2.

Fig. 24·2

In examining the causes of the tide it is usual to regard the Earth as being a smooth sphere completely covered with water. When this ideal condition has been considered allowances are made for the effects of continents; shallow water; configuration of the coasts; and friction between the water and the solid Earth's surface.

168   THE ELEMENTS OF NAVIGATION AND NAUTICAL ASTRONOMY

Let us first consider the effects of the Moon's and Sun's gravitational forces on the Earth and its waters.

4. The Moon's Effect

Although we say, in a loose way, that the Moon revolves around the Earth, the fact is that the Earth and Moon revolve, once a month, about each other.

The point about which the Earth-Moon system revolves is the common centre of gravity of the system. This is a point known as the Barycentre. The barycentre is located on a line joining the centres of gravity of the Earth and the Moon at a point about 1000 miles below the Earth's surface.

In fig. 24·3, B represents the barycentre.

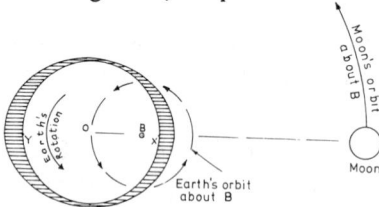

Fig. 24·3

Discounting all forces other than internal forces of the Earth-Moon system, every particle of matter within this system is acted upon by a Gravitational Force of attraction and a Centrifugal Force due to the revolution of the system about the barycentre. In these circumstances the motions around the barycentre are in equilibrium. It follows that the resultant gravitational attraction of the Earth and Moon, which tends to draw these bodies together, is exactly neutralized by the resultant centrifugal force which tends to force them apart.

Gravitational and centrifugal forces acting on particles at the Earth's and Moon's centres are exactly balanced, but all other particles are acted upon by a resultant gravitational/centrifugal force, which acts towards or away from the Moon.

On the hemisphere of the Earth under the Moon centrifugal force is less than gravitational force; and the water, therefore, being mobile, responds by piling up under the Moon at point X in fig. 24·3. On the other hemisphere centrifugal force exceeds gravitational force and the water tends to pile up at the point Y denoted in fig. 24·3. At every point on the great circle whose poles lie at X and Y, there tends to be Low Water. In other words, the resultant force acting upon the water—a force called the Moon's Tide-Raising Force—tends to produce an ellipsoidal water surface as illustrated in fig. 24·3.

We must imagine the major axis of the ellipsoid of water to be locked in the direction of the Moon from the Earth's centre. The rotation of the Earth within this ellipsoid of water results in the periodic rising and falling of sea level, which is referred to as the Lunar Tide.

The period of the lunar tide is half a lunar day. The lunar day has a variable length but its average value is 24 hr. 50 min., so that the period of the lunar tide is 12 hr. 25 min. on average.

5. The Sun's Effect

What has been said in respect of the Moon applies, in principle, to the Sun. The Earth and Sun revolve around their common centre of gravity, and the effect of gravitational and centrifugal forces give rise to the Solar Tide. The solar tide has a period of 12 hr., and it is said to be due to the Sun's Tide-Raising Force.

The solar tide-raising force is smaller than that of the Lunar: the ratio between them being about 34 : 15, or less accurately 7 : 3.

TIDES

6. The Luni-Solar Tide

The combination of the lunar and solar tides is known as the Luni-Solar Tide. At New Moon and Full Moon the tide-raising forces of both Moon and Sun act conjointly. The tides resulting on these occasions are known as Spring Tides. The term "spring" comes from the Saxon word "Springan" which means "to swell", and the term Spring Tides applies to the greatest range during the tidal cycle of a fortnight.

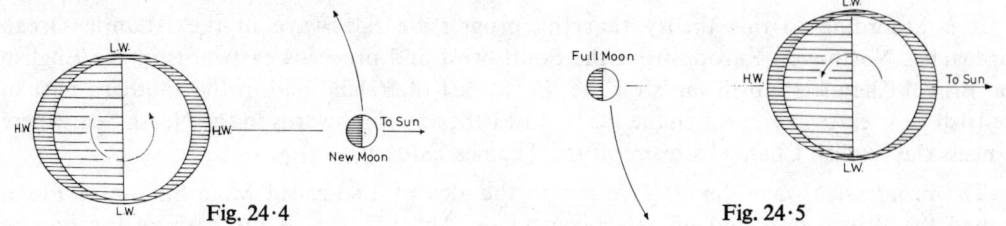

Fig. 24·4

Fig. 24·5

Fig. 24·4 illustrates the Spring Tide which occurs at the time of New Moon, when the Age of the Moon is 00 days. Fig. 24·5 illustrates the Spring Tide which occurs when the Moon is Full and its age is $14\frac{1}{2}$ days.

When the Moon and Sun are in quadrature; that is to say, when the Moon is at the First or Last Quarter, and the Age of the Moon is 07 or 21 days, respectively, the luni-solar tide is called a Neap Tide. The word "neap" is derived from the Saxon word "neafte" which means "a scarcity", and the tides that occur at the First and Third Quarters of the Moon are called Neap Tides because they are the tides of least range during the tidal cycle of a fortnight.

Fig. 24·6 illustrates conditions giving rise to neap tides.

At Springs the range of the tide is maximum. From the time of Springs to that of the following Neap tide, the range of the tide diminishes, and successive H.W.s have decreased heights, and successive L.W.s have increased heights.

At Neaps the range of the tide is minimum. From the time of Neaps to that of the following Springs the range of the tide increases, and successive H.W.s have increased heights, and successive L.W.s have decreased heights.

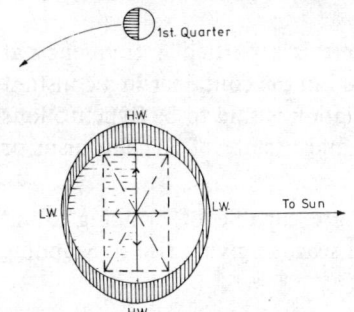

Fig. 24·6

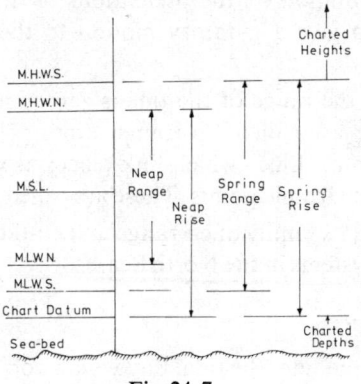

Fig. 24·7

Fig. 24·7 illustrates the various tidal levels and the more common tidal terms.

7. The Progressive Wave Theory of the Tide

According to the Progressive Wave Theory an ellipsoidal tidal wave is believed to exist only in the Southern Ocean surrounding Antarctica. The effect of the Earth's rotation is, according to this theory, responsible for the generation of branch waves, which emanate from the Southern Ocean primary tidal wave, and which progress northwards into the three oceanic gulfs of the Atlantic, Indian, and Pacific Oceans.

It is according to this theory that the progressive tide wave in the Atlantic Ocean approaches North-west Europe from the South-west and proceeds eastwards in the English and Bristol Channels; northwards off the West coast of Ireland and in the southern part of the Irish Sea; eastwards through the Pentland Firth; and southwards in the North Sea, where it meets the English Channel stream off the Thames Estuary.

The progressive wave theory gave rise to the idea of a so-called Main Stream of Flood around the British Isles and off other coastlands. This, in turn, is the basis of the Lateral System of Buoyage, in which the side on which a navigational buoy is to be passed is related to the direction of the main stream of flood in the vicinity.

8. The Standing Wave Theory of the Tide

In the Equilibrium and Progressive Wave theories of the tide, tidal phenomena are considered to be global in character. In modern ideas on tidal theory the tide is considered to be more of a local phenomenon. The rising and falling of sea level in a peripheral sea, or a bay or a gulf, is likened to the oscillation of the free surface of a fluid in a container, when a periodic force acts on the container.

For a given container of liquid having a free surface there is a particular frequency at which a periodic force having this frequency will keep the liquid in the container in a constant state of oscillation. A periodic force resulting in such an oscillation is said to be Synchronous with the Natural Frequency of the liquid in the container, and a state of Synchronism or Resonance is said to exist.

The Moon and Sun provide the periodic forces which are capable of bringing about resonance in water bodies within the ocean and its peripheral seas, so giving rise to Standing Waves of Oscillation.

The effect of the Earth's rotation on the oscillations of the water in a basin, such as the North Sea, for example, is to give a gyratory motion to the standing wave about a point called an Amphidromic Point.

At an amphidromic point the range of the tide is zero. Lines joining places at which the range of the tide is the same are called Co-Range Lines. These are roughly circular lines centred at an amphidromic point. Lines which link places at which the times of High Water are the same radiate from amphidromic points. These are called Co-Tidal Lines.

An amphidromic point and its family of co-range and co-tidal lines forms an Amphidromic System. There are three such systems in the North Sea.

9. Priming and Lagging of the Tide

If the Moon alone caused the tide, the tidal day would correspond to the lunar day, which is 24 hr. 50 min., on average, of Mean Solar Time. If this were so the times of successive a.m. and p.m. High Waters would be later each day to the extent of about 50 minutes.

TIDES

When the crest of the Solar Tide occurs before that of the Lunar Tide, the actual H.W. occurs before the time of the Moon's transit. As the Moon ages in the First and Third Quarters, the retardation of the tide is increasingly reduced. This means that the interval between the times of H.W. and the Moon's transit following the H.W. increases. When this happens the tide is said to Prime. The tide primes in the First and Third Quarters.

Fig. 24·8 illustrates the prime of the tide during the First Quarter.

During the Second and Fourth Quarters H.W. occurs after the time of the Moon's transit. When this happens, the effect on the interval between the time of the Moon's transit and that of the following H.W. is the reverse from what it is in the case of priming. In these circumstances the tide is said to Lag. Lagging is illustrated in fig. 24·9.

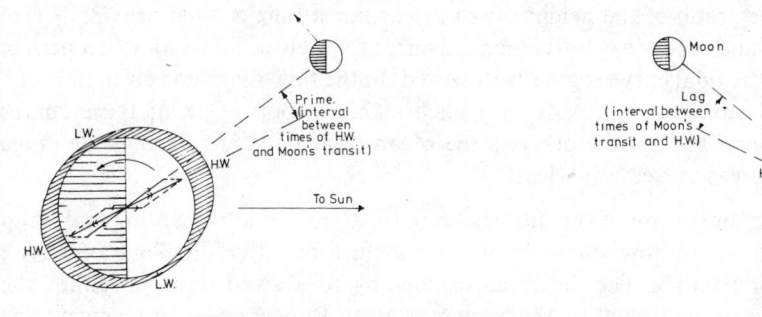

Fig. 24·8   Fig. 24·9

10. Tidal Streams

The tide causes the sea level to vary between places. This difference of levels between places gives rise to horizontal motions of water known as Tidal Streams.

Tidal streams are pronounced only in relatively shallow waters. In the open oceans conditions are not conducive to the strong development of horizontal motions of water due to the tide. In coastal waters, however, tidal streams may reach speeds of many knots.

Tidal stream information is sometimes given on charts. It is also obtainable from Tidal Stream Atlases and, of course, from Tide Tables.

It is interesting to note that tidal stream information is usually related to the time of H.W. at a given port. For the British Isles the time of H.W. at Dover is commonly used for this purpose.

11. Practical Tide Problems

The *Admiralty Tide Tables* (A.T.T.), published by the Hydrographer of the Navy, is an annual publication in three volumes. Tidal data in Volume 1 are given for European waters; in Volume 2 for the Atlantic and Indian Oceans; and in Volume 3 for the Pacific Ocean.

Each volume of the A.T.T. consists of two main parts. In Volume 1, with which alone we shall be concerned, Part 1 contains daily predictions of the times and heights of H.W.s and L.W.s for a number of places called Standard Ports. A list of standard ports appears on the inside front cover of the A.T.T. Together with the daily predictions a Tidal Diagram is provided for each standard port. By means of the appropriate diagram, the time at which the tide has a given height, or the height of the tide for any given time, for any standard port, may be found.

Part 2 of the A.T.T. contains tidal data by means of which predictions of the times and heights of H.W.s and L.W.s for a large number of places called Secondary Ports may be found. These data are in the form of Time and Height Differences which are to be applied to the tabulated time and height of High or Low Water at a particular standard port in order to find the corresponding time and height at a given secondary port.

In addition to the principal tide tables contained in Parts 1 and 2 there are several auxiliary tables to which the reader's attention is directed.

We have seen that the graph of the height of sea level against time is an approximate sine curve. At some standard and secondary ports the tidal curve for a given tide is almost a perfect sine curve, but at others the tidal curve is greatly distorted. In the tidal diagrams given in Part 1 of the A.T.T. two tidal curves are provided for each standard port. One of these curves applies to the Mean Spring Tide and the other to the Mean Neap Tide. At Springs the range of the tide is greatest, whereas at Neaps it is least.

It will be noticed on examination of the tidal diagram for Cardiff that the Spring and Neap Curves are accompanied by an abscissa scale of Time Before or After the Time of H.W.; and an ordinate scale of Factors. The factor corresponding to a given time, on either the Spring or Neap curve, when multiplied by the Spring or Neap Range, gives, respectively, the height of the sea level above that of Low Water. It will be noticed that the factor is 0 for Low Water and 1 for High Water.

On a day when the range of the tide, as found from the A.T.T., is equal or very close to the Spring or Neap Range, the required factor is lifted direct from the appropriate curve. For example, the predicted range at Cardiff for the a.m. rising tide of 5th March is 36·5 feet. This is approximately equal to the Mean Spring Range of 36·3 feet. From the Spring curve of the tidal diagram for Cardiff it should be verified that the factor corresponding to a time of 02 hr. 00 min. before H.W. is 0·73; and that the factor corresponding to a time of 03 hr. 00 min. before H.W. is 0·48.

Again, the predicted range at Cardiff for the p.m. rising tide of 27th January is 18·2 feet, this being almost the same as the Mean Neap Range of 18·4 feet. From the tidal diagram, using the Neap Curve, it will be evident that the factors corresponding to times of 01 hr. 00 min. and 01 hr. 40 min. before the time of H.W., are 0·93 and 0·82, respectively.

For ranges which are greater than the Mean Spring Range; as, for example, the rising tide at Cardiff during the morning of 20th March, when the predicted range is 40·2 feet, the Spring tidal curve is to be used for finding the required factor. For ranges which are less than the Mean Neap Range; as, for example, the falling tide at Cardiff during the morning of 26th April, when the predicted range is 15·4 feet, the Neap tidal curve is to be used for finding the required factor.

For ranges which lie between those of Mean Spring and Mean Neap the required practice is to solve the required height (or time) using each of the Spring and Neap curves, and then to interpolate between the two solutions for the predicted range.

TIDES

Example 24·1—Find the height of the sea level above that of Low Water at Cardiff for noon on 21st March.

Note—At noon the sea level is falling, therefore the time in question is AFTER that of H.W.

From A.T.T. Part 1 (see extracts):

Ht. of H.W. = 37'·7	Time of H.W. = 09 h. 28 m.
Ht. of L.W. = 0'·7	Time in Qn. = 12 h. 00 m.
Range = 37'·0	Interval = 02 h. 32 m. after H.W.

From Tidal Diagram: Mean Spring Range = 36·3 ft. Therefore use Spring Curve.

From Spring Curve:
$$\text{Factor} = 0·61$$
$$\text{Ht. of S.L. above L.W.} = 0·61 \times 37'·0 = 22'·6 \text{ ft.}$$

Answer—Height required = 22·6 feet.

Note—Having found the factor from the tidal diagram, the process of multiplication of the predicted range by the factor is facilitated by a slide rule or by means of Table 1a in the A.T.T.

Example 24·2—Find the height of the tide at Cardiff at 10 hr. 00 min. G.M.T. on 27th January.

Note—At 10 h. 00 m. the sea level is RISING, therefore the time in question is BEFORE that of H.W.

From A.T.T. Part 1:

Ht. of L.W. = 9'·4	Time of H.W. = 14 h. 48 m.
Ht. of H.W. = 27'·8	Time in Qn. = 10 h. 00 m.
Range = 18'·4	Interval = 04 h. 48 m. before H.W.

From Tidal Diagram: Mean Neap Range = 18'·4. Therefore use Neap Curve.

From Neap Curve:
$$\text{Factor} = 0·18$$
$$\text{Ht. of S.L. above L.W.} = 0·18 \times 18·4$$
$$= 3'·3$$
$$\text{Ht. of L.W. above C.D.} = 9'·4$$

$$\text{Ht. of tide at 10 h. 00 m.} = 12'·7$$

Answer—Required height = 12·7 feet.

174   THE ELEMENTS OF NAVIGATION AND NAUTICAL ASTRONOMY

Note—This is the correction to be applied to a sounding at the time in question before comparing it with charted depths. For this reason the height of the tide at any time is sometimes called the Reduction to Sounding at that time.

The alternative practical problem to that given in Examples 24·1 and 24·2, is related to finding the time at which the height of the tide is a given amount. In this case it is necessary to find the factor, not from the curve, as in the above examples, but from Table 1a. Having found the factor the required interval before or after the time of H.W. may be lifted from the tidal curve.

Example 24·3—Find the time at which the height of the tide at Cardiff during the a.m. rising tide of 18th April is 20·0 feet.

From A.T.T. Part 1:

    Ht. of L.W. = − 0′·6   Ht. of L.W. = − 0′·6
    Ht. of H.W. = +37′·7   Ht. in Qn. = +20′·0

    Range = 38′·3   Ht. above L.W. = +20′·6

From Table 1a:

From Spring Curve:

    Interval before H.W. = 02 h. 45 m.
    Time of H.W. = 08 h. 25 m.

    Time when Ht. is 20′ = 05 h. 40 m.

The above examples were such that it was not necessary to interpolate for the predicted range. The following two examples are such that it is necessary to interpolate.

Example 24·4—Find the height of the tide at Cardiff for 13 h. 30 m. G.M.T. on 24th January.

From A.T.T. Part 1:

    Ht. of H.W. = +32′·6   Time of H.W. = 11 h. 48 m.
    Ht. of L.W. = + 5′·9   Time in Qn. = 13 h. 30 m.

    Range = 26′·7   Interval = 01 h. 42 m. after H.W.

Note—The predicted range falls between those of Mean Spring and Mean Neap Ranges. Therefore it is necessary to interpolate for the predicted range.

Using Spring Curve: Using Neap Range:

    Range = 36′·3   Range = 18′·4
    Factor = 0·78   Factor = 0·81
    Ht. above L.W. = 20′·8*   Ht. above L.W. = 21′·6*

    * Using Table 1a with Predicted Range.

It is now necessary to interpolate between 20'·8 and 21'·6 (which correspond respectively to the Spring and Neap Ranges) for the predicted range. Thus:

$$\text{Ht. for } 36'\cdot 3 \text{ (Spring Range)} = 20'\cdot 8$$
$$\text{Ht. for } 18'\cdot 4 \text{ (Neap Range)} = 21'\cdot 6$$
$$\text{Ht. for } 26'\cdot 7 \text{ (predicted range)} = 20'\cdot 8 + (0\cdot 8 \times 9\cdot 6)/17\cdot 9$$
$$= 20'\cdot 8 + 0\cdot 4$$
$$\text{Ht. of S.L. above L.W.} = 21'\cdot 2$$
$$\text{Ht. of S.L. above C.D.} = 5'\cdot 9$$

$$\text{Required Height} = 27'\cdot 1$$

Answer—Required Height = 27·1 feet.

Example 24·5—Find the correction to be applied to the charted height of a tower at Cardiff on 26th January at 10 h. 00 m. G.M.T.

Note—Charted heights are given above the level of Mean High Water Springs (M.H.W.S.). These and other levels are given in Table V in A.T.T.

From A.T.T. Part 1:

Ht. of L.W. = + 8'·4	Time of H.W. = 13 h. 36 m.
Ht. of H.W. = +28'·5	Time in Qn. = 10 h. 00 m.
Range = 20'·1	Interval = 03 h. 36 m. before H.W.

From Spring Curve:	From Neap Curve:
Spring Range = 36'·3	Neap Range = 18'·4
Factor = 0·34	Factor = 0·44
Ht. above L.W. = 6'·8	Ht. above L.W. = 8'·8

$$\text{Ht. for } 36'\cdot 3 \text{ (Spring Range)} = 6'\cdot 8$$
$$\text{Ht. for } 18'\cdot 4 \text{ (Neap Range)} = 8'\cdot 8$$
$$\text{Ht. for } 20'\cdot 1 \text{ (Predicted Range)} = 6'\cdot 8 + (16\cdot 2 \times 2\cdot 0)/17\cdot 9$$
$$= 6'\cdot 8 + 1\cdot 8$$
$$\text{Ht. above L.W.} = 8'\cdot 6$$
$$\text{Ht. of L.W. above C.D.} = 8'\cdot 4$$

$$\text{Ht. of tide at 10 h. 00 m. G.M.T.} = 17'\cdot 0$$
$$\text{Ht. of M.H.W.S. (Table V)} = 37'\cdot 6$$

$$\text{Correction to charted height} = 20'\cdot 6$$

Answer—Required Correction = 20'·6.

176 THE ELEMENTS OF NAVIGATION AND NAUTICAL ASTRONOMY

Exercises on Chapter 24

1. Describe the oscillation of the sea level known as the tide.
2. Define: High Water; Low Water; Range of the Tide; Duration of the Tide; Period of the Tide.
3. Define Chart Datum.
4. Distinguish between Semi-Diurnal and Diurnal Tides.
5. Discuss the Equilibrium Theory of Tides.
6. Describe the Luni-Solar Tide. Draw a graph of the luni-solar tide for the period between New Moon and the following Full Moon.
7. What is meant by: Spring Tide and Neap Tide?
8. What is the ratio between the Tide-Raising forces of the Moon and Sun? Explain why the Moon's tide-raising force exceeds that of the Sun.
9. Discuss the Progressive Wave Theory of the Tide. What is meant by Main Stream of Flood?
10. Explain priming and lagging of tides.
11. Describe the Standing Wave Theory of Tides.
12. What are Tidal Streams? Where may information of tidal streams be found?
13. Find the height of the tide at Cardiff at 1500 hr. G.M.T. on 17th January.
14. Find the correction to apply to a sounding at Cardiff at 0500 hr. G.M.T. on 27th February.
15. Find the reduction to a sounding taken off Cardiff at noon G.M.T. on 24th February.
16. At what time during the falling a.m. tide at Cardiff was the height of the tide 19′·0 on 25th April?
17. At what time during the a.m. falling tide of 11th February was the actual depth off Cardiff 38 feet at a point where the charted depth is given as 3 fathoms?
18. Find the height of the tide at Cardiff at 06 h. 30 m. G.M.T. on 10th March.
19. Find the reduction to a sounding taken off Cardiff at 07 h. 00 m. G.M.T. on 23rd February.
20. Find the correction to be applied to a charted height of a tower at Cardiff in order to ascertain the actual height of the top of the tower above sea level at 15 h. 30 m. G.M.T. on 18th March.
21. Find the height of the tide at Cardiff at 08 h. 00 m. G.M.T. on 1st January.
22. At what time during the p.m. falling tide will the height of the tide at Cardiff on 28th April be 25·8 feet?
23. At 19 h. 30 m. G.M.T. on 23rd February a sounding taken off Cardiff gave 35′ 4″. Find the charted depth.
24. Find the time during the first falling tide of 23rd March when a vessel drawing 24′ 00″ grounded off Cardiff at a place marked on the chart as a "Drying Height 5 feet" (5).

PART 4

GENERAL ASTRONOMY

The General Astronomy dealt with in Chapters 25 to 30 inclusive, forming Part 4 of this book, serves to provide the necessary background knowledge for the study of the elements of Nautical Astronomy contained in Part 5.

CHAPTER 25

THE UNIVERSE

1. The Stars

The aggregate of all existing things, that is to say, the whole creation embracing all celestial objects and space, is known as the Universe. The universe is sometimes referred to as the Cosmos on account of its apparent perfect orderliness: the Greek word "cosmos" meaning orderly.

The materials of which the cosmos is composed is segregated into units called Island Universes or Galaxies. A Galaxy covers a vast region in which gaseous material at a very low density is interspersed with billions of Stars. Neighbouring galaxies are separated from each other by immense distances amounting to millions of Light Years, a Light Year being a unit of distance equivalent to that travelled by light in a year. The speed of light *in vacuo* is about 186,000 miles, or 300,000,000 metres, per second, so that the Light Year is a distance of considerable magnitude. Indeed the distances between stars within a galaxy are so great that it is difficult for us to imagine them—our minds being inadequate for such exercises.

A galaxy is bun-shaped, and the stars of which it is formed rotate about a central axis perpendicular to the plane of the greatest dimension of the galaxy. The galaxy to which the Sun belongs is known as the Local Galaxy: its diameter is about 100,000 light years and its maximum thickness is about 3,000 light years.

A star is a huge spherical body consisting of gaseous material at an exceedingly high temperature. It radiates electro-magnetic energy within a wide range of frequencies. Some stars are rendered visible through the agency of the electro-magnetic energy of optical frequency, or light, which they emit. The different temperatures of stars give rise to their distinctive colours: the hotter stars tend to be bluish whereas the cooler stars are reddish in colour.

Within the wide range of star-size and -type the Sun is not, in any way, outstanding. The Sun is a star of average size and of a type most commonly found in the universe. The Sun's pre-eminent importance to the human race stems from its proximity. It is our nearest star, and the Earth is one of the Sun's lesser family members.

When the night sky is viewed in the plane of the local galaxy the number of stars observable is considerably greater than when viewed in a direction perpendicular to the plane. The immense number of stars that may be seen on a dark clear night appear as a whitish belt stretching across the heavens. This belt is known as the Galactic Arch or, more familiarly, as the Milky Way.

Practically all the stars that are visible to the unaided eye belong to the local galaxy—other galaxies being so far away that only few of them can be seen without telescopic assistance. Island Universes appear as regions of diffuse light and, at one time, it was thought that they were clouds of inter-stellar gas within the local galaxy. For this reason they became known as Nebulae. To distinguish them from nebulous gas clouds within the local galaxy, island universes are sometimes called Extra-galactic Nebulae.

The stars, because of their diverse size, type and colour, have different degrees of brightness. The actual brilliance of a star at a certain specified distance is known as the star's Absolute Magnitude. The brightness of a star relative to other stars as it appears in the eye of an observer, however, is known as the star's Apparent Magnitude. The magnitude of a star is given by a number known as its Magnitude Number. The brighter stars are said to be of low magnitude and their Magnitude Numbers are small. Magnitude Number decreases as brilliance increases. Stars which are just visible to the unaided eye are said to be of the Sixth Magnitude. Such a star is reckoned to be one hundred times less bright than a star of Magnitude One.

Now the fifth root of 100 is 2·51, so that the ratio between the brightness of stars whose Magnitude Numbers differ by unity is approximately $2\frac{1}{2}$. Therefore:

A star of Magnitude 1 is $2\frac{1}{2}$ times as bright as a star of Magnitude 2

A star of Magnitude 2 is $2\frac{1}{2}$ times as bright as a star of Magnitude 3

A star of Magnitude 3 is $2\frac{1}{2}$ times as bright as a star of Magnitude 4 and so on.

Also:

A star of Magnitude 1 is $(2\frac{1}{2})^2$ times as bright as a star of Magnitude 3

A star of Magnitude 1 is $(2\frac{1}{2})^3$ times as bright as a star of Magnitude 4 and so on.

A star, the brightness of which lies between that of stars of integral magnitude numbers, has a Magnitude Number which is a decimal quantity. The star Antares, for example, has a Magnitude Number of 1·2. This means that the brilliance of Antares is less than that of a star of Magnitude 1, but exceeds that of a star of Magnitude 2. Some stars, such as for example the bright star Capella, have fractional or even negative Magnitude Numbers. The Magnitude of Capella is 0·2 and that of Sirius, the brightest of all stars not counting the Sun, is −1·6. The Apparent Magnitudes of all navigational stars are given in the *Nautical Almanac*.

The stars visible on a dark clear night are so far distant from the Earth that their real motions are not easily detected. They are, therefore, referred to as Fixed Stars. The apparent positions of the fixed stars relative to one another change extremely slowly. Nevertheless their movements are detectable from the Star Tables of the *Nautical Almanac*.

The stars are grouped into Constellations. The astronomers of Classical Times named the constellations after mythological creatures and symbols, and many of these ancient names are still in use. Stars are named after the constellations to which they belong—the constellation name being prefixed by a letter of the Greek alphabet. Most of the brighter stars have individual names as well as constellation names. The brightest star in the constellation of Taurus—the Bull— is α Taurus: it is also known as Aldebaran. It is important for a nautical astronomer to be able to recognize the navigational stars. Learning the names of the stars, and being able to identify them from a knowledge of the star patterns or constellations, are fascinating, as well as rewarding, pursuits for a student of nautical astronomy.

2. The Solar System

The Solar System comprises the Sun and those celestial bodies which revolve around it and whose movements are controlled by the Sun. The principal members of the Solar System

THE UNIVERSE

are the Sun's family of Planets. These are spherical bodies which shine by reflected light of the Sun. There are nine known planets in the Solar System. These revolve around the Sun in nearly circular Orbits, the Sun being located at the centre of the system. The orbits of the planets, in order of distance from the Sun, are illustrated in fig. 25·1.

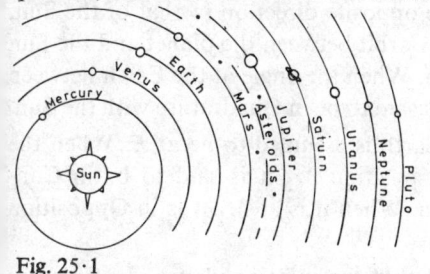

The Sun and the planets rotate about diameters which are known as the polar axes of the bodies; and the direction of rotation, in every case, is the same as that of the revolutions of the planets around the Sun.

Some of the planets have families of their own, in the form of spherical bodies called Satellites or Moons. The moons revolve around their parent planets and rotate on their polar axes in the same general direction as that of the revolution of the planets around the Sun or their rotations about their polar axes.

Fig. 25·1

In addition to the planets and satellites there are large numbers of bodies called Minor Planets or Asteroids, the orbits of which lie mainly between those of Mars and Jupiter. Other relatively unimportant members of the Solar System include Meteors and Comets, as well as a growing number of Artificial Satellites which circle the Earth or the Sun.

The Sun is an immense sphere of gaseous material at a very high temperature. It has been estimated that the temperature of the surface material of the Sun is about 6,000° Centigrade. By observing comparatively dark patches called Sun-spots the Sun may be observed to be rotating with a period of rotation of about 24½ days.

The diameter of the Sun is about 865,000 miles and it is about 750 times as massive as the rest of the Solar System together.

The planets revolve around the Sun at different rates. The Sun's gravitational force on a planet varies inversely as the square of the distance between the Sun and the planet. The nearer is a planet to the Sun the greater is the Sun's force of attraction on it. As a result planets near to the Sun travel faster than the more remote planets; and so, by greater centrifugal force, counterbalance the attraction force of the Sun.

The following table gives details of the planets.

Name	Symbol	Average Orbital Radius in 10^6 ml.	Period	Orbital Vel. in ml. per sec.
Mercury	☿	36	88 days	30
Venus	♀	67·2	225 ,,	22
Earth	⊕	92·9	365 ,,	18
Mars	♂	141·5	687 ,,	15
Asteroids	—	Various	Various	—
Jupiter	♃	483·5	12 years	8
Saturn	♄	886·5	30 ,,	5
Uranus	♅	1782	84 ,,	4
Neptune	♆	2792	164 ,,	3
Pluto	♇	3716	248 ,,	2

The planets whose orbits lie within the Earth's orbit are known as Inferior Planets. There are two such planets: Mercury and Venus. The planets whose orbits lie outside the Earth's orbit are called Superior Planets: these are Mars, Jupiter, Saturn, Uranus, Neptune and Pluto.

When a planet and the Sun lie in the same direction from the Earth the planet is said to be in Conjunction with the Sun. When a planet lies in the opposite direction to that of the Sun, that is to say, when the angle in the plane of the Earth's orbit between the planet and the Sun is 180°, the planet is said to be in Opposition to the Sun. When the angle at the Earth between the directions of a planet and the Sun is 90°, the planet is said to be in Quadrature with the Sun.

In fig. 25·2 the Earth is assumed to be at E. When the superior planet is at position M_2 it is said to be in Conjunction with the Sun. When it is at M_1 it is in Opposition to the Sun.

It will be noticed that the inferior planet illustrated in fig. 25·2 is in Conjunction with the Sun when at position V_1 or V_2. When at V_1 it is said to be at Inferior Conjunction, and when at V_2 it is at Superior Conjunction. An inferior planet can never be in Opposition to or in Quadrature with the Sun.

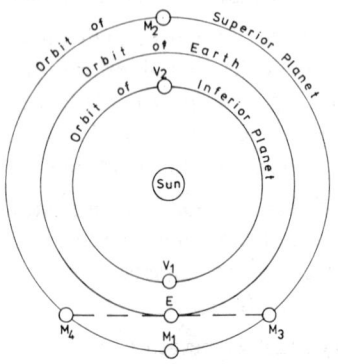

Fig. 25·2

The angle at the Earth between the Sun and any planet, measured in the plane of the Earth's orbit, is known as the Angle of Elongation of the planet. This angle is named East or West according to whether the planet is to the east or west of the Sun respectively. The angle of elongation of any superior planet may have a value of any angle up to and including 180° East or West. The maximum value of the angle of elongation of an inferior planet depends upon the radius of the planet's orbit. That for Venus is about 47° and that of Mercury is about 26°.

Inferior planets appear to oscillate to and fro about the Sun, never getting very far away from the Sun. In low and middle Latitudes they are, therefore, visible for only a relatively short duration of time after sunset or before sunrise.

Planets in or near Conjunction with the Sun are above the horizon with the Sun. They are not, therefore, visible during the hours of darkness. On the other hand, planets which are in or near Opposition with the Sun are above the horizon when the Sun is below. Such planets are, therefore, suitably placed for nautical astronomical observations.

In fig. 25·3 V_1 denotes the position of Venus when it has maximum westerly Angle of elongation. V_2 denotes Venus when it has maximum easterly elongation.

When a planet has westerly elongation it rises before the Sun and sets before the Sun. It is visible for a short period of time before sunrise but is not visible after sunset. Because of this a planet having westerly elongation is said to be a Morning Star. When a planet has easterly elongation it rises after sunrise and sets after sunset. It is therefore, visible in the evening after sunset but it is not visible before sunrise in the morning. Such a planet is called an Evening Star. It is interesting to note that ancient astronomers thought that Venus as an evening star was an entirely different body from Venus as a morning star. The evening star was known to the Romans as Hesperus and the morning star as Lucifer.

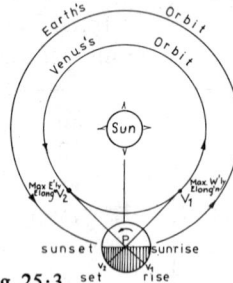

Fig. 25·3

Because a planet is a dark body which is rendered visible only by reflected sunlight, only half of its spherical surface is illuminated at any one time. When an inferior planet is at superior conjunction, its illuminated hemisphere faces the Earth. Provided that the planet does not lie on the straight line joining the Earth to the Sun, it will appear, therefore, as a shining disc. When an inferior planet is at inferior conjunction the illuminated hemisphere faces away from the Earth. It is, therefore, invisible to terrestrial observers. Because a varying amount of a planet's illuminated hemisphere is visible at the Earth—the amount depending upon the angle of elongation of the planet—inferior planets exhibit Phases similar to the phases of the Moon.

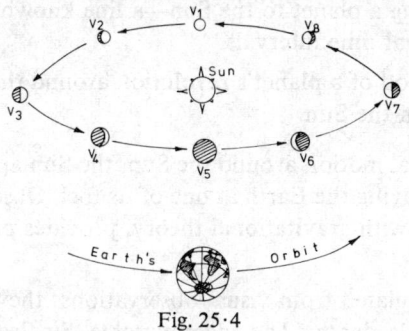

Fig. 25·4

Fig. 25·4 serves to illustrate the Phases of Venus. When Venus is at V_1, it is at superior conjunction. If it is visible in this circumstance it will appear as an illuminated disc, and is said to be Full. At positions V_3 and V_7, half of its illuminated surface is visible, and its phase is said to be Half. At V_5, when it is at inferior conjunction, it is invisible at the Earth, and it is said to be New or at Change. At positions V_2 and V_8, more than half of the illuminated hemisphere is visible, and the phase is described as Gibbous. At positions V_4 and V_6 Venus appears as an illuminated Crescent. The phases of Venus may easily be observed with good binoculars or with a ship's long glass.

At New or Change Venus is about 120 million miles nearer to the Earth than when its phase is Full. Therefore Venus appears larger when it is near Change than when it is near Full. The brilliance, or magnitude, of Venus, which is the third brightest object in the heavens excelled only by the Sun and the Moon, does not change appreciably during its period of revolution around the Sun relative to the Earth. This follows from its changing phase: Venus attaining its maximum magnitude when it is near its position of maximum elongation.

Only four of the planets are suitable for nautical astronomical purposes. These are Venus, Mars, Jupiter and Saturn, which are the Navigational Planets. Mercury, although on occasions quite bright, is too near to the Sun for it to be of value to nautical astronomers. The planets Venus, Mars and Jupiter, are sometimes visible during the hours of daylight, when they are particularly valuable for position finding when out of sight of land.

3. Kepler's Laws of Planetary Motion

The famous astronomer Johannes Kepler (1571–1630) made a close study of the motion of the planet Mars relative to the background of the fixed stars. After careful observations extending over a long period of time he concluded that the orbit of Mars is elliptical and that the Sun is located at one of the focal points of the ellipse. He also observed that the orbital velocity of Mars is greatest when Mars is nearest to the Sun and least when most remote. Similar conclusions were later extended to the other visible planets, after it was discovered that they too behaved in much the same way as Mars.

From his observations and deductions, Kepler formulated his famous laws of planetary motion:

1. Every planet revolves around the Sun in an elliptical orbit having the Sun at one focus of the ellipse.
2. The straight line joining a planet to the Sun—a line known as a radius vector—sweeps out equal areas in equal time intervals.
3. The square of the period of a planet's revolution around the Sun varies as the cube of its mean distance from the Sun.

Because of the Earth's real motion around the Sun, the Sun appears to revolve around the Earth in an elliptical orbit having the Earth at one of its foci. Observations of the Sun relative to the fixed stars, combined with gravitational theory, provides proof that the Earth revolves around the Sun.

Kepler's laws were formulated from visual observations: they were not explained mathematically until after Kepler's death. The incomparable Sir Isaac Newton (1642–1727) is credited with having provided the physical proof that the laws first enunciated by Kepler are a direct consequence of the Law of Universal Gravitation. By this law every planet is attracted by the Sun with a force known as the Solar Attraction. This force, which acts between the planet and the Sun, varies according to the Inverse Square Law. This law, expressed mathematically, is: $F \propto 1/d^2$, where F is the force and d the distance between the points at which the force emanates and acts respectively. This means that if the distance d is doubled the force would be reduced to a quarter of its initial magnitude, and that if the distance were trebled, the force would be reduced to a ninth of its initial value, and so on.

Because a planet is in motion it has a tendency to fly off tangentially to its orbit. The motion of a planet in its orbit is such that Centrifugal Force, which acts radially outwards, just balances the Solar Attraction which acts radially inwards. Centrifugal Force on a planet depends upon the planet's orbital speed, so that when a planet is nearer to the Sun than its average distance, its speed is greater than its average speed, so that increased Centrifugal Force counterbalances increased Solar Attraction.

Solar Attraction and Centrifugal Force are not the only components which influence the motions of the planets. The forces of attraction exerted on a planet by the other planets and by the satellites it may have, modify the motion. The disturbances of the planets, due to these additional factors, are known as Perturbations.

The point in a planet's orbit which is nearest to the Sun is called Perihelion, and the most remote point Aphelion.

Because the Earth's orbit is elliptical, the angular diameter of the Sun varies during the course of the year. The Sun is at perihelion in early January, and is at aphelion in early July. At perihelion the angular diameter is greatest and at aphelion it is least.

The Sun's apparent orbit around the Earth is an ellipse having the Earth at one of the foci. The points of nearest approach to, and most remote from, the Earth are known as Perigee and Apogee respectively.

THE UNIVERSE

The Moon revolves around the Earth in an elliptical orbit having the Earth at one of the foci of the ellipse. The points in the Moon's orbit which are nearest to, and most remote from, the Earth are also known as Perigee and Apogee respectively.

An ellipse may be defined as a locus of points such that the sum of the distances from any one of these points to two fixed points is a constant quantity.

Fig. 25·5 illustrates an ellipse. Each of the fixed points F_1 and F_2 is a focus of the ellipse.

$$(a+b) = (c+d) = (e+f)$$

When the foci are close together, the ellipse is more nearly circular than when the foci are widely spaced. In the limiting cases the ellipse is a perfect circle or a straight line. In the former case the foci are coincident, and in the latter they are infinitely spaced. The ratio between the difference between the greatest and least diameters of an ellipse and the greatest diameter is known as the Ellipticity of the ellipse. The ellipticity of the planetary orbits are very small fractions: that of the Earth's orbit being about 1/80. This means that the orbits are almost circular, although for diagrammatic purposes the ellipticities of planetary orbits are often exaggerated.

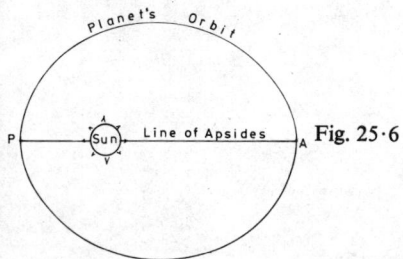

Fig. 25·5

Fig. 25·6 represents the orbit of a planet. The point P, which is the point of nearest approach to the Sun, is the planet's Perihelion. The point A denotes the planet when it is farthest from the Sun: this is its Aphelion. The straight line joining points of perihelion and aphelion of a planet is known as a Line of Apsides or Apse Line.

Fig. 25·7 serves to illustrate Kepler's Second Law. If the areas ASB and CSD are equal, the time taken for the planet to move from B to A is the same as that taken for it to move from D to C. Thus, the planet moves more rapidly in sweeping out arc DC than it does when sweeping out arc BA.

Fig. 25·6

Fig. 25·7

Exercises on Chapter 25

1. Write a short essay on The Universe.
2. Why do stars have a variety of colours?
3. Distinguish between the terms: Absolute Magnitude and Apparent Magnitude.
4. What is meant by the statement: "The magnitude of Canopus is $-0\cdot9$"?
5. How much brighter than a star of Magnitude 7 is one of Magnitude 2?
6. The magnitude of the Sun is about -26. How many stars of Magnitude 1 would be required to give the same illumination as that of the Sun?
7. What is a Constellation? Name ten constellations and the brightest star in each.

THE ELEMENTS OF NAVIGATION AND NAUTICAL ASTRONOMY

8. Name the Navigational Planets.
9. Define: Inferior Conjunction, Opposition, Quadrature.
10. Explain why planets in Conjunction with the Sun are not suitably placed for astronomical observation.
11. Using the data given in the Planet Table on page 173, compute the approximate Maximum Angle of Elongation of Venus.
12. Why is the planet Mercury not suitable for nautical astronomy?
13. Explain why Venus is sometimes a Morning Star and sometimes an Evening Star.
14. Explain why and how planets exhibit phases.
15. How may a planet be distinguished from a star?
16. State Kepler's Law of Planetary Motion.
17. Explain why the nearer planets to the Sun travel faster than more remote planets.
18. Explain why the magnitude of Venus varies only slightly during its motion around the Sun relative to the Earth even so the distance between the Earth and Venus varies considerably during this period.
19. What are planetary perturbations and their causes?
20. At what time of the year is the Sun's angular diameter greatest?
21. What is meant by: Perihelion, Aphelion, Perigee and Apogee?
22. Define: Line of Apsides.

CHAPTER 26

THE EARTH'S MOTIONS AND THE SEASONS

1. The Earth's Axial Motion

The Earth rotates on its axis in an anti-clockwise direction when viewed from above the Earth's North Pole. This means that all points on the Earth's surface, with the exception of the Earth's Poles, are continually being carried around the Earth's axis towards East, the Earth being said to spin on its axis from West to East.

The angular motion of the Earth on its axis causes the Sun and other celestial bodies to cross the sky, with apparent motions, from East to West. It is this apparent motion of the heavenly bodies due to the Earth's axial motion that provides the basis for the measurement of time. The time taken for the Earth to spin once on its axis is a natural unit of time called a Day.

The period of the Earth's spin relative to the Sun is known as a Solar Day, and the interval of time taken for any fixed star to make one apparent diurnal motion around the sky is a unit called a Star- or Sidereal Day.

2. The Earth's Orbital Motion

The Earth not only rotates on its axis; it also revolves around the Sun. Because of the Earth's revolution, and because of the greater distances of the stars from the Earth compared with the Sun's distance, the Solar and Sidereal Days are not equal in length. A detailed investigation into this important aspect of nautical astronomy will be made in Chapter 29 under the heading of Time.

The Earth revolves around the Sun at an average distance of 93,005,000 miles. When the Earth is at perihelion its distance from the Sun is about 91,000,000 miles: when at aphelion its distance is about 94,500,000 miles. The dates at which the Earth is at perihelion and aphelion become progressively later each year. In 1965 the Earth was at perihelion on January 4th, and at aphelion on July 4th: the corresponding dates in 1975 were January 5th and July 5th.

The time taken for the Earth to complete one revolution around the Sun provides a second natural unit of time called a Year.

In making one revolution around the Sun, the Earth makes $365\frac{1}{4}$ rotations on its axis with respect to the Sun, but exactly one more with respect to the fixed stars. There are, therefore, $365\frac{1}{4}$ Solar Days or $366\frac{1}{4}$ Sidereal Days in a year.

Given the average radius of the Earth's orbit, and the time taken for the Earth to make one revolution around the Sun, it may be verified that the average speed of the Earth in its orbit is about 18·5 miles per second.

3. The Celestial Sphere

Viewed from the Earth the celestial bodies appear to be projected onto the inside of a transparent sphere of infinite radius. This sphere is known as the Sky or the Celestial Sphere.

An observer has the impression that his own position occupies the centre of the celestial sphere; but, because of the Earth's relatively small size, and because of the relatively small distance (compared with the radius of the celestial sphere) between the Earth and the Sun, it is convenient, on occasions, to imagine the Earth's centre or even the Sun to occupy the central position of the sky.

The stars, because of their immense distances from the Earth, appear to maintain their positions relative to one another on the celestial sphere. On the other hand, the Sun and other members of the Solar System, because they are comparatively near to the Earth, and because of the Earth's orbital motion, change their positions relative to the fixed stars relatively rapidly.

As the Earth revolves in its orbit the Sun appears to move on the celestial sphere, against the background of the fixed stars, along a path which is in the same plane as the Earth's orbit. The complete annual path is a celestial great circle known as the Ecliptic.

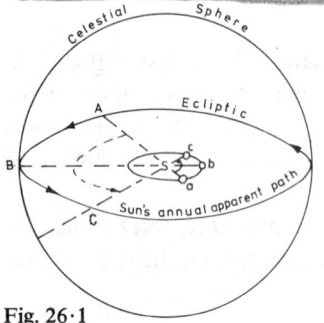

Fig. 26·1 illustrates the celestial sphere with the Sun occupying the central position. When the Earth is at position a the Sun appears to be at position A on the celestial sphere; when the Earth is at position b the Sun appears to be at position B; and when the Earth is at position c the Sun appears to be at position C. Thus, during the interval in which the Earth moves from a to b to c, the Sun appears to move from A to B to C.

Fig. 26·1

The Earth's axis is inclined to the plane of its orbit around the Sun at an angle of $66\frac{1}{2}°$. Because the Earth is a rotating body it possesses a property common to all spinning bodies known as Gyroscopic Inertia. Gyroscopic Inertia is a measure of the tendency of a spinning body to maintain its axis and plane of spin relative to space. The Earth tends to do this, and its axis tends to point in a fixed direction in space. The equator, therefore, tends to be permanently inclined to the plane of the ecliptic at an angle which is the complement of $66\frac{1}{2}°$. The great circle on the celestial sphere which is co-planar with the Earth's equator is called the Celestial Equator. The angle between the planes of the ecliptic and the celestial equator, which is $23\frac{1}{2}°$ or, more accurately $23°\ 27'$, is known as the Obliquity of the Ecliptic.

The celestial equator divides the celestial sphere into two hemispheres known, respectively, as the Northern and Southern Celestial Hemispheres.

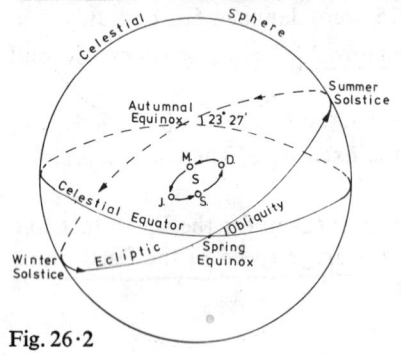

It will be noticed in fig. 26·2 that the Sun's annual apparent path—the ecliptic—cuts the celestial equator at two diametrically opposite points. When the Earth is at position M, the Sun appears to cross the celestial equator from the southern into the northern celestial hemisphere. Half a year later, when the Earth is at position S, the Sun appears to cross the celestial equator from the northern into the southern celestial hemisphere. When the Earth is at position J in its orbit, the Sun is at a position in its apparent annual orbit which is most remote from the celestial equator, and to the north of the celestial equator.

Fig. 26·2

Six months later, when the Earth is at position D, the Sun is most remote from the celestial equator but to the south.

4. The Seasons

The date when the Sun, in its annual apparent orbit around the Earth, crosses from the southern into the northern celestial hemisphere (the position occupied by the Sun when the Earth is at position M in fig. 26·2), is about March 21st each year. On this day of the year the Earth's axis lies in the plane of a great circle on the Earth which separates the Dark from the Illuminated hemispheres of the Earth. This great circle is called the Circle of Illumination; and at every point on it the Sun will appear to be rising or setting.

On March 21st every place on the Earth will be twelve hours in each of the Dark and Illuminated hemispheres of the Earth. Thus, on this day daylight and darkness have the same duration all over the Earth, and the Sun rises at 6 a.m. and sets at 6 p.m. Because day and night are equal on March 21st the point occupied by the Sun on the ecliptic on this day is called an Equinox or Equinoctial Point. From the date when the Sun occupies this equinox until about June 22nd (on which date the Earth is at position J in fig. 26·2), the plane of the Circle of Illumination swings out of alignment more and more with the Earth's axis. On June 22nd the Sun is at a point in its apparent annual orbit at which it changes its direction of motion on the celestial sphere from north-going to south-going. On this day the Earth's axis is inclined at an angle of $23\frac{1}{2}°$ to the plane of the Circle of Illumination, and the Earth's North Pole is directed towards the Sun. The Sun then appears to "stand-still" in the sky relative to the celestial equator. For this reason the point on the ecliptic occupied by the Sun on June 22nd is called a Solstice or Solstitial Point.

The interval between the dates when the Sun is at the equinox and solstice described above is known, in the northern hemisphere, as the Season of Spring, and the equinoctial and solstitial points are known, respectively, as the Vernal (or Spring) Equinox and the Summer Solstice.

From June 22nd to about September 23rd, the angle which the Earth's axis makes with the plane of the Circle of Illumination changes from $23\frac{1}{2}°$ back to $0°$. This interval is known, in the northern hemisphere, as the season of Summer. On September 23rd the Sun is at a point on the ecliptic similar to that at which it occupied six months earlier on March 21st. This is another equinox or equinoctial point. On September 23rd, which is the date of the Autumnal Equinox, the plane of the Circle of Illumination contains the Earth's axis and, as on March 21st, daylight and darkness all over the Earth are each twelve hours and the Sun rises at 6 a.m. and sets at 6 p.m.

The interval between the dates when the Sun is at the Autumnal Equinox until about December 22nd, the angle which the Earth's axis makes with the plane of the Circle of Illumination, increases again from $0°$ to $23\frac{1}{2}°$ with the Earth's North Pole, this time directed away from the Sun. This interval is known, in the northern hemisphere, as the season of Autumn.

At about December 22nd the Sun is at a point on the ecliptic at which its direction of motion relative to the celestial equator changes from south-going to north-going. This is the solstitial point known as the Winter Solstice.

During the six-month period between March 21st and September 23rd the Earth's northern hemisphere is directed towards the Sun. The northern hemisphere, therefore, receives more heat and light from the Sun during this half-year than does the southern hemisphere. In the six-month period from September 23rd to March 21st the reverse is the case.

The seasons are illustrated in fig. 26·3.

Fig. 26·3

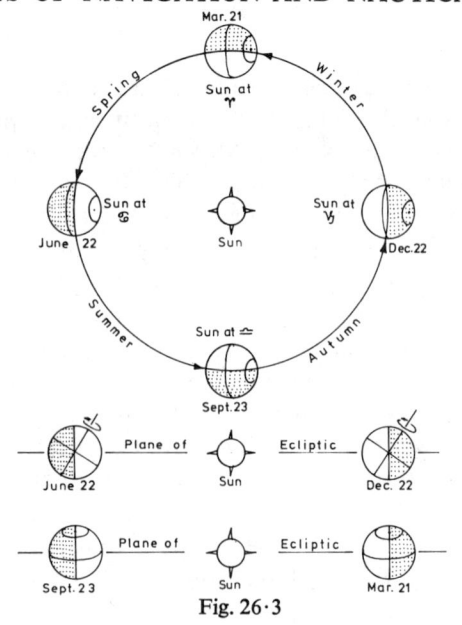

5. Unequal Lengths of Daylight and Darkness During the Year

At about June 21st, the date of the Summer Solstice, when the Earth's North Pole is tilted directly towards the Sun, every point on the Earth located to the north of the parallel of Latitude $66\frac{1}{2}°$ N. experiences total daylight.

Consider an observer to be located on the parallel of $66\frac{1}{2}°$ N. Because of the Earth's rotation he will be carried around the Earth's axis in the direction indicated in fig. 26·4. When he is at position A he is on the Circle of Illumination. The Sun, therefore, is on his horizon. Twelve hours later, after the Earth has rotated through an angle of 180°, the observer will be at position B. During the 12-hour period the Sun will have risen to reach its greatest angular distance from the horizon. This distance—and this is easily verified from fig. 26·4—is 47°. In the 12-hour period following the instant when the observer is at position B, the Sun's Altitude—as the angular distance of a celestial body from the horizon is called—decreases from 47° to 0°, when the observer will again be at position A. During the whole of the 24-hour period the observer will have been on the illuminated hemisphere, and will have experienced daylight, for the whole day. It is evident from fig. 26·4 that the same applies to every observer located to the north of Latitude $66\frac{1}{2}°$ N. The polar cap bounded by this parallel of Latitude is known on this day as the Zone of Perpetual Daylight.

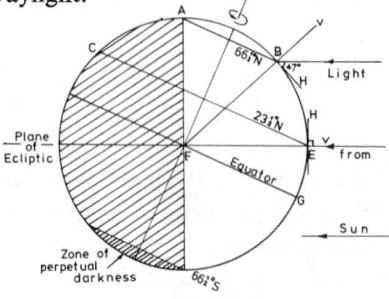

Fig. 26·4

THE EARTH'S MOTIONS AND THE SEASONS

Now consider an observer located on the parallel of Latitude of $23\frac{1}{2}°$ N. The time taken for him to be carried around the Earth's axis from position C to E (refer to fig. 26·4) is clearly 12 hours. In this interval he will travel from C to D in the dark hemisphere and from D to E in the illuminated hemisphere. A moment's consideration will show that, during the complete day, he is on the dark side of the circle of illumination for less than 12 hours and on the illuminated side for more than 12 hours. This applies to all observers located in the northern hemisphere.

It will be noticed in fig. 26·4 that the Sun is vertically overhead to an observer located at position E. This means that, in the interval during which the observer was carried around the Earth's axis from D to E, the Sun increased its altitude from 0° to 90°. It is evident that an observer on the equator experiences 12 hours' daylight and 12 hours' darkness during the course of the day. It is also evident from fig. 26·4 that the maximum altitude of the Sun at any position on the equator on June 21st is $66\frac{1}{2}°$ bearing North of an observer.

An observer located on the parallel of $66\frac{1}{2}°$ S. experiences total darkness during the whole of the day when the Sun is at the Summer Solstice. The same applies to every position to the south of that parallel of Latitude.

On June 21st all places in the southern hemisphere experience a longer period of darkness than of daylight.

At about December 22nd, the date of the Winter Solstice, when the Earth's North Pole is tilted directly away from the Sun, every point on the Earth located to the north of the parallel of Latitude $66\frac{1}{2}°$ N. experiences total darkness.

From fig. 26·5 it is readily seen that on December 22nd the Sun attains a maximum altitude of 47°, at which time it bears due North, to any observer in Latitude $66\frac{1}{2}°$ S. On this day all places on the equator have 12 hours each of daylight and darkness. All places in the southern hemisphere experience a longer period of daylight than of darkness, and all places in the northern hemisphere experience a longer period of darkness than of daylight during the course of the day.

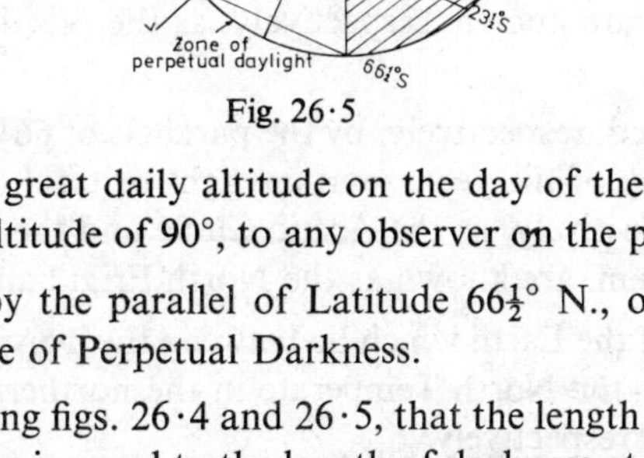

Fig. 26·5

When the Sun reaches its great daily altitude on the day of the Winter Solstice it will be vertically overhead, with an altitude of 90°, to any observer on the parallel of Latitude $23\frac{1}{2}°$ S.

The polar cap bounded by the parallel of Latitude $66\frac{1}{2}°$ N., on the date of the Winter Solstice, is known as the Zone of Perpetual Darkness.

It will be seen, by comparing figs. 26·4 and 26·5, that the length of daylight at any place on the day of the Summer Solstice is equal to the length of darkness at the same place on the day of the Winter Solstice.

At about March 21st and September 23rd, the dates respectively of the Spring and Autumnal Equinoxes, neither pole of the Earth is directed towards the Sun. The Earth's axis on these days lies in the plane of the circle of illumination, so that daylight and darkness are each 12 hours all over the Earth.

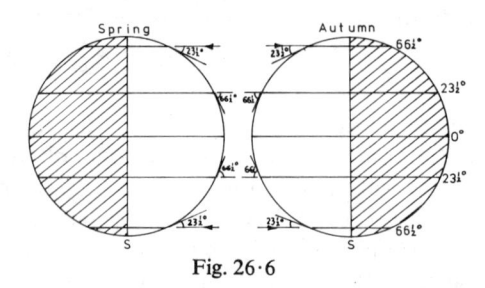

Fig. 26·6

At either equinox the altitude of the Sun at the equator reaches a maximum value of 90°. For places in other Latitudes the greatest daily altitude of the Sun is equal to the complement of the Latitude of the place. This may be verified from fig. 26·6.

6. Climatic Zones

Because of the Earth's axial tilt of $66\frac{1}{2}°$ to the plane of its orbit around the Sun, the altitude of the Sun at noon, which is the time of day at which the Sun attains its greatest daily altitude, varies throughout the year.

At the equator on June 21st, the date of the Summer Solstice, the Sun's noon-day altitude is $66\frac{1}{2}°$ bearing North of the observer. From this date to that of the Autumnal Equinox on September 23rd, the Sun's noon-day altitude increases progressively to 90°. From September 23rd to December 21st, the noon-day altitude of the Sun decreases from 90° to $66\frac{1}{2}°$; but, during this period, the Sun attains its greatest daily altitude bearing South of an observer. The variation of the Sun's noon-day altitude throughout the year is similar for all Latitudes: this being 47°, or twice the obliquity of the ecliptic.

As a consequence of the changing noon-day altitude of the Sun the Earth is divided into Climatic Zones. Between the parallels of $23\frac{1}{2}°$ N. and $23\frac{1}{2}°$ S. the Sun's altitude at noon is never less than $66\frac{1}{2}°$; and that, on at least one day of the year, the Sun's noon-day altitude is 90°. The spherical zone contained between these parallels of Latitude is known as the Torrid Zone. The northern and southern boundary parallels of the Torrid Zone are known, respectively, as the Tropic of Cancer and the Tropic of Capricorn. The parts of the Torrid Zone to the north and south of the equator are known, respectively, as the North Torrid and the South Torrid Zones.

The polar caps bounded, respectively, by the parallels of $66\frac{1}{2}°$ N. and $66\frac{1}{2}°$ S. are regions in which, on at least one day of the year, total daylight or total darkness is experienced. These parallels are known, respectively, as the Arctic Circle and the Antarctic Circle, and the caps which are bounded by them, are known as the North Frigid and the South Frigid Zones.

The spherical zones of the Earth which lie between the Torrid and Frigid Zones are known as the Temperate Zones—the North Temperate in the northern and the South Temperate in the southern hemisphere, respectively.

Fig. 26·7 illustrates the Earth's climatic zones.

THE EARTH'S MOTIONS AND THE SEASONS

7. Unequal Lengths of the Seasons

The Earth is at perihelion a fortnight or so after the date of the Winter Solstice. It will be remembered that when the Earth is nearest to the Sun its orbital speed is greatest. Thus, from the date of the Winter Solstice to that of the Spring Equinox, that is to say, during the season of northern Winter, the Earth travels faster than during the other seasons. This causes the first day of Spring in the northern hemisphere to be brought forward.

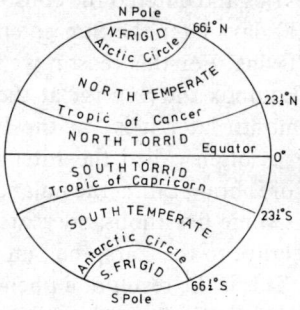

Fig. 26·7

The Summer Solstice occurs a fortnight or so before the Earth is at aphelion. The Earth, therefore, travels comparatively slowly at the time of the Summer Solstice, and during northern Summer the Earth travels more slowly than in the other seasons. This results in the first day of Autumn in the northern hemisphere being delayed.

The seasons are not, therefore, of equal length.

For the northern hemisphere:

Spring is 93 days
Summer is 94 days
Autumn is 90 days
Winter is 89 days

Because northern Winter takes place when the Earth is comparatively near to the Sun, the severity of northern Winter is mitigated. On the other hand southern Summer is warmer than it would be were the seasons of equal length.

Fig. 26·8 illustrates the unequal lengths of the seasons.

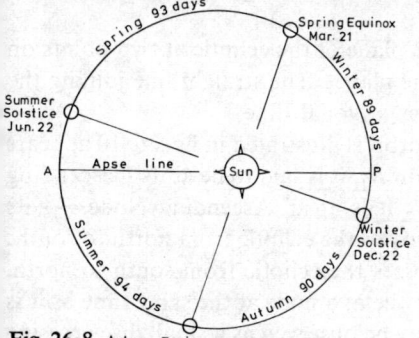

Fig. 26·8

8. The Zodiacal Belt

The Sun, in tracing out its annual apparent path across the celestial sphere, moves eastwards at the rate of 360° in a year, or about 1° per day, this motion being relative to the fixed stars. The constellations through which the Sun passes during its annual tour of the heavens are twelve in number, and they are known as the Constellations or Signs of the Zodiac.

The belt on the celestial sphere on which the Signs of the Zodiac lie extends for about 8° on each side of the ecliptic. The name given to this zone is the Zodiacal Belt.

The Ancient Egyptian astronomers are to be credited for having been the first to study, systematically, the Sun's motion relative to the background of the fixed stars. They deduced that at the time of the Spring Equinox the Sun occupied a position in the sky in the constellation of Aries the Ram. They noticed that at about 30 days later the Sun moved out of Aries and entered the constellation of Taurus the Bull, and that 30 days later the Sun entered the constellation of Gemini the Twins. For this reason we say that, on the date of the Spring Equinox the Sun is "at the First Point of Aries", and that a month later it is "at the First Point of Taurus", and so on.

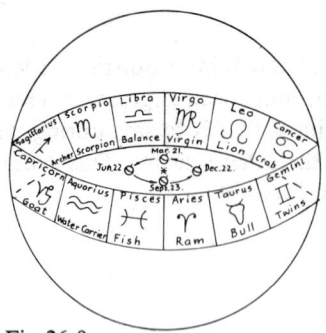

Fig. 26·9

The Spring Equinox, therefore, is often called the First Point of Aries, and the Autumnal Equinox is called the First Point of Libra—Libra, the Balance, being the sixth Sign of the Zodiac.

At the present time, which is some five thousand years after the ancient astronomers first investigated the Sun's motion relative to the stars, the Sun is not at the First Point of Aries on the first day of northern Spring. This is the result of a phenomenon known as the Precession of the Equinoxes, which we will explain in Chapter 42, which deals with the mechanical properties of Spinning bodies.

The Spring Equinox, despite the fact that it no longer coincides with the First Point of Aries, is still referred to by the latter name.

The Latin and corresponding English names of the Signs of the Zodiac are given in fig. 26·9.

The Signs of the Zodiac are easily memorized from the following ancient rhyme:

> "The ram, the bull, the heavenly twins,
> And next the crab, the lion shines,
> The virgin and the scales;
> The scorpion, archer and he-goat,
> The man who holds the watering pot,
> And fish with glittering tails."

9. Planetary Orbits

The orbits of the planets are not co-planar with the plane of the Earth's orbit nor with the planes of each other's orbits. Their orbital planes make small angles, not usually more than a few degrees, with that of the Earth's orbit. The orbits of the planets projected from the Earth onto the celestial sphere results in paths which lie near to the ecliptic. It follows that the planets lie in or near to the zodiacal belt.

The plane of a planet's orbit around the Sun cuts the plane of the ecliptic at two points on the celestial sphere which are known as the Nodes of the planet. The straight line joining the two nodes of a planet is known as a Line of Nodes or as a Nodal Line.

When the planet whose orbit is illustrated in fig. 26·10 appears to be at the ecliptic at position N_1, it is said to be at its Descending Node. When at position N_2 it is at an Ascending Node. At its descending node a planet crosses the ecliptic from north to south: at an ascending node it crosses the ecliptic from south to north.

Should an inferior planet lie at a node at the same time as it is at inferior conjunction it may be observed as a small disc crossing the face of the Sun. Such a phenomenon is known as a Transit of the planet.

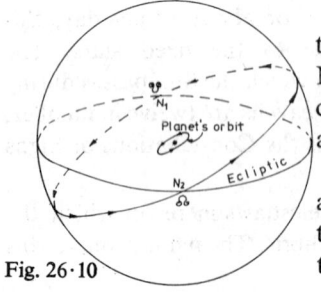

Fig. 26·10

THE EARTH'S MOTIONS AND THE SEASONS

10. Direct and Retrograde Motion

The motion of the Sun relative to the fixed stars is known as Direct Motion. Other celestial bodies which move relative to the fixed stars in this direction are said to move with Direct Motion. Celestial bodies which move relative to the fixed stars in the opposite direction are said to move with Retrograde Motion.

The apparent motion of a planet relative to the fixed stars is the resultant of its own orbital motion around the Sun and that which is imparted to it by virtue of the Earth's orbital motion. If the path of a planet is plotted on a star map, it will be found that, in some cases, the motion relative to the stars is sometimes direct and sometimes retrograde. When the motion changes from direct to retrograde and back again, the planet traces out a loop known as a Retrogressive Loop.

Exercises on Chapter 26

1. Describe the motions of the Earth. What natural units of time are derived from these motions?
2. Define: Solar Day; Sidereal Day. Compare the lengths of these units of time.
3. Draw a diagram to illustrate the Ecliptic and the Celestial Equator. Name the points of intersection of these circles.
4. Explain why the Ecliptic and the Earth's Orbit are co-planar.
5. What are the causes of the seasons? State the dates of the commencements of the seasons.
6. Describe the causes of unequal lengths of daylight and darkness during the year.
7. Explain the meaning of: Equinox; Solstice.
8. Describe the Earth's climatic zones.
9. Describe the Zodiacal Belt. Why are the planets normally found in this belt?
10. What is meant by: Ascending and Descending Nodes?
11. Show that the greatest daily altitude of the Sun on the days of the equinoxes is equal to the complement of the Latitude of the observer.
12. Describe: Direct and Retrograde Motion as these terms apply to the apparent motions of certain planets.

CHAPTER 27

DEFINING CELESTIAL POSITIONS

1. The Cartesian System of Co-ordinates

The common system of describing the position of a point in a plane is the Cartesian System, named after the French philosopher René Descartes. In this system a position is described relative to two axes of reference—usually called the x- and y-axes, respectively, which are mutually perpendicular straight lines in the plane. The axes of reference intersect each other at a point called the Origin, and distances measured x-wards from the y axis and y-wards from the x axis are called Abscissae and Ordinates respectively. The abscissa and ordinate of a given point in the plane are called the Co-ordinates of the point.

An extension of the Cartesian System is used for describing positions on a spherical surface. In this case the position is described relative to two great circles the planes of which are perpendicular to one another, and the two co-ordinates are the respective arcs of these great circles of reference. To describe a position on the surface of the spherical Earth, for example, the two great circles of reference are the Equator, from which the co-ordinate Latitude is measured; and the Greenwich Meridian from which the co-ordinate Longitude is measured. In other words the Latitude and Longitude of a point on the Earth are the co-ordinates of the points in terms of the commonly used system of defining terrestrial positions.

2. The Ecliptic System of Defining Celestial Positions

In the Ecliptic System of defining a celestial position the two co-ordinates are known, respectively, as Celestial Latitude and Celestial Longitude. The two reference great circles, in this case, are the Ecliptic, from which Celestial Latitude is measured; and the Secondary to the Ecliptic through the First Point of Aries from which Celestial Longitude is measured.

Secondaries to the Ecliptic are known as Circles of Latitude. All circles of Latitude converge to the Pole of the Ecliptic. In other words, a circle of Latitude is a semi-great circle which connects the poles of the ecliptic.

The Celestial Latitude of a point is defined as the arc of the circle of Latitude on which the point is located, measured from the ecliptic to the point. It is named North or South according as the point lies north or south of the ecliptic, respectively.

The Celestial Longitude of a point is defined as the arc of the ecliptic measured eastwards from the First Point of Aries to the circle of Latitude on which the point is located. In other words, it is the angle at the pole of the ecliptic contained between the circles of Latitude through the First Point of Aries and the point.

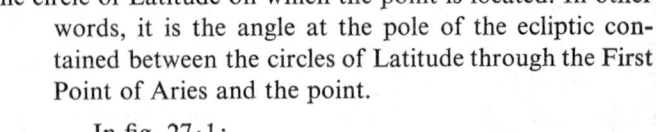

Fig. 27·1

In fig. 27·1:

Arc AX = Celestial Latitude of X

Arc ΥA = Celestial Longitude of X

Arc BY = Celestial Latitude of Y

Reflex Arc ΥPB = Celestial Longitude of Y

3. The Horizon System of Defining Celestial Positions

Only a half of the celestial sphere is visible to an observer at any instant of time. This follows because the opaque Earth itself obscures the observer's view. The great circle on the celestial sphere which divides the celestial sphere into the Visible and Invisible Hemispheres is known as the Celestial Horizon. The poles of the celestial horizon are the Zenith for the visible hemisphere, and the Nadir for the invisible hemisphere.

Because they cross the celestial horizon at 90° secondaries to the celestial horizon are called Vertical Circles. A vertical circle is defined as a semi-great circle which connects the zenith and nadir of an observer.

The co-ordinates used in the horizon system of defining celestial positions are Altitude and Bearing or Azimuth. The Altitude of a celestial point is the angle at the centre of the celestial sphere contained between the point and the horizon measured in the plane of the vertical circle on which the point lies. The altitude of the zenith is 90°, and the altitude of every point on the celestial horizon is 0°.

All points on the celestial sphere having the same altitude are located on a small circle which is parallel to the celestial horizon. Such a small circle is called a Parallel of Altitude.

The Bearing of a celestial point is the arc of the celestial horizon, or the angle at the zenith of an observer, contained between the vertical circle which lies in the North–South plane, and the vertical circle on which the point is located. The bearing of a point is usually given as an acute angle contained between the North or South point of the horizon and the direction of the point.

Because celestial meridians converge towards the celestial pole the direction of either of the Earth's Poles from any observer is in the plane of the meridian on which the observer is located. The celestial meridian which lies in the same plane as that of the observer's terrestrial meridian is known as the Observer's Celestial Meridian; or, more simply, as the Observer's Meridian. The zenith of an observer and also the celestial poles are located on the observer's celestial meridian.

The observer's celestial meridian is divided into two semi-great circles each terminating at the celestial poles. That part of the observer's celestial meridian which contains the observer's zenith is called the Observer's Upper, or Superior, Celestial Meridian. The other part, which contains the observer's nadir, is called the Observer's Lower, or Inferior, Celestial Meridian.

The observer's celestial meridian coincides with the vertical circle passing through the North and South points of his horizon. The bearing of a celestial point may, therefore, be defined as the acute angle at the zenith contained between the observer's celestial meridian and the vertical circle passing through the point.

The Azimuth of a celestial point is the angle at an observer's zenith contained between the observer's upper celestial meridian and the vertical circle on which the point is located.

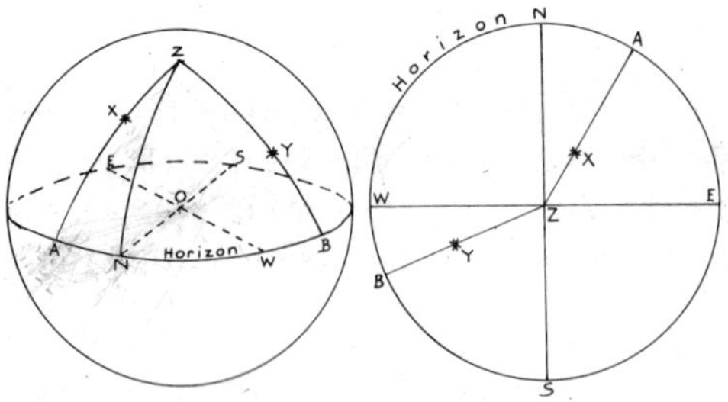

Fig. 27·2

In fig. 27·2:

AX or XOA = Altitude of X
NA or NOA = Bearing of X
NA or NZA = Azimuth of X

BY or YOB = Altitude of Y
SB or SOB = Bearing of Y
NB or NZB = Azimuth of Y

The bearing of a celestial point is named from North or South, whichever direction is nearer to that of the point itself, to East or West according as the point is to the east or the west, respectively, of the observer's celestial meridian.

4. The Celestial Equatorial System of Defining Celestial Positions

In the celestial equatorial system of defining celestial positions the co-ordinates are known as Declination and Hour Angle.

The celestial equator divides the celestial sphere into two celestial hemispheres. Every point in the Northern Celestial Hemisphere, the pole of which is the North Celestial Pole, is said to have North Declination. Every point in the South Celestial Hemisphere has South Declination.

The declination of a celestial position or point is the spherical distance of a great circle arc secondary to the celestial equator, measured from the celestial equator to the position or point. Secondaries to the celestial equator are known as Celestial Meridians. The declination of a point may, therefore, be defined as the arc of a celestial meridian intercepted between the celestial equator and the point.

All points having the same declination in either the northern or the southern celestial hemisphere lie on a small circle which is parallel to the celestial equator. Such a small circle is known as a Parallel of Declination. Parallels of Declination on the celestial sphere are analogous to parallels of Latitude on the Earth.

All celestial meridians converge towards two points on the celestial sphere known as the Celestial Poles. The North Celestial Pole is the point on the celestial sphere which is vertically above the Earth's North Pole, that is to say it is at the zenith of the North Pole. The South Celestial Pole is at the zenith of the Earth's South Pole. Celestial meridians on the celestial sphere are analogous to meridians on the Earth.

Celestial meridians are sometimes known as Hour Circles. This follows because the angle at the celestial pole at any instant, measured westwards from the observer's upper celestial meridian, to the celestial meridian of a heavenly body, is the measure of the time that has elapsed since the body was on the observer's upper celestial meridian. That is to say, it is a measure of the time since the body was at Meridian Passage.

The angle at the celestial pole measured westwards from the observer's upper celestial meridian to the Hour Circle of a celestial body is known as the Local Hour Angle (L.H.A.) of the body.

DEFINING CELESTIAL POSITIONS

The angle at the celestial pole between the Greenwich upper celestial meridian and the hour circle of a given celestial body, measured westwards from the Greenwich upper celestial meridian, is known as the Greenwich Hour Angle (G.H.A.) of the body.

The angle at the celestial pole measured westwards from the celestial meridian of the First Point of Aries to the celestial meridian of a celestial body is known as the Sidereal Hour Angle (S.H.A.) of the body.

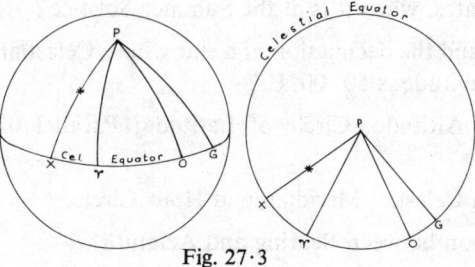

Fig. 27·3

In fig. 27·3:

$$\text{L.H.A. of star } \star = OX$$
$$\text{G.H.A. of star } \star = GX$$
$$\text{S.H.A. of star } \star = \Upsilon X$$

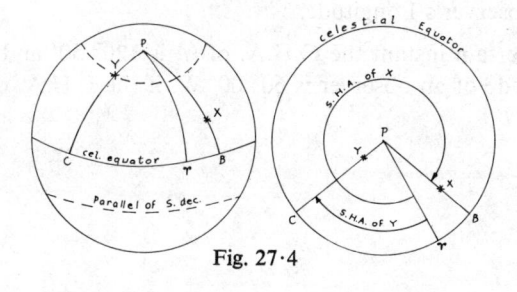

Fig. 27·4

In fig. 27·4:

$$\text{Declination of star } X = XB$$
$$\text{Declination of star } Y = YC$$
$$\text{S.H.A. of star } Y = \Upsilon C$$

The nautical astronomical problem of finding Longitude is one in which the navigator relates the position of an observed heavenly body using the co-ordinates of the Horizon System, with the body's position at the instant of the observation, using co-ordinates of the Celestial Equatorial System. A detailed discussion on this is given in Chapter 34 under the heading "The Astronomical Triangle".

Exercises on Chapter 27

1. Describe the Cartesian System of defining the position of a point in a plane, and explain how this system is adapted for describing terrestrial positions.

2. Define each of the co-ordinates used in each of the three systems employed for describing celestial positions.

3. Define: Sidereal Hour Angle; Local Hour Angle; Greenwich Hour Angle.

4. Define the position of the First Point of Aries using the Ecliptic and Celestial Equatorial Systems.
5. Compute the Celestial Latitude and the Celestial Longitude of a star whose declination is 30° 00′ N. and whose S.H.A. is 311° 15′.
6. What is the Sun's celestial position using (*a*) Ecliptic co-ordinates, and (*b*) Celestial Equatorial co-ordinates, when it is at the Summer Solstice?
7. What is the S.H.A. and the declination of a star whose Celestial Latitude is 30° 00′ N. and whose Celestial Longitude is 50° 00′ E.?
8. Define: Parallel of Altitude; Circle of Latitude; Parallel of Declination; Parallel of Latitude.
9. Distinguish between Celestial Meridian and Hour Circle.
10. Explain the distinction between Bearing and Azimuth.
11. Explain the difference between Observer's Upper Celestial Meridian and Observer's Lower Celestial Meridian.
12. Show that the celestial equator passes through the East and West points of any observer's celestial horizon.
13. Prove that for a given instant the difference between the L.H.A. of a star and its G.H.A. is a measure of the observer's Longitude.
14. Prove that if at a certain instant the G.H.A. of ♈ is 120° 50′ and the S.H.A. of a star is 15° 18′, the Longitude of an observer is 60° 00′ W. if the L.H.A. of the star is 76° 08′.

CHAPTER 28

THE APPARENT DIURNAL MOTION OF CELESTIAL BODIES

1. Diurnal Circles

Because of the axial rotation of the Earth towards East, the celestial sphere appears to revolve around the Earth towards the West. This causes the celestial bodies to perform daily circular paths around the Earth with the celestial pole as centre. These apparent daily paths are called Diurnal Circles.

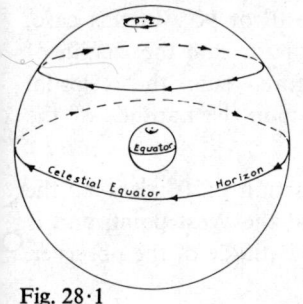

Fig. 28·1

In fig. 28·1, which illustrates diurnal circles as they appear to an observer in Latitude 90°, the celestial horizon of the observer is coincident with the celestial equator.

A heavenly body whose declination is 0° travels around the celestial horizon of an observer at either of the Earth's poles, completing its diurnal circle in the time taken for the Earth to rotate once on its axis. Other celestial bodies to the same observer maintain constant altitudes which equal their respective declinations. All stars alter their bearings or azimuths to the extent of 15° per hour, and all objects in the visible hemisphere are above the horizon for the whole day.

If the time of the year is northern Spring or Summer, the Sun would be above the horizon of an observer at the Earth's North Pole for the whole day; but, unlike the stars, its altitude would change at a slow rate equivalent to the rate of change of its declination.

Celestial bodies which are above the horizon throughout the day are known as Circumpolar Bodies. At either pole of the Earth, all bodies in the visible hemisphere are circumpolar bodies.

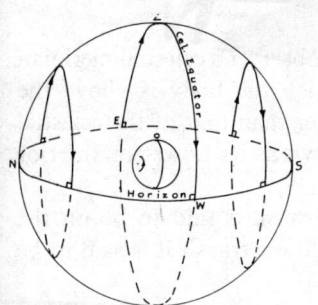

Fig. 28·2

In fig. 28·2, which illustrates diurnal circles as they appear to an observer in Latitude 0°, the celestial horizon of the observer is coincident with a celestial meridian. The diurnal circles of all celestial bodies are bisected by the observer's celestial horizon. This means that all celestial bodies to an observer on the equator are above the celestial horizon for exactly half the day and below it for the other half. At the equator all celestial bodies rise and set, so that there are no circumpolar bodies.

When a heavenly body crosses an observer's celestial meridian it is said to Culminate; to Transit; or to be at Meridian Passage. A celestial body at the point of culmination on an observer's upper celestial meridian reaches its greatest altitude for the day.

From figs. 28·3 and 28·4 it will be noticed that, to an observer in Latitude 0°, a star having a declination of 0° culminates with an altitude of 90°. The point of culmination is the zenith of the observer. Such a body crosses the sky in its diurnal path such that its bearing is always due East or due West. Its diurnal circle is a great circle on the celestial sphere which passes through the East and West points of the observer's celestial horizon. This great circle is known as the Prime Vertical Circle, or more commonly as the P.V.

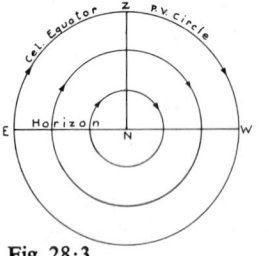

Fig. 28·3

The celestial meridian which is coincident with the horizon of an observer on the equator, cuts the observer's celestial meridian at an angle of 90°. Any celestial body on this meridian, which is known as the Six o-Clock Hour Circle, is 6 hours from the observer's celestial meridian.

It will be seen in fig. 28·4 that the interval between the instants at which a body rises and of its culmination is 6 hours. The interval from its culmination to its setting is also 6 hours.

Fig. 28·4

Fig. 28·5 illustrates diurnal circles as they appear to an observer in any Latitude other than 0° or 90°. In this case, the celestial pole lies between the horizon and the zenith of an observer. The horizon, therefore, cuts the celestial equator at an angle which depends upon the Latitude of the observer.

A celestial body having a declination of 0° rises at the East point of the horizon and sets at the West point, and is above the horizon, regardless of the Latitude of the observer, for exactly half a day.

A body which has north declination rises such that its bearing is between North and East, and it sets such that its bearing is between North and West. Such objects are above the horizon of any observer in the northern hemisphere for more than half the day.

It will be noticed in fig. 28·5 that some celestial bodies are above the horizon throughout the day. These are circumpolar bodies.

A circumpolar body crosses an observer's celestial meridian on two occasions, at both of which the body is above the horizon, during the day. The transit at which its altitude is greater than that of the celestial pole is known as its upper or Superior Transit: the other is known as its Lower or Inferior Transit.

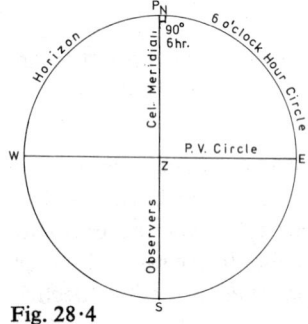

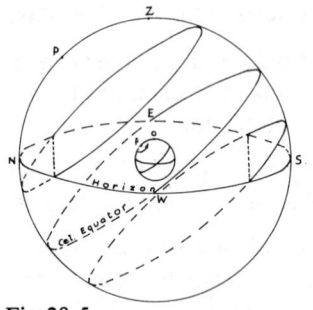

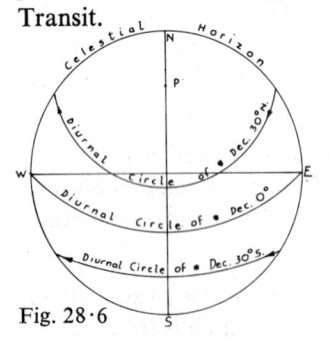

Fig. 28·5

A circumpolar body at upper transit is said to be on the meridian Above the Pole: when at lower transit it is said to be on the meridian Below the Pole.

The number of stars above the horizon at both upper and lower transits depends upon the Latitude of the observer. The greater the Latitude, the greater the number of circumpolar stars. It will be remembered that to an observer at either of the Earth's poles all celestial bodies above the horizon are circumpolar.

Fig. 28·6

THE APPARENT DIURNAL MOTION OF CELESTIAL BODIES

Fig. 28·6 illustrates diurnal circles projected onto the plane of the horizon of an observer in Latitude 50° N.

It will be noticed from figs. 28·5 and 28·6 that some celestial bodies cross the prime vertical circle during their diurnal paths, whereas others do not. The conditions necessary for a celestial body to be circumpolar, and the conditions necessary for a body to cross the prime vertical circle of an observer, will now be considered.

Fig. 28·7 serves to demonstrate the very important relationship between the altitude of the celestial pole and the Latitude of the observer.

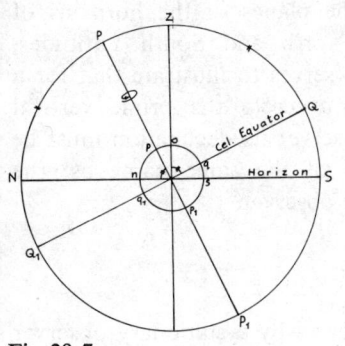

Fig. 28·7

In fig. 28·7:

$oq = QZ =$ Latitude of observer O

$PQ = 90°$

Thus: $PZ = (90° - QZ)$

$ZN = 90°$

Thus: $PN = (90° - PZ)$

$\qquad = 90° - (90° - QZ)$

$\qquad = QZ$

But: $QZ =$ Latitude of observer O

and: $PN =$ Altitude of Celestial Pole

Therefore:

$$\text{Latitude of Observer} = \text{Altitude of Celestial Pole}$$

The arc of a celestial meridian contained between a celestial body and the celestial pole is known as the Polar Distance of the body. The polar distance of a heavenly body is equal to $90° \pm$ declination of the body.

For a body to be circumpolar its polar distance must be less than the observer's Latitude. Thus, in Latitude 30° N., all stars whose declinations are greater than 60° (the complement of 30°) will be circumpolar. It follows, therefore, that for a star to be circumpolar its declination must be greater than the co-Latitude of the observer. Moreover, the Latitude and the declination must have the same name.

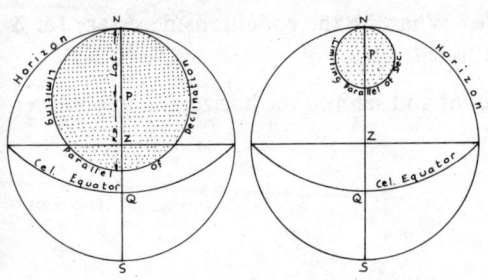

Fig. 28·8

Fig. 28·8 shows projections of the celestial sphere onto the planes, respectively, of the horizons of observers situated in the northern hemisphere. The arc NP is equal to the altitude of the celestial pole: it is, therefore, equal to the Latitude of the observer.

$PQ = 90°$

$NZ = 90°$

Therefore: $ZQ = NP =$ Latitude of Observer

For a celestial body to be circumpolar its polar distance must be less than the Latitude of the observer.

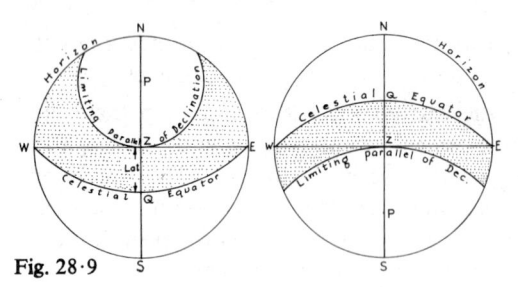

Fig. 28·9 shows projections of the celestial sphere onto the planes of the horizons of observers in North and South Latitudes, respectively. It serves to illustrate that for a celestial body to cross the prime vertical circle of an observer its declination must be less than, but of the same name as, the Latitude of the observer.

Fig. 28·9

Exercises on Chapter 28

1. Describe the apparent daily motions of a fixed star as viewed by a stationary observer located (*a*) in Latitude 90°, (*b*) at the South Pole, (*c*) on the equator, and (*d*) in Latitude 40° N.

2. What is the Latitude of an observer whose celestial horizon coincides with the ecliptic?

3. Explain why celestial bodies having North declination are above the horizon of an observer in the northern hemisphere for more than 12 hours during each day.

4. What is meant by the term Culmination? What other terms are used to denote the same phenomenon?

5. Prove that all stars having declinations greater than 36° N. are circumpolar to an observer in Latitude 54° N.

6. What are the conditions necessary for a star not to rise above the horizon of any given observer?

7. What are the conditions necessary for a star to cross the prime vertical circle of an observer?

8. Prove that the altitude of the celestial pole is equal to the latitude of the observer.

9. Explain how an observer may find his latitude from an observation of a star on his meridian above the pole.

10. Explain how the latitude of an observer may be found from the altitudes of a star above and below the pole, respectively, when the declination of the star is not known.

11. What is meant by the Six o-Clock Hour Circle? What are the conditions necessary for a celestial body to cross the Six o-Clock Hour Circle?

12. Explain why it is that all celestial bodies rise out of and set into the horizon of an observer on the equator at an angle of 90°.

CHAPTER 29

TIME

1. The Units of Time

Time, in the astronomical sense, denotes that which persists while events take place. Events are contained in time as objects are contained in space. Time exists before an event and after an event, and it measures the event as it occurs. Time is, therefore, measurable duration.

Time is normally measured by a clock the mechanism of which is adjusted so that the clock registers, in hours, minutes and seconds, the interval that has elapsed since a certain astronomical event took place.

The Earth, itself, because of its relatively uniform rate of rotation, affords the means of establishing a suitable unit of time. The Earth's real axial motion is made manifest by the apparent diurnal motion of the objects on the celestial sphere. The rotation of the Earth on its axis is almost perfectly uniform, and the time taken for it to perform one rotation is a natural unit of time known as a Day. Depending upon what celestial body is used to establish this unit of time, determines the type of day and its length. The unit derived from the apparent motion of a fixed celestial body is referred to as a Star- or a Sidereal-Day. The unit derived from the apparent diurnal motion of the Sun is known as a Sun- or Solar-Day. The unit derived from the apparent diurnal motion of the Moon is known as a Moon- or Lunar-Day.

Because of the Earth's and Moon's orbital motions the lengths of the different time units are not equal. Moreover, because of the irregularities of the Earth's and Moon's orbital motions, the lengths of the Solar and Lunar Days are not uniform.

The day commences at the instant at which the celestial body used in its determination is on the observer's celestial meridian.

The Solar Day starts when the Sun is on the observer's lower celestial meridian: a solar clock set to solar time indicates 00 h. 00 m. 00 s. at this instant.

The fixed point on the celestial sphere used for determining the Sidereal Day is the point at which the Sun, in its apparent annual path, crosses the celestial equator from the southern into the northern celestial hemisphere. This point is the Spring Equinox, more usually known as the First Point of Aries when dealing with time.

The Sidereal Day commences at the instant at which the First Point of Aries is on the observer's upper celestial meridian: a clock set to sidereal time registers 00 h. 00 m. 00 s. at this instant.

It should be noted that the celestial meridians from which solar and sidereal times are measured, respectively, are diametrically opposed to each other.

A Sidereal Day is defined as the interval which elapses between two successive transits of the First Point of Aries across the UPPER meridian of a stationary observer.

A Solar Day is defined as the interval between successive transits of the Sun across the LOWER meridian of an observer.

Although the Earth is moving in her orbit with a speed of about 18·5 miles per second, the direction of the First Point of Aries, or that of any other fixed point in space, does not change because the radius of the celestial sphere is infinite. The sidereal day, therefore, is the interval of time taken for the Earth to make exactly one rotation on its axis. Thus, for each angle of 15° through which the Earth rotates one sidereal hour passes.

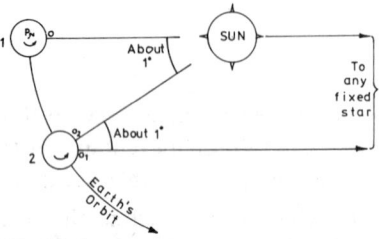

As the Earth moves in its orbit around the Sun, relative to the fixed stars, appears to move eastwards across the sky at the rate of 360/365°, or approximately 1°, per day. It follows that, in the interval between two successive transits of the Sun across the upper meridian of a stationary observer, the Earth rotates through an angle of about 361° on its axis.

Fig. 29·1

Referring to fig. 29·1: imagine the Sun and any fixed star to be on the upper meridian of the observer denoted by O. The next time that the star occupies the upper meridian of the observer occurs after the observer has been carried with the Earth to position 2, and when he is at O_1. In travelling from O to O_1 the observer has been carried around the Earth's axis through an angle of exactly 360°. Before the Sun is again on the observer's upper meridian, the Earth has to rotate through a further angle of about 1° in order to carry the observer to position O_2. A Solar Day, therefore, is a longer period of time than a Sidereal Day.

The sidereal day, assuming a uniform rate of rotation for the Earth, is a constant unit of time: but, because of the Earth's erratic orbital speed—being fastest when the Earth is at perihelion, and slowest when at aphelion—the length of the Solar Day is not uniform. In January for example, when the Earth is near perihelion, the daily angle swept out by the Earth in its orbit around the Sun is more than 1°: in July, when the Earth is near aphelion, the angle is less than 1° per day. The Solar Day, therefore, is longer in January than it is in July.

It is not convenient to use Sidereal Time in everyday affairs because the Sun governs these to a large extent. But the Sun is not perfectly suitable because the length of the Solar Day is not constant.

In order to overcome the inconvenience of the inconstancy of the Solar Day, and yet use the Sun as the basis of time-keeping, an artificial point known as the Mean Sun is used. The Mean Sun moves at the average speed of the True Sun but, instead of moving along the ecliptic, which is the path traced out by the True Sun, the Mean Sun moves in the Earth's rotation; that is to say, it moves along the celestial equator.

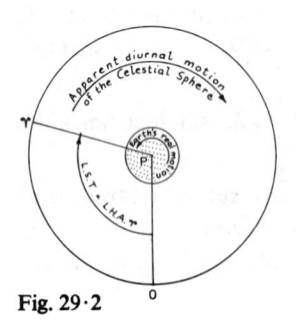

The unit of time derived from the apparent diurnal motion of the Mean Sun is known as the Mean Solar Day. The True Solar Day is usually known as the Apparent Solar Day because it is derived from the diurnal motion of the Sun which is an "apparent" motion.

Fortunately the celestial meridian of the Mean Sun is never very far from that of the True Sun. Time by the Mean Sun, therefore, is very nearly the same as time by the True Sun.

Fig. 29·2

2. Time at an Instant

As explained in Chapter 28, celestial meridians are sometimes called hour circles. The angle contained between an observer's celestial meridian and the meridian, or hour circle, passing through a given celestial point is known as the Hour Angle of that point.

The Sidereal Time at any instant for any place may be defined as the Hour Angle of the First Point of Aries. This is illustrated in fig. 29·2, in which it is seen that the Hour Angle of the First Point of Aries is a measure of the time that has elapsed since the meridian passage of the First Point of Aries. Thus, at any instant:

$$\text{Local Sidereal Time (L.S.T.)} = \text{L.H.A. of } \Upsilon$$

The Apparent Solar Day commences when the True Sun crosses an observer's lower celestial meridian. Thus, the Apparent Solar Time at any instant is defined as the angle at the celestial pole measured westwards from the observer's lower celestial meridian to the meridian, or hour circle, of the True Sun at that instant. It follows that because Solar Time is measured from the observer's lower meridian, whereas Hour Angle is always measured from an observer's upper meridian, the Hour Angle of the True Sun differs from the Solar Time by 12 hours.

The Local Mean Time at any instant is defined as the angle at the celestial pole, or the arc of the celestial equator, measured westwards from an observer's lower meridian to the meridian of the Mean Sun at the instant. It is the time that has elapsed since the Mean Sun occupied the observer's lower celestial meridian. Thus, at any instant:

$$\text{Local Mean Time (L.M.T.)} = \text{Hour Angle of the Mean Sun} \pm 12 \text{ hours}$$

$$\text{Local Apparent Time (L.A.T.)} = \text{Hour Angle of the True Sun} \pm 12 \text{ hours}$$

Or:
$$\text{L.M.T.} = \text{H.A.M.S.} \pm 12 \text{ hr.}$$

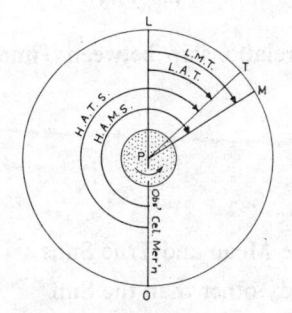

$$\text{L.A.T.} = \text{H.A.T.S.} \pm 12 \text{ hr.}$$

In fig. 29·3:
$$\text{L.M.T.} = \text{H.A.M.S.} - 12 \text{ hr.}$$

$$\text{L.A.T.} = \text{H.A.T.S.} - 12 \text{ hr.}$$

In fig. 29·4:
$$\text{L.M.T.} = \text{H.A.M.S.} + 12 \text{ hr.}$$

$$\text{L.A.T.} = \text{H.A.T.S.} + 12 \text{ hr.}$$

Fig. 29·3

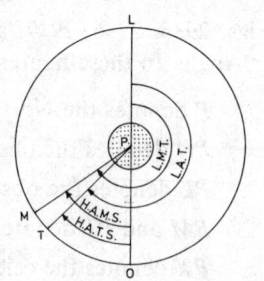

Fig. 29·4

Because the True and Mean Suns usually occupy different hour circles, L.M.T. and L.A.T. at any given instant are not the same. The L.M.T. differs from the L.A.T. by a quantity which is equivalent to the angle at the celestial pole, or the corresponding arc of the celestial equator, contained between the hour circles of the True and Mean Suns, respectively. This quantity, expressed in units of time, is called the Equation of Time. The Equation of Time is sometimes defined as the excess of L.M.T. over L.A.T. It is the amount of time that must be applied to the L.M.T. (or L.A.T.) to get the corresponding L.A.T. (or L.M.T.). If, for example, the L.M.T. is 11 h. 45 m. at the instant the L.A.T. is 11 h. 49 m., the excess of L.M.T. over L.A.T. is a negative amount, and the equation of time in this circumstance takes a minus sign and is described as −4 minutes. If, on the other hand L.M.T. is 11 h. 49 m. and the corresponding L.A.T. is 11 h. 45 m. the equation of time is +4 minutes.

$$\text{Equation of Time } (e) = \text{L.M.T.} - \text{L.A.T.}$$

or:
$$e = \text{H.A.M.S.} - \text{H.A.T.S.}$$

The equation of time is positive when the Mean Sun lies to the west of the True Sun, and it is negative when the Mean Sun lies to the east of the True Sun.

Fig. 29·5

Fig. 29·6

Fig. 29·7

Fig. 29·8

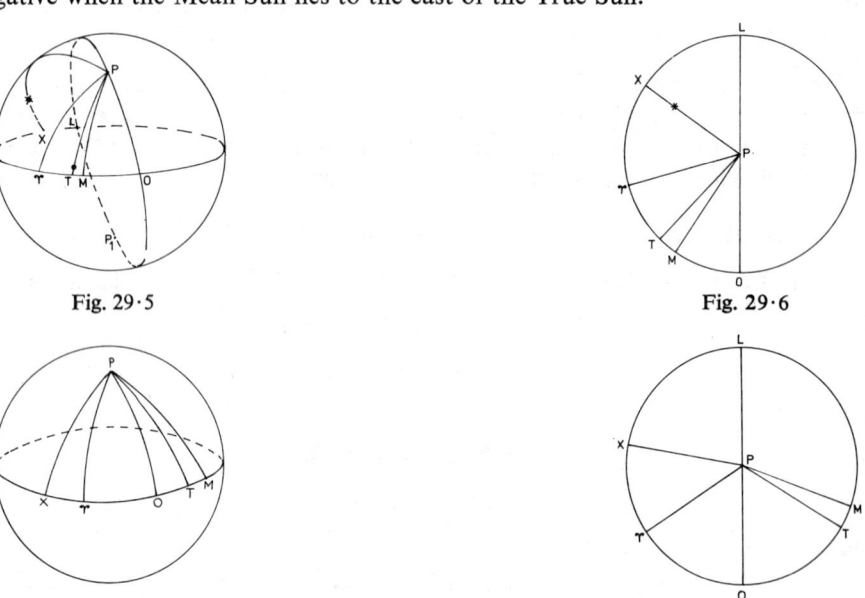

Figs. 29·5, 29·6, 29·7, and 29·8, serve to illustrate the relationship between Time and Hour Angle. In these figures.

P denotes the North celestial pole

PO denotes the observer's upper celestial meridian

PL denotes the observer's lower celestial meridian

PM and PT denote, respectively, the meridians of the Mean and True Suns

PX denotes the celestial meridian of any celestial body other than the Sun

$P\gamma$ denotes the celestial meridian of the First Point of Aries

ΥO = L.S.T.

LOM = L.M.T.

LOT = L.A.T.

OM = H.A.M.S.

OT = H.A.T.S.

$MT = e$

ΥX = S.H.A. of X

Note carefully that:

L.M.T. = H.A.M.S. $\pm$ 12 hr.

L.A.T. = H.A.T.S. $\pm$ 12 hr.

e = L.M.T. $-$ L.A.T. ($-$ve)

L.S.T. = H.A. of $X -$ S.H.A. of X

In figs. 29·7 and 29·8, the True Sun and the Mean Sun are East of the observer's upper meridian. In this case:

L.M.T. = H.A.M.S. $-$ 12 hr.

L.A.T. = H.A.T.S. $-$ 12 hr.

3. The Equation of Time

The True Sun is not a perfect timekeeper for two reasons, viz.:

1. Its motion in its apparent annual orbit around the Earth is irregular.

2. It moves in the ecliptic whereas the Earth rotates in the plane of the celestial equator.

The equation of time, therefore, is considered to be composed of two parts. The first is due to (1) above, and is known as the component due to Eccentricity; and the second, due to (2) above, is known as the component due to Obliquity.

The True Sun moves irregularly in the ecliptic. Consider a point moving in the ecliptic at a uniform rate equal to the average rate of the True Sun. This point is known as the Dynamical Mean Sun (D.M.S.). The component of the equation of time due to eccentricity is a measure of the difference between the Hour Angles of the True Sun and the Dynamical Mean Sun.

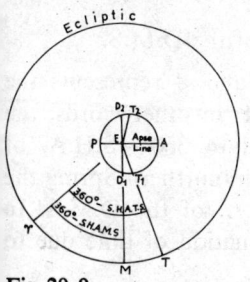

Fig. 29·9

It will be remembered that the Sun's apparent orbit around the Earth is an ellipse having the Earth at one of its foci; and that the speed of the Sun in the celestial sphere is greater when it is near perigee than when it is near apogee.

In fig. 29·9, the ellipse represents the apparent annual orbit of the Sun around the Earth. T and D represent, respectively, the True Sun and the D.M.S. P is the point in the Sun's apparent annual orbit at which it is closest to the Earth, and A represents the point at which its distance is greatest. These points are perigee and apogee respectively.

The line of apsides bisects the Sun's apparent annual orbit. The time taken by the True Sun to move from *P* to *A* is, therefore, the same as that taken for it to move from *A* to *P*. This follows, because, according to Kepler's Second Law, the times taken for the Sun's radius vector to sweep out equal areas are equal

The D.M.S. is coincident with the True Sun at perigee. Now the D.M.S. moves such that it sweeps out equal ANGLES in equal intervals. The D.M.S. moves from *A* to *P*, therefore, in the same time as that taken for the True Sun to move from *P* to *A*. Thus, the True Sun and the D.M.S. are again in coincidence at apogee.

In travelling from *P* to *A* the True Sun, which moves fastest when at perigee, moves ahead of the D.M.S., and their greatest angular separation occurs half way between perigee and apogee. This maximum separation, which amounts to about 2°, occurs in early April. From April until July—which is the time of apogee—the angular separation decreases until the time when the True Sun and the D.M.S. are in coincidence at apogee.

At apogee the True Sun moves slowest. Therefore, the D.M.S. moves ahead of the True Sun. The maximum separation occurs in early October. After this date the separation decreases until the True Sun and the D.M.S. are again in coincidence at the next perigee.

The maximum difference of time between that of the True Sun and the D.M.S. amounts to about 8 minutes. From perigee to apogee the True Sun is ahead of the D.M.S. The component of the equation of time due to eccentricity, therefore, is positive. From apogee to perigee the D.M.S. is ahead of the True Sun and this component, accordingly, is negative.

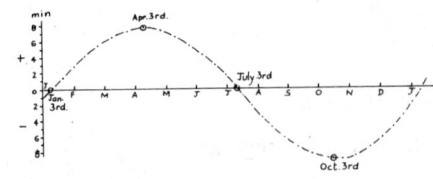

Fig. 29·10

The component of the equation of time due to eccentricity is represented graphically in fig. 29·10.

The D.M.S. increases its Celestial Longitude at a uniform rate. If the planes of the Earth's rotation and the ecliptic were coincident, the D.M.S. would be a perfect timekeeper. But because of the obliquity of the ecliptic, the point which is to afford the means of measuring time must move at a constant rate in the celestial equator. Consider a point on the celestial equator which is coincident with the D.M.S. at the Spring Equinox. Imagine this point to move such that its S.H.A. decreases uniformly at the same rate as the changing Celestial Longitude of the D.M.S. This point is known as the Astronomical Mean Sun (A.M.S.) or, more commonly, as simply the Mean Sun.

The component of the equation of time due to obliquity is a measure of the difference between the Hour Angles of the D.M.S. and the A.M.S.

Component due to Obliquity = H.A. of the A.M.S. − H.A. of the D.M.S.

In fig. 29·11 *D* represents the D.M.S., and *A* represents the A.M.S. The arc ♈*D* is equal to the arc ♈*A*; in other words, the Celestial Longitude of the D.M.S. is equal to 360° − S.H.A. of the A.M.S. It is evident from fig. 29·11 that in northern Spring the S.H.A. of the A.M.S. is less than the S.H.A. of the D.M.S. In Spring, therefore, the components of the equation of time due to obliquity is a negative quantity.

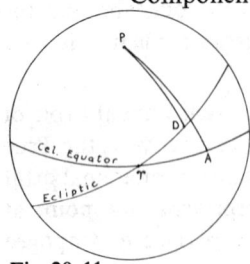

Fig. 29·11

By the time of the Summer Solstice the D.M.S. will have increased its Celestial Longitude by 90°. The A.M.S. will have decreased its S.H.A. and increased its Celestial Longitude by the same amount. At the time of the Summer Solstice the component due to obliquity is zero because the D.M.S. is coincident with the A.M.S. The maximum value of the difference in the Hour Angles of the D.M.S. and the A.M.S. occurs half way between the times of the Spring Equinox and the Summer Solstice. At that instant the difference in times by the A.M.S. and the D.M.S. is about 10 minutes.

From the date of the Summer Solstice to that of the Autumnal Equinox, the D.M.S. is ahead of the A.M.S., and the component due to obliquity is a positive quantity. From the time of the Autumnal Equinox to that of the Winter Solstice, the D.M.S. is astern, or to the East, of the A.M.S. During this part of the year the component due to obliquity is a negative quantity. From the date of the Winter Solstice to that of the Spring Equinox the component due to obliquity is positive; and from the time of the Spring Equinox to that of the Summer Solstice it is negative.

The component of the equation of time due to obliquity is shown graphically in fig. 29·12.

By combining the graphs of the two components, as shown in fig. 29·13, the equation of time for any time in the year may be found.

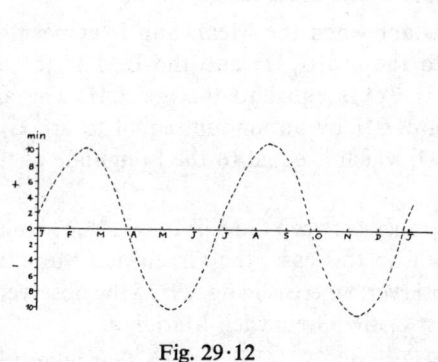

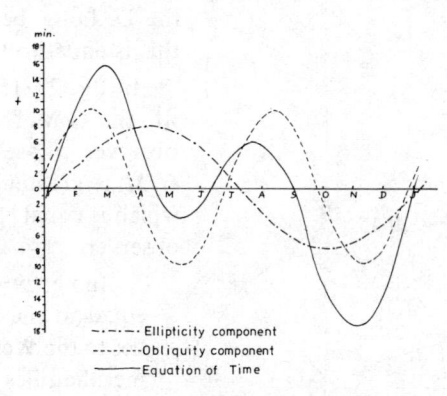

Fig. 29·12

Fig. 29·13

It will be noticed in fig. 29·13 that the equation of time is zero on four occasions each year. The dates of these times may be verified by inspection in the *Nautical Almanac*.

4. Comparison of Solar and Sidereal Time Units

The Earth makes one revolution in its orbit in the time it takes to make $365\frac{1}{4}$ rotations on its axis. During this interval the Sun revolves once in the ecliptic and, therefore, makes one apparent revolution with respect to the fixed stars. Thus, in relation to the stars, the Earth makes $366\frac{1}{4}$ rotations in the time it takes to make $365\frac{1}{4}$ rotations with respect to the Sun. Thus:

$$365\frac{1}{4} \text{ Solar Days} = 366\frac{1}{4} \text{ Sidereal Days}$$

From this relationship we see that:

$$1 \text{ Solar Day} = 1 \text{ d. } 00 \text{ h. } 03 \text{ m. } 56\cdot5 \text{ s. of Sidereal Time}$$
$$1 \text{ Sidereal Day} = 0 \text{ d. } 23 \text{ h. } 04 \text{ m. } 04\cdot1 \text{ s. of Solar Time}$$

5. Time and Longitude

The angle at the celestial pole contained between the celestial meridians of any two observers is a measure of the difference in Longitude between the positions of the observers.

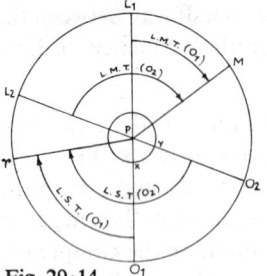

Fig. 29·14

In fig. 29·14 x and y denote two observers whose upper celestial meridians are PO_1 and PO_2, and whose lower celestial meridians are PL_1 and PL_2, respectively.

The L.M.T. at x is equal to the arc of the celestial equator L_1M. The L.M.T. at y is equal to the arc L_2M. The difference between these arcs is arc L_1L_2, which is equal to the arc O_1O_2 or arc xy. This, in turn, is equal to the difference of Longitude between x and y. Thus, the D. Long. between two observers is equal to the difference between their L.M.T.s. It may be seen from fig. 29·14 that D. Long. is also equal to the difference between the L.S.T.s at the two meridians. It is clear that time and Longitude are closely related.

Longitude is measured East or West from the Greenwich Meridian. Thus, the difference between the local time at the Greenwich Meridian and that at an observer's meridian at a given instant, is a measure, in time units, of the D. Long. between the Greenwich and observer's meridian, and this is equal to the Longitude of the observer.

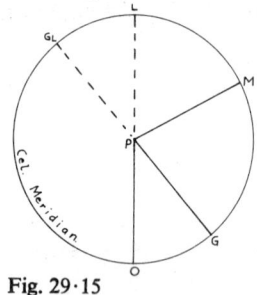

Fig. 29·15

In fig. 29·15, at the instant when the Mean Sun is at position M, the G.M.T. is equal to the arc G_LM; and the L.M.T. for an observer whose meridian is TO is equal to the arc LM. The arc G_LM is greater than the arc LM by an amount equal to arc G_LL which is equal to the arc GO, which is equal to the Longitude of the observer.

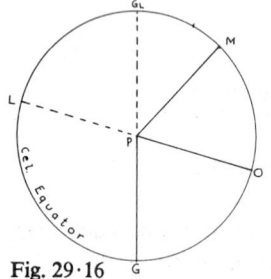

In fig. 29·16 the difference between G.M.T. and L.M.T. is also equal to the arc GO; but, in this case, the Greenwich Meridian lies to the West of the observer, whereas in fig. 29·15 the observer's meridian lies to the West of the Greenwich Meridian.

Fig. 29·16

From figs. 29·14, 29·15 and 29·16, it is seen that when the G.M.T. is greater than the L.M.T. the Longitude of the observer is named West, and that when the G.M.T. is less than the L.M.T. the Longitude of the observer is named East.

Hence the mnemonical rule:

> Longitude West Greenwich time best
> Longitude East Greenwich time least.

6. Time at Sea

The respective local times at two places which lie on different meridians differ to the extent of the D. Long. between the two places. It follows that, if it is desired to keep L.M.T. on a moving vessel it will be necessary to adjust the clock steadily as the vessel moves eastwards or westwards. For a change of each 15° of Longitude the clock will have to be altered 1 hour; for each 15′ of change of Longitude it will have to be altered by 1 minute; and for each 15″ of change of Longitude it will have to be altered by 1 second. It is not convenient, neither is it necessary, to continually change the clock time in this way.

In days gone by it was customary to set the clock, usually in the late evening, so that it registered the Apparent Time (A.T.S.) corresponding to the meridian at which the vessel was expected to reach at noon (L.A.T.) on the following day. It will be remembered that at noon (L.A.T.) the Sun attains its greatest daily altitude. If this altitude is measured it is comparatively easy for the navigator to ascertain the Latitude of his vessel. It was, therefore, convenient to have the clock set such that the navigator could be warned of the time to make his noon observation.

In many vessels a system of timekeeping known as Zone Time is used. In this system the clock is always an exact number of hours different from G.M.T. To facilitate the use of Zone Time, the Earth's surface is divided into Time Zones which are depicted on a Time Zone Chart. Time Zones are regions bounded by meridians whose Longitudes differ by 15°. The region known as Zone O is bounded by the meridians of $7\frac{1}{2}°$ E. and $7\frac{1}{2}°$ W. All vessels keeping Zone Time in this region have their clocks set to G.M.T., which is the correct Mean Time for the central meridian, which is the Greenwich Meridian. The Time Zone immediately to the east of Zone O extends from Longitude $7\frac{1}{2}°$ E. to Longitude $22\frac{1}{2}°$ E. In this region, which is known as Zone -1, vessels keeping Zone Time have their clocks set to one hour ahead of G.M.T. Thus, to find G.M.T. one hour is deducted from the clock time: hence the minus sign in front of the Zone number. Zone -2 extends from Longitude $22\frac{1}{2}°$ E. to Longitude $37\frac{1}{2}°$ E., and so on. Time Zones which lie to the west of Zone O, have positive numbers: the zone extending from $7\frac{1}{2}°$ W. to $22\frac{1}{2}°$ W. is Zone $+1$; that extending from $22\frac{1}{2}°$ W. to $37\frac{1}{2}°$ W. is Zone $+2$, and so on.

Time Zone 12 lies diametrically opposite to Zone O. It is divided into two parts by a line known as the Date Line. Ideally, the Date Line should coincide with the meridian of 180°, but its position is modified so as to avoid differences in time and date in certain island groups which straddle the 180th meridian. The part of Zone 12 which lies to the west of the Date Line is known as Zone -12, and that part which lies to the east is known as Zone $+12$.

Consider a vessel travelling westwards in the course of circumnavigating the Earth. For each 15° of Longitude made good to the West the clock will be retarded 1 hour. It is obvious that when the vessel arrives on her initial meridian, she will have changed her Longitude by 360° and her clock time will have been retarded by 24 hours. A whole day, therefore, would have been "lost" when the vessel's time is compared with the local calendar. To overcome this the date is altered by one day when crossing the Date Line, and this is the reason why this line is so named. When crossing the Date Line from West to East the date is advanced one day, and when crossing it from East to West the date is retarded one day.

7. Standard Time

When a vessel arrives in port it is often necessary for the clock to be re-set so that it corresponds to the time used by the people on shore. The times used by civil authorities are known as Standard Times, a list of which is given in the *Nautical Almanac*.

8. The Years

The time taken by the Earth to make exactly one revolution around the Sun relative to the fixed stars is known as a Sidereal Year. The sidereal year in Mean Solar units of time is 365 days 06 hours 09 minutes 09 seconds. The sidereal year is defined as the interval between successive instants when the Sun occupies the same position in the celestial sphere relative to a fixed point in space.

The time taken by the Earth to make one revolution with reference to the First Point of Aries is slightly shorter than a sidereal year. This is due to the precession of the Earth's axis which causes the equinoctial points to move westwards across the celestial sphere at an average rate of about 50″ of arc per year. This retrograde motion of the equinoxes gives rise to another unit of time known as the Tropical Year. The tropical year is defined as the interval between two successive Spring equinoxes. In Mean Solar units of time it is 365 days 05 hours 48 minutes 46 seconds.

The apse line of the Earth's orbit moves eastwards around the orbit at an average rate of about $11\frac{1}{4}″$ of arc per year. The interval between two successive perihelions, therefore, is slightly longer than a sidereal year. In Mean Solar units of time it is 365 days 06 hours 13 minutes 48 seconds. This period is known as an Anomalistic Year.

9. Co-ordinated Universal Time

The basis of the time-keeping described above is the rotation of the Earth. Solar and Sidereal Times are, therefore, described as Rotational Times. Now the speed of rotation of the Earth is not perfectly uniform: it is affected by atmospheric and oceanic phenomena, as well as by movements of material within the body of the solid Earth. So that for highly accurate time measuring the rotation of the Earth is not satisfactory.

The determination of the units of rotational times is a highly specialized branch of astronomy. Astronomical observations at a given observatory led to the determination of what astronomers call Universal Time (U.T.) Because of a short-term irregularity in the Earth's rotation, known as Polar Variation, U.T. varies slightly as between observatories, so that a correction is applied to U.T. to give a standard form of time known as U.T. 1. An empirical correction to allow for annual changes in the speed of the Earth's rotation is applied to U.T. 1 to give U.T. 2. This system of time-keeping was adopted internationally in 1956.

But even U.T. 2 is not sufficiently accurate for certain scientific purposes, so that an alternative system of time-measuring was sought.

Now the unit of time is a fundamental physical quantity, so that in 1956, the International Committee of Weights and Measures adopted as the unit of time the "Second of Ephemeris Time". Up to 1956 the unit was a second of time determined by the irregular rotation of the Earth—the unit being 1/86,400 of the Mean Solar Day. Ephemeris Time (E.T.), obtained from the orbital motion of the Earth, corresponds to U.T. 2 over a long period of time. The problem of making E.T. available led to the use of an Atomic Clock, the principle of which in no way depends on the rotation speed of the Earth.

An Atomic Clock employs energy changes within atoms to produce extremely uniform waves of electro-magnetic radiation which can be counted. In 1967 the unit of time adopted was the second of the International System of Units (S.I.). This is defined as the duration of a stated number of periods of radiation of a "caesium atom-133".

A system of time-keeping, known as Co-ordinated Universal Time (U.T.C.), has been in use since 1972. This system developed from the relating of Ephemeris Time (E.T.) to Atomic Time (A.T.), and by adjusting the atomic clock so as to remain close to U.T. 2.

In 1972 the practice was adopted of keeping U.T.C. to within about 0·5 sec. of U.T. 2 by resetting a clock on U.T.C. by precisely one second when necessary. These one-second jumps, known as Leap Seconds, are inserted at the ends of solar days; and, of course, notice well in advance is given whenever a leap second is to be introduced. Time Signals are broadcast on this system so that U.T.C. is now the reference time for common, as well as for nautical astronomical, use.

10. The Calendar

The systematic arrangement of units of time constitutes an Almanac or Calendar. The prime function of a calendar is to provide the means of recording dates of important events for the benefit of posterity. The earliest calendars were devised mainly for the purpose of determining the dates of religious feasts, a fact which resulted in a profusion of styles. The present system in general use was prescribed by Pope Gregory XIII in 1582. The Old-Style calendar in use in Britain prior to the introduction of the Gregorian Calendar in 1752, was the Julian Calendar which was invented by an Alexandrian astronomer named Sosigenes and introduced by Julius Caesar after whom it is named.

The Julian year contains $365\frac{1}{4}$ days. The 365 days were divided into 12 months each having an integral number of days. The extra quarter-days were allowed to accumulate to form a whole day which was intercalated into the month of February at four-year intervals. Thus, every fourth year contained 366 days compared with 365 days which formed the so-called Common years. Years containing 366 days were called Leap Years or Bissextile Years, and the intercalated day was called the Bissextus. Leap Years were so called because, in such a year, any given date does not advance one day on the date of the corresponding day of the preceding year, as in Common Years, but "leaps" over the additional day. Leap Years are those whose numbers are exactly divisible by four.

It is important that a calendar should be such that the seasons recur on the same dates of successive years. That this was not the case with ancient calendars resulted in them falling into disuse.

The ideal calendar year is the time taken by the Earth to revolve once in its orbit with respect to the equinoxes. This interval is the Tropical Year. The Julian Calendar was devised on the assumption that the year contained exactly 365 days 06 hours. It was, therefore, in error to the extent of about 11 minutes per year. The accumulation of this error over a period of 400 years amounts to about 72 hours or 3 days. The Gregorian Calendar took into account this error; and, accordingly, dropped 3 days every 400 years. Thus, leap years in our present calendar are those whose numbers are divisible by 4, *except* certain initial years of centuries. The initial years of centuries are not leap years unless the first two numbers of the year number are exactly divisible by 4. Thus, the years 1800 and 1900 were not leap years, but the year 2000 will be a leap year. The Gregorian Calendar is not perfect, although its error amounts to only about 2 days in 6000 years.

The first year of the Julian Calendar was made unduly long and it was known as the Year of Confusion. But the Roman writer Macrobius referred to it as the Last Year of Confusion. The first Julian Year commenced on the day of the first New Moon following the Winter Solstice of the year.

The year 1752, when the Gregorian Calendar was adopted in Britain, was made several days short—the date jumping from the 3rd to the 14th of September. This brought the next Spring Equinox to the 21st of March. It is interesting to reflect that at so recent a date—only about 8 generations ago—there were folk who genuinely thought that by altering the calendar at the time, they were being deprived of 11 days of their lives.

The ancient Roman Calendar was essentially a Lunar Calendar. The months commenced on the days of the New Moon. These days were known as the Calends, and the days of Full Moon, which occurred on the 14th or 15th day of each month, were known as the Ides.

The Moon is still used in the Ecclesiastical Calendar. The phase of the Moon is used in connection with the calculation of the date of Easter, which is the most important date in the Christian Calendar. It is from the date of Easter that the other important Christian feast days are calculated. Easter Day, in general, falls on the first Sunday after the Full Moon which follows the Spring Equinox. Easter, therefore, must fall between March 21st and April 27th. The date of the appropriate Full Moon is calculated from the Metonic Cycle, and it is sometimes different from the date of the astronomical Full Moon.

The Metonic Cycle is named after the Athenian astronomer Meton who first discovered it. It is a period of about 19 years after which the Moon's phases recur on the same day of the Solar Year. The number of the year in the Metonic Cycle is known as the Golden Number. The Golden Number may be found by adding 1 to the year number and dividing the result by 19: the remainder is the Golden Number. This is used in conjunction with the Sunday or Dominical Letter, for determining the date of Easter. If the days of the week are lettered from A to G starting with A on the first of January, the letter for Sunday will change each year. The letter is called the Sunday or Dominical Letter. Leap Years have two Sunday Letters: one for the period up to the Leap Year Day, and the other for the remainder of the year.

Exercises on Chapter 29

1. Describe how the common units of time, namely the Day and the Year, are established.
2. What angles on the celestial sphere correspond to Solar Time and Sidereal Time at a given instant?
3. When do (*a*) the Solar Day, and (*b*) the Sidereal Day, commence?
4. Why is the length of the Apparent Solar Day not constant?
5. Explain why a Solar Day is longer than a Sidereal Day. Prove that the number of Sidereal Days in a year is exactly one more than the number of Solar Days in the year.
6. Explain clearly the derivation of a Mean Solar Day.
7. Explain the effect on daily life were Sidereal instead of Solar Time used as the basis of everyday time-keeping.
8. What is the Equation of Time? Prove that the Equation of time (*e*) is given by:

$$e = \text{H.A.M.S.} - \text{H.A.T.S.}$$

9. Explain why it is necessary to know the G.M.T. and the L.M.T. for a given instant in order to find Longitude.
10. Explain each of the two components which combine to form the Equation of Time.
11. Compare the lengths of the Mean Solar Day and the Sidereal Day.
12. Explain how time is kept on board ship.
13. Explain Zone Time. If a ship is in Longitude 132° E. and her Zone Time is 1300 hr. show that the G.M.T. at the same instant is 0400 hr.

TIME

14. What is meant by Standard Time?
15. Explain the Calendar in current use.
16. Which of the following are not Leap Years: 1600, 1800, 1904, 1944, 1972? (*Answer*—1800).
17. Prove that: G.H.A.★ − L.H.A.★ = Longitude of observer, where ★ denotes any heavenly body.
18. Compare the lengths of the Sidereal Year, the Tropical Year and the Anomalistic Year, in units of Mean Solar Time.
19. Show that when a Sidereal Clock and a Mean Solar Clock, both set and keeping perfect time, register 00 h. 00 m. 00 s., that the date is September the 23rd.
20. Show that when a Mean Solar Clock registers 12 h. 00 m. 00 s. at the same instant as a Sidereal Clock registers 13 h. 00 m. 00 s., the date is October 23rd, approximately.
21. Describe the motions of the Dynamical Mean Sun and the Astronomical Mean Sun.
22. Show that the Equation of Time is zero on four days each year.
23. What effect has the equation of time on the lengths of the forenoon and afternoon?
24. Write a brief account of U.T.C.

CHAPTER 30

THE MOON

1. The Moon and its Motions

The Earth's satellite, the Moon, is the closest celestial body to the Earth. Its diameter is about 2000 miles, this being about a quarter of the Earth's diameter. The Moon has the distinction among the satellites of the Solar System of being the one nearest in size to its parent planet.

The Moon is often said to revolve around the Earth. More strictly the Earth and the Moon revolve around their common centre of gravity. This point, known as the Barycentre, lies on the straight line joining the centres of the Earth and Moon, and is located at a point about 1000 miles within the Earth.

The Moon's orbit is elliptical: its nearest approach to the Earth is about 222,000 miles, and its most remote point from the Earth lies at about 253,000 miles from the Earth. The names given to the points of nearest approach and greatest distance are Perigee and Apogee respectively.

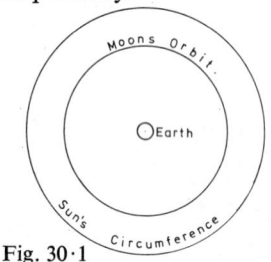

Fig. 30·1 illustrates the relative dimensions of the Earth, Moon's Orbit, and the Sun. It is interesting to note that the diameter of the Sun is considerably greater than the diameter of the Moon's orbit.

As the Moon moves in its orbit it describes a great circle on the celestial sphere against the background of the fixed stars. It moves about 13° per day to the eastwards. This comparatively rapid motion may easily be observed in a relatively short space of time. In an hour, for example, the Moon moves through an arc of the sky, relative to the stars, approximately equal to its angular diameter.

Fig. 30·1

To complete a circuit around the celestial sphere with respect to the stars, the Moon takes about 360°/13 days. The actual time is $27\frac{1}{3}$ days, and this period is known as a Sidereal Period.

The easterly motion of the Moon relative to the stars causes the Moon to rise, culminate and set, later each day to the extent of an average interval of 50 minutes of time. This interval is known as the Retardation of the Moon's Rising and Setting.

The plane of the Moon's orbit is inclined at an angle of about $5\frac{1}{4}°$ to the plane of the ecliptic. The points on the celestial sphere where the two planes intersect are referred to as the Nodes of the Moon's Orbit. The node at which the Moon is located when it crosses from south to north of the ecliptic, in so doing changing her Celestial Latitude from South to North, is known as the Ascending Node. The other node is called the Descending Node.

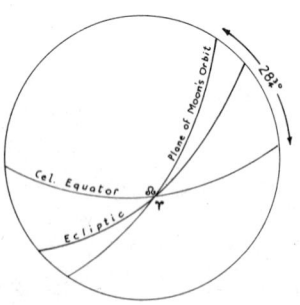

Fig. 30·2

218

THE MOON

The straight line, known as the Nodal Line, which joins the nodes of the Moon's orbit, has a retrograde motion along the ecliptic similar to the precession of the equinoxes. The nodes perform a complete revolution around the celestial sphere in 18·6 years. This is an important eclipse period known as the Saros. The effect of the nodal motion is that the limits of the Moon's declination in any given sidereal period change periodically with a period of 18·6 years. The limits of the Moon's declination during a given sidereal period are dependent upon the relative positions of the Nodes and the Equinoxes.

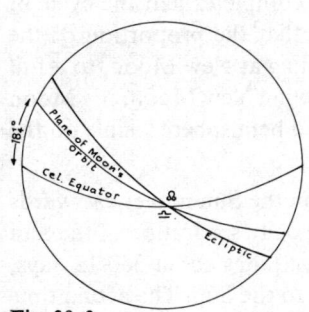

Fig. 30·3

Figs. 30·2, 30·3 and 30·4, illustrate the varying maximum values of the declination of the Moon during particular sidereal periods.

When the Ascending Node is at the Spring Equinox, as illustrated in fig. 30·2, the Moon's declination varies between $(23\frac{1}{2}+5\frac{1}{4})°$, that is $28\frac{3}{4}°$ N. and S. during the sidereal period.

When the Ascending Node coincides with the Autumnal Equinox, as illustrated in fig. 30·3, the Moon's declination varies between $(23\frac{1}{2}-5\frac{1}{4})°$, that is, $18\frac{1}{4}°$ N. and S.

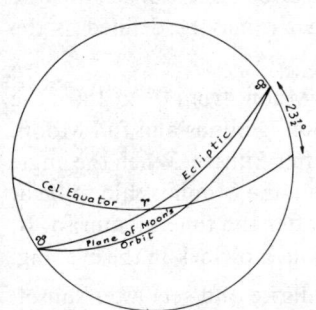

Fig. 30·4

When the nodes coincide with the Solstitial Points, as illustrated in fig. 30·4, the Moon's declinational limits are the same as those of the Sun, that is to say they are $23\frac{1}{2}°$ N. and S.

It is to be noted that, whatever may be the relative positions of the Moon's Nodes and the Equinoxes, the average rate of change of the Moon's declination is very rapid.

2. The Phases of the Moon

The Moon is rendered visible by reflected sunlight. The proportion of the Moon's illuminated hemisphere, visible at the Earth at any time, depends upon the relative positions of the Earth, Moon and Sun. The changing shapes of the part of the Moon's surface visible at the Earth are known as the Phases of the Moon.

When the Moon is in conjunction with the Sun, its illuminated hemisphere is directed away from the Earth. In this circumstance no part of the Moon's illuminated hemisphere is visible at the Earth. At this time the Moon and the Sun cross an observer's celestial meridian at the same instant of time. It follows that the time at which the Moon is in conjunction with the Sun must be 12 o'clock L.M.T. When this occurs the Moon is said to be New or at the Change, and its Age is said to be 00 d. 00 h. 00 m. 00 s.

When the Moon is in opposition with the Sun, its illuminated surface is facing the Earth. The whole of the Moon's illuminated hemisphere is, therefore, visible at the Earth. At this time the Moon appears as a disc of light and its phase is said to be Full. The Full Moon occurs when the Sun is on an observer's lower celestial meridian, the Moon being on the observer's upper meridian at the time. Full Moon, therefore, occurs at Midnight L.A.T. The angle between the Moon and the Sun, measured in the plane of the ecliptic, at the time of Full Moon, is 180°, so that the Moon rises at the time the Sun sets, and sets at the time the Sun rises.

From the time of New Moon to that of Full Moon the Moon completes half the cycle of her phases. During this time the Moon is said to Wax, meaning that the proportion of the Moon's illuminated surface visible at the Earth increases from nothing at New Moon, to a full disc at Full Moon. From the time of Full Moon to that of the following New Moon, the Moon is said to Wane, for the reason that the proportion of its illuminated hemisphere visible on the Earth diminishes from maximum to zero.

It will be remembered that because of the Earth's orbital motion the Sun moves eastwards across the celestial sphere at the rate of about 1° per day. Thus, the daily separation of the Sun and Moon amounts to about 12° per day. It follows that the Moon takes about 360/12 days, approximately, to complete a circuit of the celestial sphere relative to the Sun. The actual time is about $29\frac{1}{2}$ days—an interval known as a Lunation or Synodic Period. The Lunation is the time taken for the Moon to complete a cycle of its phases, so that it may be defined as the interval of time between two successive New Moons.

As the angle between the Moon and Sun increases after New Moon, from 0° to 180°, the Moon's visible shape appears successively as a crescentic, half Moon, gibbous and full Moon. During this half lunation the Moon rises after Sunrise and sets after Sunset. When the angle between the Moon and Sun is 90°, during the first half of a lunation, the Moon is said to be at the First Quarter. At this time the Moon rises at about 6 hours after the time of sunrise. It, therefore, crosses the upper celestial meridian of an observer at about 6 o'clock in the evening.

During the second half of a lunation the Moon rises before sunrise and sets after sunset. When the Moon is midway between Full and New, that is to say when it is at quadrature during the second half of the lunation, it is said to be at the Third Quarter. At this time the Moon rises at about six hours before sunrise, so that it crosses an observer's upper celestial meridian at about 6 o'clock in the morning.

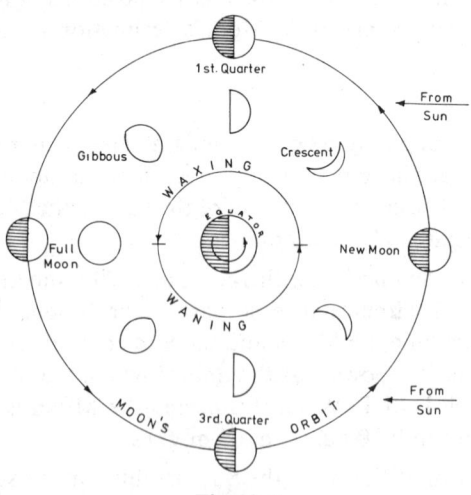

Fig. 30·5

The phases of the Moon are illustrated in fig. 30·5.

3. The Age of the Moon

The Age of the Moon is the interval of time that has elapsed since the last New Moon.

Age at New Moon = 0 days exactly
Age at 1st Qr. = 7 days approximately
Age at Full Moon = 15 days approximately
Age at 3rd Qr. = 22 days approximately

The Age of the Moon on the 1st of January is known as the Epact for the year. The Epact is used in the computation of the date of Easter. Because 12 lunations amount to 354 days, which is about 11 days short of the year, the Epact increases by 11 on successive years.

4. Winter and Summer Full Moons

Because the Sun moves in the ecliptic, successive Full Moons take place in different parts of the celestial sphere. This has an effect on the duration of Moonlight.

In Summer, when the Sun has northerly declination, the Full Moon has southerly declination. In the northern hemisphere celestial bodies which have south declination are above the horizon for less than half a day. Therefore the Summer Full Moons are below the horizon for more than half the day.

In Winter, when the Sun has southerly declination, the Full Moons have northerly declination. They are, therefore, above the horizon for more than half the day. During the long Winter nights of the northern hemisphere the Full Moon is said to "ride high in the sky".

5. Spring and Autumn Full Moons

In Spring in the Northern Hemisphere, when the Sun is near the First Point of Aries, the Full Moons are near the First Point of Libra. In Autumn, when the Sun is near Libra, the Full Moons are near Aries. In September, the Moon's declination changes most rapidly during the lunation because it crosses the ecliptic.

In Spring, when the Sun's declination changes from South to North, the Full Moon's declination changes from North to South. This has the effect of speeding up the time of Moonset. It will be remembered that celestial bodies which have South declination are above the horizon of an observer in the northern hemisphere for less than half the day, and that the greater is the South declination the shorter is the period the body is above the horizon.

In Autumn the Full Moon's declination changes from South to North as the Sun's declination changes from North to South. The time of the setting of the Full Moon is, therefore, advanced. The retardation of the Moon's rising and setting, therefore, is offset by the northerly change in its declination. Therefore, in the Autumn, the Moon when near Full rises only slightly later on successive days. The interval between the times of sunset and moonrise is relatively small for several days near the time of Full Moon. Hence, before darkness sets in, the large Moon rises, and the reflected sunlight assists the northern farmers in the labour of the harvest. For this reason the Full Moon occurring nearest to the time of the Autumnal Equinox is called the Harvest Moon.

A similar phenomenon occurs during the few days on each side of the day of Full Moon following the Harvest Moon. At this time of the year, the harvest having been gathered, the sporting activities of the farmers are assisted by moonlight, and the Full Moon following the Harvest Moon is called the Hunter's Moon.

6. Earth-Shine

The Earth as observed from the Moon passes through phases which recur every $29\frac{1}{2}$ days exactly as the Moon's phases. When the Moon is New at the Earth, the Earth is "New" at the Moon. At the time of Full Moon an observer on the Moon would observe "Full Earth". The Earth being a better reflector, area for area, than the Moon, and because of the Earth's atmosphere, the Earth is a magnificent spectacle to the intrepid space travellers who orbit the Earth or who land on the Moon.

When the Moon is crescent-shaped the portion of its surface which is not illuminated is facing the Earth. This dark hemisphere of the Moon, on those occasions when the sky is cloud-free and clear, may be observed to be slightly illuminated by sunlight which has been reflected from the Earth, and re-reflected from the Moon's surface. This illumination is called Earth-Shine, and the spectacle gives rise to the phenomenon, due to irradiation, known as "The Old Moon lying in the New Moon's arms".

7. Moon's Librations

The Moon rotates on its axis once in a sidereal period, and the speed of this rotation is virtually uniform. Because of this, the same face of the Moon is always presented to the Earth.

Now although the rotation of the Moon takes place at a uniform rate, that of the Moon's orbital motion is not uniform. In obedience to Kepler's Second Law the Moon travels fastest when it is at perigee and slowest when it is at apogee. The average orbital speed is exactly equal to the uniform rotation speed. Thus, at perigee, when the orbital speed is greater than the axial speed, a narrow strip of the Moon's surface beyond its normal western edge heaves into view. At apogee the reverse is the case and, in consequence, a part of the Moon's surface beyond its normal eastern edge becomes visible to terrestrial observers.

The Moon's axis of rotation is inclined to the plane of its orbit at an angle of about 84°. As the Moon revolves in its orbit, therefore, a continually changing aspect is presented to terrestrial observers. The Moon seems to "nod" to and fro once in a sidereal period. This allows observers on the Earth to observe parts of the lunar surface extending to as much as 6°—the complement of the angle of inclination of the Moon's spin axis to the plane of the Moon's orbit—over the Moon's poles.

These apparent irregularities in the Moon's motions are known as Librations. They result in about 59% of the Moon's surface being observable from the Earth.

8. Eclipses

An eclipse of the Sun occurs when the Moon lies on the straight line joining the Earth and Sun. Because the Moon and the Earth are opaque bodies they cast shadows away from the direction of the Sun. At a Solar Eclipse the shadow of the Moon falls on the Earth.

An eclipse of the Moon occurs when the Moon passes through the shadow of the Earth.

Were the Earth's and Moon's orbits co-planar there would be an eclipse of the Sun at every New Moon and an eclipse of the Moon at every Full Moon. But the plane of the Moon's orbit is inclined at an angle of about $5\frac{1}{4}°$ to the plane of the Earth's orbit. The declinations of the New and Full Moons are not, therefore, usually the same in magnitude as that of the Sun, so that eclipses of the Sun do not take place at every New Moon, and eclipses of the Moon do not take place at every Full Moon.

THE MOON

For an eclipse to occur the Moon must lie on the ecliptic: hence the name given to the great circle on the celestial sphere which is co-planar with the Earth's orbit. When the Moon is on the ecliptic it is at one of its nodes. So that for an eclipse to occur the Moon must be at or near a node and it must lie on the straight line joining the Earth to the Sun.

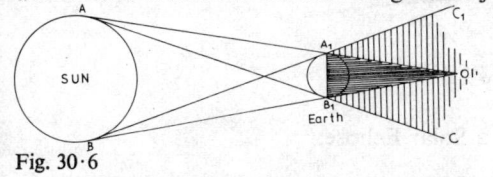

In fig. 30·6 the two tangents AA_1 and BB_1 meet at the point O, and the cone of section A_1OB_1 is a region within which no sunlight enters. This dark shadow cone is known as the Umbra.

Fig. 30·6

The tangents BA_1 and AB_1 mark out the plane section of the cone within which partial sunlight enters. This region is known as the Penumbra.

If the Moon enters the penumbra it is so slightly obscured that it is hardly noticeable. But if the Moon enters the umbra it is eclipsed. Sometimes the Moon does not pass entirely into the umbra, in which case a Partial Eclipse occurs. Should the Moon pass completely into the umbra a Total Eclipse occurs.

During a total lunar eclipse the Moon is sometimes visible as a dull red disc—like a blood-drop hanging in the night sky. This is due to sunlight being refracted on passing through the Earth's atmosphere, and the red constituents of which, being refracted to a smaller degree than the higher frequency constituents, impinge upon and illuminate the eclipsed Moon.

A lunar eclipse is visible at all places which have the Moon above the horizon during the time it is in the Earth's shadow.

Because the Moon and the Sun have approximately the same angular diameters, the length of the Moon's shadow cone is roughly equal to the radius of the Moon's orbit. Thus, during a solar eclipse, the apex of the Moon's shadow cone sweeps out a narrow belt on the Earth's surface within which the eclipse is visible. Because of the eccentricities of the Earth's and Moon's orbits, it sometimes happens that the apparent diameter of the Moon is greater than that of the Sun. This is a case of the Earth being near aphelion at the same time as the Moon is near its perigee. A solar eclipse occurring in these circumstances is known as a Total Solar Eclipse, in which the whole of the Sun's disc is obscured during the period of totality. An eclipse of the Sun which occurs when the Moon's apparent diameter is smaller than that of the Sun's, is known as Annular Eclipse. During an annular eclipse a narrow ring, or annulus, of the Sun's surface is visible during the eclipse. Such an eclipse occurs when the Moon is near apogee at the same time the Earth is near perihelion.

An eclipse of the Sun is visible only within a narrow strip of the Earth's surface, the width of the strip depending upon the relative distances of the Moon and Sun. The strip is never more than about 170 miles in width. Within the penumbra, which may extend over a circular area of the Earth of radius 2000 miles, a Partial Eclipse of the Sun occurs.

Fig. 30·7 illustrates a Total Solar Eclipse.

Fig. 30·7

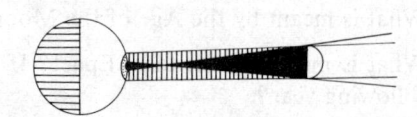

Fig. 30·8

Fig. 30·8 illustrates an Annular Eclipse.

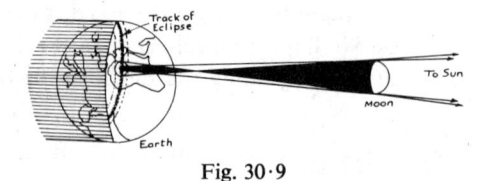

Fig. 30·9

Fig. 30·9 illustrates the track on the Earth of a Solar Eclipse.

9. Occultations

During the Moon's monthly circuit of the heavens it frequently passes over stars, and sometimes over planets. When this happens an Occultation takes place. For an occultation to occur the Moon and the occulted body must have the same celestial position.

It is interesting to observe an occultation of a star using the long glass or a good pair of binoculars. It will be seen that the star vanishes abruptly behind the edge of the Moon and reappears on the western edge, just as abruptly as it disappeared at the eastern edge some minutes before. The fact that a star disappears abruptly when being occulted by the Moon serves to show that the Moon is devoid of an atmosphere.

An occultation of a planet by the Moon is not nearly so dramatic as a star occultation: because of the planet's appreciable angular diameter it fades out gradually on being occulted.

Very occasionally a planet and a star occupy the same position in the celestial sphere. When this happens the star is said to be occulted by the planet.

Exercises on Chapter 30

1. Distinguish between a Sidereal Period and a Synodic Period of the Moon.
2. Explain a Lunation.
3. Describe the phases of the Moon and explain their cause.
4. Explain why the maximum declination of the Moon varies from one lunation to another.
5. Define Ascending Node and Descending Node.
6. In what circumstances would the Moon have a maximum declination during the lunation of $28\frac{3}{4}°$ N. and S.?
7. At what latitude would an observer be able to see the Moon on the horizon at lower meridian passage at a time when the nodes coincide with the Solstices?
8. Explain clearly why it is that the time of Moonrise occurs later each day by an amount known as the Retardation of the Moon.
9. What is meant by the Age of the Moon?
10. What is meant by the term Epact? If the epact is 4 on a certain year what is it on the following year?
11. Explain the expression: "The Full Moon rides high in Winter".

THE MOON

12. Explain the expression: "The Old Moon lies in the New Moon's arms".
13. Explain the Harvest Moon.
14. Explain Moon's Librations and indicate their effects.
15. Explain the causes of eclipses.
16. What conditions are necessary for a Solar Eclipse to occur?
17. What conditions are necessary for a Total Solar Eclipse, and for an Annular Eclipse of the Sun to occur?
18. What conditions are necessary for the period of totality of a Solar eclipse to be maximum?
19. What type of eclipse occurs if the Sun's S.H.A. and declination are equal to those of the Moon, and the angular diameters of the Moon and Sun are 15·9′ and 16·2′ respectively.
20. Explain the appearance of the Moon during some total lunar eclipses.
21. Define Occultation. Describe the occultation of a planet by the Moon.

PART 5

NAUTICAL ASTRONOMY

Nautical Astronomy deals with the problems of finding a vessel's position when out of sight of land using astronomical principles.

The astronomical data required for solving the problems of Nautical Astronomy are contained in the *Nautical Almanac*, a very important instrument of nautical astronomy and one with which the navigator should be thoroughly familiar.

The basis of Nautical Astronomy is the relationship between a celestial body's position using co-ordinates of the horizon system with the body's position at the same instant of time using the co-ordinates of the celestial equatorial system.

The altitude of a celestial body, which is a co-ordinate of the horizon system, is measured by means of a sextant. The declination of an observed body, which is a celestial equatorial co-ordinate, may be lifted from the *Nautical Almanac*. Altitude, declination and Latitude, are functions, respectively, of each of the three sides of a celestial triangle known as the Astronomical- or *PZX*-Triangle. In the general nautical astronomical problem a *PZX*-triangle has to be solved. The chapters in Part 5 of this book deal with the several aspects of this important branch of the practical work of a navigator.

The fundamental feature of modern Nautical Astronomy is the Astronomical Position Line. An astronomical position line is found from an astronomical observation or Sight, as such an observation is familiarly called. In by-gone days astronomical observations were made specifically for finding Latitude or Longitude. Nowadays, they are made essentially to determine astronomical position lines.

The astronomical methods of finding Latitude at sea have been used since the time when the Portuguese navigators of the Great Age of Discovery first applied astronomical methods for position-finding when out of sight of land.

Compared with the problem of finding Latitude at sea that of finding Longitude in early times was one of considerable complexity and difficulty. It was not until the mid-eighteenth century that a satisfactory solution of the Longitude problem was discovered. Two methods for finding Longitude at sea became available at about the same time. The method most familiar to modern navigators requires the use of an accurate mechanical timekeeper. The first successful Chronometer, which is the name given to an accurate watch used for finding Longitude at sea, was the invention of an eighteenth-century horologist named John Harrison. Although Harrison's chronometer made its appearance during the early part of the eighteenth century, almost a century was to pass before these instruments were readily available on the score of expense. During this long period the second of the two general methods for finding Longitude at sea, known as the Lunar Method, was the more commonly used method.

The obsolete lunar method required the comparison of a measured Lunar Distance with predicted Lunar Distances given in the *Nautical Almanac*. The term Lunar Distance applies to the angle contained between the Moon and the Sun or some other celestial body lying in or near to the plane of the ecliptic. The early *Nautical Almanacs* were designed essentially for the lunar method of finding Longitude at sea. Since the early part of the present century, when predicted lunar distances ceased to be given, the *Nautical Almanac* has served essentially for the purpose of solving *PZX*-triangles.

G.M.T. is available to a navigator from his chronometer, the error of which may be obtained from Radio Time Signals transmitted at specified times throughout the day. The principal tables in the *Nautical Almanac* enable a navigator to find the declination and G.H.A. of any navigational celestial body for any given instant of G.M.T. We shall see in the following chapters how position lines are obtained from the remarkable and unfailing methods of Nautical Astronomy.

CHAPTER 31

FINDING THE TRUE ALTITUDE

1. The Altitude

The True Altitude of a celestial body is an arc of a vertical circle contained between the body and the celestial horizon. A true altitude is found by applying certain Altitude Corrections to a measured altitude obtained by means of a sextant. The Sextant is the nautical astronomer's instrument for measuring arcs or angles. In Nautical Astronomy the principal arcs measured by means of a sextant are arcs of vertical circles.

The measured altitude is referred to as the Sextant Altitude. If the sextant possesses Index Error (see Chapter 43), the sextant altitude must be corrected by applying a correction for the index error to give an angle known as the Observed Altitude. This is illustrated in fig. 31·1.

The Observed Altitude of a celestial body is defined as the angle at the eye of the observer contained between the apparent directions, respectively, of the body and the visible horizon measured in the plane of the vertical circle on which the body lies. The distance of the visible horizon depends upon the height of the observer's eye above the level of the sea.

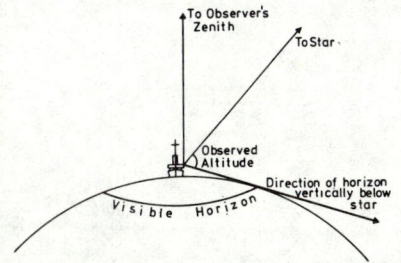

Fig. 31·1

The Visible Horizon is a small circle on the Earth's surface which bounds an observer's view in the open sea. It may be defined as a circle every point on which the sea meets the sky. The radius of the visible horizon, as seen in fig. 31·2, increases as the height of the observer's eye increases.

In fig. 31·2, B's visible horizon is a larger circle than A's visible horizon. This follows because B's eye is elevated to a greater extent than A's.

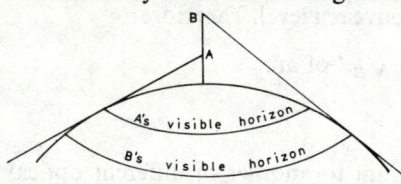

Fig. 31·2

An observer whose eye is at sea level has no visible horizon. The part of the celestial sphere which is visible to such an observer is bounded by a small circle on the celestial sphere which is parallel to the observer's celestial horizon, and on whose plane the observer's eye lies. This is called the observer's Sensible Horizon. It will be remembered that the celestial horizon is a great circle on the celestial sphere every point on which is 90° from the observer's zenith. The Earth's centre lies on the observer's celestial horizon.

2. Dip

The direction in any given vertical plane of the visible horizon is depressed below the direction of the sensible horizon in the same vertical plane by an angle which increases as the observer's height of eye increases. This angle of depression is known as Dip.

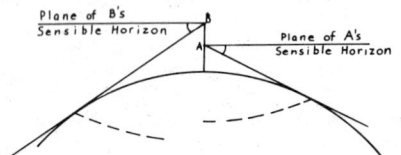

In fig. 31·3 it will be seen that the dip of B's visible horizon is greater than that of A's. This follows because B's height of eye is greater than A's.

Ignoring the effect of atmospheric refraction the dip of the visible horizon in minutes of arc is equal to the distance of the visible horizon in nautical miles. This is proved with reference to fig. 31·4.

In fig. 31·4:

AB is the vertical height of the observer's eye above sea level.

AV is the distance of the observer's visible horizon in nautical miles.

arc $AV = \theta =$ Distance of visible horizon in miles

$SBV =$ Dip in minutes

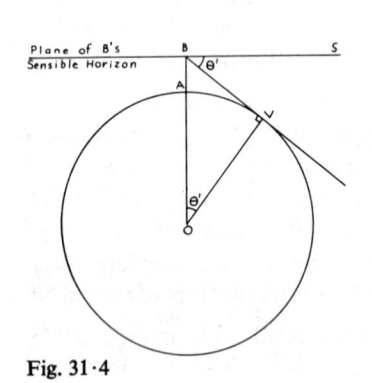

In Triangle BOV:

$BVO = 90°$

$BOV = \theta'$

$OBV = (90° - \theta')$

But:

$SBO = 90°$

Therefore:

$SBV = \theta'$

Thus:

Dip of visible horizon in mins. = Distance of visible horizon in mls.

In Part 3, Chapter 19, we proved that the distance of the theoretical horizon is equal to $1·06 \sqrt{h}$ miles, where h is the height of the observer's eye above sea level. Therefore:

Dip of Theoretical Horizon $= 1·06 \sqrt{h}\,'$ of arc.

3. Refraction

The path of a ray of light, in passing from one medium to another of different optical density, changes its direction by an angle known as Refraction. Refraction depends upon the relative optical densities of the two media and also upon the angle which the path of the ray of light makes with the common surface of the two media. When the ray strikes the surface normally; that is to say, when the angle of incidence is 90°, refraction is zero. Refraction increases as the angle of incidence decreases.

A ray of light from a celestial body, in passing through the Earth's atmosphere, is refracted so that the path of the ray of light is a slight curve. The direction of this curved path at the surface of the Earth is such that the altitude of a celestial body is apparently greater than it would be if refraction did not exist.

FINDING THE TRUE ALTITUDE

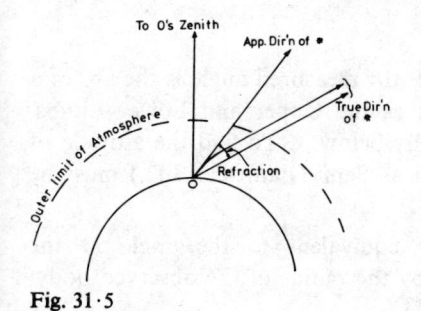

Fig. 31·5 serves to illustrate that Atmospheric Refraction is equal to the angle between the true direction of a celestial body and its apparent direction.

The value of refraction depends upon the altitude of the celestial body. It is greatest for a body whose altitude is zero; that is to say, for a body on the horizon. Its maximum value is about 33' of arc. Refraction of light from a body at the zenith; that is to say, a body whose altitude is 90°, is zero. This follows because the ray of light which enters the observer's eye from such a body, strikes the atmosphere normally.

Fig. 31·5

Refraction is affected by changes in temperature and pressure of the atmosphere. In most collections of Nautical Tables, a table giving Mean Refraction is given for a so-called Standard Atmosphere in which the sea level temperature and pressure are assumed to have specified values. In addition to this table another is given in which are tabulated corrections to be applied to the Mean Refraction for temperatures and pressures different from those for which the Mean Refractions are computed.

Certain atmospheric and sea conditions may give rise to Abnormal Refraction. If this is suspected, results of observations of celestial bodies should be treated with caution.

4. Effect of Refraction on Dip

A ray of light from the sea surface at the visible horizon is refracted such that the distance of the visible horizon is slightly greater than that of the theoretical horizon. The distance of the visible horizon, when atmospheric refraction is normal, is about one-thirteenth of the theoretical distance greater than the theoretical distance. This gives $1 \cdot 15 \sqrt{h}$ for the distance of the visible horizon. The actual dip of the visible horizon is less than the theoretical dip. When refraction is normal the dip of the visible horizon is $0 \cdot 98 \sqrt{h}$'s of arc. The effect of refraction on dip and on distance of the horizon is illustrated in fig. 31·6.

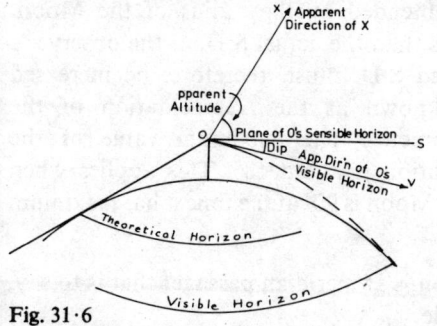

Tables of Distance and Dip of the visible horizon are calculated from the formulae stated above and are inserted in most collections of Nautical Tables. A dip table is also to be found in the *Nautical Almanac*.

The Observed Altitude of a celestial body, reduced by the dip of the visible horizon, gives the Apparent Altitude of the body. The Apparent Altitude of a celestial body is defined as the angle at the observer's eye contained between the apparent direction of the body and the plane of the observer's sensible horizon measured in the plane of the vertical circle on which the body lies.

Fig. 31·6

In fig. 31·6 *XOV* is the observed altitude of the celestial body which lies in the direction indicated. *SOV* is the actual dip of the visible horizon, and *XOS* is the apparent altitude of the body.

5. Semi-Diameter

When the altitude of the Sun or the Moon is observed, the measured angle is the arc of a vertical circle between the top or bottom edge, known as the Upper and Lower Limbs, respectively, of the body and the vertical horizon vertically below it. To find the altitude of the centre of the Sun or the Moon, a correction known as Semi-Diameter (S.D.) must be applied to the altitude of the limb.

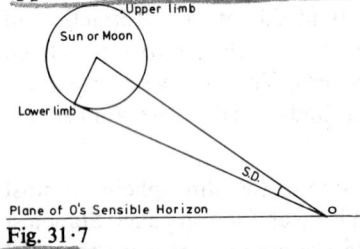

Semi-diameter is equivalent to the angle at the observer subtended by the radius of the observed body, as illustrated in fig. 31·7.

It is to be noted that when the lower limb is observed, the S.D. correction is to be added to the altitude of the limb; and that when the upper limb is observed the S.D. correction is to be subtracted from the altitude of the limb.

Fig. 31·7

The S.D. of both the Sun and Moon may be found from the daily pages of the *Nautical Almanac*. The Sun's S.D. varies between 15'·8 and 16'·3 during the course of the year: it is least when the Earth is at aphelion and greatest when the Earth is at perihelion, the dates of these events being, respectively, early July and early January. The Moon's S.D. varies between about 14'·7, when the Moon is at apogee, and about 16'·7 when the Moon is at perigee.

6. Augmentation of the Moon's Semi-Diameter

Because of the Moon's comparative nearness to the Earth, the distance of the Moon from an observer decreases appreciably as the Moon's altitude increases. This is illustrated in fig. 31·8.

Because OA in fig. 31·8 is greater than OB, the Moon's S.D. is greater when the Moon is at B than it is when the Moon is at A. The value of the Moon's S.D. tabulated in the *Nautical Almanac*, is the angle at the Earth's centre subtended by the radius of the Moon. This is always less than the actual S.D. at the observer's eye. The tabulated S.D. must, therefore, be increased by an amount known as the Augmentation of the Moon's Semi-diameter. The maximum value of the Moon's augmentation is about 18″. This applies when the altitude of the Moon is 90° at the time it has maximum S.D.

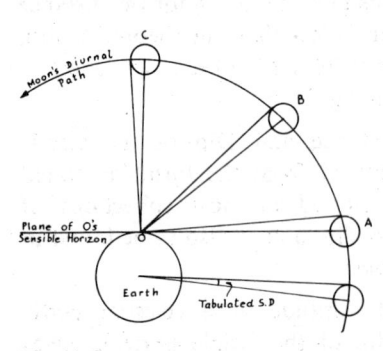

Fig. 31·8

The greatest S.D. for the day occurs when the Moon is at meridian passage; that is to say, at the instant when the Moon reaches its greatest altitude.

7. Parallax

Stars are so far distant from the Earth that their directions measured from the Earth's centre are sensibly the same as they are from any point on the Earth's surface.

FINDING THE TRUE ALTITUDE

In fig. 31·9:

Altitude of X above the sensible horizon = XOS

True Altitude of $X = XCH$

Because OX and CX are parallel: $XOS = XCH$

Therefore:

Altitude of X above sensible horizon = Altitude of X above celestial horizon.

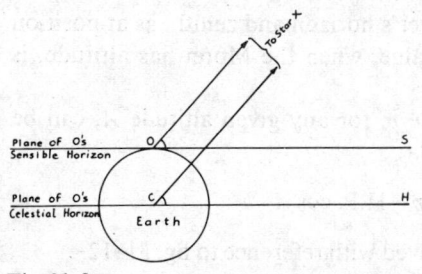

Fig. 31·9

Members of the Solar System, especially the Moon, are comparatively near to the Earth; and the respective directions of these bodies from the Earth's centre and surface are not parallel as is the case with the fixed stars.

The true altitude of any body of the Solar System is always slightly greater than the altitude of the body above the sensible horizon by an angle which depends on:

▸ (i) the distance of the body from the Earth
▸ (ii) the apparent altitude of the body.

The angle by which the true altitude exceeds the altitude of the celestial body above the sensible horizon is known as Celestial Parallax. Parallax may be defined as the angle at the centre of the observed body contained between the respective directions of the observer and the centre of the Earth.

Let M, in fig. 31·10, represent the Moon; O an observer; and C the centre of the Earth.

Altitude of M above sensible horizon = MOS

True altitude = MCH

$= MTS$

Now: $MTS - MOS = OMT$

Therefore:

Alt. above Celestial Horizon − Alt. above Sensible Horizon = Parallax.

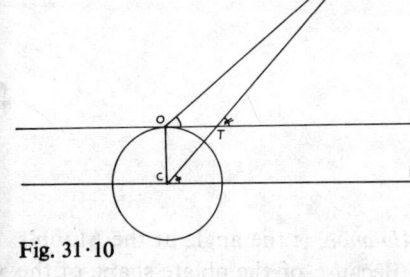

Fig. 31·10

Parallax for a given celestial body is greatest when the body is on the sensible horizon of an observer. The value at this instant is known as the Horizontal Parallax (H.P.). The H.P. of any body may readily be deduced from the Earth's radius and the distance of the body from the Earth.

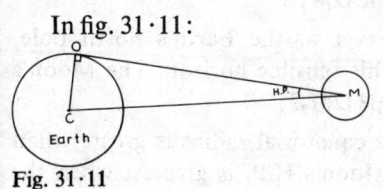

In fig. 31·11:

Radius of the Earth = OC = 4000 ml. approx.

Distance of Moon = CM = 240,000 ml. approx.

$\sin OMC = 4000/240,000$

$= 1/60 = OMC$

Fig. 31·11

Therefore:

Moon's H.P. = $1/60^c$ or $57'·3$.

When the Moon is at any position between an observer's horizon and zenith, as at position B in fig. 31·12, the parallax is less than the H.P. Its value, when the Moon has altitude, is known as Parallax-in-Altitude.

Parallax-in-Altitude p, for any given altitude A, can be found from the formula:

$$p = \text{H.P.} \cos A$$

This formula is proved with reference to fig. 31·12.

By the Plane Sine Formula:

$$\sin OBC / OC = \sin COB / CB$$
$$\sin OBC = OC/CB \cdot \sin COB$$
$$= OC/CB \cdot \sin(90° + A)$$
$$= OC/AC \cdot \sin(90° + A)$$

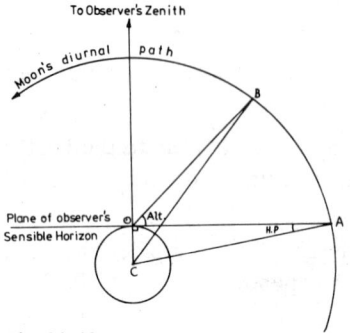

and: $\sin p = \sin \text{H.P.} \cos A$

Fig. 31·12

Since p and H.P. are small angles, this formula reduces to:

$$p' = \text{H.P.}' \cos A$$

in which A denotes the altitude above the plane of the sensible horizon.

Example 31·1—Find the parallax-in-altitude of the Moon if its apparent altitude is 60° 00′, and its H.P. is 60′·2.

$$p = \text{H.P.} \cos A$$
$$= 60'·2 \cos 60°$$
$$= 60'·2 \times \tfrac{1}{2}$$
$$= 30'·1$$

Answer—Moon's parallax-in-altitude = 30′·1.

8. Effect of Earth's Shape on Horizontal Parallax

The H.P. of the Moon, as tabulated in the *Nautical Almanac*, is the angle at the Moon's centre subtended by the equatorial radius of the Earth. Because of the oblate shape of the Earth, the Moon's H.P. for any other Latitude is slightly less than the tabulated value.

In fig. 31·13, E denotes an observer on the equator. M_1 denotes the Moon on the observer's sensible horizon. The Moon's H.P. is angle OM_1E.

P denotes an observer at the Earth's north pole, and M_2 the Moon on his sensible horizon. The Moon's H.P., in this case, is angle OM_2P.

Because the Earth's equatorial radius is greater than her polar radius, the Moon's H.P. is greatest when the Latitude of the observer is 0°, and it decreases as the Latitude increases.

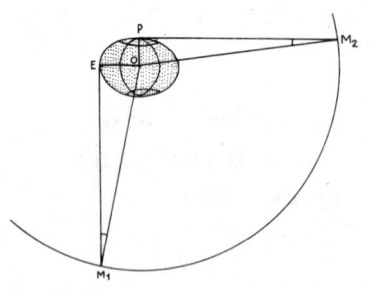

Fig. 31·13

FINDING THE TRUE ALTITUDE

The Reduction of the Moon's Horizontal Parallax for the Figure of the Earth, for all navigable Latitudes, is tabulated in most collections of Nautical Tables. An examination of such a table will show that this reduction is never more than about 12″.

9. Irradiation

Irradiation is a physiological phenomenon in which a bright object viewed against a darker background appears larger than it is; and a dark object viewed against a lighter background appears smaller. The celestial bodies viewed against a relatively dark sky appear slightly larger than they really are because of irradiation. The visible horizon, on the other hand, because it is viewed against the relatively light sky, appears to be depressed by this optical phenomenon.

When the Sun's lower limb is observed the two effects of irradiation of the Sun and the horizon, respectively, tend to neutralize each other, so that no resultant irradiation correction is necessary. When, however, the Sun's upper limb is observed the two effects combine, and the irradiation effect may be as much as a minute of arc or even more. In practice, therefore, the Sun's lower limb is to be preferred when observing the altitude of the Sun.

10. The Correction of Observed Altitudes

In the following examples altitudes are corrected using the Altitude Correction Tables given in Norie's or Burton's Nautical Tables. Many practical navigators use the tables given in the *Nautical Almanac*: readers are advised to familiarize themselves with these excellent tables.

11. Correcting Star Altitudes

Stars are so far distant from the Earth that they appear as mere pinpoints of light with no apparent size. Parallax is nil and so also is semi-diameter. The true altitude of a star is found by applying individual corrections for dip and refraction to the observed altitude.

All collections of Nautical Tables include a Star Total Correction Table which gives the combined values of dip and refraction against arguments Height of Eye and Observed Altitude. This table is invariably used in practice in preference to the individual correction tables.

Example 31·2—Find the true altitude of Aldebaran if its sextant altitude was 30° 21′·5, index error = −1′ 30″, height of eye = 45 feet. Use the individual correction tables.

$$\begin{aligned}
\text{Sextant Altitude} &= 30°\ 21'\cdot 5 \\
\text{Index Error} &= -1'\cdot 5 \\
\hline
\text{Observed Altitude} &= 30°\ 20'\cdot 0 \\
\text{Dip} &= -6'\cdot 6 \\
\hline
\text{Apparent Altitude} &= 30°\ 13'\cdot 4 \\
\text{Refraction} &= -1'\cdot 6 \\
\hline
\text{True Altitude} &= 30°\ 11'\cdot 8
\end{aligned}$$

Answer—True Altitude = 30° 11′·8.

Example 31·3—Find the true altitude of Sirius if the sextant altitude is 45° 38′·0, index error = nil, height of eye = 68 feet. Use the total correction table.

$$\begin{align}
\text{Sextant Altitude} &= 45° \ 38'·0 \\
\text{Index Error} &= 0'·0 \\
\hline
\text{Observed Altitude} &= 45° \ 38'·0 \\
\text{Total Correction} &= -9'·0 \\
\hline
\text{True Altitude} &= 45° \ 29'·0
\end{align}$$

Answer—True Altitude = 45° 29′·0.

12. Correcting Planet Altitudes

The navigational planets are comparatively close to the Earth; but, because of their relatively small size, they are often assumed to be very small discs of light with no appreciable diameters. In practice, therefore, the true altitude of a planet is often found in exactly the same way as that of a star. If, however, great accuracy is required, parallax-in-altitude and a phase correction should be applied, in addition to corrections for dip and refraction. It will not, in general, be possible to apply these additional corrections without the aid of the *Nautical Almanac*.

The phase correction for a planet arises from the centre of the illuminated portion of the planet not coinciding with the centre of the planet's disc. The correction, which normally applies only to Mars and Venus, is a function of the relative positions of the Sun and the planet, and it varies as the cosine of the angle between the vertical circle through the planet and the great circle connecting the celestial positions of the Sun and the planet.

Example 31·4—Find the true altitude of Jupiter if its sextant altitude is 24° 50′·5, index error = +1′·5, height of eye = 30 feet.

$$\begin{align}
\text{Sextant Altitude} &= 24° \ 50'·5 \\
\text{Index Error} &= +1'·5 \\
\hline
\text{Observed Altitude} &= 24° \ 52'·0 \\
\text{Dip} &= -5'·4 \\
\hline
\text{Apparent Altitude} &= 24° \ 46'·6 \\
\text{Refraction} &= -2'·1 \\
\hline
\text{True Altitude} &= 24° \ 44'·5
\end{align}$$

Answer—True Altitude = 24° 44′·5.

13. Correcting Sun Altitudes

The true altitude of the Sun is found by applying corrections for dip, refraction, semi-diameter; and, for upper limb observations, irradiation (which should be assumed to be 1'·0) as well. If great accuracy is required a correction for parallax-in-altitude is also applied; but this correction, being negligibly small, is usually ignored in practice.

Example 31·5—The sextant altitude of the Sun's lower limb was 30° 40'·5. Find the true altitude if the index error is nil, the height of eye is 40 feet, and the date is 20th June.

$$\begin{aligned}
\text{Sextant Altitude} &= 30°\ 40'\cdot 5 \\
\text{Index Error} &= 0'\cdot 0 \\
\hline
\text{Observed Altitude} &= 30°\ 40'\cdot 5 \\
\text{Dip} &= -6'\cdot 2 \\
\hline
\text{Apparent Altitude of L.L.} &= 30°\ 34'\cdot 3 \\
\text{Refraction} &= -1'\cdot 6 \\
\hline
\text{True Altitude of L.L.} &= 30°\ 32'\cdot 7 \\
\text{Semi-diameter} &= +15'\cdot 8 \\
\hline
\text{True Altitude} &= 30°\ 48'\cdot 5 \\
\end{aligned}$$

N.B.—Parallax-in-altitude for 20th June amounts to 0'·1.

If this is applied the true altitude is 30° 48'·6.

Answer—True Altitude = 30° 48'·6.

A Sun's Total Correction Table is provided in Norie's and Burton's Tables. This gives a combined correction for dip, refraction and a constant semi-diameter. The table is entered with arguments Observed Altitude and Height of Eye. To allow for the variation in the Semi-diameter throughout the year an auxiliary table, known as the Monthly Correction Table, is provided.

For upper limb observations the Sun's diameter is first subtracted from the observed altitude, and this is used as an argument, with Height of Eye, in the Sun's Total Correction Table.

Example 31·6—The sextant altitude of the Sun's lower limb on 21st January was 35° 50'·5, index error = −1'·5, height of eye = 50 feet. Find the Sun's true altitude using the total correction table from Norie's or Burton's Tables.

238   THE ELEMENTS OF NAVIGATION AND NAUTICAL ASTRONOMY

$$\begin{aligned}\text{Sextant Altitude} &= 35°\ 50'\cdot 5 \\ \text{Index Error} &= \ -1'\cdot 5\end{aligned}$$

$$\begin{aligned}\text{Observed Altitude} &= 35°\ 49'\cdot 0 \\ \text{Total Correction} &= \ +7'\cdot 9 \\ \text{Monthly Correction} &= \ +0'\cdot 3\end{aligned}$$

$$\text{True Altitude} = 35°\ 57'\cdot 2$$

Answer—True Altitude = 35° 57'·2.

Example 31·7—On 4th October the sextant altitude of the Sun's upper limb was 22° 03'·0, index error = 1'·0 off the arc, height of eye = 24 feet. Find the Sun's true altitude using the total correction table, and allowing 1'·0 for irradiation.

$$\begin{aligned}\text{Sextant Altitude} &= 22°\ 03'\cdot 0 \\ \text{Index Error} &= \ +1'\cdot 0\end{aligned}$$

$$\begin{aligned}\text{Observed Altitude} &= 22°\ 04'\cdot 0 \\ \text{Sun's diameter} &= \ -32'\cdot 0\end{aligned}$$

$$\begin{aligned}\text{Obs'd Alt. L.L.} &= 21°\ 32'\cdot 0 \\ \text{Total Correction} &= \ +8'\cdot 8 \\ \text{Monthly Corr.} &= \ +0'\cdot 3 \\ \text{Irradiation} &= \ -1'\cdot 0\end{aligned}$$

$$\text{True Altitude} = 21°\ 40'\cdot 1$$

Answer—True Altitude = 21° 40'·1.

14. Correcting Moon Altitudes

The true altitude of the Moon is found by applying to the observed altitude, corrections for dip, refraction, semi-diameter and parallax-in-altitude. In practice it is not usual to apply correction for the Reduction of the Moon's Horizontal Parallax and Augmentation of the Moon's Semi-diameter: these two minor corrections tend to neutralize each other—one being positive and the other negative.

Example 31·8—Find the true altitude of the Moon if the sextant altitude of the lower limb is 30° 40'·0, index error = 0'·5 on the arc, height of eye = 40 feet, Moon's H.P. and S.D. (from the *Nautical Almanac*) are, respectively, 56'·5 and 15'·4.

FINDING THE TRUE ALTITUDE

$$\begin{aligned}
\text{Sextant Altitude} &= 30°\,40'\cdot 0 \\
\text{Index Error} &= -0'\cdot 5 \\
\hline
\text{Observed Altitude} &= 30°\,39'\cdot 5 \\
\text{Dip} &= -6'\cdot 2 \\
\hline
\text{Apparent Altitude} &= 30°\,33'\cdot 3 \\
\text{Refraction} &= -1'\cdot 6 \\
\hline
&30°\,31'\cdot 7 \\
\text{Semi-diameter} &= +15'\cdot 4 \\
\hline
&30°\,47'\cdot 1 \\
\text{Par-in-altitude} &= +48'\cdot 9 \\
\hline
\text{True Altitude} &= 31°\,36'\cdot 0
\end{aligned}$$

Par-in-alt = H.P. cos A
= $56'\cdot 5 \times \cos 30\tfrac{1}{2}°$
= $48'\cdot 9$

Answer—True Altitude = $31°\,36'\cdot 0$.

The above method of correcting the Moon's observed altitude is rarely used at sea. A Total Correction Table, such as that given in the *Nautical Almanac*, is normally used by practical navigators.

The Total Correction Table for Moon Altitudes given in Norie's and Burton's Tables is in two parts: one for Lower Limb, and the other for Upper Limb observations. It is constructed using a constant value for the height of eye. Adjustments for heights different from that used in the construction of the table are given in an auxiliary table.

Although the Moon's H.P. and S.D. are constantly changing in magnitude, their values at all times are directly proportional to each other. The ratio between the Moon's S.D. and its H.P. is about 1 : 4 or, more accurately, 3 : 11. This is proved with reference to fig. 31·14.

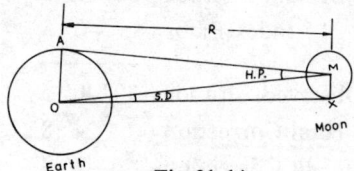

Fig. 31·14

In fig. 31·14:

Earth's radius = AO = 4000 ml. approx.

Moon's radius = MX = 1000 ml. approx.

Because angles AMO and MOX are small, we have:

AO/R = Moon's H.P. in radians

MX/R = Moon's S.D. in radians

Thus: $R = AO/\text{H.P.} = MX/\text{S.D.}$

Hence: $\dfrac{\text{Moon's S.D.}}{\text{Moon's H.P.}} = \dfrac{MX}{AO} = \dfrac{1000}{4000} = \dfrac{1}{4}$ approx.

240   THE ELEMENTS OF NAVIGATION AND NAUTICAL ASTRONOMY

This simple relationship facilitates the construction of the Moon's Total Correction Table.

Example 31·9—The sextant altitude of the Moon's upper limb was 30° 40'·0, index error = +0'·5, height of eye = 30 feet. H.P. = 60'·0, S.D. = 15'·4. Find the Moon's true altitude.

(i) Using individual corrections.
(ii) Using Total Correction Table.

(i) Using Individual Corrections:

$$\begin{aligned}
\text{Sextant Altitude} &= 30°\ 40'\cdot 0 \\
\text{Index Error} &= +0'\cdot 5 \\
\hline
\text{Observed Altitude} &= 30°\ 40'\cdot 5 \\
\text{Dip} &= -5'\cdot 4 \\
\hline
\text{Apparent Altitude} &= 30°\ 35'\cdot 1 \\
\text{Refraction} &= -1'\cdot 6 \\
\hline
&\ 30°\ 33'\cdot 5 \\
\text{Semi-diameter} &= -15'\cdot 4 \\
\hline
&\ 30°\ 18'\cdot 1 \\
\text{Parallax-in-Alt.} &= +51'\cdot 7 \\
\hline
\text{True Altitude} &= 30°\ 09'\cdot 8
\end{aligned}$$

Par-in-alt = H.P. × cos A
= 60 × cos $30\tfrac{1}{2}°$
= 51'·7

Answer—True Altitude = 30° 09'·8.

(ii) Using Total Correction Table:

$$\begin{aligned}
\text{Sextant Altitude} &= 30°\ 40'\cdot 0 \\
\text{Index Error} &= +0'\cdot 5 \\
\hline
\text{Observed Altitude} &= 30°\ 40'\cdot 5 \\
\text{Total Correction} &= +24'\cdot 8 \\
\text{Height Correction} &= +4'\cdot 4 \\
\hline
\text{True Altitude} &= 31°\ 09'\cdot 7
\end{aligned}$$

Answer—True Altitude = 31° 09'·7.

Note—When using tables such as Altitude Correction Tables, it is very important to take care over the interpolation which is sometimes necessary if accurate results are required.

15. Back Angles

When a celestial body has an altitude of more than about 60° it is possible to measure with a sextant the obtuse angle in the vertical plane between the direction of the body and that

FINDING THE TRUE ALTITUDE

of a point on the horizon which lies in a direction opposite to that of the body. This angle is called a Back Angle. Certain conditions may render it necessary or convenient to measure the back angle instead of the normally-measured "front" angle.

If the part of the horizon vertically below a body is indistinct, or if land lies in the same direction as that of the body, it may be necessary to observe the back angle instead of the altitude.

The following example illustrates how the true altitude of a celestial body is found from a back angle observation.

Example 31·10—The back angle of the Sun's upper limb was 118° 20'·0 by a sextant with no index error. Height of eye = 60 feet, Sun's S.D. = 15'·8. Find the Sun's true altitude.

Note—In a back angle observation of the Sun's upper limb the combined effect of Sun- and Horizon-irradiation may be taken as being nil.

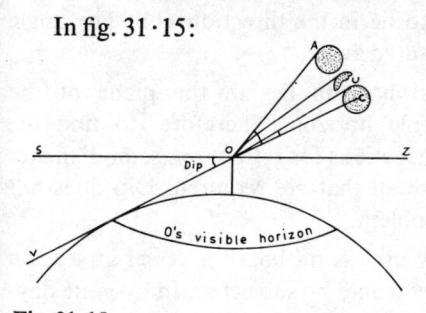

In fig. 31·15:

Observed Altitude of the Sun's U.L. = AOV
Dip = SOV
Apparent Back Angle of the Sun's U.L. = AOS
Apparent Altitude of the Sun's U.L. = AOZ
Refraction = AOU
Altitude of Sun's U.L. above Sensible Horizon = UOZ
Semi-diameter = UOC
Altitude of the Sun's centre above the S.H. = COZ

Fig. 31·15

```
Sextant Back Angle =  118° 20'·0
Index Error =              0'·0
                      ―――――――――
Observed Back Angle = 118° 20'·0
Dip =                     -7'·6
                      ―――――――――
Apparent Back Angle = 118° 12'·4
                    = 180°  0'
                      ―――――――――
Apparent Alt. of L.L. = 61° 47'·0
Refraction =             -0'·5
                      ―――――――――
                        61° 47'·1
Semi-diameter =          -15'·8
                      ―――――――――
                        61° 31'·3
Parallax =               +0'·1
                      ―――――――――
True Altitude =         61° 31'·4
```

Answer—True Altitude = 61° 31'·4.

16. The Artificial Horizon

If the visible horizon is not available because of fog or darkness, the true altitude of a celestial body may be found by employing an artificial horizon. The usual type of artificial horizon used ashore is a trough of opaque fluid, such as mercury or oil, which reflects the image of the observed body to the observer's eye. The observer measures with his sextant the angle between the true and reflected images of the body. The index error of the sextant, after being applied to the measured angle, gives an angle which is twice the apparent altitude of the observed body.

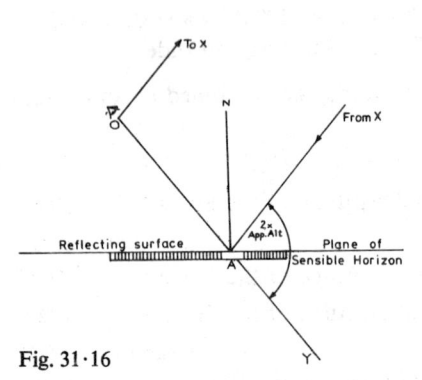

In fig. 31·16 the ray of light from the celestial body X, which strikes the reflecting surface of the artificial horizon at A, is reflected to the observer's eye at O. The incident and reflected rays make the same angle with the normal AN, and the reflected image appears to lie in the direction OY. The angle XAY is the measured angle.

The artificial horizon lies in the plane of the observer's sensible horizon. Therefore, to find the true altitude, corrections for refraction, semi-diameter and parallax, are all that are required. Dip does not enter into the problem.

Fig. 31·16

The mercury type of artificial horizon is practically useless on board a vessel at sea, on account of the inevitable vibration to which the mercury would be subjected. In by-gone days navigators often made observations ashore at places of accurately known Longitude, in order to find the G.M.T., so as to provide a means of rating their chronometers. These observations were facilitated by the mercury horizon.

17. The Bubble Attachment

Some marine sextants are fitted with an Artificial Vertical in the form of a bubble. The Bubble Attachment enables a navigator to observe visible celestial bodies at night or in thick weather when the visible horizon is not available. Results obtained from using a bubble attachment are not as accurate as those obtained from good visual observations. Nevertheless, in some circumstances, such observations may provide the navigator with worthwhile fixes so alleviating possible anxiety. A Total Correction Table for use with Bubble Sextant Observations is included in Norie's or Burton's Nautical Tables. Instructions are also given in the *Nautical Almanac* for using the Altitude Correction Tables given therein for use with bubble sextant observations.

Exercises on Chapter 31

1. Define: Visible Horizon; Sensible Horizon; Celestial Horizon. Illustrate your definitions.
2. Define: True Altitude; Apparent Altitude; Observed Altitude.
3. What is Dip? Prove that dip of the horizon in minutes of arc is equal to the distance of the horizon in nautical miles, when atmospheric refraction is ignored.
4. Prove that: Theoretical Dip $= 1\cdot06 \sqrt{h}$ where h is the observer's height of eye in feet.
5. What effect has atmospheric refraction on (i) dip of the visible horizon, (ii) distance of the visible horizon?

FINDING THE TRUE ALTITUDE

6. Compute the dip of the visible horizon for an observer whose height of eye is 406 feet.
7. Compute the distance of the visible horizon of an airman flying at 1000 feet above the sea surface.
8. Define Mean Refraction. Using the "Correction to Mean Refraction Table" in Norie's or Burton's Tables, explain how atmospheric temperature and pressure affect refraction.
9. What is the maximum value of the atmospheric refraction for a standard atmosphere? What is refraction of light from a celestial body in the zenith?
10. What is Celestial Parallax? Why is Parallax-in-Altitude less than Horizontal Parallax?
11. Prove that: $p = $ H.P. $\cdot \cos A$, where p is parallax-in-altitude, and A is altitude.
12. Explain the effect of the Earth's shape on parallax. Define Equatorial Parallax.
13. Define Reduction to the Moon's Horizontal Parallax.
14. Define Augmentation of the Moon's Semi-diameter.
15. Why do the Sun's and Moon's S.D.s vary with time?
16. State the dates on which the Sun's S.D. is greatest and least during the year.
17. When correcting a sextant altitude, using individual corrections, what should be the order of applying the corrections in the case of (i) a star, (ii) the Sun, (iii) the Moon?
18. What is the relationship between the Moon's H.P. and S.D.? Explain how this facilitates the construction of the Total Correction Table for the Moon.
19. Explain why the Sun's diameter must be subtracted from an altitude of the Sun's Upper Limb before using the Total Correction Table for the Sun.
20. Explain irradiation as it affects Sun observations of (i) the Lower Limb, (ii) the Upper Limb.
21. What is a "back angle"? State the circumstances when a back angle may be necessary.
22. Explain the artificial horizon.
23. No correction for dip is necessary when using an artificial horizon observation. Explain.
24. The sextant altitude of Antares was 35° 06'·0, index error = −1'·0. Height of eye = 45 feet. Find the true altitude.
25. The sextant altitude of Fomalhaut was 36° 46'·5, index error = +1'·5. Height of eye = 25 feet. Find the true altitude.
26. The sextant altitude of Sirius was 43° 43'·5, index error = 1'·5 off the arc. Height of eye = 50 feet. Find the true altitude.
27. The sextant altitude of Aldebaran was 50° 06'·5, index error = 2'·0 on the arc. Height of eye = 45 feet. Find the true altitude.
28. The sextant altitude of Polaris was 52° 41'·5, index error +2'·0. Height of eye = 60 feet. Find the true altitude.
29. On 16th February, the sextant altitude of the Sun's lower limb was 32° 20'·5, index error nil, height of eye = 42 feet. Find the true altitude.
30. The sextant altitude of Mars was 25° 00'·5, index error 2'·0 off the arc. Find the true altitude given the height of the observer's eye = 39 feet.

244   THE ELEMENTS OF NAVIGATION AND NAUTICAL ASTRONOMY

31. Find the true altitude of the Sun on 21st July if the sextant altitude of the upper limb was 54° 54'·0, index error −0'·5, height of eye = 60 feet.

32. The true altitude of a star was computed to be 36° 40'·0. Find the angle to set on a sextant if the index error is 1'·5 on the arc and the height of the observer's eye is 43 feet.

33. The true altitude of Canopus was computed to be 56° 40'·0. Find the angle to set on the sextant if the index error is 2'·0 off the arc and the observer's height of eye is 60 feet.

34. On 22nd March in cloudy weather, it is desired to set the sextant to the altitude of the Sun's lower limb for noon. The index error was 2'·0 off the arc; height of eye was 35 feet; and the computed true altitude was 34° 34'·0. Find the sextant altitude.

35. The sextant altitude of the Moon's lower limb was 42° 05'·5, index error nil, height of eye 56 feet and the Moon's H.P. was 56'·5. Find the true altitude.

36. The sextant altitude of the Moon's upper limb was 35° 53'·5, index error +1'·5, height of eye 45 feet. Moon's H.P. = 60'·5, Moon's S.D. 15'·2. Find the true altitude.

37. The sextant altitude of the Moon's lower limb was 46° 54'·5, index error = −2'·0. Moon's H.P. = 59'·8, Moon's S.D. = 15'·0. Height of eye = 70 feet. Find the true altitude.

38. On 5th October at 20 h. 00 m. G.M.T. the Moon's upper limb was 62° 08'·5 above the sea horizon as measured with a sextant the index error of which was 1'·5 off the arc. Find the Moon's true altitude if the height of the observer's eye was 20 feet. Moon's H.P. = 58'·4.

39. A back angle of Vega was 100° 42'·0, index error 0'·5 on the arc. Find the true altitude of the star if the observer's height of eye was 65 feet.

40. A back angle of the Sun's upper limb on 16th July was 115° 40'·0. Find the true altitude if the observer's height of eye was 30 feet.

41. Using an artificial horizon to observe Canopus the measured angle was 82° 45'·5, index error = −0'·5, height of eye = 100 feet. Find the true altitude.

42. On 18th February the vertical angle between the Sun's lower limb and its reflection in an artificial horizon on shore was 96° 00'·5, index error 2'·0 on the arc, height of eye = 65 feet. Find the true altitude.

CHAPTER 32

THE ASTRONOMICAL POSITION LINE

1. The Geographical Position of a Heavenly Body

Imagine a straight line extending from a celestial body to the centre of the Earth. The point on the Earth's surface located on this imaginary line is known as the Geographical Position (G.P.) of the celestial body. If an observer is situated at the place at which the body is in his zenith, his terrestrial position coincides with the geographical position of the observed body.

Fig. 32·1 illustrates the celestial sphere with the Earth at its centre. P denotes the celestial pole and p the Earth's North Pole. Notice that p is at the G.P. of P.

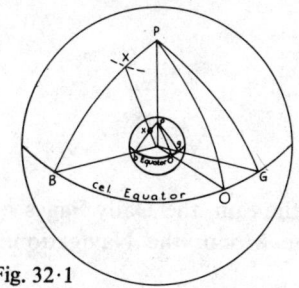

Fig. 32·1

pg represents the Greenwich Meridian
PG represents the Greenwich Celestial Meridian
po represents the Meridian of any Observer
PO represents the Observer's Celestial Meridian
X represents any celestial body
x represents the G.P. of X
px represents the Meridian of the G.P. of X.

Because the celestial meridian of X is in the same plane as the terrestrial meridian of the G.P. of X:

$$\text{arc } BX = \text{arc } bx$$

But: arc BX is the declination of X
and: arc bx is the Latitude of x
Therefore: Latitude of G.P. of X = Declination of X

From fig. 32·1 it will be seen that the angles gpb and GPB are equal.

Now: gpb is the West Longitude of the G.P. of X
and: GPB is the Greenwich Hour Angle of X
Therefore: West Longitude of G.P. of X = G.H.A. of X
= L.H.A. of X + W. Longitude of observer

The G.P. of any celestial body is, therefore, determined if the body's declination and G.H.A. are known. Provided that G.M.T. is known, the declination and G.H.A. of any navigational celestial body may be obtained from the *Nautical Almanac*.

Example 32·1—Find the G.P. of the Sun at 16 h. 00 m. G.M.T. on 30th December.

246   THE ELEMENTS OF NAVIGATION AND NAUTICAL ASTRONOMY

From *Nautical Almanac* Extracts:

$$\text{Sun's declination} = 23° 10' \cdot 3 \text{ S.}$$
$$\text{Sun's G.H.A.} = 359° 22' \cdot 5$$

Therefore: Latitude of G.P. = 23° 10'·3 S.
Longitude of G.P. = 359° 22'·5 West of Greenwich
= 00° 37'·5 East of Greenwich

Answer—Lat. G.P. = 23° 10'·3 S.
Long. G.P. = 00° 37'·5 E.

Example 32·2—Find the G.P. of the Sun at 0430 L.M.T. on 22nd September. Observer's Longitude is 82° 30'·0 W.

L.M.T. = (22) 04 h. 30 m.
Long. =   05 h. 30 m.
G.M.T. = (22) 10 h. 00 m.

From *Nautical Almanac* Extracts:

At 1000 hr. G.M.T.:   Sun's declination =   6° 49'·2 N.
Sun's G.H.A. = 150° 45'·6

Therefore: Lat. G.P. Sun =   6° 49'·2 N.
Long. G.P. Sun = 150° 45'·6 W.

Answer—Lat. G.P. = 6° 49'·2 N.
Long. G.P. = 150° 45'·6 W.

In addition to tabulations of the declination and G.H.A. of the Sun, the Daily Pages of the *Nautical Almanac* give the declinations and G.H.A.s of the Moon; the Navigational Planets; and of the First Point of Aries (♈).

Example 32·3—Find the G.P. of the Moon at 1930 L.M.T. on 22nd September. Observer's Longitude is 52° 30'·0 E.

L.M.T. = (22) 19 h. 30 m.
Long. =   03 h. 30 m. E.
—————————
G.M.T. = (22) 16 h. 00 m.
—————————

From *Nautical Almanac* Extracts:

At 1600 hr. G.M.T. on 22nd September:

Moon's declination = 15° 04'·1 S.
Moon's G.H.A. = 299° 59'·6

Therefore: Lat. G.P. = 15° 04'·1 S.
Long. G.P. = 299° 59'·6 W.
= 60° 00'·4 E.

Answer—Lat. G.P. = 15° 04'·1 S.
Long. G.P. = 60° 00'·4 E.

Example 32·4—Find the G.P. of Mars at 1800 hr. G.M.T. on 23rd September.

THE ASTRONOMICAL POSITION LINE

From *Nautical Almanac* Extracts:

At 1800 hr. G.M.T. on 23rd September:

$$\text{Declination of Mars} = 18°\,26'\cdot6\,\text{N}.$$
$$\text{G.H.A. of Mars} = 213°\,13'\cdot3$$

Therefore:
$$\text{Lat. G.P. Mars} = 18°\,26'\cdot6\,\text{N}.$$
$$\text{Long. G.P. Mars} = 213°\,13'\cdot3\,\text{W}.$$
$$= 146°\,46'\cdot7\,\text{E}.$$

Answer—Lat. G.P. = 18° 26'·6 N.
Long. G.P. = 146° 46'·7 E.

To facilitate the problem of finding the G.P. of a navigational star, the G.H.A. of the First Point of Aries (G.H.A. ♈) is used. It is demonstrated in fig. 32·2 that the G.H.A. of a given star is equivalent to the sum of the G.H.A. ♈ and the Sidereal Hour Angle of Aries (S.H.A. ♈).

In fig. 32·2:

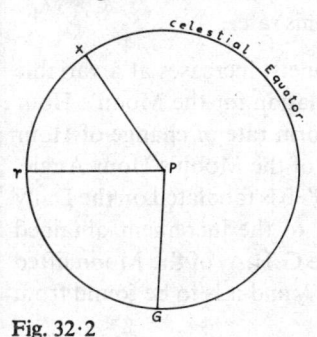

P represents the Celestial Pole
PG represents the Celestial Meridian of Greenwich
P♈ represents the Celestial Meridian of ♈
PX represents the Celestial Meridian of a star ✱

$$\text{arc } G\,♈ = \text{G.H.A. } ♈$$
$$\text{arc } ♈\,X = \text{S.H.A. } ✱$$
$$\text{arc } GX = \text{G.H.A. } ✱$$

But: $GX = G♈ + ♈X$

Fig. 32·2   Thus: G.H.A. ✱ = G.H.A. ♈ + S.H.A. ✱

Example 32·5—Find the G.P. of Aldebaran at 1600 hr. G.M.T. on 24th September.

From the *Nautical Almanac* Extracts:

At 1600 hr. G.M.T. on 24th September:

$$\text{Declination of Aldebaran} = 16°\,25'\cdot6\,\text{N}.$$
$$\text{G.H.A. } ♈ = 242°\,58'\cdot7$$
$$\text{S.H.A. Aldebaran} = 291°\,36'\cdot6$$

$$\text{Sum} = 534°\,35'\cdot3$$
$$360°\,00'\cdot0$$

$$\text{G.H.A. Aldebaran} = 174°\,35'\cdot3$$

Therefore:
$$\text{Lat. G.P. Aldebaran} = 16°\,25'\cdot6\,\text{N}.$$
$$\text{Long. G.P. Aldebaran} = 174°\,35'\cdot3\,\text{W}.$$

Answer—Lat. G.P. = 16° 25'·6 N.
Long. G.P. = 174° 35'·3 W.

248   THE ELEMENTS OF NAVIGATION AND NAUTICAL ASTRONOMY

In the above examples 32·1 to 32·5 inclusive, G.H.A.s and declinations were lifted direct from the appropriate Daily Page of the *Nautical Almanac*, these quantities being tabulated for each integral hour of G.M.T.

When the G.M.T. is not an integral number of hours, recourse must be made to the Increment and Interpolation Table provided in the *Nautical Almanac*. Increments to the G.H.A.s of the Sun, Aries and the Moon, for each minute and second from 00 m. 00 s. to 60 m. 00 s., are tabulated in these tables.

The Hour Angle of the Mean Sun increases uniformly at the rate of 15° 00'·0 or 900'·0 per hour of Mean Solar Time. The Hour Angle of the True Sun increases at a variable rate, but the variation of this rate from its average rate of increase (which is equal to the rate of increase of the Mean Sun's Hour Angle), is very small. Allowance is made for this variation in the tabulated values of the Sun's G.H.A. The increment for the Sun's G.H.A. is, therefore, computed on the assumption that the True Sun's rate of change of Hour Angle is exactly 900'·0 per hour.

The Hour Angle of the First Point of Aries increases at the uniform rate of 15° 02'·5 per Mean Solar Hour, and the Increment Table for ♈ is computed for this rate.

The Hour Angle of the Moon, as well as that of any of the planets, increases at a variable rate, and it is necessary to allow for these variations in the interpolation for the Moon's Hour Angle. The Increment Table for the Moon is computed for a uniform rate of change of Hour Angle of 14° 19'·0 per hour. This is the minimum rate of increase of the Moon's Hour Angle. The excess of the Moon's hourly increase in hour angle over 14° 19'·0 is tabulated on the Daily Pages of the *Nautical Almanac* as "v". An additional correction to the increment obtained from the Moon's Increment Table is, therefore, to be applied to the G.H.A. of the Moon lifted from the Daily Page. This correction is known as the "v correction", and it is to be found from the Interpolation Tables.

The increment for the G.H.A. of a planet is found by using the Sun's Increment Table, and applying a "v correction" as in the case of the Moon. "v" for a planet, which is tabulated on the Daily Pages of the *Nautical Almanac*, is the excess, positive or negative, of the planet's hourly motion over 15° 00'·0.

When the G.M.T. is not an integral number of hours, and it is required to find the declination of the Sun, Moon, or a navigational Planet, it is necessary to apply a correction to the declination lifted from the Daily Page. To facilitate interpolating between tabulated values of the declination corresponding to G.M.T.s which embrace the given G.M.T., the differences between adjacent tabulated values of the declination are given. These differences are designated "d". The correction to apply to a tabulated value of declination found from the Daily Page is easily found from the Interpolation Tables, using as arguments "d" and the difference between the given G.M.T. and the lower tabulated G.M.T. The correction is known as the "d correction".

The quantity "d" is the mean hourly change in the declination of Moon, Sun or Planet. It is tabulated at one-hourly intervals for the Moon and at daily intervals for the Sun and Planets.

Example 32·6—Find the G.P. of the Sun at 14 h. 40 m. 23s. on September 22nd.

THE ASTRONOMICAL POSITION LINE

From the Daily Page of the *Nautical Almanac* Extracts:

G.H.A. at 1400 G.M.T. = 350° 55'·6   v = 12'·7   Dec. = 11° 37'·8 S. d = 8.
Increment for 40 m. 23 s. = +9° 57'·7   Incr. = −5'·5
"v" correction = +   8'·8   Dec. = 11° 32'·3 S.
                    ─────────
                    361° 02'·1
                    360° 00'·0
                    ─────────
G.H.A. at 14 h. 40 m. 23 s. = 001° 02'·1

Answer—Lat. G.P. = 11° 32'·3 S.
         Long. G.P. = 01° 02'·1 W.

Example 32·8—Find the G.P. of Jupiter at 07 h. 42 m. 51 s. G.M.T. on 24th September.

From the Daily Page of the *Nautical Almanac* Extracts:

G.H.A. at 0700 G.M.T. = 256° 15'·2   v = 2'·0   Dec. = 11° 39'·5 S. d = 0'·2
Increment for 42 m. 51 s. = 10° 42'·8   Incr. = +0'·1
v correction =   +1'·5   Dec. = 11° 39'·6 S.
                        ─────────
                        266° 59'·5
                        360° 00'·0
                        ─────────
                         93° 00'·5

Therefore: Lat. of G.P. of Jupiter = 11° 39'·6 S.
                 Long. of G.P. of Jupiter = 93° 00'·5 E.

Answer—Lat. of G.P. = 11° 39'·6 S.
         Long. G.P. = 93° 00'·5 E.

2. Circles of Equal Altitude

Imagine an observer to be at the G.P. of a star at a certain instant of time. The true altitude of the star at the instant would be 90° 00'·0. Now suppose a second observer, to observe the same star at the same instant of time, and to find the true altitude to be 80° 00'·0. It follows that the second observer's zenith is 10° 00'·0 distant from that of the first observer, and that the second observer must be located on a small circle of radius 10° 00'·0 and centred at the first observer's position. Such a circle on the Earth's surface, at which at a given instant the altitude of a given celestial body is the same at all points on it, is called a Circle of Equal Altitude.

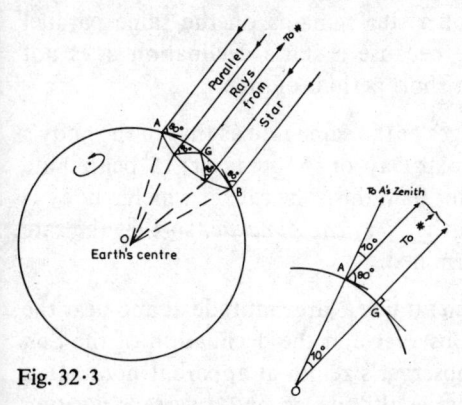

The radius of a circle of equal altitude in nautical miles is equal to the number of minutes of arc in the angular distance of the observed celestial body from the observer's zenith. That is to say, it is equal to the complement of the altitude of the body, an angle known as the body's Zenith Distance.

Fig. 32·3

Fig. 32·3 illustrates a circle of equal altitude of radius 10° or 600 miles.

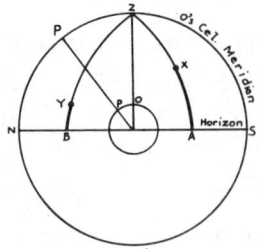

250   THE ELEMENTS OF NAVIGATION AND NAUTICAL ASTRONOMY

Fig. 32·4

Fig. 32·4 represents the celestial sphere projected onto the plane of the celestial meridian of an observer whose zenith is at Z. Fig. 32·5 represents the celestial sphere illustrating the same conditions as in fig. 32·4, but projected, in this case, onto the plane of the observer's celestial horizon.

In figs. 32·4 and 32·5:

AX = Altitude of X; ZX = Zenith Distance of X
BY = Altitude of Y; ZY = Zenith Distance of Y

If the zenith distance of a celestial body is small, and its declination also is small, the corresponding circle of equal altitude may be plotted on a Mercator Chart as a circle centred at the G.P. of the observed body. The observer's position may be fixed on such a circle, for which reason it is known as a Position Circle.

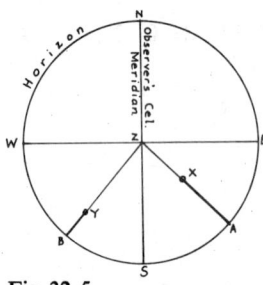

Fig. 32·5

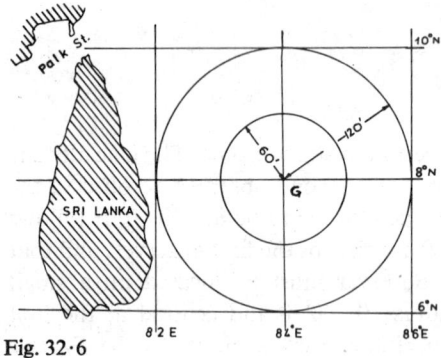

Fig. 32·6 represents a portion of a Mercator Chart. Suppose the G.P. of the Sun at a given instant to be at G in Latitude 08° 00'·0 N., Longitude 84° 00'·0 E. If the True Altitude of the Sun at the instant is 89° 00'·0, the observer must lie on a circle of equal altitude centred at G and of spherical radius 1° or 60 miles.

The Latitude of the G.P. of a celestial body changes if the declination of the body changes. The G.P. of a star remains on the same parallel of Latitude because a star's declination does not change over short periods of time.

Fig. 32·6

The Longitude of the G.P. of a celestial body changes at the same rate as that of the body's Hour Angle. For the Mean Sun this is 360° per Mean Solar Day, or 15° per hour; 15' per minute or 15" per second of time. For a star it is slightly greater than this. The rate at which a body is changing its Hour Angle may readily be found from the *Nautical Almanac*, so that the rate of change of Longitude of a body's G.P. is readily determined.

It happens, when located in the tropics, that the Sun attains a large altitude at and near the time of its meridian passage. If the Latitude of the observer and the declination of the Sun are equal in name and magnitude, the Sun is at the observer's zenith at apparent noon. It is possible in a circumstance when the Sun has a very large altitude, to find a vessel's position very easily and very speedily in the following manner.

THE ASTRONOMICAL POSITION LINE

The Sun's altitude is measured with a sextant and the chronometer time of the observation noted, and the G.M.T. found. With the G.M.T. the declination and the G.H.A. of the Sun may be lifted from the *Nautical Almanac* and the Sun's G.P., therefore, may be plotted on the chart. A position circle is then drawn centred at the plotted G.P. and with radius in miles equal to the zenith distance of the Sun in minutes of arc. The Sun's altitude is again observed two or three minutes after the first observation, and a second position circle is drawn on the chart. The vessel's position is then fixed at the intersection of the two position circles appropriate to the vessel's D.R. position, thus removing the ambiguity arising for the circles intersecting at two positions.

The timing of the observations must be such that the two position circles intersect at a good angle of cut. Three observations produce three position circles which intersect at a common point so resolving any ambiguity. If the interval between observations is more than a few minutes it may be necessary to transfer the first position circle for the run between the sights.

This method of finding a vessel's position has severe limitations. It fails at times when considerable error would result from the distortion of the plotted circles of equal altitude, which should be assumed to be circles on a Mercator Chart only in cases in which the zenith distance of the observed body is small and the Latitude of the G.P. of the body is also small. The method does not, therefore, solve the general problem in Nautical Astronomy, in which the zenith distance of the observed body is generally many tens of degrees, and the radius of the corresponding circle of equal altitude many hundreds, or even thousands, of miles.

Any small arc of a circle of equal altitude when plotted on a Mercator Chart may be assumed to be a straight line lying at right angles to the direction of the observed body from the observer's position. Such a line is known as an Astronomical Position Line.

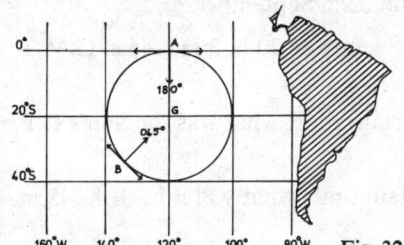

In fig. 32·7 an observer located at A observes the body whose G.P. is at G to bear due South. The position line obtained from this observation lies 090°–270°. An observer at B lies on a position line the direction of which is 135°–315°. This is so because the G.P. of the observed body bears 045° from the observer.

Fig. 32·7

Suppose that at a certain time an observer ascertains the zenith distances of two celestial bodies A and B whose G.P.s are located at a and b respectively. Fig. 32·8 illustrates the two circles of equal altitude corresponding to the two observations. The observer's position, which must lie on both circles, must be at X or Y. No difficulty is experienced in deciding which of these two possible positions is the position of the observer.

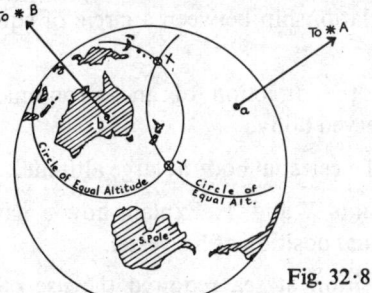

The general problem in Nautical Astronomy involves:

(i) Finding the latitude and longitude of a point on a circle of equal altitude and:

Fig. 32·8

(ii) Finding the direction of the circle at this point.

With this information a position line is determined, and the vessel's position is located on it.

The various methods of ascertaining astronomical position lines will be investigated in the following chapters.

Exercises on Chapter 32

1. Define: Geographical Position of a Heavenly Body.
2. Prove that the Latitude of the G.P. of a heavenly body is equal to the declination of the body; and that the Longitude of the G.P. of the body is equal to its G.H.A.
3. Using the *Nautical Almanac* Extracts, find the G.P. of the Sun:
    (i) at 10 h. 00 m. 00 s. G.M.T. on September 24th.
    (ii) at 09 h. 00 m. L.M.T. on September 22nd. Longitude 60° W.
    (iii) at 21 h. 00 m. L.M.T. on September 23rd. Longitude 45° W.
4. Using the *Nautical Almanac* Extracts, find the G.P. of the Sun:
    (i) At 23 h. 42 m. 25 s. G.M.T. on 12th June.
    (ii) At 02 h. 43 m. 56 s. L.M.T. on 14th June. Longitude 30° 15′ E.
5. At what position on the Earth will the Sun be at the zenith at 12 h. 00 m. 00 s. G.M.T. on 16th June?
6. At what position on the Earth will the Sun be at the zenith at 16 h. 13 m. 40 s. L.M.T. on 14th June, given that the Longitude of the observer is 37° 50′ E.?
7. Find the G.P. of Alioth at 16 h. 40 m. 36 s. on 15th June.
8. Find the G.P. of Hamal at 20 h. 41 m. 45 s. on 17th June.
9. Find the G.P. of Schedar at 07 h. 40 m. 45 s. L.M.T. on 16th June given that the observer's Longitude is 29° 58′ W.
10. Find the G.P. of the Moon at 06 h. 41 m. 37 s. G.M.T. on 22nd September.
11. At what position on the Earth will the Moon be at the zenith at 20 h. 36 m. 46 s. G.M.T. on 22nd September?
12. If the Sun's G.P. is 23° 16′·0 N., 165° 42′·0 E. at a certain time, what was the Sun's G.P. 01 h. 15 m. 20 s. before this time?
13. If the Sun's G.P. is 23° 20′·0 S., 19° 22′·0 W. at a certain time, what will it be 01 h. 43 m. 54 s. later?
14. Define: Circle of Equal Altitude. Explain why the spherical radius of a circle of equal altitude is equal to the zenith distance of the observed body.
15. Define: Astronomical Position Circle. What is the relationship between a circle of equal altitude and an astronomical position line?
16. Define: Astronomical Position Line. Explain why the direction of an astronomical position line is at right angles to the bearing of the observed body.
17. Explain a simple method of fixing from observations of a celestial body at large altitude.
18. Two astronomical position circles intersect at two points *X* and *Y*. Explain how a navigator may determine which of these positions is the actual position of his vessel.
19. A fifteenth-century method of finding a vessel's position at sea required the use of a terrestrial globe and two altitude observations of stars. Explain this method.
20. Devise a method of fixing a vessel at sea using a globe and two observations of the Sun, given the date and the interval of time between the observations.

CHAPTER 33

MERIDIAN ALTITUDE OBSERVATIONS

1. Latitude by Meridian Altitude

A star, in performing its apparent diurnal motion, attains its greatest altitude at the instant it bears due North or due South on the upper celestial meridian of a stationary observer. This greatest daily altitude is called the Meridian Altitude. The meridian altitude is named North or South according as the star bears, respectively, North or South from the observer. By subtracting the meridian altitude from 90° 00′ the Meridian Zenith Distance (M.Z.D.) is found. The M.Z.D. is named according to the direction of the zenith from the observed body. Thus, if the meridian altitude is named North the meridian zenith distance is named South, and *vice versa*.

A position line obtained from an observation of a celestial body at meridian passage lies along a parallel of Latitude; its direction, therefore, is 090°–270°. A meridian altitude observation enables an observer to ascertain the exact Latitude of his vessel from a single observation.

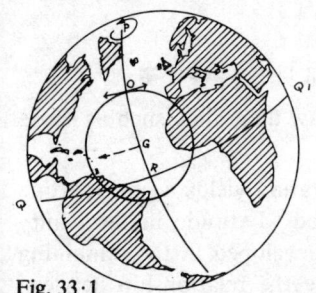

In fig. 33·1, which represents the Earth, P is the North Pole and QQ_1 is the equator. Imagine G to be the geographical position of a star which bears due South of an observer at O. The observer's position lies on a circle of equal altitude of radius GO. This radius, in angular units, is equal to the M.Z.D. of the star.

The latitude of G is equivalent to the declination of the star. If the declination is known, the Latitude of the observer may readily be found. The Latitude of the observer is equal to the arc RO, and this is the sum of the arcs OG and GR. Arc OG is equal to the M.Z.D. of the star, and arc GR is equal to the declination of the star. Thus:

Fig. 33·1

Latitude of Observer = M.Z.D. ✸ + Declination ✸

This relationship is illustrated in figs. 33·2 and 33·3.

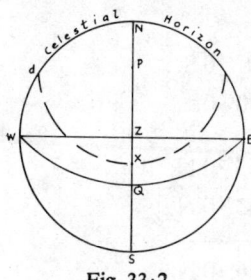

Fig. 33·2

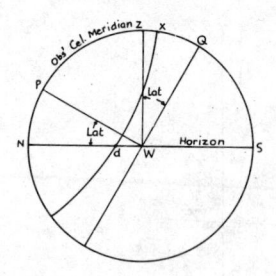

Fig. 33·3

Fig. 33·2 is a projection of the celestial sphere onto the plane of the celestial horizon of an observer whose zenith is projected at Z. Fig. 33·3 is a projection of the celestial sphere on the plane of the observer's celestial meridian. From both figures it may readily be seen that:

$$\text{Latitude of Observer} = \text{Altitude of Celestial Pole}$$
$$= NP$$
$$= ZQ$$
$$= ZX + XQ$$
$$= \text{M.Z.D.} \star + \text{Declination} \star$$

In the above example the Latitude of the observer is the SUM of the M.Z.D. and the declination of the observed body. In some cases the Latitude is found by taking the DIFFERENCE between the M.Z.D. and declination of the observed body. In all cases the Latitude is a combination of the M.Z.D. and the declination.

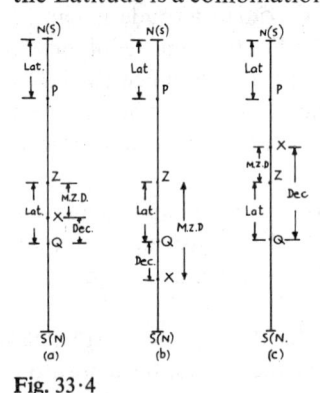

Fig. 33·4, in which *NZS* represents the observer's celestial meridian on the plane of his celestial horizon, illustrates the relationship between Latitude of Observer, and M.Z.D. and Declination of an observed body.

Referring to fig. 33·4:

In case (i): Latitude = M.Z.D. $\star$ + Declination $\star$
In case (ii): Latitude = M.Z.D. $\star$ − Declination $\star$
In case (iii): Latitude = Declination $\star$ − M.Z.D. $\star$

In every case:

$$\text{Latitude} = \text{M.Z.D.} \star \pm \text{Declination} \star$$

Fig. 33·4

When solving meridian altitude problems it is advisable to draw diagrams, such as those in fig. 33·4, to assist in the solutions.

It should be noted that, essentially, a meridian altitude observation yields a position line which lies 090°–270° along the parallel of the computed, or "Observed", Latitude, in the vicinity of the meridian of the observer. This important matter will be developed in the remaining pages of this chapter. The following examples serve to illustrate the relationship between Latitude of Observer, M.Z.D. and Declination of an observed celestial body.

Example 33·1—The True Meridian Altitude (T.M.A.) of Alpheratz, whose declination is 28° 43'·0 N., is 62° 07'·0 S. Find the Latitude of the observer.

In fig. 33·5:

(SX) = T.M.A.	= 62° 07'·0 S.	
	90° 00'·0	
(ZX) = M.Z.D.	= 27° 53'·0 N.	
(QX) = Declination	= 28° 43'·0 N.	
(ZQ) = Latitude	= 56° 36'·0 N.	

Fig. 33·5

Answer—Latitude = 56° 36'·0 N.

MERIDIAN ALTITUDE OBSERVATIONS

Example 33·2—The T.M.A. of Betelguese (declination 07° 24'·0 N.) was 42° 10'·0 N. Find the Latitude of the observer.

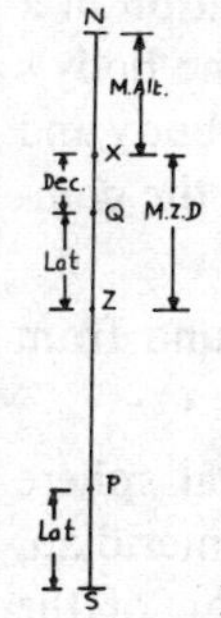

In fig. 33·6:

(NX) = T.M.A.   = 42° 10'·0 N.
                     90° 00'·0
                     ─────────
(ZX) = M.Z.D.   = 47° 50'·0 S.
(QX) = Declination = 7° 24'·0 N.
                     ─────────
(ZQ) = Latitude   = 40° 26'·0 S.

Fig. 33·6

Answer—Latitude = 40° 26'·0 S.

Example 33·3—Compute the T.M.A. of Antares whose declination is 26° 17'·0 S., if the observer's Latitude is 30° 20'·0 N.

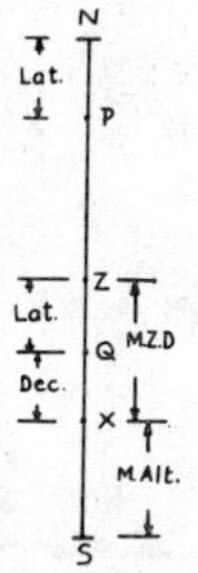

In fig. 33·7:

(ZQ) = Latitude   = 30° 20'·0 N.
(QX) = Declination = 26° 17'·0 S.
                     ─────────
(ZX) = M.Z.D.   = 56° 37'·0 N.
                     90° 00'·0
                     ─────────
(SX) = T.M.A.   = 33° 23'·0 S.

Fig. 33·7

Answer—True Meridian Altitude = 33° 23'·0 S.

Example 33·4—The T.M.A. of a star, observed by an observer in Latitude 40° 55'·0 N. was 54° 22'·0 S. Find the declination of the star.

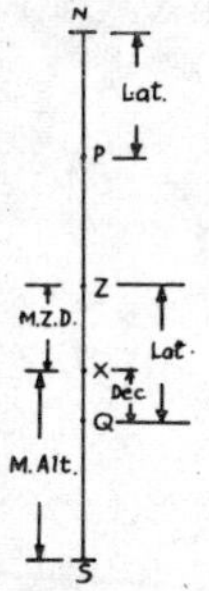

In fig. 33·8:

(SX) = T.M.A.   = 54° 22'·0 S.
                     90° 00'·0
                     ─────────
(ZX) = M.Z.D.   = 35° 38'·0 N.
(ZQ) = Latitude   = 40° 55'·0 N.
                     ─────────
(QX) = Declination = 05° 17'·0 N.

Fig. 33·8

Answer—Declination = 05° 17'·0 N.

2. Latitude by Meridian Altitude of a Body at Lower Meridian Passage

In the above examples, we considered cases in which the observed body is at Upper Meridian Passage. We shall now discuss the meridian altitude observation of a celestial body at Lower Meridian Passage.

A circumpolar body is above the horizon at its lower transit, at which time its altitude is least for the day and is less than the altitude of the celestial pole. A body at lower transit is said to be "On the Meridian Below the Pole". To find the Latitude from an observation of a body on the meridian below the pole, the body's true altitude is simply added to the body's polar distance, the polar distance being the arc of a great circle contained between the body and the celestial pole. In the case of a circumpolar body, the polar distance is equal to the complement of the declination of the body.

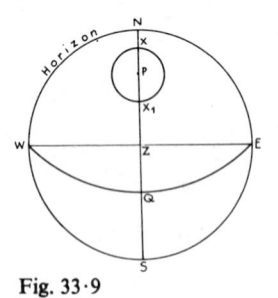

Fig. 33·9 and fig. 33·10 illustrate how Latitude is found from an observation of a body at lower meridian passage.

Fig. 33·9 and fig. 33·10 are projections of the celestial sphere onto the planes of the observer's celestial horizon and meridian, respectively. In fig. 33·9, the small circle is the projection of the diurnal circle of a circumpolar star. When this star is at lower meridian passage, at X, its true altitude is NX. This arc added to PX gives NP, the altitude of the celestial pole, which is equal to the Latitude of the observer.

Fig. 33·9

$$PX = PX_1$$
$$= 90° - QX_1$$
$$= 90° - \text{Declination of } X$$
$$= \text{Polar Distance of } X$$

Therefore:

$$\text{Latitude} = NP$$
$$= NX + PX$$
$$= \text{True Altitude of } X + \text{Polar Distance of } X$$

Fig. 33·10

Example 33·5—The true altitude of Kochab (declination 74° 26'·0 N.) on the meridian below the pole, was 20° 40'·0 N. Find the Latitude of the observer.

In fig. 33·11:

$NX =$ True Altitude $= 20° 00'·0$ N.

$PX =$ Polar Distance $= 15° 34'·0$

$NP =$ Latitude $\quad = 36° 14'·0$ N.

Fig. 33·11

Answer—Latitude $= 36° 14'·0$ N.

3. Effect of Observer's Motion on Meridian Altitude Observations

When a celestial body is at upper meridian passage it attains its greatest altitude for the day. When a celestial body is at lower transit its altitude is least for the day. This applies strictly to a stationary observer observing a celestial body whose declination is constant. We shall now investigate the effect of the observer's motion over the Earth's surface.

It is not uncommon practice, when observing the meridian altitude of a body for Latitude, to measure the maximum altitude of the body and to take this as the meridian altitude. This

MERIDIAN ALTITUDE OBSERVATIONS

practice may introduce appreciable and unnecessary error. When a celestial body is near the meridian its rate of change of altitude is usually very small. An observer on a fast moving vessel heading northerly or southerly, is increasing or decreasing his distance from the G.P. of the observed body. Therefore, the body, when at meridian passage, is changing its altitude at a rate which is equal to the rate at which the observer is changing his Latitude.

If a vessel is travelling towards the G.P. of a celestial body the altitude of the body continues to increase after it has culminated until the rate of change of the observer's Latitude is equal to the rate of change of the body's altitude. At this instant the body is said to "Dip". If a vessel is travelling away from the G.P. of an observed body the body dips before the time of its meridian passage.

Not only does a vessel's northerly or southerly motion cause the maximum altitude to occur at a time different from that of its meridian altitude; any change in the declination of the observed body produces the same effect. To a stationary observer observing the meridian altitude of a body which transits North of the observer's zenith, northerly change in the body's declination results in the maximum altitude occurring AFTER meridian altitude. Southerly change in the body's declination results in the maximum altitude occurring BEFORE meridian altitude. For an observer observing the meridian altitude of a body which transits south of the observer's zenith the reverse applies.

4. G.M.T. of Sun's Meridian Passage

To ensure, when observing for Latitude, that the correct meridian altitude is measured, the G.M.T. of the meridian passage of the body should be computed and the observation made precisely at that time. We shall now examine the method of finding the time of meridian passage of a celestial body.

When the True Sun is on an observer's upper celestial meridian its L.H.A. is 00 h. 00 m. 00 s. The observer's Longitude applied to this gives the G.H.A. of the True Sun corresponding to its meridian passage at the observer. Now the G.H.A. of the True Sun is tabulated against G.M.T. in the *Nautical Almanac* so that if G.H.A. is known the G.M.T. is determined.

Example 33·6—Find the G.M.T. of the Sun's meridian passage across the meridian of 101° 24′ W. on June 16th.

$$\text{L.H.A.T.S. at meridian passage} = 00° 00′·0$$
$$\text{Longitude} = 101° 24′ \text{ W.}$$

$$\text{G.H.A.T.S. at meridian passage} = 101° 24′·0$$

$$\text{G.M.T.} = 18 \text{ h. } 00 \text{ m.} \rightarrow \text{G.H.A.} = 89° 52′$$
$$\text{Increment} = 46 \text{ m.} \leftarrow \text{Increment} = 11° 32′$$

$$\text{Required G.M.T.} = 18 \text{ h. } 46 \text{ m.}$$

Answer—Required G.M.T. = 18 h. 46 m.

An alternative, and more practical, method is to apply the Longitude direct to the time of the Sun's meridian passage given at the foot of the right-hand Daily Page of the *Nautical Almanac*. This time is strictly the G.M.T. of the True Sun's meridian passage at Greenwich, but it may be taken as the L.M.T. of the Sun's meridian passage over any meridian without introducing material error.

Example 33·7—Find the G.M.T. of the Sun's meridian passage over the meridian of 120° 00'·0 W. on 17th June.

From *Nautical Almanac* Extracts:

L.M.T. of Sun's meridian passage =	(17)	12 h. 00 m.
Longitude =		08 h. 00 m. W.
G.M.T. of Sun's meridian passage =	(17)	20 h. 00 m.

Answer—G.M.T. = 20 h. 00 m.

Example 33·8—Find the G.M.T. of the Sun's meridian passage over the meridian of 90° 00'·0 E. on 31st December.

From *Nautical Almanac* Extracts:

L.M.T. of Sun's meridian passage =	(31)	12 h. 03 m.
Longitude =		06 h. 00 m. E.
G.M.T. of Sun's meridian passage =	(31)	06 h. 03 m.

Answer—G.M.T. = 06 h. 03 m.

5. G.M.T. of Moon's Meridian Passage

The G.M.T. of the upper and lower meridian passages of the Moon over the meridian of Greenwich may be found from the right-hand Daily Pages of the *Nautical Almanac*. It will be remembered that the interval between successive transits of the Moon over the meridian of a stationary observer is always more than 24 hours of Mean Solar Time. For example, the Moon was on the upper meridian of Greenwich on 16th June at 11 h. 18 m. G.M.T. On the 17th June the Moon was at upper meridian passage at Greenwich at 12 h. 10 m. G.M.T. The interval between these times of transit is 24 h. 52 m. The excess of the Lunar Day over the Mean Solar Day is, therefore, 52 m.

Suppose that an observer is located in Longitude 90° W., the Moon will cross his meridian at a quarter of a lunar day after it has crossed the Greenwich meridian. It follows, therefore, that on the 16th of June the Moon will cross the meridian of 90° W. (90/360 of 24 h. 52 m.), which is 06 h. 13 m., after it has crossed the Greenwich meridian. In other words the Moon will cross the meridian of 90° W. on 16th June at 17 h. 31 m. G.M.T. This time is 06 h. 13 m. later than 11 h. 18 m., which is the G.M.T. of the Moon's upper meridian passage at Greenwich on 16th June.

On 17th June the Moon will cross the meridian of 90° E. (270/360 of 24 h. 52 m.), that is, 18 h. 47 m. after it has crossed the Greenwich meridian on 16th June; or (90/360 of 24 h. 52 m.) before it will cross the Greenwich meridian on 17th June.

MERIDIAN ALTITUDE OBSERVATIONS

In practice, when it is necessary to find the G.M.T. of the Moon's meridian passage, it is usual first to find the L.M.T. of the Moon's meridian passage at the observer and then, by applying the Longitude in time, to find the G.M.T. of meridian passage.

The L.M.T. of the Moon's meridian passage is found by applying a Correction for Longitude to the tabulated G.M.T. of the Moon's meridian passage at Greenwich. The Correction for Longitude is a fraction of the excess of the Lunar Day over 24 hours, and it is proportional to the observer's Longitude.

In the interval between successive transits of the Moon, the Moon passes over 360° of Longitude. The L.M.T. of the Moon's meridian passage to an observer in any West Longitude is later than the G.M.T. of its meridian passage at Greenwich by an amount which is equal to:

W. Long./360 × Excess of Lunar Day over 24 hours.

Similarly for any East Longitude the L.M.T. of the Moon's meridian passage occurs earlier than its passage across the Greenwich meridian by an amount which is equal to:

E. Long./360 × Excess of Lunar Day over 24 hours.

Example 33·9—Find the G.M.T. of the Moon's upper meridian passage over the meridian of 60° W. on 31st December.

From *Nautical Almanac* Extracts:

G.M.T. of mer. pass. at Greenwich = (31)	04 h. 04 m.
Corr. for Long. (60/360 × 48 m.) =	+08 m.
L.M.T. of mer. pass. at Observer = (31)	04 h. 12 m.
Longitude in time =	04 h. 00 m. W.
G.M.T. of mer. pass. at Observer = (31)	08 h. 12 m.

Answer—G.M.T. = 08 h. 12 m.

Example 33·10—Find the G.M.T. of the Moon's upper meridian passage over the meridian of 36° 00′ E. on 23rd September.

From *Nautical Almanac* Extracts:

G.M.T. of mer. pass. at Greenwich = (23)	20 h. 56 m.
Corr. for Long. (36/360 × 46 m.) =	−8 m.
L.M.T. of mer. pass at Observer = (23)	20 h. 48 m.
Longitude in time =	02 h. 24 m. E.
G.M.T. of mer. pass. at Observer = (23)	18 h. 24 m.

Answer—G.M.T. = 18 h. 24 m.

Should the local time of the Moon's meridian passage be near midnight on any day, the next meridian passage will occur at a very early time during the morning of the next but one day. For example, the Moon is on the meridian of Greenwich at 23 h. 08 m. on 31st May.

260 THE ELEMENTS OF NAVIGATION AND NAUTICAL ASTRONOMY

The following meridian passage at Greenwich takes place at 00 h. 07 m. on 2nd June. In other words there was no meridian passage at Greenwich on the 1st of June. This does not mean that there has not been a meridian passage over other meridians on that date. For instance, the Moon will cross certain meridians East of the Greenwich meridian provided that the Correction for Longitude is greater than 07 minutes. All meridians to the East of that of Longitude (360/59 × 7 m.), that is 42° E. approximately, will experience a meridian passage of the Moon on 1st June.

Example 33·11—Find the G.M.T. of the Moon's upper meridian passage over the meridian of 120° 00′ W. on 1st July.

From *Nautical Almanac* Extracts:

G.M.T. of Moon's mer. pass. at Greenwich on 30th June = 23 h. 48 m.
G.M.T. of Moon's mer. pass. at Greenwich on 1st July = ————
G.M.T. of Moon's mer. pass at Greenwich on 2nd July = 00 h. 43 m.

G.M.T. of mer. pass. at Greenwich = (30) 23 h. 48 m.
Correction for Longitude (120/360 × 55 m.) = +18 m.

L.M.T. of mer. pass. at Observer = (1) 00 h. 06 m.
Longitude in time = 08 h. 00 m. W.

G.M.T. of mer. pass. at Observer = (1) 08 h. 06 m.

Answer—G.M.T. = (1) 08 h. 06 m.

Example 33·12—Find the G.M.T. of the Moon's upper meridian passage over the meridian of 60° 00′ E. on 1st June.

From *Nautical Almanac* Extracts:

G.M.T. of Moon's mer. pass. at Greenwich on 31st May = 23 h. 08 m.
G.M.T. of Moon's mer. pass. at Greenwich on 1st June = ————
G.M.T. of Moon's mer. pass. at Greenwich on 2nd June = 00 h. 07 m.

G.M.T. of mer. pass. at Greenwich = (2) 00 h. 07 m.
Long. Corr. (60/360 × 60 m.) = −10 m.

L.M.T. of mer. pass. at Observer = (1) 23 h. 57 m.
Longitude in time = 04 h. 00 m. E.

G.M.T. of mer. pass. at Observer = (1) 19 h. 57 m.

Answer—G.M.T. = (1) 19 h. 57 m.

6. G.M.T. of Planet's Meridian Passage

The G.M.T. of a planet's meridian passage at Greenwich is given to the nearest minute of time, for every third day throughout the year, on the left-hand Daily Pages of the *Nautical*

MERIDIAN ALTITUDE OBSERVATIONS

Almanac. The tabulated time is the approximate L.M.T. of the planet's meridian passage over every other meridian. If it is required to find the exact L.M.T. (to the nearest minute of time) of meridian passage of a planet, it is necessary to apply a Correction for Longitude to the G.M.T. of meridian passage at Greenwich, as in the case for the Moon.

Example 33·13—Find the G.M.T. of the upper meridian passage of Jupiter over the meridian of 90° 00′ W. on 14th June.

From *Nautical Almanac* Extracts:

```
        G.M.T. of mer. pass. at Greenwich = (14)   19 h. 52 m.
            Corr. for Long. (90/360 × 4 m.) =       − 01 m.
                                                   ─────────
        L.M.T. of mer. pass. at Observer = (14)    19 h. 51 m.
                         Longitude in time =        06 h. 00 m. W.
                                                   ─────────
        G.M.T. of mer. pass. at Observer = (15)    01 h. 51 m.
                                                   ─────────
```
Answer—G.M.T. = (15) 01 h. 51 m.

Example 33·14—Find the G.M.T. of the upper meridian passage of Saturn over the meridian of 90° 00′ E. on 15th June.

From *Nautical Almanac* Extracts:

```
        G.M.T. of mer. pass. at Greenwich = (15)   23 h. 53 m.
              Long. Corr. (90/360 × 6 m.) =         +01 m.
                                                   ─────────
        L.M.T. of mer. pass. at Observer = (15)    23 h. 54 m.
                         Longitude in time =        06 h. 00 m. E.
                                                   ─────────
        G.M.T. of mer. pass. at Observer = (15)    17 h. 54 m.
                                                   ─────────
```
Answer—G.M.T. = (15) 17 h. 54 m.

Observation Notes relating to the planets, appropriate for the whole year, together with a diagram illustrating the relative positions of the planets and the Sun, are given on pages 8 and 9 of the *Nautical Almanac*.

7. G.M.T. of Star's Meridian Passage

When a celestial body is on the upper meridian of an observer, its L.H.A. is 00 h. 00 m. 00 s. Therefore, the L.S.T.—which is equivalent to the L.H.A. of ♈—at the time of a star's meridian passage is equal to (360° − S.H.A. of the star).

Fig. 33·12, which represents the celestial sphere projected onto the plane of the celestial equator serves to show the relationship between S.H.A. ✱ and L.H.A. of ♈ when the star is on the observer's upper celestial meridian.

The L.H.A. of ♈ at the time of a star's meridian passage may be found by subtracting the star's S.H.A. from 360° 00′. By applying the Longitude of the observer to the L.H.A. of ♈, the G.H.A. of ♈ is found. The G.M.T. corresponding to any given G.H.A. of ♈ may be found by interpolation using the *Nautical Almanac*.

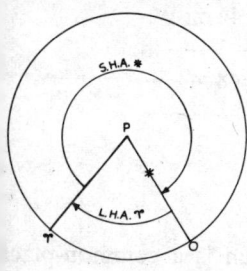

Fig. 33·12

Example 33·15—Find the G.M.T. of the upper meridian passage of Achernar over the meridian of 11° 00′ W. on 15th June.

$$\text{S.H.A. of Achernar} = 335° 58'$$
$$360° 00'$$

$$360° - \text{S.H.A.} = 24° 02' = \text{L.H.A. mer. pass.}$$
$$\text{Longitude} = 11° 00' \text{ W.}$$

$$\text{G.H.A. mer. pass.} = 35° 02'$$

From *Nautical Almanac* Extracts:

G.M.T. when G.H.A. is 23° 06′ = (15)	08 h. 00 m. 00 s.
Increment for (35° 02′ − 23° 06′) =	47 m. 44 s.
G.M.T. of star's mer. pass. = (15)	08 h. 47 m. 44 s.

Answer—G.M.T. = (15) 08 h. 47 m. 44 s.

In practice, the G.M.T. of a star's meridian passage may be found by applying (360° − S.H.A. star) to the G.M.T. of meridian passage of the First Point of Aries. The G.M.T. of the meridian passage of Aries is tabulated every third day on the left-hand Daily Pages of the *Nautical Almanac*.

The G.M.T. of the passage of Aries over the meridian of Greenwich is nearly the same as the L.M.T. of its passage over every other meridian. (It will be remembered that the sidereal day is 4 minutes shorter than a solar day.) Therefore, the approximate L.M.T. of a star's meridian passage may be found by applying (360° − S.H.A. ✶) to the G.M.T. of meridian passage of Aries found from the *Nautical Almanac*.

Example 33·15, using this method, is solved as follows:

From *Nautical Almanac* Extracts:

G.M.T. of mer. pass. of Aries at Greenwich = (15)	06 h. 28 m.
(360° − 335° 58′) = 24° 02′ =	01 h. 36 m.
Approx. L.M.T. mer. pass. at Observer = (15)	08 h. 04 m.
Longitude in time =	00 h. 44 m. W.
Approx. G.M.T. mer. pass. at Observer = (15)	08 h. 48 m.

Answer—G.M.T. = (15) 08 h. 48 m.

8. Position Line from Meridian Altitude Observation

Finding Latitude from a meridian altitude observation of the Sun is a common-place task of practical navigation. Occasionally a navigator observes the Moon, and sometimes the

MERIDIAN ALTITUDE OBSERVATIONS

planet Venus or Jupiter (these planets, when suitably placed, being visible during the daytime), when at meridian passage, in order to find his Latitude. It is seldom, however, that the meridian altitude of a star is observed at sea. Problems involving star meridian altitudes are largely of academic interest only.

The following examples illustrate how the Latitude and position line are deduced from meridian altitude observations.

Example 33·16—16th June in E.P. Latitude 36° 05′ N., Longitude 16° 00′ E. Find the G.M.T. at which the star Fomalhaut is at meridian passage. If the meridian altitude observed was 24° 18′·0 S., and the height of the observer's eye was 50 feet, find the Latitude and position line.

From *Nautical Almanac* Extracts:

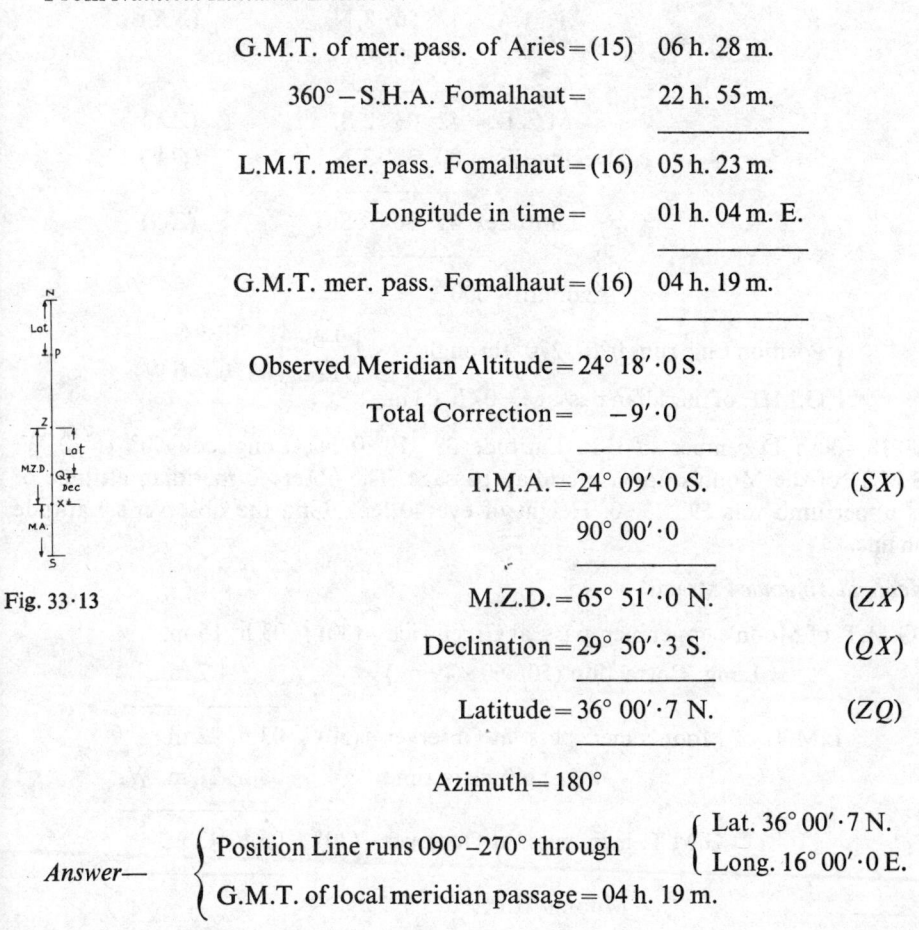

$$
\begin{array}{rl}
\text{G.M.T. of mer. pass. of Aries} = (15) & 06\,\text{h. }28\,\text{m.} \\
360° - \text{S.H.A. Fomalhaut} = & 22\,\text{h. }55\,\text{m.} \\ \hline
\text{L.M.T. mer. pass. Fomalhaut} = (16) & 05\,\text{h. }23\,\text{m.} \\
\text{Longitude in time} = & 01\,\text{h. }04\,\text{m. E.} \\ \hline
\text{G.M.T. mer. pass. Fomalhaut} = (16) & 04\,\text{h. }19\,\text{m.} \\
\end{array}
$$

Observed Meridian Altitude =	24° 18′·0 S.	
Total Correction =	−9′·0	
T.M.A. =	24° 09′·0 S.	(SX)
	90° 00′·0	
M.Z.D. =	65° 51′·0 N.	(ZX)
Declination =	29° 50′·3 S.	(QX)
Latitude =	36° 00′·7 N.	(ZQ)
Azimuth =	180°	

Fig. 33·13

Answer— { Position Line runs 090°–270° through { Lat. 36° 00′·7 N. Long. 16° 00′·0 E.
G.M.T. of local meridian passage = 04 h. 19 m.

Example 33·17—16th June in E.P. Latitude 41° 30′·0 S., Longitude 03° 00′·0 W. Find the G.M.T. of the meridian passage of Mars. If the sextant altitude of Mars at meridian passage was 48° 02′·0 bearing North, find the Latitude and position line. Index error 1′·0 off the arc. Height of eye 30 feet.

264   THE ELEMENTS OF NAVIGATION AND NAUTICAL ASTRONOMY

From *Nautical Almanac* Extracts:

$$\begin{aligned}
\text{Approx. L.M.T. mer. pass. of Mars} = (16) \quad &06\text{ h. }51\text{ m.} \\
\text{Longitude in time} = \quad &12\text{ m. W.} \\
\hline
\text{Approx. G.M.T. mer. pass. of Mars} = (16) \quad &07\text{ h. }03\text{ m.}
\end{aligned}$$

$$\begin{aligned}
\text{Sextant Meridian Altitude} &= 48°\ 02'\cdot 0\ \text{N.} \\
\text{Index Error} &= \quad +1'\cdot 0 \\
\hline
\text{Observed Meridian Altitude} &= 48°\ 03'\cdot 0\ \text{N.} \\
\text{Total Correction} &= \quad -6'\cdot 2
\end{aligned}$$

T.M.A. = 47° 56'·8 N.	(*NX*)
90° 00'·0	
M.Z.D. = 42° 03'·2 S.	(*ZX*)
Declination = 00° 32'·3 N.	(*QX*)
Latitude = 41° 30'·9 S.	(*ZQ*)

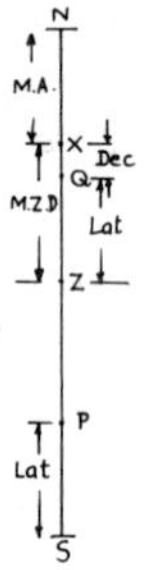

Azimuth = 000°

Fig. 33·14

Answer— { Position Line runs 090°–270° through { Lat. 41° 30'·9 S.
              G.M.T. of meridian passage = 07 h. 03 m.   Long. 03° 00'·0 W.

Example 33·18—30th December in E.P. Latitude 39° 10'·0 N., Longitude 50° 15'·0 W. Find the G.M.T. of the Moon's upper meridian passage. The observed meridian altitude of the Moon's upper limb was 59° 23'·0. Height of eye 40 feet. Find the observer's Latitude and position line.

From *Nautical Almanac* Extracts:

$$\begin{aligned}
\text{G.M.T. of Moon's upper mer. pass. at Greenwich} = (30) \quad &03\text{ h. }15\text{ m.} \\
\text{Long. Correction }(50/360 \times 49\text{ m.}) = \quad &+7\text{ m.} \\
\hline
\text{L.M.T. of Moon's mer. pass. at Observer} = (30) \quad &03\text{ h. }22\text{ m.} \\
\text{Longitude in time} = \quad &03\text{ h. }21\text{ m. W.} \\
\hline
\text{G.M.T. mer. pass. at Observer} = (30) \quad &06\text{ h. }43\text{ m.}
\end{aligned}$$

Moon's H.P. = 57'·7

Moon's declination at 06 h. 00 m. = 08° 42'·2 N.   d = 9'·7
Increment for 43 m. =   7'·0

Moon's declination at 06 h. 43 m. = 08° 49'·2 N.

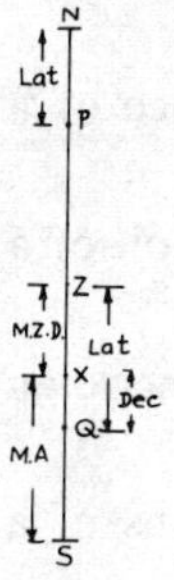

MERIDIAN ALTITUDE OBSERVATIONS

Observed Meridian Altitude Moon's U.L. = 59° 23′·0 S.
Total Correction (+3·3+3·6) =   +6′·9

(SX)   T.M.A. = 59° 29′·9 S.
           90° 00′·0

(ZX)   M.Z.D. = 30° 30′·1 N.

(QX)   Declination = 08° 49′·2 N.

Fig. 33·15   (ZQ)   Latitude = 39° 19′·3 N.

Azimuth = 180°

Answer—
$\begin{cases} \text{Position Line runs 090°–270° through} \\ \text{G.M.T. Moon's transit} = 06 \text{ h. } 43 \text{ m.} \end{cases}$
$\begin{cases} \text{Lat. } 39° \ 19'\cdot3 \text{ N.} \\ \text{Long. } 50° \ 15'\cdot0 \text{ W.} \end{cases}$

Example 33·19—30th December in E.P. Latitude 26° 35′·0 N. Longitude 55° 15′·0 W. Find the G.M.T. of the Sun's upper meridian passage. If the observed meridian altitude of the Sun's lower limb was 40° 02′·5 find the Latitude and position line. Height of eye 45 feet.

From *Nautical Almanac* Extracts:

L.M.T. of Sun's transit = (30)   12 h. 02 m.
Longitude in time =   03 h. 41 m. W.

G.M.T. of Sun's transit = (30)   15 h. 43 m.
Declination at 15 h. 00 m. G.M.T. = 23° 10′·5 S.   d = 0′·2
Increment for 43 m. =   −0′·1

Declination for 15 h. 43 m. G.M.T. = 23° 10′·4 S. 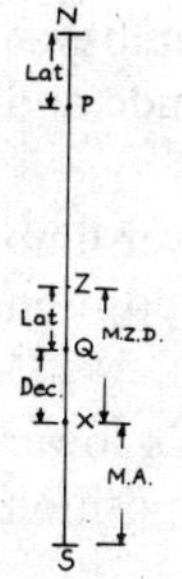
Observed Meridian Altitude of Sun's L.L. = 40° 02′·5 S.
Total Correction (+8′·3+0′·5) =   +8′·8

(SX)   T.M.A. = 40° 11′·3 S.
           90° 00′·0

(ZX)   M.Z.D. = 49° 48′·7 N.

(QX)   Declination = 23° 10′·4 S.

(ZQ)   Latitude = 26° 38′·3 N.

Fig. 33·16

Azimuth = 180°

Answer—
$\begin{cases} \text{Position Line runs 090°–270° through} \\ \text{G.M.T. of transit} = 15 \text{ h. } 43 \text{ m.} \end{cases}$
$\begin{cases} \text{Lat. } 26° \ 38'\cdot3 \text{ N.} \\ \text{Long. } 55° \ 15'\cdot0 \text{ W.} \end{cases}$

Exercises on Chapter 33

1. What is the relationship between Observer's Latitude, Meridian Zenith Distance of a celestial body, and Declination of the body?
2. Explain clearly how an observer may find his Latitude from a sextant observation of a celestial body on his upper celestial meridian.
3. Explain how Latitude is found from an observation of a celestial body at lower meridian passage.
4. Explain how an observer may find his Latitude by making sextant observations of a circumpolar body of unknown declination.
5. Distinguish between Meridian Altitude and Maximum Altitude. Why is it important to observe the meridian altitude and not the maximum altitude when observing for Latitude?
6. Find the G.M.T. of the Sun's meridian passage:
    (i) in Long. 30° E. on 16th June
    (ii) in Long. 150° W. on 17th June
    (iii) in Long. 170° E. on 22nd September.
7. Find the G.M.T. of the Moon's meridian passage:
    (i) in Long. 30° W. on 16th June
    (ii) in Long. 155° E. on 23rd September
    (iii) in Long. 90° E. on 31st December.
8. Find the G.M.T. of the meridian passage of:
    (i) Venus in Long. 70° W. on 23rd September
    (ii) Mars in Long. 90° E. on 16th June.
9. Find the G.M.T. of the upper meridian passage of:
    (i) Achernar in Long. 25° W. on 15th June
    (ii) Fomalhaut in Long. 20° E. on 17th June
    (iii) Alpheratz in Long. 90° W. on 30th December.
10. Find the observer's Latitude if the true meridian altitude of Hamal was 50° 10′ S.
11. Find the Latitude of the observer if the true meridian altitude of Bellatrix was 35° 08′ N.
12. 16th June in estimated Long. 78° 45′ E. the meridian altitude of the Sun's lower limb was 32° 48′·0 N. Index error 1′·0 on the arc. Height of eye 45 feet. Find the Latitude and position line.
13. 23rd September in estimated Long. 29° 45′ W. the meridian altitude of Sun's lower limb was 45° 00′·5 S. Index error −1′·0. Height of eye 50 feet. Find the Latitude and position line.
14. 16th June in estimated Longitude 160° 45′ W. the meridian altitude of the Sun's lower limb was 63° 15′·0 S. Index error nil. Height of eye 30 feet. Find the Latitude and position line.
15. 1st January in Long. 156° 00′ E. the meridian altitude of the Sun's lower limb was 80° 20′·0 N. Index error 0′·5 off the arc. Height of eye 40 feet. Find the Latitude and position line.
16. 16th June in estimated Long. 53° 15′ W., the meridian altitude of the Moon's lower limb was 51° 22′·0 S. Index error nil. Height of eye 30 feet. Find the Latitude and position line.

MERIDIAN ALTITUDE OBSERVATIONS

17. 17th June in estimated Long. 70° 00′ W., Altair bearing due South had an observed altitude of 48° 20′·0. Index error was −1′·0. Height of eye 30 feet. Find the Latitude and position line.

18. 31st December during evening twilight the meridian altitude of Achernar was 62° 20′·0 S. Index error nil. Height of eye 50 feet. Find the Latitude and position line given that the estimated Longitude was 75° 00′ E.

19. 15th June in estimated Longitude 152° 00′ E. the meridian altitude of Avior was 26° 03′·5 S. Index error 1′·0 on the arc. Height of eye 30 feet. Find the Latitude and position line.

20. 12th June in estimated Long. 120° 00′ W., the meridian altitude of Arcturus was 41° 05′·0 S. Index error nil. Height of eye 25 feet. Find the Latitude and position line.

21. 30th December during morning twilight in estimated Longitude 23° 00′ W. Pollux bore 000° and had an observed altitude of 61° 45′·0. Index error 1′·5 off the arc. Height of eye 35 feet. Find the Latitude and position line.

CHAPTER 34

THE ASTRONOMICAL TRIANGLE AND SIGHT REDUCTION

1. Introduction

An observation, or Sight, of a celestial body enables a navigator to plot an astronomical position line on his chart. The direction of an astronomical position line is at right angles to that of the observed body at the time of the sight.

A position line obtained from a sight of a celestial body on an observer's meridian lies 090°–270° along a parallel of Latitude. The meridian altitude problem for obtaining an East–West position line is relatively simple, as we have seen from Chapter 33. The problem of finding a position line from a sight of a celestial body which lies out of the observer's celestial meridian involves solving a spherical triangle by a process known as Sight Reduction. The spherical triangle solved in reducing a sight is a celestial triangle known as the Astronomical- or *PZX*-Triangle.

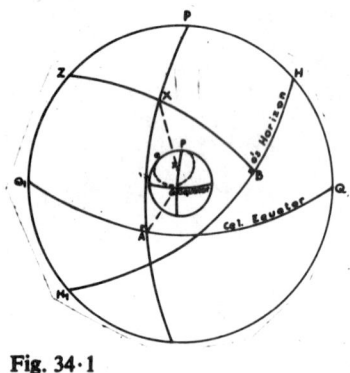

Fig. 34·1 illustrates a typical *PZX*-triangle. In fig. 34·1, *p* represents the Earth's North Pole and *P* the celestial pole. *o* represents the observer and *Z* his zenith. *x* represents the geographical position of a star or other celestial body denoted by *X*. The great circles QQ_1 and HH_1 represent, respectively, the celestial equator and the celestial horizon of the observer.

The spherical triangle, the three angles of which are *P*, *Z* and *X*, respectively, is the Astronomical Triangle. Notice that the spherical triangle *pxo* on the Earth is geometrically similar to the astronomical triangle *PZX*. Referring to fig. 34·1:

Fig. 34·1

PH = Altitude of Celestial Pole
   = Observer's Latitude

But since: $ZH = 90°$
Therefore: PZ = Observer's co-Latitude
   AX = Declination of X
And since: $AP = 90°$
Therefore: PX = Polar Distance of X
   BX = Altitude of X
But since: $BZ = 90°$
Therefore: ZX = Zenith Distance of X.

268

THE ASTRONOMICAL TRIANGLE AND SIGHT REDUCTION

The three sides of the *PZX*-triangle are:

PZ = Co-Latitude of an observer whose zenith is at Z

ZX = Zenith Distance of the Observed Body

PX = Polar Distance of the Observed Body.

The three angles of the *PZX*-triangle are:

P = Hour Angle of Observed Body at observer whose zenith is at Z

Z = Azimuth of Observed Body to an observer whose zenith is at Z

X = Parallactic Angle (of relatively minor account in practical Nautical Astronomy).

The zenith distance of the observed body X, in minutes of arc, is equal to the great circle distance, in nautical miles, between the observer and the geographical position of the observed body. The distance, xo in fig. 34·1, is the radius of a circle of equal altitude of the body X. A small arc of this circle, when projected onto a Mercator Chart or plotting sheet, is a Position Line somewhere on which the navigator may fix his vessel's position.

Fig. 34·2 illustrates a typical *PZX*-triangle projected onto the plane of the celestial equator. The direction of P from o, which denotes the observer on the Earth, is N. θ° W., where θ is the azimuth of the body indicated by the angle *PZX* in the Astronomical Triangle. The direction of the circle of equal altitude at o, denoted by the pecked line, is at right angles to the direction of x, the G.P. of X, at o. This direction is determined if the azimuth of X is known.

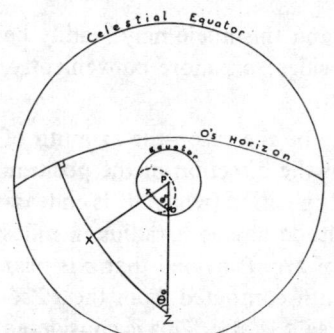

Fig. 34·2

In our discussion of the *PZX*-triangle above, we have made no mention of the important fact that an observer at sea is generally unaware of his precise position. It is not possible, therefore, for him to form a *PZX*-triangle such as that illustrated in figs. 34·1 and 34·2. This gives rise to a problem in computing a position line when the observer knows neither his Latitude nor Longitude. We shall see later how this problem is overcome.

There are two general methods of computing a position line from an astronomical observation. These are known, respectively, as the Longitude Method and the Intercept Method.

2. The Longitude Method

The essence of the Longitude Method of sight reduction is the computation of the Hour Angle of an observed body at a position where a circle of equal altitude is intersected by a parallel of chosen Latitude. By comparing the computed Hour Angle with the Greenwich Hour Angle of the body at the time of the observation, the Longitude of the position—the so-called Calculated Longitude—is ascertained. The azimuth of the observed body is then found, and the required position line is plotted on the chart at right angles to the azimuth and through a position having the Chosen Latitude and the Calculated Longitude. The principle of the method is illustrated in fig. 34·3.

In fig. 34·3, the point c lies on the chosen parallel of Latitude. The zenith at c is denoted by Z_c.

$cp = PZ_c =$ Co-Latitude of c

$px = PX =$ Polar Distance of X

$xc = XZ_c =$ Zenith Distance of X at c, or at any other point on the circle of equal altitude through c.

Hence, a PZX-triangle is formed from the sides PZ_c, PX, and XZ_c.

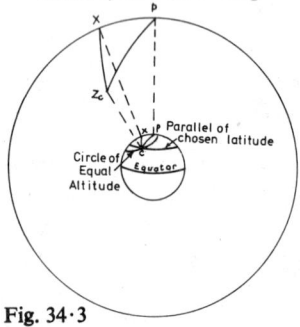

The angle XPZ_c in the PZX-triangle formed is the Hour Angle of the observed body X at any point on c's meridian. This angle is readily found by solving the PZX-triangle given the three sides. The Longitude of the meridian of c is then easily found by comparing the Calculated Hour Angle with the Greenwich Hour Angle at the time of the observation. The G.M.T. of the observation is, of course, obtained from the chronometer the error of which is known. The G.M.T. also allows the observer to lift the declination of the observed body from the *Nautical Almanac*.

Fig. 34·3

The angle XZ_cP is the azimuth of the observed body at c, and this angle may readily be found by computing the PZX-triangle given the initial three sides; or, more conveniently, by the spherical sine formula using the computed Hour Angle.

If the observer's position is not at c—and this will generally be the case—the azimuth of the observed body (which is necessary for the determination of the direction of the position line) will generally be different from what it is at c. The observer's position (which he is endeavouring to discover) does, however, lie on a circle of equal altitude having a radius in miles equal to the zenith distance of the observed body in minutes of arc. Provided that c is near to the actual but unknown position of the observer, the azimuth computed from the PZX-triangle may be taken to be the same at the observer's position as it is at c. This is considered in relation to fig. 34·4.

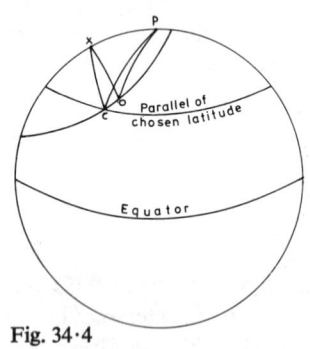

Suppose that o in fig. 34·4 represents an observer's actual position, and that c is the position computed in the way we have described above. The radius of the circle of equal altitude through o and c is usually in the order of many hundreds or even thousands of miles. The direction of x from o is different from that at c; but, provided that c and o are close to each other, the directions of ox and cx are so slightly different from each other that, bearing in mind that an azimuth to the nearest half a degree or so is sufficiently accurate for nautical astronomical purposes, they may be considered to be the same without introducing error.

Fig. 34·4

The chosen Latitude must, of necessity, be near to the actual, although unknown, Latitude of the observer at the time of the sight. In this circumstance the resulting position line may be projected onto the chart as a straight line at right angles to the calculated azimuth and through

THE ASTRONOMICAL TRIANGLE AND SIGHT REDUCTION

a position the Latitude of which is equal to the chosen Latitude, and the Longitude of which is equal to the calculated Longitude.

The Longitude Method is summarized as follows:

1. The altitude of a celestial body is measured with a sextant and the G.M.T. of the sight noted.

2. The measured altitude is corrected to give a True Altitude, which, subtracted from 90° gives the zenith distance of the observed body, or the side ZX of the PZX-triangle.

3. The declination of the observed body is lifted from the *Nautical Almanac*. By combining the declination with 90° (subtracting it from 90° when the Latitude and declination have the same name, and adding it to 90° when the Latitude and declination have different names) the polar distance of the observed body is found, and this gives side PX of the PZX-triangle.

4. A Latitude near to the actual but unknown Latitude of the vessel is chosen. This is subtracted from 90° to give the co-Latitude and hence the side PZ of the PZX-triangle.

5. Angle P of the PZX-triangle is computed using a direct trigonometrical method, or, more practically, by using a Short-Method or Inspection Table (see Chapter 38).

6. The computed angle P—Hour Angle of observed body—is compared with the Greenwich Hour Angle of the body at the time of sight. The difference between the G.H.A. (obtained from the *Nautical Almanac*) and the computed H.A. is the Longitude of a point on the chosen parallel through which to project the required position line.

7. Angle Z of the PZX-triangle is found—almost invariably by inspection from Azimuth Tables. The direction of the required position line is at right angles to the azimuth, and the position line is projected in this direction through a point having a Latitude equal to the chosen Latitude and a Longitude equal to the calculated Longitude.

3. General Remarks on the Longitude Method

When the chosen Latitude and the observed body's declination are both North or both South, that is to say, when they have the same name, the polar distance of the body is found by subtracting the declination from 90° 00′.

When the observed body is West of the observer, the Hour Angle of the body is less than 180° 00′.

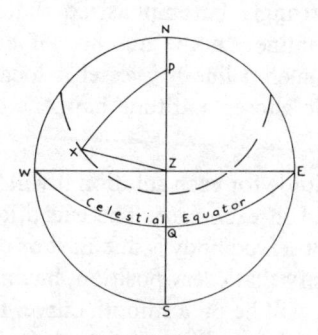

Fig. 34·5

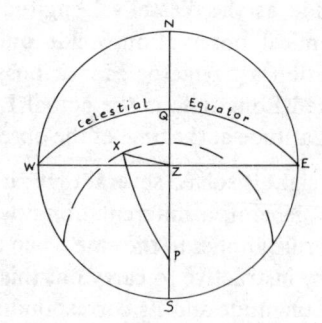

Fig. 34·6

Figs. 34·5 and 34·6 illustrate typical *PZX*-triangles projected onto the plane of an observer's celestial horizon. In fig. 34·5 the observer's Latitude and the body's declination are both North: in fig. 34·6 they are both South. In both figs. the body is West of the meridian in which case the angle *P* is less than 180° 00′.

When a celestial body is East of the meridian of an observer its Hour Angle is greater than 180°. In these cases the angle *P* of the *PZX*-triangle is found by subtracting the Hour Angle from 360° 00′.

When the chosen Latitude and the declination of an observed body have different names, the polar distance of the body is found by adding the declination to 90° 00′.

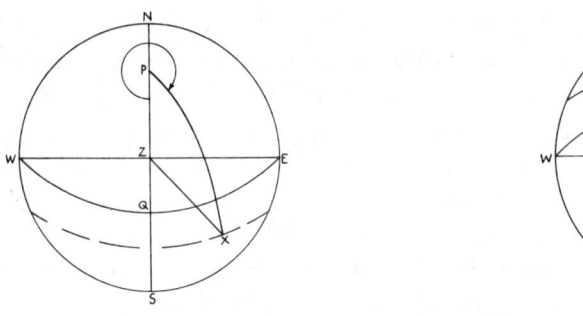

Fig. 34·7   Fig. 34·8

Figs. 34·7 and 34·8 illustrate typical *PZX*-triangles in which the declination of the observed body *X* and the Latitude of an observer have different names. In both figs. the Hour Angle is greater than 180°—the observed body lying East of the meridian—and the angle *P*, therefore, is obtained by subtracting the Hour Angle from 360° 00′.

In practice, when computing a *PZX*-triangle using a direct method of spherical trigonometry, it is usual to choose a Latitude corresponding to the Latitude by estimation in order to determine the side *PZ* of the *PZX*-triangle. When using Short-Method or Inspection Tables, it is usually necessary to choose a Latitude having an integral value.

When a sight has been reduced and a calculated Longitude found, it is important to realize that the calculated Longitude is not necessarily the Longitude of the observer at the time of the observation. It is not uncommon to hear inexperienced navigators refer to the Calculated Longitude as the Vessel's Longitude: it cannot too strongly be emphasized that a single astronomical observation yields only a single position line. In the absence of additional information a navigator cannot possibly tell where on such a line his vessel is located. The calculated Longitude is the actual Longitude only if the chosen Latitude happens to be the actual Latitude at the time of the observation.

If a sight is solved several times using a different Latitude for each solution it will be found that, in general, a different Longitude will be computed in each case. The exception to the general rule applies to the case when the azimuth of the observed body is due East or due West. It is very instructive to carry out this exercise and to verify that every position, having a computed Longitude and its corresponding chosen Latitude, will lie on a smooth curve, the direction at every point on which is equal to the azimuth at the position of the point. Provided

THE ASTRONOMICAL TRIANGLE AND SIGHT REDUCTION

that the zenith distance of the observed body is large and that the azimuth of the body is not too near due North or due South, the "smooth curve" will be projected on the Mercator Chart as a straight line.

4. The Longitude Method in Practice

The following examples are solved using a direct method popularized by P. L. H. Davis of the *Nautical Almanac* Office. Davis is credited with having introduced the arrangement of the Haversine Table in which Natural and Logarithmic Haversines are tabulated abreast of one another, thus facilitating computation using the so-called Haversine Method.

Example 34·1—17th June in D.R. position Lat. 40° 02'·0 N., Long. 54° 00'·0 W., at about 0800 hr. on board, the observed altitude of the Sun's lower limb was 38° 20'·0. Height of eye 26 feet. Chronometer time 11 h. 32 m. 26 s. Chronometer error 08 m. 40 s. slow on G.M.T. Using the Longitude method of sight reduction ascertain the position line on which the observer is located at the time of the observation.

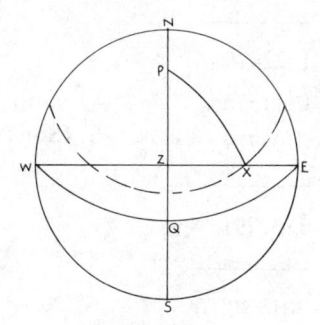

Fig. 34·9

Fig. 34·9 illustrates the *PZX*-triangle for this problem.

The first step in this problem is to ascertain whether the G.M.T. of the observation is 11 h. or 23 h. remembering that the dial of a chronometer extends only to 12 hr. This is the reason why the time on board, namely 0800 hr. approx., is given.

Approx. L.M.T. = 08 h. 00 m.	Chron. Time = 11 h. 32 m. 26 s.
Approx. Long. W. = 03 h. 36 m.	Chron. Error = +8 m. 40 s.
Approx. G.M.T. = 11 h. 36 m.	G.M.T. (17th) = 11 h. 41 m. 06 s.
Dec. at 11 h. G.M.T. = 23° 22'·4 N.	Obs. Altitude = 38° 20'·0
Increment for 41 m. = +0'·1	Total Corr. = +9'·8
Dec. at 11 h. 41 m. 06 s. = 23° 22'·5 N.	True Altitude = 38° 29'·8
90° 00'·0	90° 00'·0
Polar Distance (*PX*) = 66° 37'·5	Zen. Distance (*ZX*) = 51° 30'·2

274 THE ELEMENTS OF NAVIGATION AND NAUTICAL ASTRONOMY

Latitude = 40° 02′·0 N.	G.H.A. at 11 h. = 344° 49′·7
90° 00′·0	Incr. for 41 m. 06 s. = 10° 16′·5
co-Lat. (PZ) = 49° 58′·0	G.H.A. = 355° 06′·2
(PX) = 66° 37′·5	Calc. H.A. = 301° 30′·0
(PX∼PZ) = 16° 39′·5	Calc. Long. = 53° 36′·2 W.

hav $P = \{\text{hav } ZX - \text{hav } (PZ \sim PX)\}$ cosec PZ cosec PX

nat hav $ZX = 0\cdot 18877$
nat hav $(PZ \sim PX) = 0\cdot 02098$

nat hav $\theta = 0\cdot 16779$

log hav $\theta = \bar{1}\cdot 22476$
log cosec $PZ = 0\cdot 11596$ Azimuth (from Azimuth Tables) = 090°
log cosec $PX = 0\cdot 03719$ (Sun is on P.V.)

log hav $P = \bar{1}\cdot 37791$

Calc. H.A. = 301° 30′·0

Position Line runs 000°–180° through Lat. 40° 02′·0 N., Long. 53° 36′·2 W.

Example 34·2—24th September in D.R. position Lat. 50° 20′·0 N., Long. 10° 30′·0 W. at about 3 p.m. on board, the sextant altitude of the Sun's lower limb was 25° 26′·5. Index error 0′·5 off the arc. Height of eye 40 feet. Chronometer time 3 h. 40 m. 20 s. Chronometer error 00 m. 22 s. slow on G.M.T. Ascertain, using the Longitude method, the position line on which the vessel was located at the time of sight.

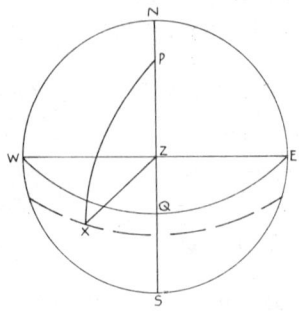

Fig. 34·10

THE ASTRONOMICAL TRIANGLE AND SIGHT REDUCTION

Fig. 34·10 illustrates this problem.

Approx. L.M.T. = 15 h. 00 m.	Chron. Time = 15 h. 40 m. 20 s.
Approx. Long = 00 h. 42 m.	Chron. Error = +00 m. 22 s.
Approx. G.M.T. = 15 h. 42 m.	G.M.T. (24th) = 15 h. 40 m. 22 s.

N.B.—Sun is West of meridian

Dec. at 15 h. = 00° 25'·1 S.   d = 1'·0

Increment = +0'·7

Dec. = 00° 25'·8 S.

90° 00'·0

PX = 90° 25'·8

Lat. = 50° 20'·0 N.

90° 00'·0

PZ = 39° 40'·0

PX = 90° 25'·8

$(PX - PZ)$ = 50° 45'·8

Sext. Alt. = 25° 26'·5

Index Er. = +0'·5

Obs'd Alt. = 25° 27'·0

Total Cor. = +8'·0

True Alt. = 25° 35'·0

90° 00'·0

ZX = 64° 25'·0

hav P = {hav ZX − hav $(PX - PZ)$} cosec PZ cosec P

nat hav ZX = 0·28409

nat hav $(PX - PZ)$ = 0·18374

nat hav θ = 0·10035

log hav θ = $\bar{1}$·00152

log cosec PZ = 0·19496

log cosec PX = 0·00001

log hav P = $\bar{1}$·19649

Calc. H.A. = 46° 42'·3

G.H.A. at 15 h. = 46° 58'·3

Increment = 10° 10'·5

G.H.A. = 57° 08'·8

Calc. H.A. = 46° 42'·3

Calc. Long. = 10° 26'·5 W.

Azimuth (From Azimuth Tables) = 233½°

Position Line runs 143½°–323½° through Lat. 50° 20'·0 N., Long. 10° 26'·5 W.

Example 34·3—23rd September in D.R. position Lat. 45° 10'·0 S. Long. 118° 00'·0 W. the sextant altitude of the star Alphard was 32° 30'·0. Index error +1'·0. Height of eye 30 feet. Chronometer time 13 h. 45 m. 10 s. Chronometer error 02 m. 05 s. fast on G.M.T. Ascertain the position line on which the observer is located.

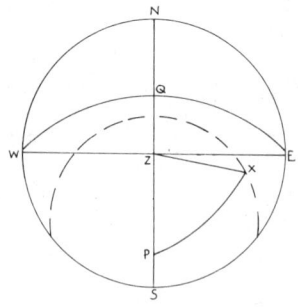

276 THE ELEMENTS OF NAVIGATION AND NAUTICAL ASTRONOMY

Fig. 34·11

Fig. 34·11 illustrates this problem.

Chron. Time = 13 h. 45 m. 10 s.	S.H.A. Alphard = 218° 36'·7
Chron. Error = −02 m. 05 s.	360° 00'·0
G.M.T. (23rd) = 13 h. 43 m. 05 s.	φ = 141° 23'·3
	= 9 h. 25 m.

G.M.T. of Transit of ♈ = (22) 23 h. 46 m.
φ = 09 h. 25 m.
G.M.T. of Transit of ✹ = (23) 09 h. 11 m.

Therefore the ✹ is East of the meridian and its H.A. is > 180°.

Lat. = 45° 10'·0 S.	Sext. Altitude = 32° 30'·0
PZ = 44° 50'·0	Index Error = +1'·0
Dec. = 8° 28'·7 S.	Obs. Altitude = 32° 31'·0
PX = 81° 31'·3	Total Corr. = −6'·9
(PX − PZ) = 36° 41'·3	True Altitude = 32° 24'·1
	ZX = 57° 35'·9

$$\text{hav } P = \{\text{hav } ZX - \text{hav}(PX - PZ)\} \operatorname{cosec} PX \operatorname{cosec} PZ$$

nat hav ZX = 0·23208	G.H.A. ♈ at 13 h. = 196° 52'·2
nat hav (PX − PZ) = 0·09920	Increment = 10° 48'·0
nat hav θ = 0·13288	G.H.A. ♈ = 207° 40'·2
log hav θ = 1̄·12347	S.H.A. ✹ = 218° 36'·7
log cosec PZ = 0·15178	G.H.A. ✹ = 426° 16'·9
log cosec PX = 0·00477	Calc. H.A. = 308° 14'·1
log hav P = 1̄·28002	Calc. Long. = 118° 02'·8 W.
Calc. H.A. = 308° 14'·1	

Azimuth (From Azimuth Tables) = 113°

Position Line runs 023°–203° through Lat. 45° 10'·0 S., Long. 118° 02'·8 W.

Example 34·4—24th September in D.R. position Lat. 35° 10'·0 N. Long. 165° 00'·0 E., during morning twilight at about 0550 hr. on board, the observed altitude of Hamal was

THE ASTRONOMICAL TRIANGLE AND SIGHT REDUCTION

39° 55'·0. Height of eye 25 feet. Chronometer time 6 h. 44 m. 10 s. Chronometer error 01 m. 53 s. fast on G.M.T. Ascertain the position line on which the observer is located.

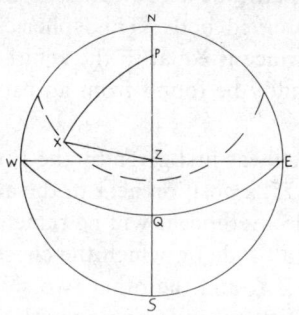

Fig. 34·12

Fig. 34·12 illustrates this problem.

Approx. Local Time = (24) 05 h. 50 m.	S.H.A. ✸ = 328° 46'·9
Approx. Longitude = E. 11 h. 00 m.	360° 00'·0
Approx. G.M.T. = (23) 18 h. 50 m.	φ = 31° 13'·1
	= 02 h. 05 m.
Chron. Time = 18 h. 44 m. 10 s.	G.M.T. of transit of ♈ = 23 h. 50·7 m.
Chron. Error = −1 m. 53 s.	φ = 02 h. 05 m.
G.M.T. (23rd) = 18 h. 42 m. 17 s.	G.M.T. of transit of ✸ = 01 h. 55·7 m.
	(24th)

Therefore, ✸ is West of meridian and H.A. is < 180°.

Lat. = 35° 10'·0 N.   Observed Altitude = 39° 55'·0
PZ = 54° 50'·0   Index Error = −6'·0
Dec. = 23° 16'·1 N.   True Altitude = 39° 49'·0
PX = 66° 43'·9   ZX = 50° 11'·0
(PX − PZ) = 11° 53'·9

hav P = {hav ZX − hav (PX − PZ)} cosec PZ cosec PX

nat hav ZX = 0·17983   G.H.A. ♈ at 18 = 272° 04'·5
nat hav (PX − PZ) = 0·01075   Increment = 10° 36'·0
nat hav θ = 0·16908   G.H.A. ♈ h. = 282° 40'·5
log hav θ = 1̄·22811   S.H.A. ✸ = 328° 46'·9
log cosec PZ = 0·08752   G.H.A. ✸ = 251° 27'·4
log cosec PX = 0·03684   Calc. H.A. = 56° 39'·2
log hav P = 1̄·35247   Calc. Longitude = 165° 11'·8 E.
Calc. H.A. = 56° 39'·2

Azimuth (From Azimuth Tables) = 273°

Position Line runs 003°–183° through Lat. 35° 10'·0 N., Long. 165° 11'·8 E.

5. The Intercept Method

The radius of a circle of equal altitude is equivalent to the great circle distance between an observer and the G.P. of an observed body. This spherical distance in minutes of arc (or nautical miles) on the Earth's surface is equal to the zenith distance of the observed body in minutes of arc, and this may readily be found from an astronomical observation or "sight" of the body.

In the Intercept Method, which was first given by the French Naval Officer, Captain (later Admiral) Marcq St. Hilaire in 1875, a position near to the actual but unknown position of the vessel is chosen. In the Longitude Method, it will be remembered, only a Latitude is chosen. The radius of the circle of equal altitude on which the chosen position lies is computed using the angle P opposite to the side ZX, and the other two sides, viz., PZ and PX, of the PZX-triangle. This computed Zenith Distance is compared with the True Zenith Distance obtained from the altitude observation of the body. The difference between the computed and observed zenith distances is known as the "intercept". The chosen position is plotted on the chart, and the intercept is projected in a direction from the plotted chosen position corresponding to the azimuth of the observed body. The position line is then projected through the end of the intercept in a direction at right angles to the azimuth of the body. Fig. 34·13 serves to illustrate the principle of the Intercept or "Marcq St. Hilaire Method" of sight reduction.

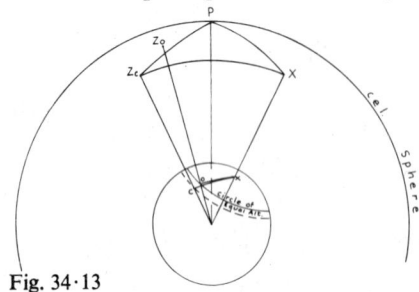

Fig. 34·13

Referring to fig. 34·13: o denotes an observer and Z_o his zenith. c is a chosen position which is near to o; and x is the G.P. of an observed celestial body. Because the distance oc is small (never more than 40 or so miles in practice), and the distance from o (or c) to x is great (usually in the order of many hundreds, or even thousands of miles), the direction of x from o is almost the same as that of x from c. No error is introduced in practice by assuming these directions to be the same. It is to be noted that the direction of x from o (or c) is equivalent to the azimuth Z in the PZX-triangle.

The Intercept Method is summarized as follows:

1. The altitude of a celestial body is observed and the G.M.T. of the observation noted.

2. The zenith distance of the observed body at the observer's actual, but unknown, position is found by subtracting the true altitude of the body from $90° 00'·0$.

3. A position (Latitude and Longitude) near to the observer's position is chosen.

4. The Latitude of the chosen position subtracted from $90° 00'·0$ gives the side PZ_c of the astronomical triangle which has to be solved.

5. The Longitude of the chosen position allows the observer to find the angle XPZ_c of the astronomical triangle. This angle is found simply by combining the Longitude of the chosen position with the G.H.A. of the observed body determined by means of the *Nautical Almanac*.

6. The declination of the body at the time of the observation is found from the *Nautical Almanac*, and the polar distance gives the side PX of the astronomical triangle.

7. The side XZ_c of the astronomical triangle is computed using the angle P and the adjacent sides PZ_c and PX.

8. The azimuth of the body is found, usually from Azimuth Tables.

THE ASTRONOMICAL TRIANGLE AND SIGHT REDUCTION

9. The position line is plotted on the chart through the end of the intercept which is found by taking the difference between the Computed Zenith Distance and the Observed Zenith Distance. The direction of the position line is at right angles to the azimuth.

Great care must be taken to plot the intercept in the correct direction. If the Computed Zenith Distance (C.Z.D.) is greater than the Observed Zenith Distance (O.Z.D.) the observer must be nearer to the G.P. of the observed body than is the Chosen Position. In this case the intercept is named TOWARDS. If, on the other hand, the C.Z.D. is less than the O.Z.D. the observer must be farther from the G.P. of the observed body than is the Chosen Position. In this case the intercept is named AWAY. The following examples illustrate this.

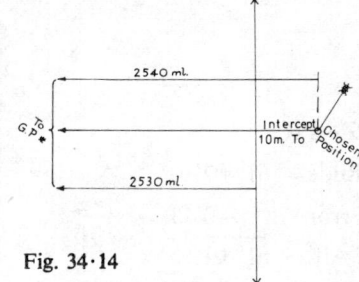

Example 34·5—The O.Z.D. of a star was 42° 20'·0 and the C.Z.D. was 42° 30'·0. The azimuth of the star was 240°. Plot the position line.

Fig. 34·14 illustrates that the intercept in this case is named TOWARDS.

Example 34·6—The Sun's O.Z.D. was 20° 00'·0 and the C.Z.D. was 19° 50'·0. The azimuth was 330°. Plot the position line.

Fig. 34·15 illustrates that, in this case, the intercept is named AWAY.

Fig. 34·14

It is clear from examples 34·5 and 34·6, that the intercept is named AWAY when the O.Z.D. is greater than the C.Z.D., and that it is named TOWARDS when the C.Z.D. is greater than the O.Z.D. Most practical navigators use a mnemonic for finding, without effort, whether an intercept is AWAY or TOWARDS. A common mnemonic is FOG meaning "From" (or AWAY) when the **O**.z.d. is **G**reater than the C.Z.D. Mnemonics, or "Donkey's Bridges" as they are sometimes called, often serve useful purposes, but it is always a source of satisfaction to understand the basis of the mnemonic.

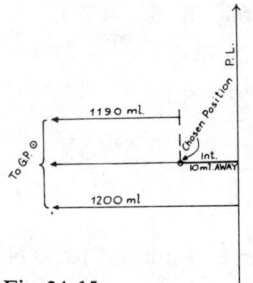

Fig. 34·15

6. The Intercept Method in Practice

Example 34·7—13th June during evening twilight at about 1950 hr., the sextant altitude of the star Denebola was 61° 02'·0. Index error 0'·5 on the arc. Height of eye 20 feet. Chronometer time 22 h. 43 m. 50 s. Chronometer error 00 m. 05 s. fast on G.M.T. Using Latitude 36° 10'·0 N., Longitude 44° 00'·0 W., ascertain the position line on which the observer was located using the Intercept Method.

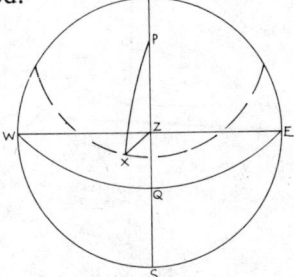

Fig. 34·16

280 THE ELEMENTS OF NAVIGATION AND NAUTICAL ASTRONOMY

Chron. Time = 22 h. 43 m. 50 s.	G.H.A. ♈ at 22 h. = 231° 42'·2
Chron. Error = −00 m. 05 s.	Increment = 10° 58'·0
G.M.T. (13th) = 22 h. 43 m. 45 s.	G.H.A. ♈ = 242° 40'·2
	S.H.A. ✳ = 183° 15'·8
Latitude = 36° 10'·0 N.	G.H.A. ✳ = 65° 56'·0
PZ = 53° 50'·0	Longitude = 44° 00'·0 W.
Declination ✳ = 14° 48'·3 N.	H.A. ✳ = 21° 56'·0
PX = 75° 11'·7	
(PX − PZ) = 21° 21'·7	

hav ZX = hav P sin P sin PZ sin PX + hav (PZ ∼ PX)	
log hav P = $\bar{2}$·55859	Sext. Altitude = 61° 02'·0
log sin PZ = $\bar{1}$·90704	Index Error = −0'·5
log sin PX = $\bar{1}$·98634	Observed Alt. = 61° 01'·5
log hav θ = 2·$\bar{4}$5097	Total Corr. = −5'·0
nat hav θ = 0·02824	True Alt. = 60° 56'·5
nat hav (PX − PZ) = 0·03435	O.Z.D. = 29° 03'·5
nat hav ZX = 0·06259	C.Z.D. = 28° 58'·5
ZX = 28° 58'·5	Intercept = 5'·0 AWAY

Azimuth (from Azimuth Tables) = 228½°

Position Line runs 138½°–318½° through a point 5·0 miles 048½° from Latitude 36° 10'·0 N., Longitude 44° 00'·0 W.

Example 34·8—31st December at about 0800 hr. on board, the sextant altitude of the Sun's lower limb was 36° 07'·0. Index error −1'·0. Height of eye 76 feet. Chronometer time 3 h. 47 m. 10 s. Chronometer error 05 m. 06 s. fast on G.M.T. Ascertain the position line on which the observer was located using the Intercept Method and a chosen position in Latitude 40° 05'·0 N., Longitude 63° 30'·0 E.

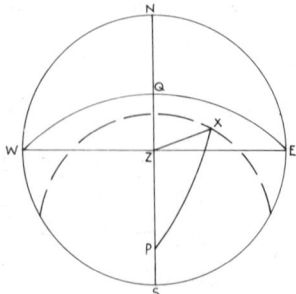

Fig. 34·17

Fig. 34·17 illustrates Example 34·8.

THE ASTRONOMICAL TRIANGLE AND SIGHT REDUCTION

Approx. Local Time = (31st) 08 h. 00 m.		G.H.A.T.S. at 03 h. = 224° 19′·2	
Longitude E. =	04 h. 14 m.	Increment =	10° 31′·0
Approx. G.M.T. = (31st) 03 h. 46 m.		G.H.A.T.S. = 234° 50′·2	
Chron. Time =	03 h. 47 m. 10 s.	Longitude =	63° 30′·0 E.
Chron. Error =	−05 m. 06 s.	H.A. =	298° 20′·2
G.M.T. (31st) =	03 h. 42 m. 04 s.	P =	61° 39′·8
Dec. (03 h.) = 23° 08′·5 S.	d = 0′·2	Latitude =	40° 05′·0 S.
Increment =	−0′·1	PZ =	49° 55′·0
Dec. = 23° 08′·4 S.		(PX − PZ) =	16° 56′·6
PX = 66° 51′·6			

$$\text{hav } ZX = \text{hav } P \sin PZ \sin PX + \text{hav } (PX \sim PZ)$$

log hav P =	$\bar{1}$·41941	Sext. Altitude = 36° 07′·0	
log sin PZ =	$\bar{1}$·88372	Index Error =	−1′·0
log sin PX =	$\bar{1}$·96357	Obs. Altitude = 36° 06′·0	
log hav θ =	$\bar{1}$·26670	Total Corr. =	+6′·5
nat hav θ =	0·18480	True Altitude = 36° 12′·5	
nat hav (PX − PZ) = 0·02170		O.Z.D. = 53° 47′·5	
nat hav ZX = 0·20650		C.Z.D. = 54° 03′·4	
ZX =	54° 03′·4	Intercept =	15′·9 TOWARDS

Azimuth (From Azimuth Tables) = 091½°

Position Line runs 001½°–181½° through a point 15·9 miles 091½° from Lat. 40° 05′·0 S., Long. 63° 30′·0 E.

Example 34·9—30th December at about 3 p.m. on board, the sextant altitude of the Sun's lower limb was 23° 25′·0. Index error 1′·0 off the arc. Height of eye 80 feet. Chronometer time 01 h. 34 m. 50 s. Chronometer error 08 m. 05 s. slow on G.M.T. Find, using the Intercept Method, the position line on which the observer was located, the chosen position being Lat. 35° 10′·0 N., Long. 161° 15′·0 W.

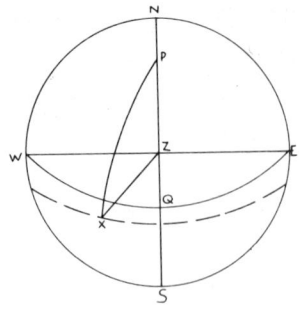

282 THE ELEMENTS OF NAVIGATION AND NAUTICAL ASTRONOMY

Fig. 34·18

Fig. 34·18 illustrates Example 34·8.

Approx. Local Time = (30th) 15 h. 00 m.
Longitude = 10 h. 45 m.
Approx. G.M.T. = (31st) 01 h. 45 m.

Chron. Time = 01 h. 34 m. 50 s.	G.H.A.T.S. at 01 h. = 194° 19'·8
Chron. Error = +08 m. 05 s.	Increment = 10° 43'·8
G.M.T. (31st) = 01 h. 42 m. 55 s.	G.H.A.T.S. = 205° 03'·6
Dec. (01 h.) = 23° 08'·8 S. d = 0'·2	Longitude = 161° 15'·0
Increment = −0'·1	H.A. = 33° 48'·6
Dec. = 23° 08'·7 S.	Lat. = 35° 10'·0 N.
PX = 113° 08'·7	PZ = 54° 50'·0
	$(PX-PZ)$ = 58° 18'·7

hav ZX = hav P sin PZ sin PX + hav $(PX-PZ)$

log hav P = $\bar{2}$·92712	Sextant Alt = 23° 25'·0
log sin PZ = $\bar{1}$·91248	Index Error = +1'·0
log sin PX = $\bar{1}$·96358	Obs. Altitude = 23° 26'·0
log hav θ = $\bar{2}$·80318	Total Corr. = +5'·4
nat hav θ = 0·06355	True Altitude = 23° 31'·4
nat hav $(PX \sim PZ)$ = 0·23736	O.Z.D. = 66° 28'·6
nat hav ZX = 0·30091	C.Z.D. = 66° 32'·2
ZX = 66° 32'·2	Intercept = 2'·8 TOWARDS

Azimuth (from Azimuth Tables) = 214°

Position Line runs 124°–304° through a point 2·8 miles 214° from Latitude 35° 10'·0 N., Long. 161° 15'·0 W.

7. The Modified Formula

In the Longitude Method of sight reduction the angle P is found from the formula:

$$\text{hav } P = \frac{\text{hav } ZX - \text{hav } (PX \sim PZ)}{\sin PX \sin PZ}$$

This formula is simplified by a modification illustrated in fig. 34·19.

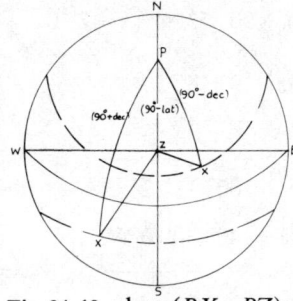

Referring to fig. 34·19:

$$PZ = (90° - NP) = (90° - \text{Lat.})$$

Therefore:

$$\sin PZ = \cos \text{Lat.}$$
$$PX = (90° \pm QY) = (90° \pm \text{dec.})$$

Therefore:

$$\sin PX = \cos \text{dec.}$$

Fig. 34·19

$$\text{hav } (PX \sim PZ) = \text{hav } \{(90° \pm \text{dec.}) \sim (90° - \text{Lat.})\}$$
$$= \text{hav } (\text{Lat.} \sim \text{dec.}) \text{ when Lat. and dec. have different names}$$
$$= \text{hav } (\text{Lat.} + \text{dec.}) \text{ when Lat. and dec. have the same name}$$

In general, therefore:

$$\text{hav } (PX \sim PZ) = \text{hav } (\text{Lat.} \pm \text{dec.})$$

The Haversine Formula given above may, therefore, be modified to:

$$\text{hav } P = \frac{\text{hav } ZX - \text{hav } (\text{Lat.} \pm \text{dec.})}{\cos \text{Lat.} \cos \text{dec.}}$$

or:

$$\text{hav } P = \{\text{hav } ZX - \text{hav } (\text{Lat.} \pm \text{dec.})\} \sec \text{Lat.} \sec \text{dec.}$$

By transposition, we have:

$$\text{hav } ZX = \text{hav } P \cos \text{Lat.} \cos \text{dec.} + \text{hav } (\text{Lat.} \pm \text{dec.})$$

this being the corresponding formula for use with the Intercept Method of sight reduction.

The following examples illustrate the use of the Modified Haversine Formula.

Example 34·10—23rd September, during evening twilight the observed altitude of Jupiter West of the meridian was 16° 42'·0. Height of eye 26 feet. Chronometer time 20 h. 44 m. 10 s. Chronometer error 07 m. 04 s. fast on G.M.T. Using the modified Haversine Formula and the Intercept Method, ascertain the position line on which the observer was located given the chosen position in Lat. 35° 20'·0 N., Long. 40° 30'·0 W.

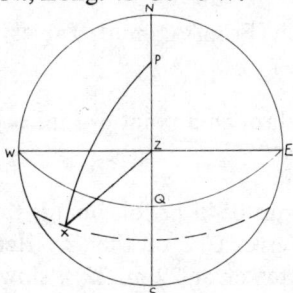

Fig. 34·20

Fig. 34·20 illustrates Example 34·10.

284 THE ELEMENTS OF NAVIGATION AND NAUTICAL ASTRONOMY

Approx. Local Time = (23rd) 18 h. 00 m.   Chron. Time = 20 h. 44 m. 10 s.

Approx. Longitude = W.   2 h. 42 m.   Chron. Error =   −07 m. 04 s.

Approx. G.M.T. = (23rd) 20 h. 42 m.   G.M.T. (23) = 20 h. 37 m. 06 s.

Dec. at 20 h. = 11° 37′·5 S.   d = 0′·2   G.H.A. at 20 h. = 90° 53′·4   v = 2′·0

Increment =   +0′·1   Increment = 9° 16′·5

                                                     v corr. =   +1′·3

Dec. = 11° 37′·6 S.

                                             G.H.A. Jupiter = 100° 11′·2

Latitude = 35° 20′·0 N.   Longitude = 40° 30′·0 W.

(Lat. + dec.) = 46° 57′·6   H.A. = 59° 41′·2

hav ZX = hav P cos Lat. cos dec. + hav (Lat. + dec.)

log hav P = $\bar{1}$·39381   Obs. Altitude = 16° 42′·0

log cos Lat. = $\bar{1}$·91158   Total Corr. =   −8′·4

log cos dec. = $\bar{1}$·99100

                              True Altitude = 16° 33′·6

log hav θ = $\bar{1}$·29639

                              O.Z.D. = 73° 26′·4

nat hav θ = 0·19788   C.Z.D. = 73° 20′·2

nat hav (1+d) = 0·15875   Intercept =   6′·2 AWAY

nat hav ZX = 0·35663

ZX = 73° 20′·2

Azimuth (From Azimuth Tables) = 243°

Position Line runs 133°–313° through a point 6·2 miles 043° from Latitude 35° 20′·0 N., Longitude 40° 30′·0 W.

Example 34·11—13th June at about 0545 hr. on board, the sextant altitude of the Moon's upper limb was 57° 01′·5. Index error 0′·5 on the arc. Height of eye 30 feet. Chronometer time 15 h. 14 m. 20 s. Chronometer error 29 m. 22 s. slow on G.M.T. Using the modified formula and the Intercept Method ascertain the position line on which the observer was located given the chosen position in Lat. 40° 20′·0 N., Long. 150° 00′·0 W.

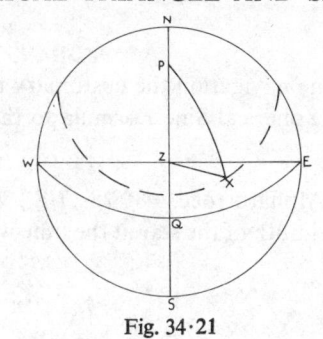

THE ASTRONOMICAL TRIANGLE AND SIGHT REDUCTION

Fig. 34·21

Fig. 34·21 illustrates Example 34·11.

Approx. Local Time = (13) 05 h. 45 m.	Chron. Time = 15 h. 14 m. 20 s.
Longitude W. = 10 h. 00 m.	Chron. Error = +29 m. 22 s.
Approx. G.M.T. = (13) 15 h. 45 m.	G.M.T. (13th) = 15 h. 43 m. 42 s.

Dec. at 15 h. = 13° 50'·4 N. d = 7'·0	G.H.A. = 89° 10'·0 v = 13'·2
Incr. = +5'·1	Incr. = 10° 25'·6
Dec. = 13° 55'·5 N.	v corr. = +9'·6
Lat. = 40° 30'·0 N.	G.H.A. = 99° 45'·2
	Long. W. = 150° 00'·0
(Lat. − dec) = 26° 24'·5	H.A. = 309° 45'·2

hav ZX = hav P cos Lat. cos dec. + hav (Lat. − dec.)

log hav P = $\bar{1}$·25589	Sext. Altitude = 57° 01'·5
log cos Lat. = $\bar{1}$·81106	Index Error = −0'·5
log cos dec. = $\bar{1}$·38139	
	Obs. Altitude = 57° 01'·0
log hav θ = $\bar{2}$·44834	Total Corr. = +8'·8
nat hav θ = 0·02808	True Altitude = 57° 09'·8
nat hav (1 − d) = 0·05217	
	O.Z.D. = 32° 50'·2
nat hav ZX = 0·08025	C.Z.D. = 32° 54'·8
ZX = 32° 54'·8	Intercept = 4'·6 TOWARDS

Azimuth (From Azimuth Tables) = 106°

Position Line runs 016°–196° through a position 4·6 miles 106° from Latitude 40° 20'·0 N., Longitude 150° 00'·0 W.

8. The Azimuth

In the early days of position line navigation the customary method of finding the azimuth was to apply the relatively simple Spherical Sine Formula to the *PZX*-triangle. The following example illustrates the method.

Example 34·12—The altitude of Alphard (dec. = 8° 28'·7 S.) was 32° 24'. The observer was in Latitude 45° 10'·0 S. Find the azimuth of the star at the time when its H.A. was 308° 14'.

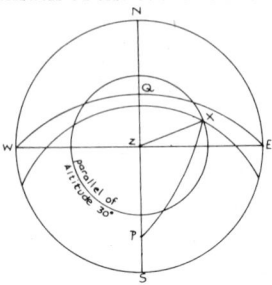

Fig. 34·22

Referring to fig. 34·22:

$$\sin Z = \sin PX \sin P / \sin ZX$$

or:

$$\sin Z = \cos \text{dec.} \sin P \sec \text{alt.}$$

$$\log \cos \text{dec.} = \bar{1}\cdot 99525$$
$$\log \sin P = \bar{1}\cdot 89515$$
$$\log \sec \text{alt.} = 0\cdot 07349$$
$$\log \sin Z = \bar{1}\cdot 96389$$

Therefore:
$$Z = \text{S. } 113° \text{ E.}$$

N.B.—Care was necessary in using this formula because the sine of an angle is equal to the sine of its complement; and, in the example given, it is not an uncommon mistake to call the azimuth 67° instead of its complement, viz. 113°.

This example should be compared with Example 34·3.

Answer—Azimuth = S. 113° E.

At the time when position line navigation was discovered iron, and later steel, vessels were becoming increasingly common. The difficulties associated with the magnetic compass on board iron and steel vessels paved the way for the introduction of pre-computed Azimuth Tables of the Davis and Burdwood type which are triple-entry tables giving azimuths against Latitude, hour angle and declination. The Davis and Burdwood Tables, although designed specifically for checking magnetic compasses, were readily adapted for position line navigation. They are still commonly used today for this purpose as well as for checking compasses. Although straightforward in use care is important in the necessary interpolation.

Another type of azimuth table is the *ABC* table which we shall discuss in detail in Chapter 38.

THE ASTRONOMICAL TRIANGLE AND SIGHT REDUCTION

9. Compass Error by Azimuth

Compass Error is found by comparing an observed Compass Bearing of a body whose True Bearing is known. In the days before pre-computed Azimuth Tables, the true bearing of a celestial body was computed by the observer. This required the solution of the *PZX*-triangle for the angle *Z*. Celestial bodies are more conveniently observed for azimuth when their altitudes are not great. Azimuth Mirrors and other devices used for observing azimuths or bearings are subject to error when observing bodies at altitudes in excess of about 30°.

True Azimuths at the present time are almost always obtained from Azimuth Tables or *ABC* Tables.

10. Compass Error by Amplitude

The angle at the observer's position between the East or West point of his celestial horizon and the direction of a heavenly body at rising or setting is known as the body's Rising- or Setting-Amplitude. To facilitate finding compass error from an observation of a body at rising or setting, Amplitude Tables are included in collections of tables such as those of Norie's or Burton's. Amplitude Tables are double-entry tables giving amplitudes against arguments Latitude and declination. Amplitudes are readily computed by applying Napier's Rules to the *PZX*-triangle, which is a quadrantal triangle in cases in which the True Altitude of the body is 0°. Fig. 34·23 illustrates the so-called Amplitude Formula.

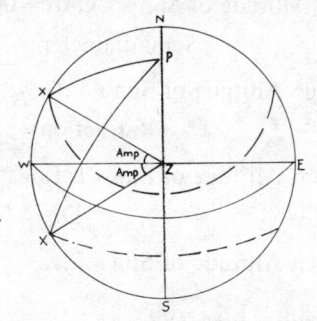

Fig. 34·23

In fig. 34·23:

$$ZX = 90°$$
$$PZ = \text{co-Latitude of observer}$$
$$PX = 90° \pm \text{declination of observed body}$$
$$PZX = 90° \pm \text{amplitude of observed body} = (90° \pm A°)$$

By Napier's Rules:

$$\cos PX = \cos(90° \pm A) \cos(90° - PZ)$$

That is: $\sin \text{dec.} = \sin A \cos \text{Lat.}$

or: $\sin A = \sin \text{dec.} \sec \text{Lat.}$

Amplitudes are always named according to the declination of the observed body, illustrated in Example 34·13.

288   THE ELEMENTS OF NAVIGATION AND NAUTICAL ASTRONOMY

Example 34·13—Find the Sun's rising amplitude when its declination is 20° S., and the observer's Latitude is 40° N.

$$\sin \text{Amp.} = \sin \text{dec.} \sec \text{Lat.}$$

$$\log \sin \text{dec.} (20°) = \bar{1} \cdot 53405$$

$$\log \sec \text{Lat.} (40°) = 0 \cdot 11574$$

$$\log \sin \text{Amp.} = \bar{1} \cdot 64979$$

$$\text{Rising Amplitude} = \text{E. } 26\tfrac{1}{2}° \text{ S.}$$

Answer—Amplitude = E. $26\tfrac{1}{2}°$ S.

It is important to bear in mind that when observing the Sun or Moon in an amplitude observation that the centre of the observed body should be on the celestial horizon of the observer, and to remember that the visible horizon does not coincide with the celestial horizon. The effects of refraction, parallax and dip, results in the Sun's and Moon's lower limb having an altitude of several minutes of arc at the time its centre is on the celestial horizon. The following examples demonstrate this.

Example 34·14—Find the observed altitude of the Sun's lower limb at a time when its semi-diameter is 16′ and the dip of the horizon is 5′.

$$\begin{aligned}
\text{True Altitude of Sun's Centre} &= 00° \ 00' \\
\text{Semi-diameter} &= \ -16' \\
\text{True Altitude of Sun's L.L.} &= \ -16' \\
\text{Refraction} &= \ +33' \\
\text{Apparent Altitude of Sun's L.L.} &= \ +17' \\
\text{Dip} &= \ +5' \\
\text{Observed Altitude of Sun's L.L.} &= \ +22'
\end{aligned}$$

Answer—Observed Altitude of Sun's L.L. = 00° 22′.

Example 34·15—Find the observed altitude of the Moon's lower limb at a time when the Moon's H.P. and S.D. are, respectively, 64′ and 16′, and the dip of the horizon is 5′.

$$\begin{aligned}
\text{True Altitude of Moon's centre} &= \ \ 00° \ 00' \\
\text{Semi-diameter} &= \ \ -16' \\
\text{True Altitude of Moon's L.L.} &= -00° \ 16' \\
\text{Parallax} &= -01° \ 04' \\
\text{Refraction} &= \ \ +33' \\
\text{Apparent Altitude of Moon's L.L.} &= \ \ -47' \\
\text{Dip} &= \ \ +5' \\
\text{Observed Altitude of Moon's L.L.} &= \ \ -42'
\end{aligned}$$

Answer—Observed Altitude of Moon's L.L. = −42′.

THE ASTRONOMICAL TRIANGLE AND SIGHT REDUCTION

It is particularly important when using amplitude tables to make the observation at the precise time at which the body's centre is on the celestial horizon, especially when the Latitude is high. The reason for this stems from the fact that the higher the Latitude the sharper the angle a body's diurnal circle makes with the horizon. In these circumstances the rate at which a body is changing its azimuth is relatively large, and a small change in altitude results in a large change in azimuth. In practice it is better to treat all observations made to find compass error as straightforward azimuth problems and to ignore the amplitude table.

Exercises on Chapter 34

1. Explain the term "sight" used in nautical astronomy.
2. Explain carefully the sides and angles of a typical *PZX*-triangle.
3. Explain the principles of astronomical position line navigation.
4. Describe Captain Sumner's discovery.
5. Analyse the Longitude Method of obtaining a position line, and explain the circumstances in which the method falls down.
6. Explain carefully the intercept method of sight reduction.
7. Explain the basis of naming the intercept TOWARDS or AWAY.
8. Demonstrate that the intercept method is superior to the Longitude method of sight reduction.
9. Explain why a single sight yields only a position line and not a position.
10. Show how the haversine formula applied to the *PZX*-triangle for finding (i) angle *P*, (ii) side *ZX*, is modified by using Latitude and declination instead of sides *PZ* and *PX* respectively.
11. Explain the Amplitude Table and its construction.
12. What precautions are necessary when making observations of (i) Azimuths, and (ii) Amplitudes?
13. Explain why great care is necessary when making amplitude observations in high Latitudes.
14. 16th June, at about 0730 hr. on board, the observed altitude of the Sun's lower limb was 38° 05'·0. Height of eye 36 feet. Chronometer time 11 h. 30 m. 05 s. Chronometer error 8 m. 12 s. slow on G.M.T. Using Latitude 40° 00'·0 N., Longitude 53° 30'·0 W. reduce the sight using the intercept method, and ascertain the position line on which the vessel was located.
15. 22nd September, during the afternoon, the observed altitude of the Sun's lower limb was 26° 30'·0. Height of eye 40 feet. Chronometer time 15 h. 35 m. 20 s. Chronometer error 4 m. 02 s. slow on G.M.T. Using Lat. 50° 00'·0 N., Long. 10° 00' W., reduce the sight using the intercept method, and ascertain the position line on which the vessel was located.
16. 15th June, at about 1400 hr. on board, the observed altitude of the Sun's lower limb was 12° 30'·0. Height of eye 35 feet. Chronometer time 0 h. 59 m. 20 s. Chronometer error 19 m. 10 s. fast on G.M.T. The sight was reduced using Lat. 45° 10'·0 S., Long. 29° 30'·0 E., find the intercept and position line.

290   THE ELEMENTS OF NAVIGATION AND NAUTICAL ASTRONOMY

17. 30th December, a.m. on board, the observed altitude of the Sun's lower limb was 44° 46′·0. Height of eye 63 feet. Chronometer time 4 h. 22 m. 10 s. Chronometer error 20 m. 05 s. slow on G.M.T. The sight was reduced using Lat. 35° 10′·0 S., Long. 59° 30′·0 E. Ascertain the position line.

18. 22nd September, a.m. on board, the observed altitude of Alphard, east of the meridian, was 30° 12′·0. Height of eye 45 feet. Chronometer time 1 h. 35 m. 20 s. Chronometer error 5 m. 02 s. slow on G.M.T. The sight was reduced using Lat. 44° 50′·0 S., Long. 120° 00′·0 W. Find the intercept and position line.

19. 14th June, during morning twilight, the observed altitude of Mars was 64° 02′·0. Index error 0′·5 off the arc. Height of eye 26 feet. Chronometer time 10 h. 55 m. 23 s. Chronometer error 13 m. 04 s. fast on G.M.T. Using Lat. 25° 10′·0 S., Long. 50° 00′·0 W., find the intercept and position line. Verify that the Longitude method of sight reduction breaks down for this problem.

20. 22nd September, during evening twilight, the observed altitude of Jupiter was 19° 32′ 0. Height of eye 29 feet. Chronometer time 7 h. 01 m. 23 s. Chronometer error 18 m. 14 s. fast on G.M.T. Reduce the sight using Lat. 45° 00′·0 N., Long. 25° 00′·0 W., and ascertain the intercept and position line.

21. 14th June, during morning twilight, the observed altitude of Altair was 31° 44′·0. Height of eye 44 feet. Chronometer time 08 h. 00 m. 00 s. Chronometer error 16 m. 04 s. fast on G.M.T. Using Lat. 25° 00′·0 S., Long. 32° 00′·0 W., find the intercept and position line.

22. 15th June, during morning twilight, the observed altitude of Achernar was 51° 47′·0. Height of eye 49 feet. Chronometer time 7 h. 40 m. 03 s. Chronometer error 2 m. 20 s. slow on G.M.T. Using Lat. 27° 00′·0 S., Long. 28° 00′·0 W., find the intercept and position line.

23. 23rd September, during morning twilight, the observed altitude of Regulus was 40° 40′·0. Height of eye 36 feet. Chronometer time 5 h. 40 m. 20 s. Chronometer error 4 m. 15 s. fast on G.M.T. Using Lat. 26° 00′·0 N., Long. 165° 00′·0 W., ascertain the position line using the intercept method.

24. 23rd September, p.m. on board, the observed altitude of the Moon's upper limb was 12° 11′·3. Height of eye 40 feet. Chronometer time 18 h. 40 m. 20 s. Chronometer error 01 m. 02 s. slow on G.M.T. Reduce the sight using the intercept method using Lat. 54° 00′·0 N., Long. 15° 00′·0 W., and ascertain the intercept and position line.

25. 31st December, a.m. on board, the observed altitude of the Moon's upper limb was 35° 48′·7. Height of eye 40 feet. Chronometer time 8 h. 36 m. 02 s. Chronometer error 00 m. 04 s. slow on G.M.T. Using Lat. 48° 00′·0 N., Long. 30° 00′·0 W., reduce the sight using the intercept method and ascertain the intercept and position line.

Note—It would be a useful and illuminating exercise to solve questions 14 to 25 inclusive using the Longitude method of sight reduction, and to verify that the same position line is found in every case.

CHAPTER 35

POSITION LINES AND PLOTTING CHART

1. Introduction

Not uncommonly, a navigator observes the Sun in the morning and solves his sight using the Longitude Method. It is important to realize, as we have stressed in Chapter 34, that the calculated Longitude is not generally the observer's actual Longitude at the time at which the observation is made. The calculated Longitude is the observer's actual Longitude only if the Latitude used in the computation is the actual Latitude of the observer at the time of the sight. Because, in general, the Latitude of a vessel is not known, the Longitude of the vessel cannot be determined from a single sight. It is important to appreciate that the result of a morning Sun-sight is NOT a Longitude but a POSITION LINE. The Longitude calculated is that of a point located on the position line which also lies on the parallel of the Latitude used in reducing the sight; and the direction of the position line is at right angles to the bearing of the observed body at the time of the sight.

Referring to fig. 35·1, suppose an observation of the Sun bearing 070° gives a calculated Longitude $A°$ when Latitude $a°$ is used in the computation; then the calculated Longitude would be $B°$ had Latitude $b°$ been used; Longitude $C°$ had Latitude $c°$ been used; and so on. Notice that, in every case, although a different Longitude is computed, the same position line is determined.

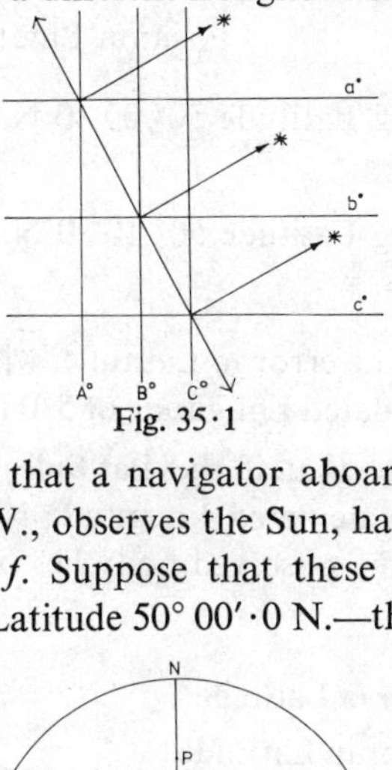

Fig. 35·1

Referring to fig. 35·2, suppose that a navigator aboard a stationary vessel in Latitude 50° 00'·0 N., Longitude 30° 00'·0 W., observes the Sun, having North declination, when it is successively at positions d, e and f. Suppose that these observations are reduced by the Longitude Method using, in turn, Latitude 50° 00'·0 N.—the true Latitude of the vessel—and Latitude 50° 10'·0 N.

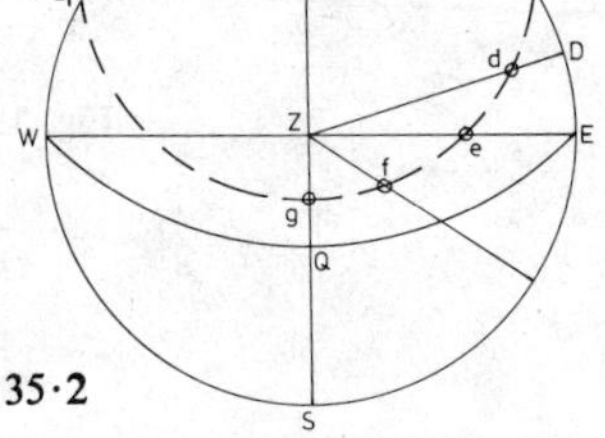

Fig. 35·2

292 THE ELEMENTS OF NAVIGATION AND NAUTICAL ASTRONOMY

(i) *Sun at Position d:*

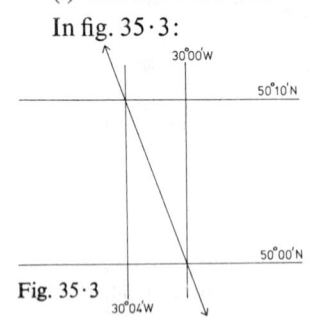

In fig. 35·3: Azimuth $NZD = 077°$

Position Line: Direction $= 347°–167°$

Using Latitude 50° 00'·0 N., the calculated Longitude $=$ 30° 00'·0 W.

Using Latitude 50° 10'·0 N., the calculated Longitude $=$ 30° 04'·0 W.

The error in Latitude, which is 10'·0, produces an error in Longitude of 2·5 miles, or 4'·0 W.

Fig. 35·3

(ii) *Sun at Position e:*

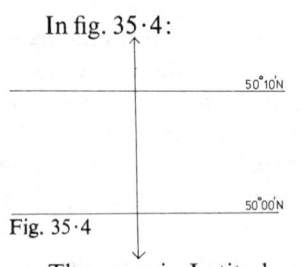

In fig. 35·4: Azimuth $NZE = 090°$ (Sun on Prime Vertical Circle)

Position Line: Direction $= 000°–180°$

Using Latitude 50° 00'·0 N., the calculated Longitude $=$ 30° 00'·0 W.

Using Latitude 50° 10'·0 N., the calculated Longitude $=$ 30° 00'·0 W.

Fig. 35·4

The error in Latitude, which is 10'·0, produces no error in the calculated Longitude.

(iii) *Sun at Position f:*

In fig. 35·5: Azimuth $NZF = 118°$

Position Line: Direction $= 028°–208°$

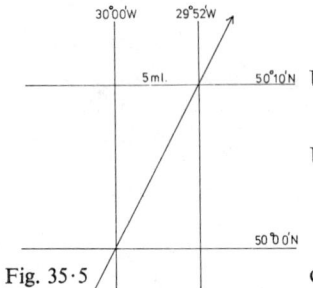

Using Latitude 50° 00'·0 N., the calculated Longitude $=$ 30° 00'·0 W.

Using Latitude 50° 10'·0 N., the calculated Longitude $=$ 29° 52'·0 W.

Fig. 35·5

The error in Latitude, which is 10'·0, produces an error in the calculated Longitude of 5·0 miles, or 8'·0 E.

The above examples illustrate that if the Latitude used in the computation is not the True Latitude of the vessel the calculated Longitude is NOT the vessel's actual Longitude, except in the special case when the observed body lies on the observer's prime vertical circle at the time of the observation.

2. Error in Longitude due to Error in Latitude

Error in Longitude due to error in Latitude may be found as follows:

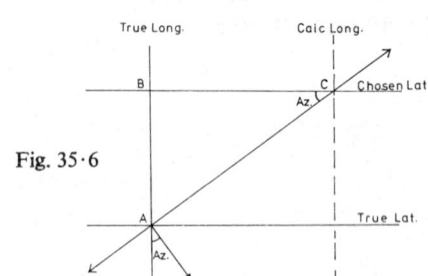

Fig. 35·6

POSITION LINES AND PLOTTING CHART

Referring to fig. 35·6: In triangle ABC:

$$\text{Error in Latitude} = AB$$
$$\text{Corresponding Error in Departure} = BC$$
$$\text{Azimuth} = BCA$$

Now: $\qquad\qquad\qquad\qquad\qquad BC = AB \cot BCA$

That is: $\quad$ Error in Departure = Error in Latitude . cot Azimuth

But: $\quad$ Error in Departure = Error in Longitude . cos Latitude

Therefore: $\quad$ Error in Longitude = Error in Latitude . cot Azimuth . sec Latitude

From this formula it may be seen that, for any given Latitude, the error in Longitude consequent upon an error in Latitude varies as the cotangent of the azimuth. Thus, error in Longitude increases from zero when the azimuth is 90° to infinity when the azimuth is 0°. It is for this reason that when making an observation for Longitude, optimum conditions prevail when the observed body is on the observer's prime vertical circle; or, in other words, when the observed body bears 090° or 270°. It is for this reason that in days before the advent of astronomical position line navigation, navigators made a point of observing the Sun when on, or nearest to, the prime vertical circle.

It may also be seen from the formula that error in Longitude varies as the secant of the Latitude, so that the error increases with the Latitude; being least when the Latitude is 0° (secant 0° = 1), and infinity when the Latitude is 90° (secant 90° = ∞).

A table, given in Norie's and Burton's Nautical Tables, gives the error in Longitude consequent upon an error of one minute in Latitude, for all values of Latitude and azimuth. This table is known as the Longitude Correction Table. The Longitude Correction may also be obtained from the *A B C* Table described in Chapter 38.

3. Effects of Errors in Altitude

Any error in an altitude used in calculating a position line affects the position through which to plot the position line. If the altitude is in error and too great, the position line will be displaced towards the direction of the observed body. If, on the other hand, the altitude is in error and too small, the position line will be displaced away from the direction of the observed body.

Referring to fig. 35·7, suppose that the true position of a vessel is represented as being at A and that, due to an error in altitude amounting to AB, a false position line is drawn through B. The position line passes through the meridian of the vessel at C. Therefore, the error in Latitude due to an error of AB in the altitude is AC.

The position line passes through the parallel of the vessel's Latitude at D, so that the error in departure, due to the error AB in altitude, is AD.

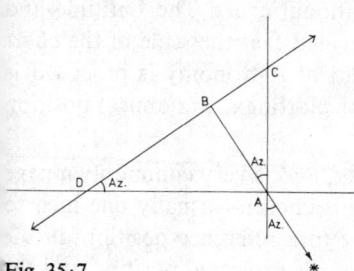

Fig. 35·7

In the triangles ABC and ABD which, if the distance AB is small, may be considered to be plane right-angled triangles, the angles BDA and BAC are each equal to the azimuth of the body at the time of the observation.

In triangle ABC: $\qquad AB = AC \cos CAB$

Thus: Error in Altitude = Error in Latitude . cos Azimuth

Or: Error in Latitude = Error in Altitude . sec Azimuth

These formulae indicate that Error in Latitude is greatest when secant Azimuth is greatest; that is to say, when the Azimuth is 90°. The Error is least when the Azimuth is 0°; that is, when the observed body is on the meridian. Now the secant of 0° is unity, so that, for a body at meridian passage, an error in altitude produces an equal error in Latitude.

In triangle ABD $\qquad AD = AB \operatorname{cosec} ADB$

Thus: Error in Departure = Error in Altitude . cosec Azimuth

Or: Error in Altitude = Error in Departure . sin Azimuth

But: Error in Departure = Error in Longitude . cos Latitude

Thus: Error in Altitude = Error in Longitude . cos Latitude . sin Azimuth

Or: Error in Longitude = Error in Altitude . sec Latitude . cosec Azimuth

This formula indicates that, for any error in altitude for a given Latitude, the error in Longitude is greatest when the cosecant of the azimuth is greatest. This occurs when the azimuth is 90°, that is to say when the body is on the observer's prime vertical circle. The error decreases as the azimuth decreases from 90° to 0°.

It may also be seen from the formula that, for any given azimuth, error in Longitude due to error in altitude is greatest when secant Latitude is greatest. This occurs when the Latitude is 90°.

Error in altitude may arise from the sextant used in taking the altitude not being in correct adjustment. It may also result through faulty refraction and dip corrections.

Any error in altitude which affects all position lines equally may readily be eliminated or reduced in cases in which the position lines are obtained from observations of four stars one in each quadrant of the compass. If, for example, stars bearing 000°, 090°, 180° and 270°, are observed, the four false position lines will intersect to form a square; If the same error affects all four position lines the true position of the vessel is then at the centre of the square.

4. The Plotting Chart

The most convenient way of ascertaining a vessel's position from astronomical observations, is to plot the resulting position lines on a large-scale navigational chart. The Latitude and Longitude of the vessel may then be lifted from the chart provided that the scale of the chart is sufficiently large. The sea-chart in use at times when Nautical Astronomy is practised is usually a small-scale chart: it is, therefore, quite unsuitable for plotting astronomical position lines. In these circumstances resort is made to a Plotting Sheet.

A plotting sheet is simply a sheet of blank or squared paper, and is very commonly a page of the navigator's workbook. A convenient scale of distance is chosen—usually one inch to represent ten miles—and the position lines are drawn relative to a reference position on the plotting sheet. The D. Lat. and the Departure between the vessel's projected position and the reference position may readily be measured. Departure is then converted into D. Long. usually by means of the traverse table, and the vessel's position found by applying the D. Lat. and D. Long. to the reference position.

POSITION LINES AND PLOTTING CHART

The method of finding a vessel's position by means of a plotting sheet is superior to those in which tabular methods are used. The latter, often used mechanically, tend to cause the user to lose sight of the principles of the problem. Moreover, the degree of reliability of the observed position is better assessed from the plotting sheet on which the problem is clearly displayed, than is the case when a tabular method is employed.

The following examples illustrate the use of the plotting sheet.

Example 35·1—Using position Lat. 31° 23'·0 S., Long. 49° 43'·0 W. to reduce his sights, a navigator's observation of Canopus bearing 136° gave an intercept of 5'·0 TOWARDS, and that of α Pavonis bearing 220° gave an intercept of 8'·5 AWAY. Find the vessel's position by plotting.

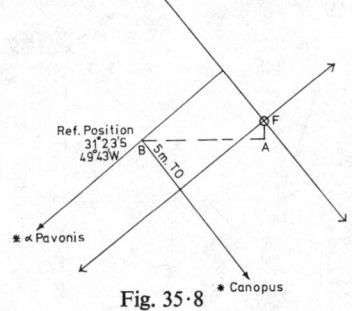

Fig. 35·8

In fig. 35·8:

$$\text{Departure } (AB) = 10'·0 \text{ E.}$$
$$\text{D. Long. (from Trav. Tables)} = 11'·7 \text{ E.}$$

Ref. Position: (B) Lat.	31° 23'·0 S.	Long.	49° 43'·0 W.
(AF) D. Lat.	2'·7 N.	D. Long.	11'·7 E.
Vessel's Position (F) Lat.	31° 20'·3 S.	Long.	49° 31'·3 W.

Answer—Vessel's Position: Lat. = 31° 20'·3 S., Long. = 49° 31'·3 W.

Example 35·2—An observation of the Sun bearing 120°, at 0930 hr. gave a calculated Longitude of 32° 10'·0 W., using Lat. 40° 00'·0 N., to reduce the sight. The vessel travelled on a course of 292° (T) at a speed of 12·0 knots until noon, when the Latitude by observation of the Sun on the meridian gave a Latitude of 40° 25'·0 N. Find the Longitude of the vessel at noon.

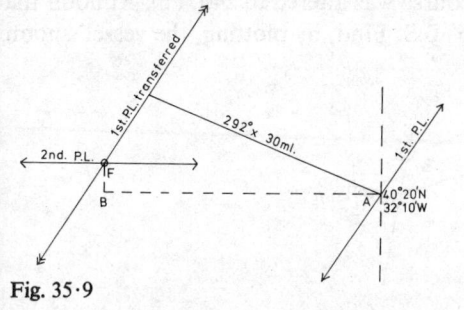

Distance travelled = 2·5 × 12·0
= 30·0 miles

From fig. 35·9:

Departure (AB) =   31'·4
D. Long. =   41'·0 W.
Long. A = 32° 10'·0 W.
Long. F = 32° 51'·0 W.

Fig. 35·9

Answer—Longitude at noon = 32° 51'·0 W.

296   THE ELEMENTS OF NAVIGATION AND NAUTICAL ASTRONOMY

Note—An alternative solution to Example 35·2 involves plotting the transferred position line through a position obtained by applying the D. Lat. and D. Long.—found from the Traverse Tables—corresponding to the course and distance made good (292° (T)×30·0 miles) to the position through which the first position line is plotted in the above solution. By so doing the necessity of plotting the first position line and the run is obviated.

Example 35·3—Using Lat. 50° 45′·0 N., Long. 30° 10′·0 W., the intercept obtained from an observation of the Sun bearing 040° (T) was zero. The vessel then travelled on a course of 213° (T) for a distance of 40·0 miles, when a second observation of the Sun bearing 075° (T) gave an intercept of 5·0 miles TOWARDS. Find by plotting the vessel's position at the time of the second observation, given that the position used in reducing the second sight was Lat. 50° 00′·0 N., Long. 31° 00′·0 W.

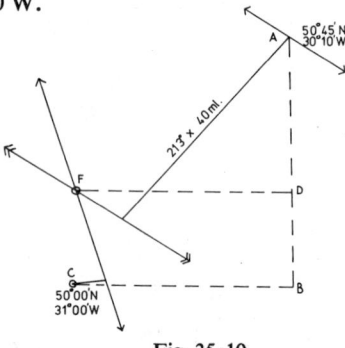

Fig. 35·10

D. Lat. (*AB*)= 45′ S.	Dep. (*AF*)= 32′·8
D. Long. (*AC*)= 50′ W.	D. Long. (*AF*)= 51′·0 W.
Dep. (*AC*)= 32·0 miles	D. Lat. (*AF*)= 25·0
Lat. *A*= 50° 45′·0 N.	Long. *A*= 30° 10′·0 W.
D. Lat. *AF*= 25′·0 S.	D. Long. *AF* = 51′·0 W.
Lat. *F*= 50° 20′·0 N.	Long. *F*= 31° 01′·0 W.

Answer—Lat. = 50° 20′·0 N., Long. = 31° 01′·0 W.

Example 35·4—At 0800 hr., using Lat. 20° 00′·0 S., Long. 170° 00′·0 E., an observation of the Sun bearing 075° gave an intercept of 8·5 miles AWAY. The vessel travelled on a course of 275° (T) at a speed of 8·0 knots. At 1100 hr. the course was altered to 240° (T). At noon the Latitude by meridian altitude of the Sun was 20° 15′·0 S. Find, by plotting, the vessel's noon position.

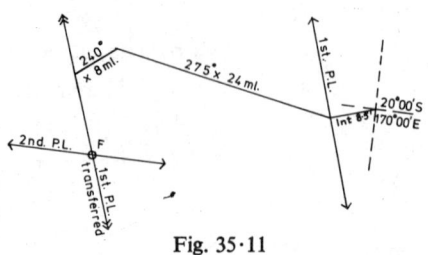

Fig. 35·11

POSITION LINES AND PLOTTING CHART

From fig. 35·11:

$$\text{Dep. } AF = 37 \cdot 0 \text{ miles}$$
$$\text{D. Long. } AF = 39' \cdot 5 \text{ W.}$$
$$\text{Long. } A = 170° \ 00' \cdot 0 \text{ E.}$$
$$\text{Long. } F = 169° \ 20' \cdot 5 \text{ E.}$$

Answer—Noon Position: Lat. = 20° 15′·0 S., Long. = 169° 20′·5 E.

Example 35·5—At 0600 hr. using Lat. 46° 45′·0 N., Long. 51° 45′·0 W. an observation of a star bearing 045° gave an intercept of 5·0 miles AWAY. After travelling for 22·0 miles on a course of 250° (T) Cape Race (Lat. 46° 39′·0 N., Long. 53° 04′·0 W.) bore 032° (T). Find the vessel's position at the time at which the bearing of Cape Race was observed.

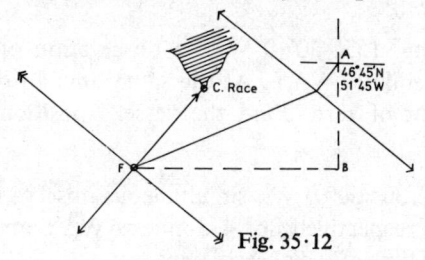

Fig. 35·12

From fig. 35·12

$$\text{D. Lat. } AF = 17' \cdot 5 \text{ S.}$$
$$\text{Dep. } AF = 18' \cdot 5 \text{ miles}$$
$$\text{D. Long. } AF = 26' \cdot 8 \text{ W.}$$

Lat. A = 46° 45′·0 N.	Long. = 51° 45′·0 W.
D. Lat. AF = 17′·5 S.	D. Long. = 26′·8 W.
Lat. F = 46° 27′·5 N.	Long. = 52° 11′·8 W.

Answer—Lat. = 46° 27′·5 N., Long. = 52° 11′·8 W.

Exercises on Chapter 35

1. Explain why it is impossible to find a vessel's position when out of sight of land from a single astronomical observation.

2. Is it possible to find an observer's Longitude from a single observation? Explain your answer.

3. Discuss the factors which influence the error in Longitude consequent upon an error in Latitude when the Longitude Method is used for sight reduction.

4. Prove: Error in Long. = Error in Lat. . cot Az. . sec Lat.

5. Explain how the Longitude Correction Table is computed.

6. Explain clearly the effect on an astronomical position line due to an error in altitude.

7. Prove: Error in Lat. = Error in Alt. . sec Az.

298 THE ELEMENTS OF NAVIGATION AND NAUTICAL ASTRONOMY

8. Prove: Error in Long. = Error in Alt. . sec Lat. . cosec Az.

9. Describe the use of a simple plotting sheet. How does a plotting chart differ from a navigational chart when used for plotting astronomical position lines for fixing?

10. Using Lat. 40° 05′·0 N., for reducing an observation of the Sun bearing 285° by the Longitude Method, the calculated Longitude was 46° 15′·0 W. At the same time the Latitude by meridian altitude of Venus was found to be 40° 10′·0 N. Find the vessel's position.

11. Using Lat. 25° 10′·0 S., Long. 120° 33′·0 W., the intercept obtained from an observation of a star bearing 065° was 6·0 miles TOWARDS. At the same time an observation of a star bearing 134° gave an intercept of 4·5 miles TOWARDS. Find the vessel's position at the time of the observations.

12. Using Lat. 44° 10′·0 N., Long. 155° 30′·0 W., an observation of the Moon bearing 254° gave an intercept of 5·5 miles AWAY. At the same time an observation of a star bearing 105° gave an intercept of zero. Find the vessel's position at the time of the observations.

13. Using Lat. 60° 30′·0 S., Long. 30° 00′·0 W., simultaneous observations of stars bearing 158° and 020° gave intercepts respectively, of 4·5 miles AWAY and 3·0 miles AWAY. Find the vessel's position at the time of the observations.

14. The calculated Longitude obtained from an observation of the Sun which bore 085° was 43° 06′·0 W. The latitude used in reducing the sight was 50° 45′·0 N. The vessel travelled on a course of 280° (T) for a distance of 30·0 miles, when a second observation of the Sun which bore 165° (T) gave an intercept of 4·0 miles AWAY, the position used for reducing the sight being Lat. 51° 00′·0 N., Long. 43° 30′·0 W. Find the vessel's position at the time of the second observation.

15. Using Lat. 40° 30′·0 S., the calculated Longitude obtained from an observation of the Sun bearing 105° was 124° 20′·0 W. The vessel travelled on a course of 090° (T) for a distance of 20·0 miles, and then on a course of 140° (T) for a distance of 15·0 miles. At the end of this run the Latitude by meridian altitude of the Sun was 40° 30′·0 S. Find the Longitude of the vessel at noon.

16. At 1000 hr. when the log registered 46·0, the Sun was observed bearing 156° (T) and the intercept, using Lat. 40° 00′·0 S., Long. 00° 00′·0. to reduce the sight was 5·0 miles AWAY. At 1400 hr. when the log registered 86·0, an observation of the Sun bearing 210° gave an intercept of 4·0 miles TOWARDS, using Lat. 40° 00′·0 S., Long. 00° 30′·0 W. Find the vessel's position at 1400 hr. given that the course made good between the times of the observations had been 275° (T).

17. At 0300 hr. a point of land in Lat. 37° 00′·0 N., Long. 08° 54′·0 W. bore 040° (T). The vessel travelled for a distance of 25·0 miles on a course of 195° (T), during which time the current was estimated to have set 270° (T) for a distance of 5·0 miles. At the end of this run, an observation of a star bearing 040° gave an intercept of 5·0 miles AWAY, using Lat. 36° 20′·0 N., Long. 09° 00′·0 W. to reduce the sight. Find the vessel's position at the time of the observation of the star.

CHAPTER 36

CELESTIAL BODIES NEAR THE MERIDIAN

1. General Remarks

It is sometimes useful to know what stars are within a given interval from their times of meridian passage.

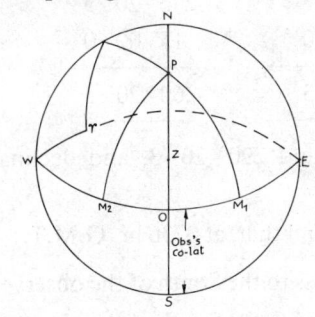

Fig. 36·1

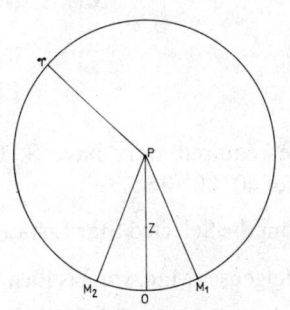

Fig. 36·2

In figs. 36·1 and 36·2, which are projections of the celestial sphere onto the planes of an observer's celestial horizon and celestial equator, respectively, PO represents the upper celestial meridian of the observer, and $P\Upsilon$ represents the celestial meridian of the First Point of Aries. The arc $O\Upsilon$ is the Local Hour Angle of Υ. This may readily be found by applying the observer's Longitude to the G.H.A. of Υ which is tabulated against G.M.T. in the *Nautical Almanac*. The L.H.A. of Υ subtracted from 360° 00' gives the Sidereal Hour Angle of the Observer's Upper Celestial Meridian.

Suppose that the celestial meridian PM_1 lies 15° to the East of the observer's upper celestial meridian, and that the celestial meridian PM_2 lies 15° to the West of the observer's celestial meridian. All celestial bodies which lie within these two celestial meridians have Sidereal Hour Angles within 15° of that of the observer's upper celestial meridian.

All celestial bodies which lie between the celestial meridian PM_1 and the observer's upper celestial meridian have S.H.A.s which are less than the S.H.A. of the observer's upper celestial meridian. On the other hand, all celestial bodies which lie between the observer's upper celestial meridian and the celestial meridian PM_2 have S.H.A.s which are more than the S.H.A. of the observer's upper celestial meridian.

It may be seen from fig. 36·1 that, for a celestial body to be above the horizon at the time of its upper transit, its declination, if of opposite name to that of the observer's Latitude, must be less than the co-Latitude of the observer. For a celestial body to be above the horizon at the time of its lower transit, its polar distance must be less than the observer's latitude. (Refer to Part 4, Chapter 28.)

The following examples illustrate a method of finding what stars are within a given interval of time of meridian passage.

300 THE ELEMENTS OF NAVIGATION AND NAUTICAL ASTRONOMY

Example 36·1—What visible stars are less than 1 hour from the upper celestial meridian of an observer in Lat. 50° 00′ N., Long. 30° 00′ W. at 0736 G.M.T. on 22nd September.

$$\begin{aligned}
\text{G.H.A. } \Upsilon \text{ at 07 hr. G.M.T.} &= 105°\ 38'·2 \\
\text{Increment for 36 m.} &= 09°\ 01'·5 \\
\text{G.H.A. } \Upsilon \text{ at 0736 hr. G.M.T.} &= 114°\ 39'·7 \\
\text{Longitude} &= 30°\ 00'·0\ \text{W.} \\
\text{L.H.A.} &= 84°\ 39'·7 \\
&\ \ 360°\ 00'·0
\end{aligned}$$

S.H.A. Obs. Upper Mer. =	275° 20′·3	275° 20′·3
Interval 1 hour =	15° 00′·0	15° 00′·0
Limiting S.H.A.s =	290° 20′·3	260° 20′·3

The required stars have S.H.A.s between 260° 20′·3 and 290° 20′·3, and declinations North of 40° 00′·0 S.

From the Selected Star List in the *Nautical Almanac*, we find that, at 0736 hr. G.M.T.:

Betelgeuse (Mag. var.) is East of the meridian and will cross to the South of the observer.
Alnilam (Mag. 1·8) is West of the meridian and has crossed to the South of the observer.
Elnath (Mag. 1·8) is West of the meridian and has crossed to the South of the observer.
Bellatrix (Mag. 1·7) is West of the meridian and has crossed to the South of the observer.
Capella (Mag. 0·2) is West of the meridian and has crossed to the North of the observer.
Rigel (Mag. 0·3) is West of the meridian and has crossed to the North of the observer.

Example 36·2—What visible navigational stars are within 1½ hours of the lower celestial meridian of an observer in Lat. 50° 00′·0 S., Long. 34° 15′·0 W. at 2040 hr. G.M.T. on 30th December.

$$\begin{aligned}
\text{G.H.A. } \Upsilon \text{ at 20 hr. G.M.T.} &= 38°\ 45'·0 \\
\text{Increment for 40 m.} &= 10°\ 01'·6 \\
\text{G.H.A. } \Upsilon \text{ at 2040 hr. G.M.T.} &= 48°\ 46'·6 \\
\text{Longitude} &= 34°\ 15'·0\ \text{W.} \\
\text{L.H.A.} &= 14°\ 31'·6 \\
&\ \ 360°\ 00'·0
\end{aligned}$$

S.H.A. Observer's U. Mer. =	345° 28′·4	
	180° 00′·0	180° 00′·0
S.H.A. Observer's L. Mer. =	165° 28′·4	165° 28′·4
Interval 1½ hours =	+22° 30′·0	−22° 30′·0
Limiting S.H.A.s =	187° 54′·4	142° 58′·4

CELESTIAL BODIES NEAR THE MERIDIAN

The required stars have S.H.A.s between 187° 58'·4 and 142° 58'·4, and declinations South of 40° 00'·0 S.

From the Selected Star List, we find that at 2040 hr. G.M.T.:

Hadar (Mag. 0·9) is West of the Observer's lower meridian and will cross South of the observer.

Gacrux (Mag. 1·6) is East of the observer's lower meridian and has crossed South of the observer.

Acrux (Mag. 1·1) is East of the observer's lower meridian and has crossed South of the observer.

2. Position Line from Ex-Meridian Observation

The term Ex-Meridian applies to an altitude observation of a body which is relatively near to the observer's celestial meridian. Such an observation enables an observer to determine his Latitude provided that he knows his approximate Longitude. In practice, because an observer does not know his Longitude when making an ex-meridian observation, the Latitude he finds from his sight is not necessarily his actual Latitude. It is the Latitude of a point on an astronomical position line, the Longitude of the point being that used in the reduction of the ex-meridian sight.

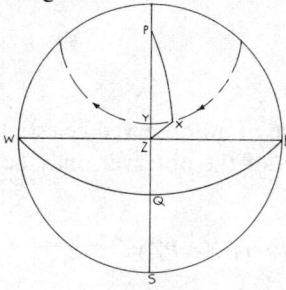

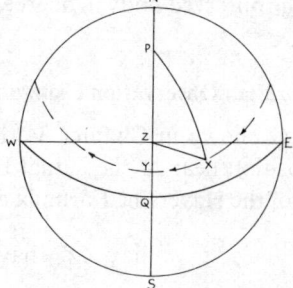

Fig. 36·3   Fig. 36·4

Figs. 36·3 and 36·4 illustrate the celestial sphere, and depict typical PZX-triangles projected onto the plane of the celestial horizon of an observer whose zenith is projected at Z. Assume that the declination of the body X remains constant during the time it takes to move from its present position X to position Y on the observer's celestial meridian. The arc ZY is the Meridian Zenith Distance (M.Z.D.) of the body; and, because PY and PX are equal, we have:

$$\text{M.Z.D.} = (PX \sim PZ)$$

The Spherical Haversine Formula applied to the PZX-triangle contains the term $(PX \sim PZ)$, thus:

$$\text{hav } P = \frac{\text{hav } ZX - \text{hav } (PX \sim PZ)}{\sin PX \sin PZ}$$

Transposing, we have:

$$\text{hav } (PX \sim PZ) = \text{hav } ZX - \text{hav } P \sin PX \sin PZ$$

Or:

$$\text{hav M.Z.D.} = \text{hav } ZX - \text{hav } P \sin PX \sin PZ$$
$$\text{hav M.Z.D.} = \text{hav } ZX - \text{hav } P \cos \text{Lat.} \cos \text{Dec.}$$

This formula is known as the Ex-Meridian Haversine Formula. It may be used for finding a position line provided that the Hour Angle (P), and the Latitude and declination are within certain limits.

The second term in the right-hand side of the ex-meridian formula is a small quantity if the angle P is a small angle. This term contains the factor cos Lat. Now the observer's Latitude is not known: the problem in hand being one in which a Latitude is to be computed. Provided that the Latitude used in the computation approximates to the actual but unknown Latitude of the vessel, there will be only a small error introduced into the term hav P cos Lat. cos dec. It follows, therefore, that the M.Z.D. may be calculated; and, by applying to the M.Z.D. the declination of the observed body, a Latitude may be found. The required position line passes through a point having this computed Latitude and a Longitude equal to that used for finding the angle P which figured in the computation. The direction of the position line is at right angles to the bearing of the observed body at the time of the observation.

The following practical rules should be observed when using the ex-meridian formula:

1. The numerical difference between the Latitude of the observer and the declination of the observed object should not exceed 4°.

2. The Hour Angle (Angle P) in minutes of time should not exceed the zenith distance of the observed body in degrees.

3. Ex-Meridian Observation Using the Intercept Method

We have shown in Chapter 34 that the zenith distance of an observed celestial body at a chosen position near to the actual but unknown position of the observer, may be computed by means of the Haversine Formula as follows:

$$\text{hav } ZX = \text{hav } P \sin PZ \sin PX + \text{hav } (PX \sim PZ)$$

or:

$$\text{hav M.Z.D.} = \text{hav H.A.} \cos \text{Lat.} \cos \text{dec.} + \text{hav (Lat.} \pm \text{dec.)}$$

The Calculated Zenith Distance, when compared with the Observed Zenith Distance, yields an intercept. The observer's position lies on a position line which is drawn through the end of the intercept in a direction at right angles to the bearing of the body at the time of the observation.

The Azimuth is usually found by means of Azimuth Tables. When the azimuth is a small angle—and this is generally the case in ex-meridian observations—it may be ascertained by applying the Spherical Sine Formula to the PZX-triangle and modifying it as follows:

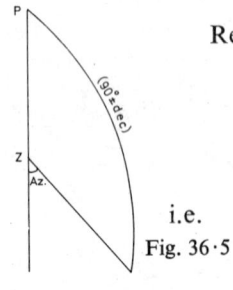

Referring to fig. 36·5:

$$\frac{\sin Z}{\sin PX} = \frac{\sin P}{\sin ZX}$$

$$\sin Z = \sin P \sin (90° \pm \text{dec}) \text{ cosec } ZX$$

i.e.

$$\sin Z = \sin \text{H.A.} \cos \text{dec. sec alt.}$$

Fig. 36·5

This formula is general regardless of the value of Z, but when a celestial body is near meridian passage it bears almost due North or due South, and the angle Z is a small angle. Therefore:

$$\sin Z = Z \text{ in radians}$$

Therefore: $\qquad Z^c = \text{H.A.}^c \cos \text{dec. sec alt.}$

And: $\qquad Z° = \text{H.A.}° \cos \text{dec. sec alt.}$

Or: $\qquad Z° = \dfrac{\text{H.A. in mins. of time}}{4} \cos \text{dec. sec alt.}$

In the case of the Sun, the cosine of whose declination is always nearly unity, its approximate azimuth may be found from the formula:

$$\text{Sun's Azimuth when near Meridian} = \dfrac{\text{H.A. in mins.}}{4} \sec \text{alt.}$$

The Traverse Table may be used to find the Sun's azimuth in these circumstances. Enter with (H.A. in mins./4) in the Distance column on the page corresponding to Altitude as a course angle. The azimuth is then lifted from the D. Lat. column.

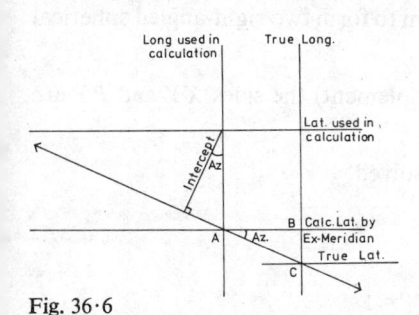

Fig. 36·6 illustrates the relationship between the ex-meridian method and the intercept method of obtaining a position line.

The D. Lat. between the Latitude used in the ex-meridian formula and that used in the intercept formula is given by:

$$\text{D. Lat.} = \text{Intercept sec Azimuth.}$$

If the Longitude used in the ex-meridian formula is not the observer's actual Longitude at the time of the observation, the observer's actual Latitude differs from the Latitude computed using the ex-meridian formula by an amount which depends upon the error in Longitude and the azimuth of the observed body. Because the position line lies nearly East–West, any such error is small.

Fig. 36·6

Referring to fig. 36·6: if the error in Longitude corresponding to the error in Departure AB is e, then:

$$\text{Error in Latitude} = BC$$
$$= AB \tan \text{Azimuth}$$
$$= e \cos \text{Lat. tan Azimuth}$$

4. Ex-Meridian Observation Using Napier's Rules

The Ex-Meridian Haversine Formula is valid only for observations of celestial bodies near an observer's UPPER celestial meridian. When a celestial body is near LOWER meridian passage its M.Z.D. is the SUM of PX and PZ. Now the quantity $(PX+PZ)$ does not appear in the Haversine Formula, so that the Ex-Meridian Haversine Formula cannot be used for solving ex-meridian problems involving bodies near the observer's lower celestial meridian.

The intercept method, in contrast, is valid for all ex-meridian sights. Another method which holds good for ex-meridian observations of bodies near lower as well as upper meridian passage, requires the division of the *PZX*-triangle into two right-angled spherical triangles and solving both using Napier's Rules.

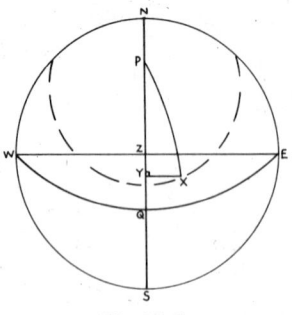

Fig. 36·7 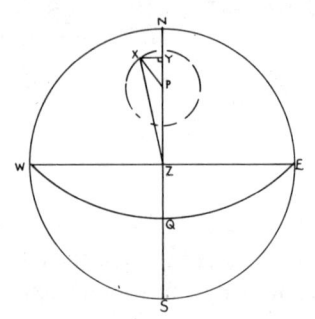   Fig. 36·8

Figs. 36·7 and 36·8 illustrate typical *PZX*-triangles. In fig. 36·7 the body *X* is near to the observer's upper celestial meridian. In fig. 36·8 the body *X* is near to the observer's lower meridian.

In solving the ex-meridian problem by Napier's Rules, a perpendicular great circle is dropped from the body onto the observer's celestial meridian to form two right-angled spherical triangles *PYX* and *ZYX*.

In triangle *PYX*: using side *PX* and angle *P* (or its supplement) the sides *XY* and *PY* are solved.

In triangle *ZYX*: using sides *XY* and *ZX* the side *ZY* is solved.

In triangle *PYX*:
$$\sin XY = \sin P \cos \text{dec.}$$
$$\tan PY = \cos P \cot \text{dec.}$$

In triangle *ZYX*:
$$\cos ZY = \cos ZX \sec XY$$

The Latitude, which is the complement of *PZ*, may then be found as follows:
$$\text{Lat.} = 90° - (PY - ZY)$$
Or:
$$\text{Lat.} = 90° - (ZY - PY)$$

5. Reduction to the Meridian

When a celestial body's Hour Angle is small, and Latitude and declination are within certain limits, the M.Z.D. may be found by applying a small correction to the ex-meridian zenith distance. This small correction is known as the Reduction to the Meridian. In practice, ex-meridian problems are invariably solved by means of Ex-Meridian Tables. There are numerous such tables, many of which give the reduction, against Latitude, declination and hour angle as arguments.

The ex-meridian tables given in Norie's and Burton's Nautical Tables appear to be the most popular in use amongst Merchant Naval officers. These tables are explained with reference to fig. 36·9.

CELESTIAL BODIES NEAR THE MERIDIAN

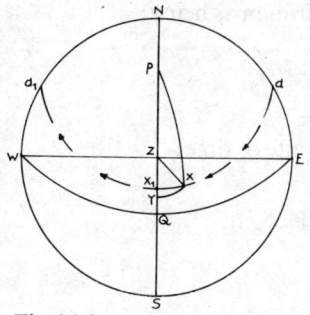

Fig. 36·9

Fig. 36·9 illustrates a typical PZX-triangle projected onto the plane of the celestial horizon of an observer whose zenith is Z. dd_1 represents the diurnal circle of a body which is at X when observed in order to find a position line using the ex-meridian method. The body is on the meridian at position X_1. Arc ZY is equal to arc ZX, so that the correction to apply to the ex-meridian zenith distance (ZX) to obtain the meridian zenith distance (ZX_1) is equal to the arc X_1Y. This correction is the reduction to the meridian denoted by r.

$$ZX = ZX_1 + X_1Y$$
$$= (PX_1 - PZ) + r$$
$$= (\text{Lat.} - \text{dec.}) + r$$
$$= (L - D) + r$$

Applying the Spherical Cosine Formula to the PZX-triangle, we have:

$$\cos ZX = \cos PZ \cos PX + \sin PZ \sin PX \cos P$$
$$\cos[(L-D)+r] = \sin L \sin D + \cos L \cos D \cos P$$
$$= \sin L \sin D + \cos L \cos D (1 - 2\sin^2 P/2)$$
$$= (\sin L \sin D + \cos L \cos D) - 2 \cdot \cos L \cdot \cos D \cdot \sin^2 P/2$$
$$= \cos(L-D) - 2\cos L \cos D \sin^2 P/2 \quad \ldots \ldots \ldots \quad (I)$$

Now:
$$\cos[(L-D)+r] = \cos(L-D)\cos r - \sin(L-D)\sin r$$
$$= \cos(L-D)[1 - 2\sin^2 r/2] - \sin(L-D)\sin r$$

If r is small,

then: $\sin r/2 = r/2$ radians

and $\sin r = r$ radians

so that:
$$\cos[(L-D)+r] = \cos(L-D)(1 - r^2/2) - \sin(L-D) r \quad \ldots \ldots \ldots \quad (II)$$

Equating (I) and (II), we have:
$$\cos(L-D)(1 - r^2/2) - \sin(L-D) r = \cos(L-d) - 2\cos L \cos D \sin^2 P/2$$

Thus:
$$r \sin(L-D) = \cos(L-D)(1 - r^2/2) - \cos(L-D) + 2\cos L \cos D \sin^2 P/2$$
$$= \cos(L-D)(1 - r^2/2 - 1) + 2\cos L \cos D \sin^2 P/2$$
$$= -\cos(L-D) r^2/2 + 2\cos L \cos D \text{ hav } P$$

And, $$r = \frac{2 \cos L \cos D \text{ hav } P}{\sin(L-D)} - \cot(L-D) r^2/2$$

The reduction to the meridian is, therefore, a combination of two parts. The second part, namely: $-\cot(L-D) r^2/2$, is a very small quantity if P is small. In this circumstance it may be ignored without introducing material error.

As a first approximation we may consider the reduction to the meridian as being:

$$r \text{ radians} = \frac{2 \cos L \cos D \text{ hav } P}{\sin (L-D)}$$

If we now express r in seconds of arc and consider P to be one minute of time:

$$r \text{ sec of arc} = \frac{2 \cos L \cos D \text{ hav } 1^m}{\sin (L-D)} \times 60 \times 3438$$

That is:

$$r'' = 1 \cdot 9635 \frac{\cos L \cos D}{\sin (L-D)}$$

This quantity is the change in a body's altitude in seconds of arc during one minute of time to or from the instant of meridian passage. The quantity is referred to as A in Norie's Ex-Meridian Tables, and as F in Burton's. So that:

$$A \text{ (or } F) = 1 \cdot 9635 \cdot \frac{\cos L \cos D}{\sin (L-D)}$$

Values of A are given for all values of Latitude and declination in Ex-Meridian Table 1.

The motion in altitude during a short interval before or after the time of meridian passage is considered to be one of uniform acceleration. The rate of change of altitude at the instant of meridian passage is zero, so that if the change in altitude in one minute from or to the instant of meridian passage is A'', the rate of change of altitude one minute from the time of meridian passage must be $2 \cdot A''$ per minute, and the average rate is A'' per minute. From the relationship in Dynamics, $s = \frac{1}{2} ft.^2$, we have:

$$\text{Change in altitude in } t \text{ mins.} = At^2$$

Ex-Meridian Table 2 provides the result of multiplying A by t^2. The table is entered with A and the Hour Angle of the observed body. This is the required First Correction, which must be applied to the ex-meridian zenith distance to obtain the M.Z.D.

If t is large, then it becomes necessary to apply the second part of the reduction, namely: $-\cot (L-D) r^2/2$. This is referred to as the Second Correction. It is tabulated in Ex-Meridian Table 3 against Altitude and First Correction.

To obtain the position line by ex-meridian observation, the azimuth of the body for the time of observation must be found. $A B C$ Tables or Davis' or Burdwood's Tables are used for finding the azimuth. The position line runs at right angles to the azimuth of the observed body.

6. Concluding Remarks on the Ex-Meridian Problem

The above discussion on the ex-meridian problem is largely of academic interest, although the problem is still considered by some navigators as being worth preserving for practical use. Moreover, it still figures occasionally in examinations designed to test nautical astronomical knowledge and principles. But it must be admitted that the method is quite redundant: the Intercept Method of sight reduction is universal in its application and this method is preferable **to all other methods of sight reduction.**

CELESTIAL BODIES NEAR THE MERIDIAN

Exercises on Chapter 36

1. Explain carefully how you would find the L.M.T. at which a given star is within a given interval of time of its upper meridian passage.

2. What navigational stars are above the horizon and within 1 hour East of the upper celestial meridian of an observer in Lat. 30° 00'·0 N., Long. 30° 00'·0 W. at 05 h. 36 m. on 30th December?

3. What navigational stars will be above the horizon of an observer in Lat. 40° 00'·0 S., Long. 60° 00'·0 E. and within 45 minutes of upper meridian passage and to the West of the observer's celestial meridian at 20 h. 43 m. on 1st January?

4. What navigational stars are above the horizon and within 1 hour of the upper celestial meridian of an observer in Lat. 35° 00'·0 N., Long. 45° 00'·0 W., at 05 h. 36 m. on 22nd September?

5. What navigational stars are within 1 hour of lower meridian passage and above the horizon of an observer in Lat. 70° 00'·0 S., Long. 60° 00'·0 W. at 21 h. 36 m. L.M.T. on 31st December?

6. What navigational stars are within half an hour of the lower meridian and are above the horizon of an observer in Lat. 60° 00'·0 N., Long 60° 00'·0 W. at 05 h. 37 m. on 23rd September?

7. Describe how the Ex-Meridian Haversine Formula is derived from the Haversine Formula for finding P in the PZX-triangle.

8. Explain how an ex-meridian may be solved by using Napier's Rules.

9. Compare the Intercept Method with the Ex-Meridian Method for finding an astronomical position line.

10. Explain the First and Second Corrections given in the Ex-Meridian Tables contained in Norie's and Burton's Tables.

CHAPTER 37

THE POLE STAR

1. Latitude from Observation of Polaris

In Chapter 28 it is demonstrated that:

Latitude of Observer = Altitude of Celestial Pole

Were a star located at the celestial pole its true altitude would equal the observer's Latitude. No star of declination 90° 00'·0 exists, but there is a bright star which is located very near to the North celestial pole. This star, of magnitude 2·2, is α *Ursa Minoris* or Polaris.

The declination of Polaris is about 89° and its Sidereal Hour Angle is about 330°. These values change comparatively rapidly on account of the precession of the equinoxes. For 1975, the mean values of the declination and S.H.A. of Polaris were, respectively, 89° 09'·2 N. and 327° 30'·0.

The diurnal circle of Polaris is a very small circle, centred at the North celestial pole, having a radius of about 1°, this being the approximate value of the Polar Distance of Polaris. It follows that the measure of the true altitude of Polaris is always within about 1° of that of the observer's Latitude.

By applying a correction to the True Altitude of Polaris, an observer readily may find his Latitude. This method of finding Latitude is one of the earliest astronomical methods used by seamen; and, as far back as the time of the Great Discoveries of the fifteenth century, the Portuguese and Spanish navigators were provided with information to enable them to find Latitude at sea from an observation of the Pole Star.

The Pole Star Tables given in the *Nautical Almanac* enable a navigator to find Latitude and azimuth of Polaris provided that the Latitude does not exceed about 65° N. Between the equator and the parallel of about 8° N., Polaris is too near the horizon to make it a suitable body for altitude observations.

In fig. 37·1, which is a projection of the celestial sphere onto the plane of the horizon of an observer whose zenith is projected at Z:

N., E., S. and W., are the Cardinal Points of the Horizon

P is the North Celestial Pole

X is Polaris

Y is a point on the Observer's Celestial Meridian having the same altitude as that of Polaris

WQE is the celestial equator

♈ is the First Point of Aries

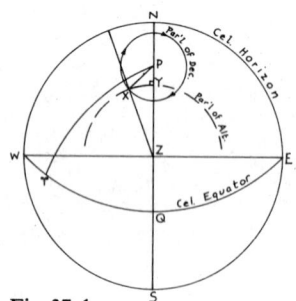

Fig. 37·1

YPX is the L.H.A. of Polaris.

Latitude of Observer = Altitude of Celestial Pole

$= NP$

$= NY - PY$

= True Altitude of Polaris − Correction

When the L.H.A. of Polaris is more than 18 h. 00 m. or less than 06 h. 00 m., the correction is to be subtracted from the True Altitude of Polaris to give the Latitude. This is the case illustrated in fig. 37·1. On the other hand, when the L.H.A. of Polaris is more than 06 h. 00 m. but less than 12 h. 00 m. the correction is to be added to the True Altitude of Polaris to give the Latitude.

Because arc PY is small, about 1°, the triangle PXY may be assumed to be a plane triangle right-angled at Y. On this assumption the correction to be applied to the True Altitude of Polaris (arc NY) to find the Latitude (arc NP) is given by:

$$PY = PX \cos YPX$$

That is: Correction = Polar Distance of Polaris . cos L.H.A. of Polaris.

To allow for the fact that the triangle PXY is not plane, the actual correction to be applied to the True Altitude is given by the formula:

$$\text{Correction} = p \cdot \cos h + p/2 \cdot \sin p \cdot \sin^2 h \cdot \tan l$$

where: $p =$ Polar Distance of Polaris

$h =$ L.H.A. of Polaris

and $l =$ Latitude of Observer.

When the L.H.A. of Polaris is 000°; that is to say, when Polaris is at upper meridian passage, the L.H.A. ♈ is about 29°, which is (360° 00′ − S.H.A. of Polaris). In this circumstance the correction to apply to the True Altitude of Polaris to obtain the Latitude is equal to the Polar Distance of Polaris, and the correction is to be subtracted.

When the L.H.A. of Polaris is 180° 00′; that is to say, when Polaris is at lower meridian passage, the L.H.A. of ♈ is about 209°. In this circumstance the correction is equal to the Polar Distance of Polaris which is to be added to the True Altitude of Polaris to find the Latitude of the observer.

2. The Pole Star Altitude Tables

The Pole Star Tables for Latitude are in three parts. The quantities extracted from the three parts are denoted by a_0, a_1 and a_2, for the first, second and third parts, respectively.

The quantity a_0 is tabulated against L.H.A. ♈. This is used as an argument partly to obviate the necessity of finding the L.H.A. of Polaris, which latter quantity determines the value of a_0. The correction is computed from the formula given above, using a Latitude of 50° and mean values of declination and S.H.A. of Polaris for the year for which the tables apply. The computed value is adjusted by the addition of a constant so that it is always a positive quantity regardless of the L.H.A. of Polaris.

310   THE ELEMENTS OF NAVIGATION AND NAUTICAL ASTRONOMY

The quantity a_1 is tabulated against L.H.A. ♈ and Latitude. The excess (positive or negative) of the value of the second term of the formula, viz. $p/2 \cdot \sin p \sin^2 h \tan l$ over its mean value for Latitude 50°, is computed. This excess is increased by a constant to make a_1 always positive.

The quantity a_2 is tabulated against L.H.A. ♈ and the date. The correction to the first term of the formula, viz. $-p \cos h$, for the actual celestial position of Polaris from the mean position used in computing a_0 and a_1, is computed and increased by a constant to give a_2 which is always positive.

The sum of the three constants is exactly 1° 00'·0, so that Latitude from an observation of Polaris is given by:

$$\text{Latitude} = \text{True Altitude of Polaris} - 1° 00' + a_0 + a_1 + a_2$$

Example 37·1—13th June in approximate Long. 30° 00'·0 W., the observed altitude of Polaris was 46° 10'·0. The chronometer time of the observation was 07 h. 35 m. 40 s. Chronometer error 2 m. 25 s. slow on G.M.T. Height of eye 40 feet. Find the observer's Latitude.

$$
\begin{array}{rl}
\text{Chron. Time} = & 07 \text{ h. } 35 \text{ m. } 40 \text{ s.} \\
\text{Error} = & +2 \text{ m. } 25 \text{ s.} \\
\hline
\text{G.M.T.} = (13) & 07 \text{ h. } 38 \text{ m. } 05 \text{ s.} \\
\text{G.H.A. ♈ at 07 h.} = & 6° \ 05'·2 \\
\text{Increment} = & 9° \ 32'·8 \\
\hline
\text{G.H.A. ♈} = & 15° \ 38'·0 \\
\text{Longitude W.} = & 30° \ 00'·0 \\
\hline
\text{L.H.A. ♈} = & 345° \ 38'·0 \\
\text{Obs. Alt.} = & 46° \ 10'·0 \\
\text{Total corr.} = & -7'·2 \\
\hline
\text{True Alt.} = & 46° \ 02'·8 \\
& -1° \ 00'·0 \\
\hline
& 45° \ 02'·8 \\
a_0 = & 00° \ 18'·4 \\
a_1 = & +0'·6 \\
a_2 = & +0'·2 \\
\hline
\text{Latitude} = & 45° \ 22'·0 \text{ N.}
\end{array}
$$

Answer—Latitude Observer = 45° 22'·0 N.

When observing Polaris, the Latitude obtained from the Pole Star Tables is the Latitude of the observer only if the Longitude used in the finding L.H.A. ♈ is the observer's actual Longitude. In general, the exact Longitude of the observer at the time of an observation is not known, so that the result of a Pole Star Altitude observation is not a Latitude but a Position Line, the direction of which is at right angles to the bearing of Polaris at the time of the observation. The position line passes through a point the Latitude of which is that found from the tables, and the Longitude of which is that used in finding L.H.A. ♈.

The azimuth of Polaris may readily be found from the Pole Star Azimuth Table.

THE POLE STAR

3. The Pole Star Azimuth Table

The Pole Star Azimuth Table is entered with L.H.A. ♈ and Latitude as arguments. Azimuths are given to the nearest 0°·1, this being more than sufficiently accurate for all practical purposes.

Fig. 37·2 serves to illustrate the method used for computing Pole Star Azimuths.

In fig. 37·2:   Azimuth of Polaris = PZX

                                          = arc NH

By the Parallel Sailing Formula:

                                   $NH = XY$ sec HX

But $XY = PX$ sin P

Therefore: Azimuth = PX sin P sec HX

Or: Azimuth = Polar Dist. sin P sec Alt.

Now   Polar Distance = 1° (nearly)

and   Altitude = Latitude (nearly)

Therefore: Azimuth = 1° sin L.H.A. Polaris sec Lat.

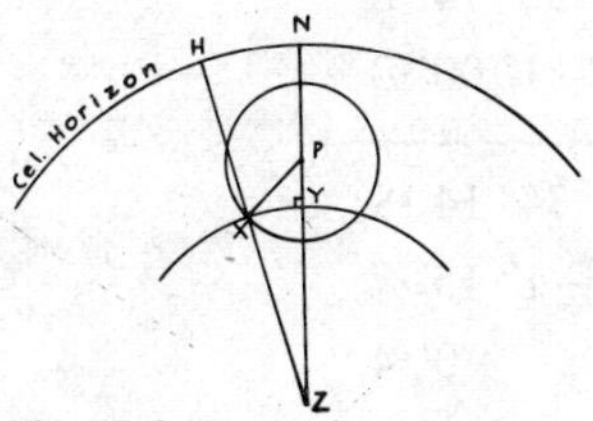

Fig. 37·2

This formula indicates that the azimuth is zero when the L.H.A. of Polaris is 000° or 180°, and that it is maximum when the L.H.A. of Polaris is 090° or 270°. In other words the azimuth of Polaris is zero when the L.H.A. of ♈ is about 29° or 209°, and it is maximum when the L.H.A. of ♈ is 109° or 299°.

The Pole Star Azimuth Table provides an easy means for checking the compass error, especially in low Latitudes, in which circumstances Polaris has a small altitude and it is ideally placed, therefore, for azimuth observations.

Example 37·2—13th June in Lat. 40° 00'·0 N., Long. 150° 00'·0 W., at 11 h. 42 m. G.M.T., the Pole Star bore 004¼ (C). The variation was 6½° W. Find the deviation for the heading of the vessel at the time of the observation.

                 G.H.A. ♈ at 11 h. = 66° 15'·1

                         Increment = 10° 31'·7

                         G.H.A. ♈ = 76° 46'·8

                       Longitude = 150° 00'·0

                          L.H.A. ♈ = 286° 46'·8

                     True Azimuth = 001¼°

                     Obs. Azimuth = 004¼°

                                Error = 3° W.

                           Variation = 6½° W.

                          Deviation = 3½° E.

Answer—Deviation = 3½° E.

Example 37·3—22nd September in D.R. position Lat. 35° 20'·0 N., Long. 30° 00'·W., the sextant altitude of Polaris was 35° 20'·5. Index error 1'·8 off the arc. Height of eye 40 feet. Chronometer time 18 h. 38 m. 46 s. Chronometer error 2 m. 23 s. slow on G.M.T. Find the position line corresponding to the observation.

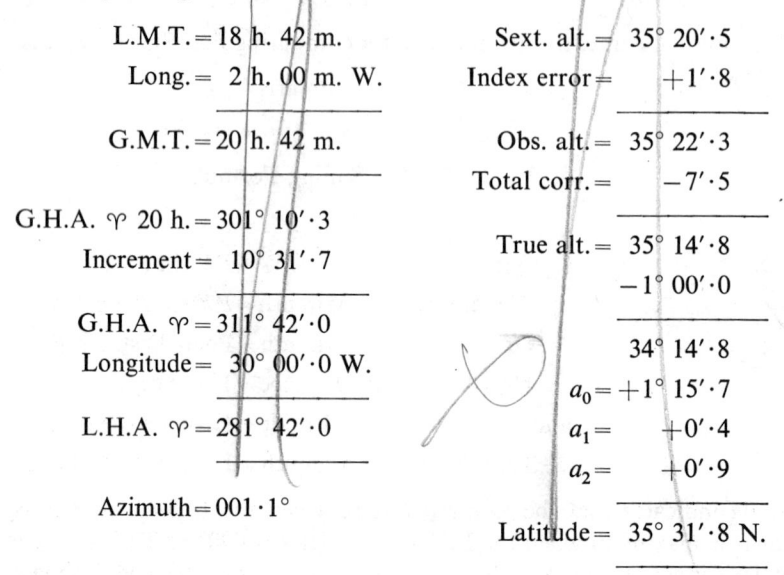

L.M.T. =	18 h. 42 m.	Sext. alt. =	35° 20'·5
Long. =	2 h. 00 m. W.	Index error =	+1'·8
G.M.T. =	20 h. 42 m.	Obs. alt. =	35° 22'·3
		Total corr. =	−7'·5
G.H.A. ♈ 20 h. =	301° 10'·3	True alt. =	35° 14'·8
Increment =	10° 31'·7		−1° 00'·0
G.H.A. ♈ =	311° 42'·0		34° 14'·8
Longitude =	30° 00'·0 W.	a_0 =	+1° 15'·7
L.H.A. ♈ =	281° 42'·0	a_1 =	+0'·4
		a_2 =	+0'·9
Azimuth = 001·1°		Latitude =	35° 31'·8 N.

Answer—Position Line 091°–271° through Lat. 35° 31'·8, Long. 30° 00'·0 W.

Exercises on Chapter 37

1. Prove that the Latitude of an observer is equal to the altitude of the celestial pole.

2. Explain why the true altitude of Polaris is always within about 1° of the Latitude of an observer.

3. Describe carefully the construction of the Pole Star Tables.

4. 12th June in approx. Long. 29° 00'·0 W. the observed altitude of Polaris was 46° 14'·0. The chronometer time of the observation was 07 h. 40 m. 20 s. Chronometer error 3 m. 22 s. slow on G.M.T. Height of eye 30 feet. Find the observer's Latitude.

5. 15th June in approx. Long. 175° 00'·0 W. the observed altitude of Polaris was 43° 25'·0. The chronometer time was 19 h. 38 m. 20 s. and the chronometer was 4 m. 13 s. slow on G.M.T. Height of eye 50 feet. Find the observer's Latitude.

6. 23rd September in approx. Long. 33° 00'·W. the observed altitude of Polaris was 39° 25'·0. The chronometer time was 05 h. 47 m. 12 s. and the chronometer was 4 m. 04 s. fast on G.M.T. Height of eye was 40 feet. Find the position line on which the observer was located at the time of the observation.

7. 24th September in approx. Long. 75° 00'·0 W. the observed altitude of Polaris was 28° 10'·5. The chronometer time was 11 h. 42 m. 20 s. and the chronometer error was 1 m. 02 s. slow on G.M.T. Height of eye 20 feet. Ascertain the position line on which the observer was located at the time of the observation.

THE POLE STAR

8. 12th June in approx. position Lat. 50° 00'·0 N., Long. 30° 00'·0 W., the compass bearing of Polaris was 008°·5 when the G.M.T. was 07 h. 44 m. 00 s. Find the compass error for the heading of the vessel at the time of observation.

9. 23rd September in approx. position Lat. 42° 00'·0 N., Long. 150° 00'·0 E., the compass bearing of Polaris was 014°·0 when the chronometer time was 08 h. 43 m. 00 s. Chronometer error 1 m. 05 s. fast on G.M.T. Find the deviation for the present heading of the vessel given that the variation is 10°·0 W.

10. 14th June in Latitude 35° 00'·0 N., Long. 65° 00'·0 W., the compass bearing of Polaris was 015°·0 when the chronometer time was 08 h. 14 m. 20 s. Chronometer error 29 m. 04 s. slow on G.M.T. Find the deviation for the present heading of the vessel given that the variation is 12°·0 W.

CHAPTER 38

NAUTICAL ASTRONOMICAL TABLES

1. Introduction

The tables we shall describe in this chapter are some of these designed to facilitate the solution of spherical triangles—principally the *PZX*-triangles in sight reduction.

2. The Principles of A B C Tables

The *A B C* Tables found in Burton's, Norie's and other Nautical Table collections, may be used to give a direct solution to any one of any four adjacent parts of a spherical triangle, provided that the other three of the adjacent parts are known. The principle of the tables is based on the Four Parts Formula of spherical trigonometry. This formula is discussed in Chapter 6.

The navigational uses of the *A B C* Tables include:

1. Finding the azimuth of a celestial body.

2. Finding great circle courses.

In addition to these uses the tables are extensively used for finding the error in Longitude consequent upon an error in Latitude used in the Longitude Method of sight reduction.

The arguments used in the *A B C* Tables are Hour Angle, Latitude and Declination. These, together with Azimuth, are involved in the principal type of problem which is solved by means of the tables.

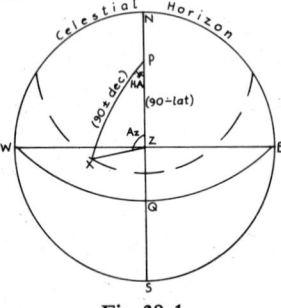

Fig. 38·1

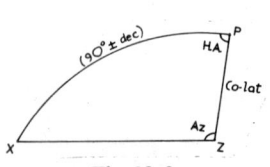

Fig. 38·2

Figs. 38·1 and 38·2 illustrate typical *PZX*-triangles with four adjacent parts marked.

By the Four Parts Formula:

$$\cos P \cos PZ = \sin PZ \cot PX - \sin P \cot Z$$

NAUTICAL ASTRONOMICAL TABLES

Dividing throughout by $\sin P \sin PZ$, we have:

$$\frac{\cos P \cos PZ}{\sin P \sin PZ} = \frac{\sin PZ \cot PX}{\sin P \sin PZ} - \frac{\sin P \cot Z}{\sin P \sin PZ}$$

From which:
$$\cot P \cot PZ = \operatorname{cosec} P \cot PX - \cot Z \operatorname{cosec} PZ$$

Or: $\quad\quad \underbrace{\cot \text{H.A.} \tan \text{Lat.}}_{1} = \underbrace{\operatorname{cosec} \text{H.A.} \cot (90° \pm \text{Dec.})}_{2} - \underbrace{\cot \text{Az.} \sec \text{Lat.}}_{3}$

Expression 1 is tabulated for all combinations of H.A. and Latitude, as *A*-Correction.

Expression 2 is tabulated for all combinations of H.A. and Declination as *B*-Correction.

Expression 3 is tabulated for all combinations of Latitude and Azimuth as *C*-Correction.

It follows that:
$$A = B - C$$
and
$$C = B - A$$

(i) *When the H.A. is less than* 90° *and the Latitude and Declination have the Same name:*

B is a positive quantity because $\cot (90° - \text{Dec.})$ and cosec H.A. are both positive quantities.

A is a positive quantity because cot H.A. and tan Lat. are both positive quantities.

In these circumstances, therefore:
$$C = +B - (+A)$$
That is:
$$\underline{C = B - A}$$

(ii) *When the H.A. is less than* 90° *and the Latitude and Declination have Opposite names:*

B is a negative quantity because $\cot (90° + \text{Dec.})$ is negative and cosec H.A. is positive.

A is a positive quantity because cot H.A. and tan Lat. are both positive quantities.

In these circumstances, therefore:
$$C = -B - (+A)$$
Or:
$$\underline{C = -B - A}$$

(iii) *When the H.A. is more than* 90° *and the Latitude and Declination have the Same name:*

B is a positive quantity because $\cot (90° - \text{Dec.})$ and cosec H.A. are both positive quantities.

A is a negative quantity because cot H.A. is negative and tan Lat. is positive.

In these circumstances, therefore:
$$C = +B - (-A)$$
Or:
$$\underline{C = +B + A}$$

316   THE ELEMENTS OF NAVIGATION AND NAUTICAL ASTRONOMY

(iv) *When the H.A. is more than 90° and the Latitude and Declination have Opposite names:*

B is a negative quantity because cosec H.A. is positive and cot $(90°+\text{Dec.})$ is negative.

A is a negative quantity because cot H.A. is negative and tan Lat. is positive.

In these circumstances, therefore:

$$C = -B - (-A)$$

Or:
$$C = -B + A$$

From (i), (ii), (iii) and (iv) it will be seen that, in general:

$$C = \pm(A+B)$$

When using $A\ B\ C$ Tables care must be taken to interpolate carefully and also to apply the correct sign to each of the tabulated values of the A, B and C, Corrections.

3. Uses of the A B C Tables

(i) *To find Error in Longitude due to Error in Latitude:*

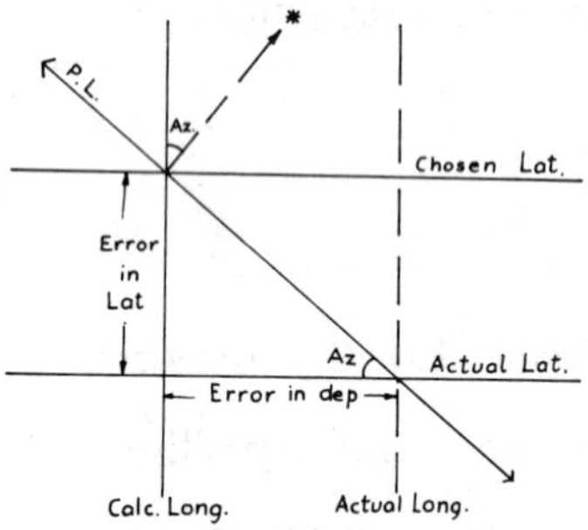

Referring to fig. 38·3: it will be seen that:

Error in Departure = Error in Latitude cot Azimuth

But: Departure = D. Long. cos Lat.

Therefore:

Error in Long. cos Lat. = Error in Lat. cot Az.

And: Error in Long. = Error in Lat. cot Az. sec Lat.

But: cot Az. sec Lat. = C-correction

Therefore:

Error in Long. = Error in Lat. C-correction

Fig. 38·3

It follows that the C-correction is numerically equal to the Error in Latitude consequent upon an Error of one minute in Latitude.

Because $(A \pm B) = C$; therefore, $(A \pm B)$ is also equal to the Error in Longitude consequent upon an Error of one minute in Latitude.

In general, Table C is entered with $(A \pm B)$ and Latitude, in order to find the Azimuth of the observed celestial body.

Example 38·1—A chosen Latitude of 50° 00'·0 N. was used to reduce a sight using the Longitude Method. The Sun's declination was 20° 00'·0 N., the calculated H.A. was 315° and the calculated Longitude was 30° 00'·0 W. The vessel travelled on a course of 240° for 32 ml., at the end of which the Sun was observed on the meridian and the Latitude by meridian altitude was found to be 49° 35'·0 N. Find the Longitude of the vessel at noon.

NAUTICAL ASTRONOMICAL TABLES

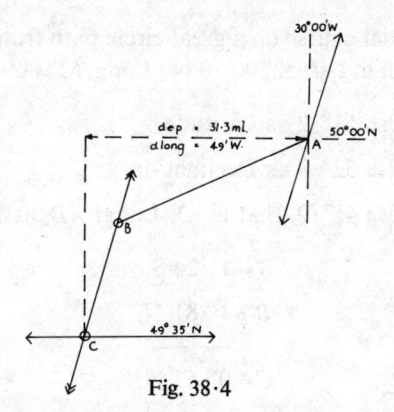

Fig. 38·4

The Longitude of the vessel at noon, found by scale drawing as illustrated in fig. 38·4, is 30° 49'·0 W.

The usual method employed at sea to find the noon Longitude using the *A B C* Tables is as follows:

$$A = 1' \cdot 19 \text{ S.}$$
$$B = 0' \cdot 51 \text{ N.}$$
$$\overline{C = 0' \cdot 68 \text{ S.}}$$

Azimuth = 113°·6

Lat. (*A*) = 50° 00'·0 N.	Long. = 30° 00'·0 W.
D. Lat. = 16'·0 S.	D. Long. = 43'·0 W.
Lat. (*B*) = 49° 44'·0 N.	Long. = 30° 43'·0 W.
Lat. (*C*) = 49° 35'·0 N.	Error = 06'·0 W.
Error = 9·0' S.	Long. (*C*) = 30° 49'·0 W.
C-correction = ×0·68	
Error in Long. = 6'·0 W.	

Answer—Noon Longitude = 30° 49'·0 W.

(ii) *To find Courses in Great Circle Sailing*:

Suppose that it is necessary to find the initial course of the great circle track from *A* to *B* as illustrated in fig. 38·5:

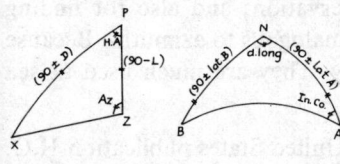

Fig. 38·5

Consider the initial Latitude as *Latitude* in Table *A*
Consider the final Latitude as *Declination* in Table *B*
Consider D. Long. between *A* and *B* as H.A. in Tables *A* and *B*
The *Azimuth*, found in Table *C*, is the required initial course.

Example 38·2—Find the initial course on a great circle path from a position in Lat. 51° 10′ N., Long. 10° 00′ W. to a position in Lat. 52° 00′·0 N., Long. 55° 00′·0 W.

$$\text{Use } 51°·2 \text{ as Latitude}$$
$$\text{Use } 52°·0 \text{ as Declination}$$
$$\text{Use } 45°·0, \text{ that is, D. Long. } AB, \text{ as H.A.}$$

$$A = 1'·24 \text{ S.}$$
$$B = 1'·81 \text{ N.}$$
$$\overline{}$$
$$C = 0'·57 \text{ N.}$$
$$\overline{}$$

Initial Course = N. 70°·4 W.

Answer—Initial Course = 289°·6.

4. Inspection Tables

The only part of a *PZX*-triangle which can be obtained from direct measurement is the zenith distance. It is for this reason that there are two, and only two, general methods for finding an astronomical position line. These are the Longitude and Intercept Methods, respectively.

The computation involved when reducing a sight using a direct method in which a *PZX*-triangle is solved by the common trigonometrical tables, is relatively complex and time-consuming. The chance, therefore, of the computer making clerical errors is considerable. For this and other reasons many attempts have been made to shorten the process of sight reduction, and numerous tables are available for this purpose.

The term Inspection Tables applies to tables from which certain unknown parts of the astronomical triangle corresponding to certain known parts, may be lifted. Apart from perhaps some interpolation, no computations are involved when solving a *PZX*-triangle using inspection tables.

The principal disadvantage of inspection tables used in Nautical Astronomy for solving Hour Angle or Altitude is the great bulk and the consequent expense of the tables.

The Altitude-Azimuth Tables by Burdwood in which azimuths are given against arguments Latitude, Declination and Altitude; and the Time-Azimuth Tables by Davis in which azimuths are given against arguments Latitude, Declination and Hour Angle; are examples of Inspection Tables in which only a relatively coarse degree of accuracy—to the nearest tenth of a degree—is required. These tables, as well as serving for azimuths are also useful for finding approximate star altitudes for setting the sextant prior to making an observation; and also for finding great circle courses in which problems great circle courses are analogous to azimuths. Because of their low degree of accuracy azimuth tables are not bulky. They are much used at sea because of their handiness.

A comprehensive set of altitude and azimuth tables is the United States publication H.O. 214. Each volume of H.O. 214 covers a range of 10 degrees. They are designed primarily for marine navigation, and they are applicable to observations of all astronomical navigational bodies.

H.O. 214 were regrouped into six volumes, each covering a range of Latitude of 15 degrees, and published by the Hydrographic Department of the British Admiralty in 1951, as H.D. 486.

H.D. 486 is entered with Latitude, Declination and Local Hour Angle of the observed body as arguments, and the respondents are Altitude to the nearest $0'·1$ and Azimuth to the nearest $0°·1$. Each whole degree of Latitude covers 24 large pages—12 for cases in which Latitude and declination have the same name, and 12 for cases in which the Latitude and declination have different names.

Another very useful inspection table is that known as H.O. 249, published by the United States Hydrographic Office. These tables, similar to H.O. 214, provide pre-computed altitudes and azimuths, correct to the nearest minute and degree of arc respectively, for each of 38 stars. The arguments are L.H.A. and Latitude, each to an integral number of degrees. Although the accuracy of altitude and azimuth in H.O. 249 is not nearly so great as that in H.O. 214, they are valuable tables for star sight reduction.

A magnificent set of Inspection Tables, under the title *Sight Reduction Tables for Marine Navigation* was published by the U.S. Naval Oceanographical Office as Pub. No. 229 in 1970. In the following year the tables, in identical format, were published by the British Ministry of Defence as H.D. Publication N.P. 401. In all respects the new work is a veritable *tour de force* as far as inspection tables for marine use are concerned.

Each of the six volumes of N.P. 401 contains two eight-degree zones of Latitude, and there is an overlap of one degree between consecutive volumes. Within each of the 12 zones of eight degrees of Latitude, the main argument is Local Hour Angle (L.H.A.) between $0°$ and $360°$. To each integral degree of L.H.A. $(=P°)$ in the range $0°$–$90°$ there corresponds an opening of two facing pages. The left-hand page covers the case of Latitude and declination of the same name. The right-hand page contains on the upper portion, the tabulations for L.H.A. $=P$ and L.H.A. $=(36°-P)$, and declinations of contrary names; and on the lower portion the tabulations for supplementary L.H.A.s $(180°-P)$ and $(180°+P)$, and declinations of the same name as Latitude. The two portions on the right-hand page are separated in each column, by a horizontal rule which, together with the vertical lines separating the columns, form a stepped configuration across the page. This separating line is called the "Contrary-Same" or "C-S" line. The horizontal segments of the C-S line indicate the degree of declination in which the horizon (alt. $=0°$) occurs. Altitudes on one side of the line are positive, that is to say they are above the horizon; and on the other side they are negative or below the horizon. The advantage of this interesting arrangement is that the reduction of a sight taken within a particular zone of Latitude, in either hemisphere, is determined uniquely by the value of the L.H.A.

The respondents are altitude (denoted by H_c) to $0'·1$; azimuth (denoted by Z) to $0°·1$; and, in smaller type, the difference (denoted by d) between the tabular altitude for a given declination entry and that for a declination of one degree higher.

Rules are given at each opening for converting tabulated azimuths to true azimuths, and an interpolation table, in four pages occupying the insides of the covers and adjacent pages, facilitates finding the correction to the tabulated altitude for the odd minutes of declination. This table is in two parts: one giving the correction for the tens of minutes in the excess of the actual declination over the entering declination, and the other giving the correction for the units of declination. The interpolation table also provides for second-difference corrections and an indication of when a second-difference correction should be applied is provided by printing the "altitude-difference", d in *italics* followed by a dot (·) for easy recognition.

5. Short Method Tables

The PZX-triangle may be divided into two right-angled spherical triangles by dropping a perpendicular great circle from one of the vertices of the triangle onto the opposite side or side produced. This is the basis of the so-called Short Method Tables.

Short method tables, in respect of the time taken for sight reduction and bulk of tables, lie between inspection tables and the trigonometrical tables used in direct methods.

The principal disadvantage of short method tables is the relative complexity of the rules for using them. Before it is possible to gain full advantage from a short method table, considerable practice is necessary in the use of the table and in the handling of the rules. In other words, complete familiarity with the table is essential for optimum use.

6. Ogura's Table

S. Ogura of the Imperial Japanese Navy first published his ingeniously contrived short method table in 1920. Ogura's method is the basis of many short method tables, and it is primarily for this reason that we shall discuss the method in some detail.

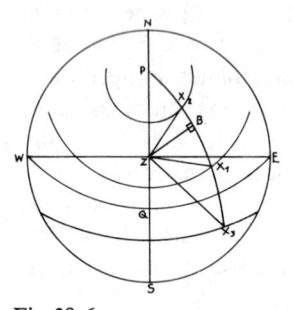

Fig. 38·6

The PZX-triangle is divided by dropping a perpendicular great circle from Z onto the side PX or PX produced. The right-angled spherical triangle PZB is known as the Time Triangle, and the right-angled spherical triangle BZX is known as the Altitude Triangle.

Fig. 38·6 illustrates the three cases, viz.:

Case 1—Latitude (φ) and Declination (d) of same name, and $\varphi < d$ in which case: $BX_1 = d - K$, where K is the declination of the point B.

Case 2—Latitude and Declination of the same name but $\varphi > d$ in which case: $BX_2 = K - d$.

Case 3—Latitude and Declination having opposite names in which case: $BX_3 = K + d$.
In all cases:
$$BX = K \pm d \quad \ldots \ldots \ldots \ldots \ldots \ldots \quad (1)$$
Also, in all cases:
$$BX = (90° \pm d) - PB \quad \ldots \ldots \ldots \ldots \ldots \quad (2)$$

Equating equations (1) and (2), we get:
$$K \pm d = 90° \pm d - PB$$
$$K = 90° - PB$$

NAUTICAL ASTRONOMICAL TABLES

In the Time Triangle (PZB), by Napier's Rules:

$$\sin co . P = \tan PB . \tan co . PZ$$

i.e. $\qquad \cos P = \tan PB . \cot PZ$

i.e. $\qquad \tan PB = \cos P . \tan PZ$

Inverting: $\qquad \cot PB = \sec P . \cot PZ$

$$\tan(90 - PB) = \sec P . \cot PZ$$

i.e. $\qquad \tan K = \sec P . \tan \varphi \quad \ldots \ldots \ldots \ldots \quad (3)$

In the Time Triangle (PZB) by the Spherical Sine Rule:

$$\sin BZ = \sin P . \sin PZ$$

i.e. $\qquad \sin BZ = \sin P . \cos \varphi$

Inverting: $\qquad \operatorname{cosec} BZ = \sin P . \cos \varphi \quad \ldots \ldots \ldots \ldots \quad (4)$

In the Altitude Triangle (BZX), by Napier's Rules:

$$\sin co . ZX = \cos BZ . \cos BX$$

$$\cos ZX = \cos BZ . \cos(K+d)$$

Inverting: $\qquad \sec ZX = \sec BZ . \sec(K+d) \quad \ldots \ldots \ldots \quad (5)$

Values of K are pre-computed from equation (3) for integral degrees of Latitude (φ) and P (hour angle).

Values of BZ are computed from equation (4), and values of log sec BZ, denoted by A, are tabulated for integral degrees of Latitude (φ) and P (hour angle).

Ogura's tables are sometimes known as A and K Tables. The procedure for using them is as follows:

1. Choose a Latitude, an integral number of degrees nearest to the estimated Latitude of the ship.

2. Apply to the G.H.A. of the observed body for the time of the observation, a Longitude so as to make P an integral number of degrees.

3. Enter the table with φ and P and extract K.

4. Apply d to K to find ($K \pm d$)
(+ when φ and d have different names, $\sim$ when φ and d have the same name).

5. Enter table with φ and P and extract A.

6. Enter the log secant table with ($K \pm d$). Add the log secant of ($K \pm d$) to A. This gives the log secant of the calculated zenith distance (from equation 5).

7. Compare the C.Z.D. with the O.Z.D. to find the intercept.

8. Use the $A\ B\ C$ table or Azimuth tables to find the azimuth of the observed body and hence the direction of the position line.

The A and K Tables afford a rapid means of finding the C.Z.D.—only three tabular entries being required.

322 THE ELEMENTS OF NAVIGATION AND NAUTICAL ASTRONOMY

The azimuth may be found by combining angle PZB in the Time Triangle with angle BZX in the Altitude Triangle.

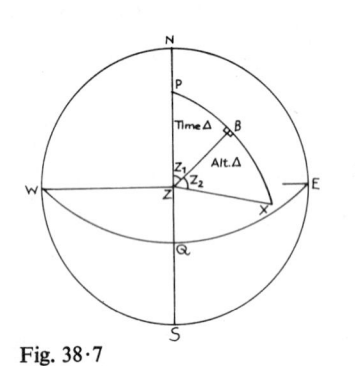

Referring to fig. 38·7

In the Time Triangle:

By Napier's Rules:

$$\cos PZ = \cot PZB \cdot \cot P$$

i.e. $\quad \cot Z_1 = \cos PZ \cdot \tan P$

Inverting: $\quad \tan Z_1 = \operatorname{cosec} \varphi \cdot \cot P \quad \ldots \ldots \quad (1)$

In the Altitude Triangle:

By Napier's Rules:

$$\sin BZ = \cot BZX \cdot \tan BX$$

i.e. $\quad \cot Z_2 = \sin BZ \cdot \cot BX$

Inverting: $\quad \tan Z_2 = \operatorname{cosec} BZ \cdot \tan (K+d) \ldots \quad (2)$

Fig. 38·7

Z_1 and Z_2 may be computed from equations (1) and (2) respectively, and tabulated against φ and P. The azimuth of the observed body is then found by combining Z_1 and Z_2.

The Altitude and Azimuth Table in Norie's Tables are based on Ogura's A and K tables, but they provide for azimuths as well.

Ogura's method of dividing the PZX-triangle into two right-angled spherical triangles has been used in tables by Dreisonstok; Smart and Shearm; Comrie (in Hughes' Sea and Air Navigation Tables); Gingrich; and Myerscough and Hamilton.

Exercises on Chapter 38

1. Show that, in solving an astronomical position line, an error in Latitude produces an error in Longitude proportional to the secant of the Latitude and the cotangent of the azimuth of the observed body.

2. Using the Four Parts Formula of spherical trigonometry, prove that:

    $\cot AZ \cdot \sec \text{Lat.} = \operatorname{cosec} \text{H.A.} \cot (90 \pm \text{dec.}) - \cot \text{H.A.} \tan \text{Lat.}$

3. What factors determine the signs (or names) of the A, B and C corrections? Show that, in all cases:

    $C = \pm(A+B)$

4. Explain how the A B C Tables are used to find:
    (i) the azimuth of a heavenly body
    (ii) a great circle course
    (iii) the Longitude correction factor.

5. Distinguish between Inspection- and Short Method-Navigation Tables.

6. Explain clearly the construction of Ogura's Short Method Table.

CHAPTER 39

RISING AND SETTING PHENOMENA

1. Introduction

When a celestial body is on an observer's celestial horizon its Local Hour Angle may readily be computed by Napier's Rules because the zenith distance of the body is 90° 00'.

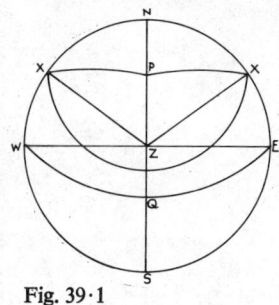

Fig. 39·1 illustrates the visible celestial hemisphere of an observer whose zenith is projected at Z. P is the projection of the North celestial pole and WQE that of the celestial equator. X is a celestial body at rising or setting. Applying Napier's Rules to the spherical triangle PZX we have:

$$\sec P = -\cot \text{dec.} \cot \text{Lat.}$$

Fig. 39·1

Example 39·1—Compute the L.H.A. of the planet Venus when it sets at a date when its declination is 20° 00' N. given the Latitude of the observer as 42° 00' N.

$\sec P = -\cot \text{dec.} \cot \text{Lat.}$

dec. = 20° 00' log cot = 0·43893 (+)

Lat. = 42° 00' log cot = 0·04556 (+)

$P =$ log sec = $\overline{0·48449}$ (−)

Therefore: $P = 180° - 70° 52'$

              $= 109° 08'$

Answer—L.H.A. = 109° 08'.

2. Sunrise and Sunset

If the L.H.A. of the Sun for the time of sunrise and sunset is known, the L.A.T. of sunrise or sunset may be found from the relationship between Apparent Solar Time and Sun's Hour Angle, thus:

L.A.T. of Sunset = L.H.A. of the Sun at sunset + 12 hr.

L.A.T. of Sunrise = L.H.A. of the Sun at sunrise − 12 hr.

It should be noted that if the Latitude of the observer and the declination of the Sun have the same Name, the Sun rises before 0600 hr. L.A.T. and sets after 1800 hr. L.A.T. This is the case illustrated in fig. 39·1. On the other hand, if the Latitude of the observer and the Sun's declination have different names, the Sun rises after 0600 hr. L.A.T. and sets before 1800 hr. L.A.T.

The L.A.T. of sunset or sunrise may be found from Davis's or Burdwood's Azimuth Tables. The times of sunrise and sunset given in these tables are those at which the Sun's centre is on the celestial horizon of an observer. The terms Theoretical-Sunrise and -Sunset are used to denote these events. The terms Visible-Sunrise and -Sunset are the times at which the Sun's upper limb is on an observer's visible horizon.

Corresponding theoretical and visible sunsets or sunrises differ on account of dip, refraction, and semi-diameter. The observed altitude of the Sun's upper limb at the time of visible sunrise or sunset is 00° 00′·0, whereas the true altitude of the Sun's centre at the time of theoretical sunrise or sunset is 00° 00′·0.

It is an easy matter to compute the true altitude of the Sun's centre for the time of visible sunrise or sunset. The following example serves to illustrate this.

Example 39·2—Find the true altitude of the Sun's centre at the time of visible sunset on 13th June, if the observer's height of eye is 36 feet.

$$
\begin{aligned}
\text{Obs. Alt. of Sun's upper limb} &= 00° \ 00'·0 \\
\text{Dip} &= -05'·8 \\
\hline
\text{App. Alt. of Sun's upper limb} &= -00° \ 05'·8 \\
\text{Refraction} &= -33'·0 \\
\hline
\text{True Alt. of Sun's upper limb} &= -00° \ 38'·8 \\
\text{Semi-diameter} &= -15'·8 \\
\hline
\text{True Alt. of Sun's centre} &= -00° \ 54'·6 \\
\end{aligned}
$$

Answer—True Alt. = −00° 54′·6.

In the circumstances applicable to example 39·2 the Sun's centre is nearly a whole degree, or double the Sun's angular diameter, below the observer's celestial horizon at the instant when its upper limb touches the observer's visible horizon. It follows that the Sun's centre is about 1° above the visible horizon when its centre is on the celestial horizon. It is important, when observing the amplitude of the Sun for checking the compass, to ensure that the observation is made at the instant when the Sun's centre is on the celestial horizon. Attention was drawn to this important matter in chapter 34, to which the reader is referred.

The G.M.T.s of sunrise and sunset at the Greenwich meridian for Latitudes at 10°-, 5°-, and 2°-intervals, are given on the right-hand daily pages of the *Nautical Almanac*. These times are approximately equal to the L.M.T.s of sunrise and sunset for all meridians other than the Greenwich meridian. To find the G.M.T. of sunrise or sunset the observer's Longitude must be applied to the time lifted from the *Nautical Almanac*. A table is provided in the *Nautical Almanac* to facilitate interpolating for Latitude.

3. Moonrise and Moonset

Theoretical Moonrise and Moonset occur when the Moon's centre is on the observer's celestial horizon. Visible Moonset and Moonrise occur when the Moon's upper limb is on the observer's visible horizon. Theoretical and visible moonrise or moonset differ on account of dip, refraction, semi-diameter and horizontal parallax.

Example 39·3—Find the observed altitude of the Moon's upper limb at the time of theoretical moonrise in Lat. 40° 00′ N., Long. 30° 00′ W. on 24th September, if the height of the observer's eye is 60 feet.

From *Nautical Almanac* Extracts:

$$\text{L.M.T. of Moonrise at Long. } 0° = 16 \text{ h. } 03 \text{ m.}$$
$$\text{Long. correction} = +03 \text{ m.}$$
$$\text{G.M.T. of Moonrise in Long. } 30° \text{ W.} = 16 \text{ h. } 06 \text{ m.}$$

$$\text{H.P.} = -55'\cdot 1$$
$$\text{S.D.} = +15'\cdot 1$$
$$\text{Ref.} = +33'\cdot 0$$
$$\text{Dip} = +7'\cdot 5$$
$$\text{Total Correction} = +0'\cdot 5$$
$$\text{True Alt. Moon's centre} = 00° \text{ } 00'\cdot 0$$
$$\text{Total Correction} = +0'\cdot 5$$
$$\text{Observed Alt. Moon's U.L.} = 00° \text{ } 00'\cdot 5$$

Answer—Obs. Alt. = $00° \text{ } 00'\cdot 5$.

In the circumstances applicable to example 39·3, the Moon's upper limb just touches the visible horizon at the instant it is on the observer's celestial horizon. It follows that the correct conditions for observing the Moon's amplitude are that its upper limb should be in contact with the observer's visible horizon.

The times of Moonrise and Moonset may be found from the tables given on the right-hand pages of the *Nautical Almanac*. The tabulated times are G.M.T.s of the phenomena at the Greenwich meridian. They are given for 10°-, 5°-, and 2°-intervals of Latitude, and tables are provided for interpolating for Latitude and Longitude.

4. Twilight

Owing to atmospheric refraction and the reflection of light from particles in the upper atmosphere, sunlight is received after the time of sunset or before the time of sunrise so long as the Sun is not more than about 18° below the observer's horizon. Sunlight received at a place in these circumstances is called Twilight.

Civil Twilight begins in the morning and ends in the evening when the Sun is 6° below the horizon. The light received from the Sun when it is between 6° and 12° below the horizon is known as Nautical Twilight. During the period of nautical twilight bright stars are visible and the horizon is clear enough for stellar observations. The light received from the Sun when it is between 12° and 18° below the observer's horizon is known as Astronomical Twilight.

The times (G.M.T. at Greenwich meridian or L.M.T. at other meridians) of the beginning and end of Nautical- and Civil-Twilight are given on the right-hand daily pages of the *Nautical Almanac*.

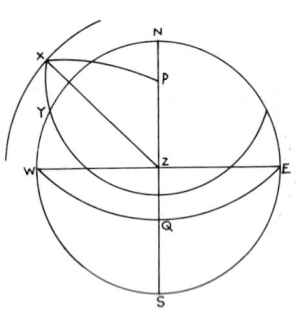

326 THE ELEMENTS OF NAVIGATION AND NAUTICAL ASTRONOMY

Fig. 39·2

In fig. 39·2 Y represents the Sun when it is on the celestial horizon of an observer whose zenith is projected at Z, and X represents the Sun when it is 18° below the observer's horizon. The time at which the period of astronomical twilight ends may be found by calculating the angle XPZ in the spherical triangle PZX in which ZX is equal to (90°+18°), or 108°; PZ is the co-Latitude of the observer, and PX is the polar distance of the Sun.

If, in the calculation, the log haversine of XPZ is more than 10·0, no triangle can be formed, in which case twilight lasts all night.

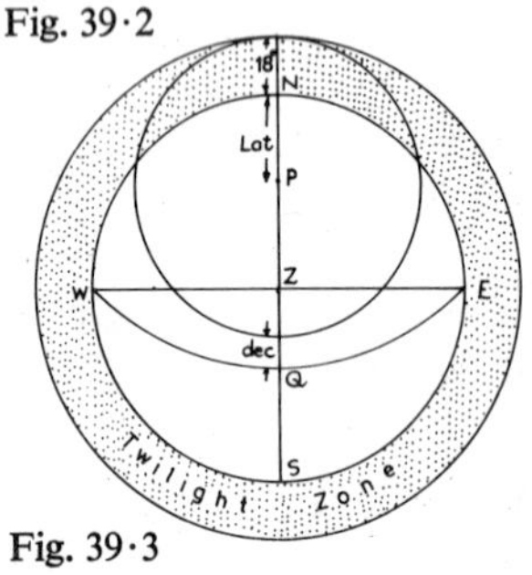

Fig. 39·3

Fig. 39·3 serves to illustrate the conditions in which twilight lasts all night. For twilight to last all night, as illustrated in fig. 39·3, the Sun's diurnal path must lie within the Twilight Zone, this being a spherical belt the limits of which are the observer's horizon and a small circle parallel to the horizon lying 18° below the observer's horizon.

For twilight to last all night (Lat.+18°) must be less than (90°−dec.).

In the limiting case:

$$(90° - \text{dec.}) = (\text{Lat.} + 18°)$$

That is: $$\text{Lat.} = (72° - \text{dec.})$$

or: $$(\text{Lat.} + \text{dec.}) = 72°$$

Therefore, for twilight to last all night, the Latitude of the observer and the Sun's declination must be of the same name and the observer's Latitude must be not less than the sum of 72° and the Sun's declination. In other words (Lat.+dec.) must be not less than 72°.

The Sun's maximum declination is 23° 27′. Therefore the lowest Latitude at which twilight can last all night is (72° 00′ − 23° 27′), which is 48° 33′ N. or S.

In low Latitudes, where the diurnal paths of celestial bodies cut the horizon at large angles, the Sun is in the twilight zone for a much shorter duration, than in high Latitudes.

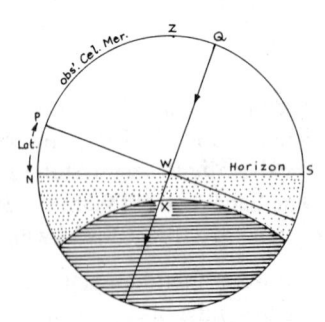

Fig. 39·4

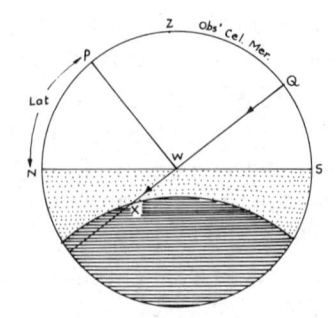

Fig. 39·5

Figs. 39·4 and 39·5 serve to illustrate that twilight is of shorter duration in the tropics than in higher Latitudes. In fig. 39·4 WX represents the path of the Sun when its declination is 00° 00′, at a place in Latitude θ where θ is equal to arc NP. The angle XPW is a measure of the interval during which the Sun is in the twilight zone.

In fig. 39·5, the interval during which the Sun is in the twilight zone is angle XPW. Clearly this angle is greater than the corresponding angle in fig. 39·4, because the observer's Latitude is greater. Therefore, the Sun passes through the twilight zone more obliquely in the higher than in the lower Latitude.

5. The Midnight Sun

If the Sun's diurnal circle does not cross the observer's horizon, the observer will experience daylight for the whole twenty-four hours of the day, and the Sun will be at lower meridian passage above the horizon. The limiting Latitude for the Sun to be visible at midnight on a given day of the year is (90°−dec.). At the limiting Latitude the Sun will graze the horizon at midnight. Thus, on 12th June, when the Sun's declination is 23° N., the Sun will graze the horizon, at midnight, of an observer in Lat. 67° N. In Latitudes higher than this the Sun's altitude at midnight on this day will be (Lat.−67°). Thus, in Latitude 77° N. on 12th June, the Sun's lower meridian altitude will be 10° bearing North.

Exercises on Chapter 39

1. Define: Theoretical Sunrise; Visible Sunrise.

2. Derive a formula for finding the time of Sunrise in terms of the observer's Latitude and the Sun's declination.

3. In what circumstances will sunrise occur (*a*) before 0600 hr. L.A.T., (*b*) after 0600 hr. L.A.T.?

4. Explain why the Sun's lower limb is about a semi-diameter above the visible horizon when the Sun's centre is on the observer's celestial horizon.

5. Explain why the Moon's upper limb should be just in contact with the visible horizon when her amplitude is observed.

6. Define: Civil Twilight; Nautical Twilight; Astronomical Twilight.

7. In what circumstances will twilight be experienced all night in Latitude 60° 00′ N.?

8. Prove that for twilight to last all night:

$$(\text{Lat.} + 18°) \not< (90° - \text{dec. of Sun})$$

9. Explain the circumstances in which the Sun will be above the horizon of an observer at midnight.

10. Why is northern Norway sometimes called The Land of the Midnight Sun?

11. Explain why the duration of twilight on any day of the year increases with Latitude.

12. Show that the least duration of twilight occurs at the equator on 21st March of any year.

CHAPTER 40

RATE OF CHANGE OF HOUR ANGLE, AZIMUTH AND ALTITUDE

1. Rate of Change of Hour Angle

For a stationary observer the Hour Angle of the Mean Sun changes at the rate of 15° per hour. That is to say, at a rate of 900′ per hour or 15′ per minute of time.

The S.H.A. of the Mean Sun decreases uniformly at the rate of 360° per year; that is at a rate of 2·46′ per hour. Therefore, the Hour Angle of the First Point of Aries, and of any fixed star, changes at a rate of (900+2·46)′ per hour, or very nearly 902·5′ per hour.

The Hour Angle of the True Sun changes at an irregular rate. The average rate of change of the Hour Angle of the True Sun is, of course, equal to the uniform rate of change of the Mean Sun's Hour Angle. The variation from the mean rate of change of the Hour Angle of the True Sun is so small, that the hourly change in his Hour Angle may be taken as 900′.

The rate of change of the Hour Angle of the Moon or a planet, which may readily be found for any time from the corresponding daily page of the *Nautical Almanac*, is very irregular.

If h' per hour is the rate of decrease of the S.H.A. of a celestial body, the rate of change of its Hour Angle is $(902·5-h)'$ per hour.

Any motion of an observer over the Earth's surface, unless it be along a meridian, will tend to cause the rate of change of the Hour Angle of a celestial body to increase if the observer's motion is easterly, and to decrease if his motion is westerly.

If x' per hour is the rate of change of an observer's Longitude towards the West, then:

Rate of change of H.A. of any celestial body $=(902·5-h-x)'$ per hour.

2. Rate of Change of Azimuth

Fig. 40·1 is a projection of the celestial sphere onto the plane of the celestial horizon of an observer whose zenith is projected at Z. N., E., S., and W., are the cardinal points of the horizon, and WQE is the projection of the celestial equator. X is a celestial body whose present altitude is CX and whose present azimuth is NZX.

Let the celestial body X in fig. 40·1 change its Hour Angle by 1 minute of time—this being represented by the arc AB. During this time the altitude of the body changes by an amount equal to WY, and its azimuth changes by an amount equal to the CD. Because the change in Hour Angle is small, triangle WXY may be regarded as being a plane triangle right-angled at W.

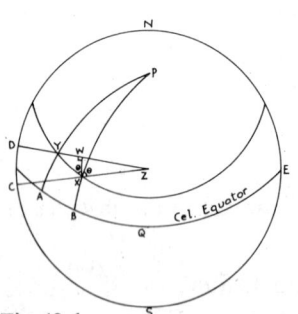

Fig. 40·1

RATE OF CHANGE OF HOUR ANGLE, AZIMUTH AND ALTITUDE

Now, $\qquad WXZ = 90°$
and $\qquad YXP = 90°$
Now $\qquad YXW = 90° - WXP$
and $\qquad PXZ = 90° - WXP$
Therefore: $\qquad YXW = PXZ$

The angles YXW and PXZ are denoted in fig. 40·1 by θ. θ is called the Parallactic Angle of the astronomical triangle-PZX.

Rate of Change of Azimuth of $X = CD'$ per minute of time
$\qquad = XW \cdot \sec CX \; '/\text{min}.$
$\qquad = XY \cdot \cos θ \cdot \sec CX \; '/\text{min}.$
$\qquad = AB \cdot \cos BX \cdot \cos θ \cdot \sec CX \; '/\text{min}.$

Now $\qquad$ arc $AB = 1$ minute of time $= 15'$ of arc
$\qquad BX =$ Declination of X
Therefore: $\qquad CX =$ Altitude of X

Rate of Change of Azimuth $= 15 \cdot \cos$ dec. $\sec$ Alt. $\cos θ \; '/\text{min}.$

From this formula it may readily be seen that for any given declination, the rate of change of a body's azimuth is greatest when its altitude is greatest. It may also be seen that the rate of change of the azimuth of a heavenly body is zero when its parallactic angle is 90°. When the parallactic angle is 0°, the body is on the observer's meridian and its rate of change of azimuth is greatest because its altitude is greatest.

The rate of change of azimuth of a celestial body may be expressed in terms of the body's Azimuth and Hour Angle as follows.

By applying the Spherical Sine Formula to the PZX-triangle, we have:
$\qquad \sin PX = \sin$ Az. cosec H.A. $\sin ZX$
or: $\qquad \cos$ dec. $= \sin$ Az. cosec H.A. $\cos$ Alt.

By substituting for cos dec. in the formula derived above, we get:

Rate of Change of Azimuth $= 15 \cdot \sin$ Az. cosec H.A. $\cos$ Alt. sec Alt. $\cos θ \; '/\text{min}.$
$\qquad = 15 \cdot \sin$ Az. cosec H.A. $\cos θ \; '/\text{min}.$

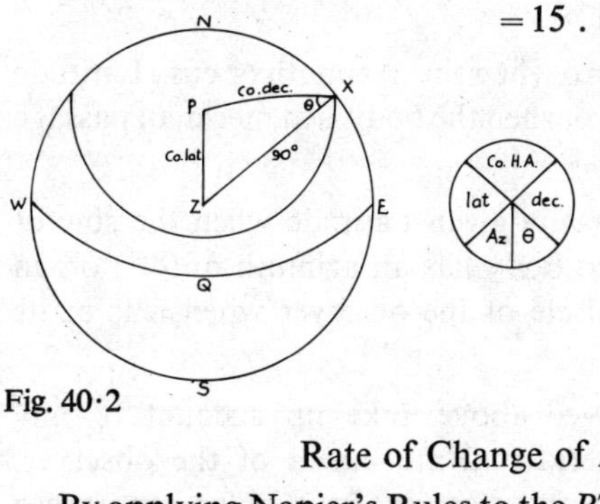

Fig. 40·2

When the altitude of a body is 0°, its rate of change of azimuth is proportional to the sine of the Latitude of the observer irrespective of the declination of the body. This may be proved with reference to fig. 40·2.

The secant of 0° is unity. Therefore, for a body on the horizon, we have, by substituting sec 0° in the first formula derived:

Rate of Change of Azimuth $= 15 \cdot \cos$ dec. $\cos θ \; '/\text{min}.$

By applying Napier's Rules to the PZX-triangle illustrated in fig. 40·2 we have:
$\qquad \sin$ Lat. $= \cos$ dec. $\cos θ$

By substituting sin Lat. for cos dec. cos θ we have:

Rate of Change of Azimuth of a celestial body on the horizon $= 15 \sin$ Lat. $'/\text{min}.$

330 THE ELEMENTS OF NAVIGATION AND NAUTICAL ASTRONOMY

This rate of change of azimuth is independent of the declination of the body. It follows that all celestial bodies at rising or setting change their azimuths at a rate which is the same for all bodies—this rate being proportional to the sine of the Latitude of the observer.

3. Rate of Change of Altitude

Referring to fig. 40·1 the change in the altitude of the body X during the interval of one minute is equivalent to WY. It follows that the rate of change of altitude is $WY\,'$ per minute.

$$\text{Rate of Change of Altitude} = WY\,'/\text{min.}$$
$$= XY \cdot \sin\theta\,'/\text{min.}$$
$$= AB \cdot \cos\text{dec.} \sin\theta\,'/\text{min.}$$

By applying the Spherical Sine Formula to the triangle-PZX, we have:

$$\sin\theta = \frac{\sin PZ \cdot \sin PZX}{\sin PX}$$

$$= \frac{\cos\text{Lat.} \sin\text{Az.}}{\cos\text{dec.}}$$

Substituting this value for $\sin\theta$ in the above formula, we have:

$$\text{Rate of change of altitude} = AB \cdot \cos\text{dec.} \; \frac{\cos\text{Lat.} \sin\text{Az.}}{\cos\text{dec.}}$$

$$= AB \cdot \cos\text{Lat.} \sin\text{Az.} \,'/\text{min.}$$

For a stationary observer the Mean Sun changes its altitude at the rate of $15 \cdot \cos\text{Lat.} \sin\text{Az.}\,'/\text{minute}$.

This rate is very nearly the same for the True Sun. In practice the same formula is used for finding the rate of change of a star's altitude, although the exact formula, which applies to any celestial body, is:

$$\text{Rate of change of Altitude} = \frac{(902\cdot 5 - h - x)}{60} \cdot \cos\text{Lat.} \sin\text{Az. in }\,'/\text{min.}$$

The rate of change of altitude of a heavenly body is zero when $\sin$. Azimuth or $\cos$. Latitude is zero. In other words the rate of change of altitude is zero when the body is at meridian passage or when the observer's Latitude is 90° North or South.

The maximum rate of change of altitude occurs for any given Latitude when the sine of the Azimuth is greatest. This occurs when the observed body has an azimuth of 90°; or, in cases when the body never crosses the prime vertical circle of the observer when it is at its limiting azimuth.

The formulae for rate of change of altitude, derived above, take no account of any declinational movement of the observed body, or meridianal movement of the observer. Changes of declination of the body or of Latitude of the observer are of particular importance when the body is at or near meridian passage.

When a celestial body is at meridian passage its rate of change of altitude due to its changing hour angle is zero. The body's altitude may change when the body is at meridian passage on account of changes in the body's declination or the observer's Latitude. Because

RATE OF CHANGE OF HOUR ANGLE, AZIMUTH AND ALTITUDE

of these factors the maximum daily altitude of a celestial body does not necessarily occur when the body is at meridian passage. Consequently the meridian altitude is not necessarily the maximum altitude. Maximum altitude and meridian altitude are the same only if the observer is not changing his Latitude at the time of the observation, and the body is not changing its declination.

When a body's geographical position and the observer's position are closing, the body's altitude at the time of meridian passage is increasing. In these circumstances maximum altitude occurs after meridian altitude.

When a body's geographical position and the observer's position are opening, the body's altitude at the time of meridian passage is decreasing. In these circumstances maximum altitude occurs before meridian altitude.

The rate at which a body is changing its declination may be found from the appropriate daily page of the *Nautical Almanac*. The following examples show how a body's rate of change of altitude may be found for the time of the body's meridian passage.

Example 40·1—Find the rate of change of the Sun's altitude when it is on the observer's upper celestial meridian of an observer in Latitude 30° 00′ N. on board a vessel steaming 000° (T) at 20·0 knots. The date is 22nd September.

From *Nautical Almanac* Extracts:

Declination of the Sun at Apparent Noon; that is, at 1153 L.M.T. or 1353 G.M.T. = 00° 22′·6 N.

The rate of change of the Sun's declination at this time is 01′·0 per hour towards South.

The ship's rate of change of Latitude is 20′·0 per hour towards North.

The Geographical position of the Sun and the observer's position are, therefore, opening at the rate of (20·0+1·0)′ per hour. That is at 21′·0 per hour.

At Apparent noon, therefore, the rate of change of the Sun's altitude is 21′·0 per hour, and the altitude is decreasing. Meridian altitude occurs, therefore, after maximum altitude.

Example 40·2—23rd September. Find the rate of change of the Moon's altitude when at upper meridian passage to an observer in Latitude 40° 00′ N. on board a vessel travelling 180° (T) at 15·0 knots. The observer's Longitude is 30° 00′ W.

L.M.T. meridian passage at Long. 0° = (23) 20 h. 56 m.
Longitude correction 30/360 of 46 m. =   +4 m.
L.M.T. meridian passage at 30° W. = (23) 21 h. 00 m.
Longitude in time =   2 h. 00 m.
G.M.T. meridian passage at ship = (23) 23 h. 00 m.

Moon's declination = 11° 21′·8 S. Changing at the rate of 8′·0 per hour towards North. Rate of change of Latitude is 15′·0/hour towards South. Therefore, the geographical position of the Moon and the observer's position are closing at the rate of (8·0+15·0)′ per hour. That is, at 23′·0 per hour.

332   THE ELEMENTS OF NAVIGATION AND NAUTICAL ASTRONOMY

At the time of the Moon's meridian passage the rate of change of the Moon's altitude is 23'·0 per hour, and the altitude is increasing. Meridian altitude occurs, therefore, before maximum altitude.

4. Interval Between Maximum and Meridian Altitudes of the Sun

Let the combined rates of change of an observer's Latitude and observed body's declination be y' per hour. Maximum altitude occurs when y' per hour is equal numerically to the rate of change of altitude due to the body's changing hour angle.

Let the rate of change of the observer's longitude be x' per hour towards West. In other words x is positive or negative when the observer is travelling westwards or eastwards respectively.

$$\text{Rate of change of Sun's altitude} = (900-x) \cos \text{Lat.} \sin Az. \text{ '/hour} \quad \ldots \quad (I)$$

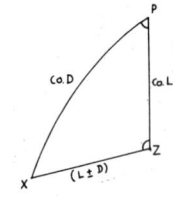

Applying the Spherical Sine Formula to the PZX-triangle illustrated in fig. 40·3, we have:

$$\sin Z = \frac{\cos \text{dec.} \sin P}{\sin ZX}$$

Fig. 40·3

Now when a celestial body is near meridian passage its zenith distance is very nearly equal to its M.Z.D.

When Latitude and declination have the same name:

$$\text{M.Z.D.} = (\text{Lat.} \sim \text{dec.})$$

When Latitude and declination have different names:

$$\text{M.Z.D.} = (\text{Lat.} + \text{dec.})$$

In general, therefore:

$$\text{M.Z.D.} = (\text{Lat.} \pm \text{dec.}), \text{ so that:}$$

$$\sin Z = \frac{\cos \text{dec.} \sin P}{\sin (\text{Lat.} \pm \text{dec.})}$$

Substituting for $\sin Z$ in formula (I), we have:

$$\text{Rate of change of Sun's altitude} = (900-x) \cos \text{Lat.} \frac{\cos \text{dec.} \sin P}{\sin (\text{Lat.} \pm \text{dec.})} \text{'/hr.}$$

$$= \frac{(900-x) \cos \text{Lat.} \cos \text{dec.} \sin P}{\sin (\text{Lat.} \pm \text{dec.})} \text{'/hr.}$$

Since P is a small angle:

$$\sin P = P \text{ radians}$$

Therefore:

$$\text{Rate} = \frac{(900-x) \cos \text{Lat.} \cos \text{dec.} P^c}{\sin (\text{Lat.} \pm \text{dec.})} \text{'/hr.}$$

RATE OF CHANGE OF HOUR ANGLE, AZIMUTH AND ALTITUDE

Let the interval between the times of maximum and meridian altitudes be t seconds of time. In other words, let the angle ZPX at the time of maximum altitude be t seconds or $t/3600$ hours. In this case:

$$P = \frac{t \cdot (900-x)}{3600 \cdot 3438} \text{ radians, so that:}$$

$$\text{Rate} = \frac{(900-x)\cos \text{Lat.}\cos \text{dec.}}{\sin(\text{Lat.} \pm \text{dec.})} \cdot \frac{t \cdot (900-x)}{3600 \cdot 3438} \text{ '/hr.}$$

$$= \frac{t \cdot (900-x)^2 \cdot \cos \text{Lat.} \cos \text{dec.}}{3600 \cdot 3438 \cdot \sin(\text{Lat.} \pm \text{dec.})} \text{ '/hr.}$$

$$= \frac{t \cdot 900^2 (1-x/900)^2 \cdot \cos \text{Lat.} \cos \text{dec.}}{3600 \cdot 3438 \cdot \sin(\text{Lat.} \pm \text{dec.})} \text{ '/hr.}$$

Now, $(1-x/900)^2 \simeq (1-x/450)$

and, $\sin(\text{Lat.} \pm \text{dec.}) = \sin \text{Lat.} \cos \text{dec.} + \cos \text{Lat.} \sin \text{dec.}$

Therefore:

$$\text{Rate} = \frac{t \cdot 900^2 (1-x/450) \cos \text{Lat.} \cos \text{dec.}}{3600 \cdot 3438 (\sin \text{Lat.} \cos \text{dec.} \pm \cos \text{Lat.} \sin \text{dec.})} \text{ '/hr.}$$

$$= \frac{t(1-x/450)}{15 \cdot 28 (\tan \text{Lat.} \pm \tan \text{dec.})} \text{ '/hr.}$$

At the time of maximum altitude the rate of change of the Sun's altitude is equal to y' per hour, so that:

$$y = \frac{t(1-x/450)}{15 \cdot 28 (\tan \text{Lat.} \pm \tan \text{dec.})}$$

and

$$t = \frac{15 \cdot 28 (\tan \text{Lat.} \pm \tan \text{dec.})}{(1-x/450)} \cdot y$$

$$= 15 \cdot 28 \, y \, (\tan \text{Lat.} \pm \tan \text{dec.}) \, (1-x/450)^{-1}$$

Now, $(1-x/450)^{-1} \simeq (1+x/450)$,

so that: $t = 15 \cdot 28 \, y \, (\tan \text{Lat.} \pm \tan \text{dec.}) \, (1+x/450)$ seconds

From this relatively simple formula, the interval t, between the times of meridian and maximum altitudes may easily be found.

Example 40·3—Find the interval between the times of maximum and meridian altitudes of the Sun on 23rd September, to an observer in Lat. 50° 00′ N., Long. 45° 00′ W., on board a vessel travelling 030° (T) at the rate of 18·0 knots.

Rate of change of observer's Latitude = 15′·6 to the North

Rate of change of observer's Longitude = 9·0 miles/hr.

= 14′·0/hr. to the East

From *Nautical Almanac* Extracts:

Sun's Declination at 1153 L.M.T. or 1453 G.M.T. = 0° 01'·8 S.

Rate of change of Sun's Declination = 1'·0 towards the South

$$x = -14'\cdot 0 \text{ per hour}$$
$$y = (15\cdot 6 + 1\cdot 0)' \text{ or } 16'\cdot 6 \text{ per hour opening}$$

Meridian altitude occurs AFTER maximum altitude.

$$t = 15\cdot 28 \, y \, (\tan \text{Lat.} + \tan \text{dec.}) \, (1 - x/450)$$
$$= 15\cdot 28 \, . \, 16\cdot 6 \, (\tan 45° + \tan 0° \, 01'\cdot 8) \, (1 - 14/450)$$
$$= 244 \text{ seconds or 4 minutes approx.}$$

Answer—Interval = 4 minutes approx.

Exercises on Chapter 40

1. Explain why the Rate of change of the Sun's Hour Angle is 900' per hour, and that of a fixed star is about 2'·5 per hour faster.

2. Explain why westerly movement of the observer causes the rate of change of a celestial body's Hour Angle to be smaller than that for a stationary observer in the same Latitude.

3. What is the rate of change of a star's Hour Angle to an observer in Latitude 60° N. travelling due East at the rate of 20·0 knots?

4. Derive a formula for finding the rate of change of the Sun's azimuth in terms of the declination and altitude of the Sun and the parallactic angle.

5. Explain why, for any given declination of the Sun, the rate of change of the Sun's azimuth is greatest when its altitude is greatest.

6. Explain why an object's movement at the instant the parallactic angle is 90° is directly towards the observer's zenith.

7. Derive a formula for finding the rate or change of the Sun's azimuth in terms of the Sun's Azimuth, Hour Angle, and Parallactic Angle.

8. Show that the rate of change of azimuth of any celestial body at rising or setting is proportional to the sine of the Latitude of the observer.

9. Show that:

    Rate of change of Sun's Altitude to a stationary observer = 15 . cos Lat. sin Az. '/min.

10. Explain why the rate of change of a celestial body's altitude is zero when it is at meridian passage.

11. Explain clearly why the times of maximum and meridian altitudes are not generally coincident to an observer travelling northwards or southwards.

12. Find the interval between maximum and meridian altitudes of the Sun on 24th September, to an observer in Lat. 42° 00' N., Long. 60° 30' W. travelling 165° (T) at the rate of 20·0 knots.

13. Find the interval between maximum and meridian altitudes of the Sun on 22nd September, for an observer travelling 340° (T) at 15·0 knots, given the observer's position: Lat. 20° 00' N., Long. 15° 00' W.

PART 6

THE INSTRUMENTS OF NAVIGATION

The present-day navigator is faced with a bewildering selection of instruments designed to assist him in bringing his vessel safely to harbour.

The principal instruments of navigation are chart, compass and log. Charts, and their complementary Sailing Directions, are discussed in Part 3 under the general heading of Coastal Navigation. In this part we shall describe the magnetic and gyroscopic compass; sextant and chronometer; sounding instruments and logs; radio direction finding; radar; hyperbolic aids to navigation; and, finally, satellite and inertial systems of navigation.

The usefulness of a tool is usually a function of the craftsmanship of the user. For this reason a navigator should aim to be a complete master of every instrument designed for his specific use. There is much for the beginner to learn—but equally as much satisfaction, in being an expert, to be gained.

CHAPTER 41

THE MAGNETIC COMPASS

1. Magnetism

A magnet is a mass of iron, steel, or other magnetic material, which has the property of attracting like masses with a force known as Magnetic Force. Natural magnets are found in many regions of the Earth, but the strength of attraction or power of these is usually very meagre. Comparatively strong magnets may be made using the magnetic properties of an electric current. These are known as Artificial Magnets.

The strength of a magnet is concentrated at two points which are called the Poles of the magnet. In an artificial magnet, which is usually a bar or needle of steel, the poles are located near the ends. If a magnet is suspended at its centre of gravity, it will tend to align itself in a definite direction relative to the Earth's surface. This direction is roughly in the North–South plane. The end which tends to point northwards is known as the North-Seeking end. The other is known as the South-Seeking end.

The ends of magnets used on board in connection with the magnetic compass, are usually painted Red at the North-seeking end and Blue at the South-seeking end. For this reason seamen almost always refer to the poles of a magnet by the terms Red and Blue.

It will be found that if two artificial magnets are brought together a force of attraction exists between poles of opposite colour, and a force of repulsion exists between poles of the same colour. A magnet's Field is the region around a magnet within which the force emanating from the magnet may influence magnetic material. The field of a magnet may be explored by means of a small Exploring Magnet. If the alignments of the needle of the exploring magnet at several points are linked, it will be found that Curves or Lines of Magnetic Force emerge from the Red end and enter the Blue end of the magnet whose field is being explored. The direction of a line of force is regarded conventionally as the direction towards which the Red end of the exploring magnet points. Thus we say that the lines of force of a magnet emanate from the Red end and enter the Blue end of a magnet.

The fundamental law of magnetism is:

Like poles repel each other, and
Unlike poles attract each other.

2. Terrestrial Magnetism

Observations of the alignment of a freely suspended magnet at many places on the Earth's surface indicate that the Earth itself has the properties of a huge natural magnet. The lines of force of the Earth-magnet cut the Earth's surface at different angles at different places. The angle at which the Earth's lines of force cut the Earth's surface at any place is called the angle of Dip for the place. The angle of dip may be found by means of a Dipping Needle which is simply a magnet suspended so that it is free to incline. At places where the Red end of a dipping needle tilts downwards Dip is named Positive. Where the Blue end dips downwards Dip is named Negative.

At a position where the dip is $-90°$, the lines of force of the Earth-magnet leave the Earth's surface vertically. The position at which this occurs—a position known as the South Magnetic Pole—is in the vicinity of South Victoria Land in Antarctica.

At a position where the dip is $+90°$, the lines of force of the Earth-magnet enter the Earth vertically. The position of this point—the North Magnetic Pole—is in the region of Hudson Bay in North America.

Girdling the Earth in an approximate great circle path which is inclined at about 10° to the geographical equator is a line on which the dip at every point is 0°. This line is called the Magnetic Equator.

Lines on the Earth joining places of equal dip approximate to small circles which are parallel to the Magnetic Equator. These lines of equal dip are called Isoclinic Lines. The line of zero dip is called the Aclinic Line. This, of course, is the Magnetic Equator. All places at which the dip is negative are said to be in the South Magnetic Hemisphere and to have South Magnetic Latitude. All places at which the dip is positive are said to be in the North Magnetic Hemisphere and to have North Magnetic Latitude.

3. Variation

A freely suspended magnetic needle, when under the influence of the Earth's magnetic field alone, settles in a vertical plane which is known as that of the Magnetic Meridian. The horizontal direction towards the North Magnetic Pole is called Magnetic North, and the opposite direction is called Magnetic South. The horizontal angle between the directions of True North and Magnetic North at any place is called the Variation of the place. Variation is named, conventionally, according as the direction of Magnetic North lies to the right or left of that of True North. When the direction of Magnetic North lies to the right of that of True North, variation is named East. In other cases it is named West. The magnitude of variation may be anything between 0° and 180° E. and W.

Lines linking places at which the variation is the same are called Isogonic Lines or Isogonals. The variation changes not only from place to place but also from time to time. The value for any given epoch may be found from an isogonic chart. Information on the isogonic chart enables the navigator to bring the variation up to date.

4. The Magnetic Compass

There are two types of magnetic compasses known, respectively, as the Dry Card and the Liquid compasses.

The Dry Card Compass consists of a system of magnetised needles slung, by means of silk threads, from a graduated Compass Card of about ten inches in diameter. The needles lie parallel to the plane of the North and South graduations on the card. At the centre of the card is a Cap in which is fitted a Jewel Bearing. The compass card rests on a hard metal sharply pointed Pivot. The point of support is above the centre of gravity of the system so that the card always remains in the horizontal plane and is unaffected by dip.

The card is made of very thin rice paper and is stiffened by a thin aluminium ring around its circumference, which provides the means for supporting the needle system. The concentration of mass at the edge of the card ensures that the period of the vibration of the card is large and that there will be little chance, therefore, of a state of synchronism existing between the oscillating card and the rolling of the ship.

The mass of the card and the needle system is very small—only a fraction of an ounce—hence the friction between the cap and pivot, the magnitude of which depends upon the resultant force acting, is very small. In a perfect magnetic compass the friction between cap and pivot is virtually nil, in which case the North–South axis of the compass card lies fixed in the plane of the magnetic meridian provided that the needle is under the influence of the Earth's magnetism alone.

Painted on the bowl which houses the compass card is a Lubber Line against which the compass course of the ship is read. The bowl is supported on Gymbals within the compass Binnacle.

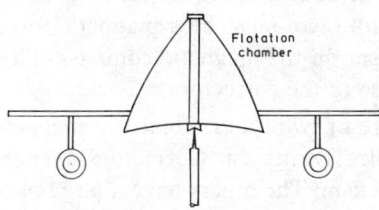

In a Liquid Compass the card is housed in a bowl of liquid of low freezing point. The buoyancy of the compass card, which is provided by means of a copper Flotation Chamber, is adjusted so that the card almost floats. The weight of the card on the pivot, therefore, is very small, so that friction between the cap and the pivot is almost eliminated. The needles in a liquid compass are encased in brass to prevent them from rusting.

Fig. 41·1

Fig. 41·1 shows a section through a liquid compass card.

5. Ship Magnetism

The magnetism of a piece of magnetic material may be Permanent or Temporary. The permanent magnetism of a mass of magnetic material is due to a magnetic property known as Hard Iron (H.I.). The temporary magnetism of a piece of magnetic material is due to its Soft Iron (S.I.) property. It should be noticed that H.I. and S.I. are names of Properties of magnetic material and that they have no direct connection with the physical properties of hardness of material.

A piece of unmagnetised magnetic material, when placed in a magnetic field, becomes magnetised by a phenomenon known as Induction. On removing the material from the inducing field only a proportion of the magnetism it acquired by induction remains: a proportion is always lost. That which is retained is due to the H.I. property of the material: that which is lost is due to its S.I. property.

A vessel is made of magnetic material, and every part of the structure of a vessel has both H.I. and S.I. properties simultaneously. The vessel acquires permanent magnetism during the time she is being constructed. This is due to the magnetising influence of the Earth. It must not be thought that a vessel's permanent magnetism is simply explained, especially in these times when prefabrication of vessels plays an important part during the period of building. As a result of the vessel's permanent magnetism a permanent magnetic field acts at the position of the magnetic compass, and this field may cause deviation. This deviation is often regarded as being due to the existence of a Red and Blue pole within the vessel. This idea should be used reservedly as it may give rise to false ideas on magnetism of vessels.

In addition to the permanent magnetism which causes a permanent magnetic field to act at the compass position another magnetic field acts, this being due to the S.I. property of the vessel. This induced field changes in magnitude and direction relative to the fore-and-aft line of the vessel with every change in the vessel's course and geographical position. The permanent field, on the other hand, never varies in magnitude or direction relative to the vessel's fore-and-aft line.

6. Simple Ideas on Compass Adjustment

The term Compass Adjustment is something of a misnomer. The compass, as such, requires no adjustment: it is as near perfect as is humanly possible, when it is supplied to a vessel. The aim of a Compass Adjuster is to place correctors in the vicinity of the compass so that their combined magnetic effect neutralises the magnetic field of the vessel at the compass position. If he is successful the compass will be influenced by the Earth's magnetic field only and it will always indicate Magnetic North and no deviation will appear on any heading. In practice perfection is seldom attained, and the best that can be done is to reduce the magnitude of deviations to small proportions. When deviation appears there must be a magnetic force acting in addition to the force due to the Earth's magnetism on the magnetic compass. This force is the resultant of the vessel's magnetism and that due to the correctors.

The correcting devices used by the compass adjuster are of two types. Some are magnets having a large proportion of H.I. property. These are called Permanent Corrector Magnets and they are used to neutralise the ship's permanent magnetism. The others have a large proportion of S.I. property. These are used to neutralise the effects of the ship's induced magnetism, and they are called the Soft Iron Correctors.

The Soft Iron Correctors are named after famous scientists who promoted the science of ship magnetism in the early days of its development. The spheres which are usually to be found placed athwart the compass on the binnacle are named after Lord Kelvin who first suggested their use. The Soft Iron Corrector placed in a brass case usually on the fore side of the binnacle is known as the Flinders' Bar after Captain Matthew Flinders the famous sailor scientist of the early nineteenth century.

For convenience of adjustment the permanent and induced magnetism of a vessel are resolved into fore-and-aft, athwartship, and keelward components.

The forces due to the vessel's permanent magnetism which act in the fore-and-aft, athwartship, and keelward directions at the Red end of the magnetic compass, are called respectively forces P, Q, and R.

Force P is eliminated by means of a permanent magnet placed fore and aft in the binnacle beneath the compass. Force Q is eliminated by means of an athwartship magnet placed in the binnacle beneath the compass. Force R, together with certain of the component forces of the Induced Magnetism of the vessel, is eliminated by means of a vertical magnet placed in the binnacle beneath the compass.

The resolution of the induced magnetism of a vessel is very much more complex than that of her permanent magnetism. Whereas three components are all that are necessary to deal with the permanent magnetism of a vessel, in the most complex cases her induced magnetism requires no less than nine components.

The purpose of Kelvin's spheres is to neutralise certain horizontal (fore-and-aft and athwartship) components of the vessel's induced magnetism. The purpose of the Flinders'

Bar is to neutralise certain horizontal forces which result from induction in vertical components of the vessel's induced magnetism.

Although collectively the component forces of a vessel's magnetism are complex, the process of neutralising it at the compass position is relatively simple. As an illustration of the procedure let us suppose that the field due to the vessel's induced magnetism is properly neutralised by the Soft Iron correctors and that the permanent magnetism of the vessel acts in the fore-and-aft line towards the stern.

When the vessel is heading 000° or 180° by Compass, her permanent magnetism acts in line with the Earth's magnetic field, so that the vessel's magnetism does not produce a deviating force. On every other compass course the vessel's magnetic force acts across the magnetic meridian and it is, therefore, a deviating force. The maximum deviation occurs on 090° or 270° by compass.

When compensating any component of a vessel's magnetism a compass adjuster normally sets the course on which the deviation due to that particular component is maximum. Thus, in the case under consideration, the course would be set to 090° (C.) or 270° (C.).

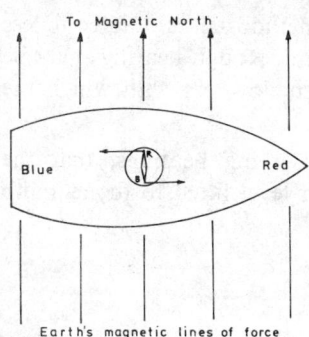

Fig. 41·2 illustrates the magnetic couple due to the vessel's permanent magnetism which acts on the magnetic compass when the course is 090° (C.). On this course the maximum West deviation occurs. To eliminate the vessel's fore-and-aft permanent magnetism, the compass adjuster removes the West deviation, which he finds by comparing the Magnetic and Compass Bearings of a distant shore mark or of the Sun or other celestial body, by placing a fore-and-aft magnet (in the binnacle beneath the compass) with its Red end pointing aft.

Fig. 41·2

To eliminate deviation due to the athwartship component of the vessel's permanent magnetism the course is set to 000° (C.) or 180° (C.), and the deviation appearing would be eliminated by means of an athwartship magnet placed in the binnacle beneath the compass.

7. The Azimuth Mirror

Compass Bearings of terrestrial or celestial bodies are observed by means of an Azimuth Mirror. This instrument which, when in use, fits over the compass card, consists of a stand on which is mounted a glass prism. The prism may be rotated about a horizontal axis by means of a screw with a milled head on which is engraved an arrow. The stand of the azimuth mirror is provided with a Spirit Level and a Shadow Pin, and also with a Sighting Tube in which is housed a Magnifying Lens.

There are two methods of observing bearings with an azimuth mirror. They are known respectively as the Arrow Up and Arrow Down methods.

The Arrow Up method is usually employed when observing heavenly bodies. The prism is adjusted by means of the milled head screw until the light from the body is reflected to the observer's eye. This is denoted by E in fig. 41·3. The observer's eye is placed so that the compass card is seen through the magnifying lens in the sighting tube. A metal pointer P indicates the required bearing when it is in alignment with the image of the observed object.

Great care must be taken to ensure that the image of the observed body is in coincidence with the pointer: if they are not an error in the bearing will result. Such an error increases as the altitude of the body increases. For this reason it is advisable not to observe bodies whose altitudes are greater than about 40°, when bodies at smaller altitudes are available.

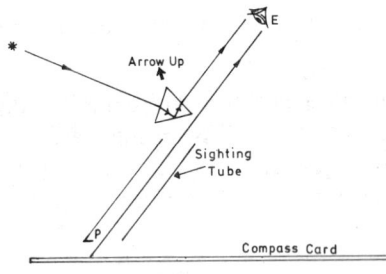

Fig. 41·3

The Arrow Down method is used for observing terrestrial bodies or celestial bodies near the horizon. With this method the observer sights the object over the top of the prism, and adjusts the prism until the compass card is reflected to his eye. The Arrow Down method is illustrated in fig. 41·4.

The Arrow Down method may be used for observing celestial bodies which are well above the horizon, and it is to be preferred to the Arrow Up method when the body is indistinct.

Bearings using the two methods should agree unless the prism is out of adjustment. Rough bearings may be observed by means of the shadow pin with which the instrument is provided.

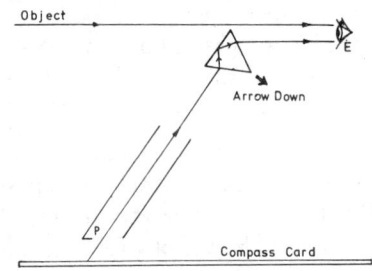

Fig. 41·4

It is essential, when observing bearings, that the azimuth mirror is perfectly horizontal. The purpose of the spirit level fitted to the azimuth mirror stand, is to ensure that this is so.

Exercises on Chapter 41

1. What is an artificial magnet? Describe how an artificial magnet may be made.
2. What is the fundamental law of magnetism?
3. Describe a Dry Card Compass.
4. Describe a Liquid Compass such as that found in a life-boat.
5. Describe carefully the binnacle of a magnetic compass.
6. Write a short essay on Terrestrial Magnetism.
7. Define: Magnetic Poles; Magnetic Equator.
8. What are: (*a*) Isoclinic Lines, (*b*) Isogonic Lines?
9. Define: Magnetic Dip.
10. Describe how a vessel acquires her magnetism.
11. Distinguish between permanent magnetism and induced magnetism of a vessel.
12. Describe the permanent and Soft Iron correctors used for neutralising a vessel's magnetism at the position of the magnetic compass.
13. Describe an azimuth mirror.
14. Describe the two methods of using an azimuth mirror.
15. How may the accuracy of an azimuth mirror be checked?

CHAPTER 42

THE GYROSCOPIC COMPASS

1. Introduction

The magnetic compass, which has served the mariner for no less than half a millennium, and which has provided him in the past with his only sure guide of direction when out of sight of land, is now being superseded by the gyroscopic compass—a marvellous product of engineering skill in which navigators have complete confidence. An important feature of the gyro compass is that it indicates the direction of True North.

The ancestor of the complex gyro compass is a simple device, invented by Foucault of pendulum fame, to demonstrate the Earth's rotation. Foucault named the instrument a Gyroscope which means to "view a spin".

At the beginning of this century, when it became possible to spin a gyro wheel electrically, attention was directed to the possibility of developing a gyroscopic compass for use on board ship. First in the field was the German scientist Dr. Anschutz, who introduced his first successful gyro compass in 1908. Three years later, in 1911, the American scientist Dr. Elmer Sperry, patented his gyro compass. The British reply to the Anschutz and Sperry compasses was invented by Dr. S. G. Brown in 1916. Since then other makes of gyro compass have appeared, notably the Arma and Plath compasses.

The usefulness of a gyro compass is not confined to directional properties only: Steering and Bearing Repeater Compasses; Course Recorder, and Automatic Helmsman may be operated from a Master Gyro Compass.

The principal disadvantage of the gyro compass is that it requires an electrical supply, a failure in which requires the use of the stand-by magnetic compass.

2. The Principles of the Gyroscope

A device known as a mechanical or model gyroscope is used to demonstrate gyroscopic principles. It is a wheel mounted in such a way that its axle may be set in any desired direction relative to its housing. Fig. 42·1 illustrates a mechanical gyroscope.

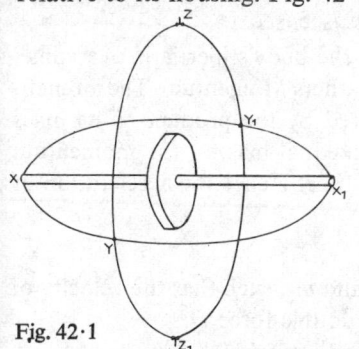

A mechanical gyroscope is said to have three Axes or Planes of Freedom. Fig. 42·1 illustrates the three axes of freedom, which are:

XX_1—which allows freedom of the wheel to spin on its axle

YY_1—which allows freedom of the axle to incline to the horizontal

ZZ_1—which allows freedom of the axle to turn in azimuth.

The centre of gravity of a mechanical gyroscope should be at the centre of gravity of the wheel and friction at all bearings should be as small as possible.

Fig. 42·1

A gyro wheel, in common with all spinning bodies, possesses the properties known as Gyroscopic Inertia and Precession.

Gyroscopic Inertia is the property of a spinning body by which it tends to maintain its axis or plane of spin. Precession is the seemingly paradoxical movement of the axis of a spinning body when a mechanical couple is applied to it.

To understand these properties it is necessary to understand elementary mechanics. The following brief notes should be familiar.

The Mass of a body is a measure of the quantity of matter in it, as indicated by the acceleration imparted to it by a given Force. The unit of mass is the Pound.

A body's Inertia is a measure of the tendency it has to maintain its present state of rest or uniform motion in a straight line. The amount of inertia a body possesses is dependent upon its mass and velocity. Newton's First Law of Motion, often referred to as the Law of Inertia, states that: "every body tends to maintain its present state of rest or uniform motion in a straight line and will do so if no force acts on it".

Anything that changes, or tends to change, a body's state of rest or uniform motion in a straight line, is known as a Force. A force is a vector quantity in that it has both magnitude and direction. A vector quantity is one that may be represented by a straight line, the direction of which represents the direction of the vector and the length of which represents its magnitude. Examples of kinds of force are:

(i) Gravitational Force which is the force the Earth exerts on bodies on or near its surface. The gravitational force acting on a body is equivalent to the Weight of the body. The unit of gravitational force is the downward force exerted by the Earth on a mass of one pound—a unit known as a Pound Weight.

(ii) Frictional Force, which is the force of resistance to motion due to the contact of two surfaces moving relatively to one another.

(iii) Magnitude Force, which is the force emanating from a pole of a magnet, which may change the motion of a mass of magnetic material.

(iv) Muscular Force.

(v) Electrical Force.

Velocity is a vector quantity and is defined as the rate of change of position in a given direction.

A body Accelerates when its velocity changes. Suppose a body starts from rest and moves with an average velocity of 1 ft./second during the first second; 2 ft./second during the second second; 3 ft./second during the third; and so on. The body is said to accelerate at the rate of one foot per second per second, an acceleration written as 1 ft./sec./sec.

When a force is applied to a body which is free to move, the body's inertia is overcome, and it moves with ever-increasing velocity. As it accelerates it gathers Momentum. The momentum of a body is the quantity of motion it possesses, measured by the product of its mass and velocity. If equal forces are applied simultaneously to unequal masses, the momentum gathered by each mass is the same after any given interval of time. Hence the acceleration of the larger mass is less than that of the smaller.

Force is measured by rate of change of momentum.

If forces P and Q act simultaneously on equal masses m_1 and m_2, such that the velocity of m_1 is double that of m_2 after a given interval, then force P is double force Q.

If forces P and Q act simultaneously on masses m and $2m$ such that the velocity of each mass at any instant is the same, force P is half force Q.

THE GYROSCOPIC COMPASS

The relation between force, mass and acceleration, is given in Newton's Second Law of Motion, which is:

"Rate of change of momentum is proportional to the applied force."

In other words:

    Applied force $\propto$ Rate of change of momentum

                      $\propto$ Rate of change of mass $\times$ velocity

                      $\propto$ Rate of change of velocity $\times$ mass

                      $\propto$ Acceleration $\times$ mass.

The unit of force is chosen such that it is the force required to give unit acceleration to a unit mass.

Hence, if appropriate units are used:

    Applied Force = Acceleration $\times$ Mass

or: $F = a \cdot m$

A body spins, or tends to spin, when a Mechanical Couple is applied to it. A couple is formed when two equal opposite forces, acting in the same plane, are separated by a perpendicular distance known as the Arm of the Couple. The tendency of a body to spin when a couple is applied to it is dependent upon the magnitude of the forces and the arm of the couple. A measure of this tendency is called the Moment of the Couple. The moment of a couple is found by taking the product of one of the forces and the arm of the couple. This may readily be proved by the Principle of Moments.

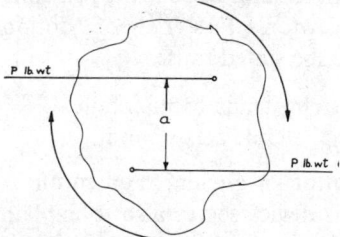

Fig. 42·2 illustrates a couple the Moment of which is $P \cdot a$ ft. lbs.

The quantity of motion possessed by a body moving in a circular path is called the body's Angular Momentum. The angular momentum of a spinning body is dependent upon its mass, its angular velocity, and the distribution of its mass.

Fig. 42·2

When a body which is free to move is acted upon by a mechanical couple its angular velocity increases and it is said to move with Angular Acceleration.

A spinning body whose mass is concentrated near its axis of spin has less angular momentum than a spinning body of the same total mass and the same angular velocity but whose mass is concentrated away from the axis of spin.

Newton's Second Law of Motion applied to spinning bodies may be stated thus:

"The rate of change of angular momentum is proportional to the applied couple."

In other words:

    Applied Couple $\propto$ Rate of change of angular momentum

                      $\propto$ Rate of change of angular velocity $\times I$

                      $\propto$ angular acceleration $\times I$

346   THE ELEMENTS OF NAVIGATION AND NAUTICAL ASTRONOMY

If appropriate units are used:

$$\text{Applied Couple} = \text{Angular Acceleration} \times I$$
$$L = \varphi \cdot I$$

The quantity I is known as the Moment of Inertia of the spinning body. It may be expressed in the form:

$$I = m \cdot k^2$$

where m is the mass of the body and k is a quantity called the Radius of Gyration of the body. The radius of gyration may be defined as the radius of motion of a particle having a mass equal to that of the spinning body.

If, through some internal cause, one particle of a revolving system changes its position and/or its angular velocity such that its contribution to the total momentum of the system changes, then one or more of the other particles of the revolving system will change position and/or angular velocity in such a way that the total angular momentum of the system remains the same.

This property of a revolving system of masses is implicit in a principle known as the Principle of the Conservation of Angular Momentum, which is:

"Every system of revolving masses, about a given axis, tends to maintain its angular momentum, and will do so if no external couple acts upon it."

Examples of revolving systems to which this principle applies are: the Solar System; the Earth-Moon system; the governor on an engine; and a pendulum.

The axis of revolution of a system or the spin axis of a rotating body, tends to maintain a fixed direction in space. If this axis changes its direction the angular momentum of the system or the body would change about the original axis, and this cannot be so if the principle of the conservation of angular momentum is to be obeyed. Newton's First Law of Motion applied to spinning bodies or revolving systems may, therefore, be stated thus:

"The axis of a spinning body, or of a revolving system, tends to maintain its present direction in space, and will do so if no external couple acts upon it."

An instructive illustration of the principle of the conservation of angular momentum is a piece of circus equipment known as a Performing Disc. This device serves also to explain the phenomenon of precession. A performing disc consists of a large platform of wood which is made to rotate about a vertical axis. A clown, by changing his position on the disc, is able to change the speed of the disc. When at the centre of the disc, the clown contributes no angular momentum to the system. When he moves from the centre towards the circumference of the disc he contributes an increasing amount of angular momentum to the system. The disc's contribution is, therefore, reduced, the reduction being effected by it slowing down. When the clown moves towards the centre of the disc, its angular velocity increases.

Fig. 42·3 illustrates a performing disc the moment of the couple causing rotation is $P \cdot a$ lbs. ft. When the clown C_1 moves towards the centre of the disc, the speed of rotation of the disc increases. When clown C_2 moves towards the edge of the disc, the speed of rotation of the disc decreases.

Fig. 42·3

From these facts the two following rules are deducted:

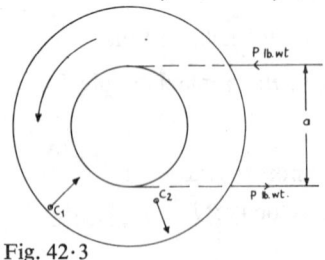

THE GYROSCOPIC COMPASS

1. When a member of a revolving system moves towards the axis of revolution an apparent couple takes effect which acts *with* the couple causing the revolution of the system.
2. When a member of a revolving system moves away from the axis of revolution an apparent couple takes effect which acts *against* the couple causing the revolution of the system.

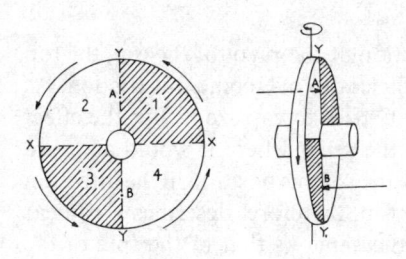

The cause of precession may be explained thus: Imagine a couple acting at A and B on the spinning wheel illustrated in fig. 42·4. The couple acts about the axis XX. If the wheel were not spinning the couple would tend to rotate the wheel about the axis XX.

Every particle in quadrant 1 is receding from the axis of rotation of the couple and, therefore, an apparent couple acts such that every particle in this quadrant tends to move in a direction opposite to that in which force A acts.

Fig. 42·4

Every particle in quadrant 2 is moving towards XX and, therefore, an apparent couple acts such that every particle in quadrant 2 tends to move in the direction in which force A acts.

Every particle in quadrant 3 is receding from XX and, therefore, an apparent couple acts such that every particle in this quadrant tends to resist force B.

Every particle in quadrant 4 is moving towards the axis XX and, therefore, an apparent couple acts such that every particle in quadrant 4 tends to act with the force B.

All particles in quadrants 1 and 4 tend to move in the same direction. That is in the direction of force B.

All particles in quadrants 2 and 3 tend to move in the same direction, which is the direction in which force A acts.

If the wheel is perfectly free to move it will rotate about the axis YY. This movement is called Precession. Note that the wheel tends to set its plane of spin in the plane of the applied couple.

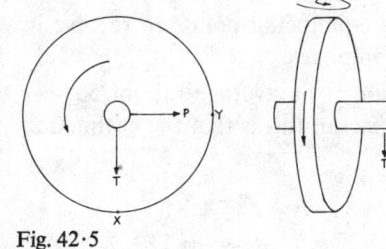

Imagine the wheel illustrated in fig. 42·5 to be spinning in the plane of the paper. The rotary force T is one of the forces of a couple, known as a torque. This rotary force tends to cause X to move into the plane of the paper. Were the wheel not spinning, this is what T would do. But, because the wheel is spinning, the effect of T is for the point Y to move into the plane of the paper.

Fig. 42·5

N.B. Y is 90° from X in the direction of spin of the wheel.

Examples of Gyroscopic Inertia

(i) *Deck Quoit*—The deck quoit must be given an initial spin to that it will remain horizontal during its flight.

(ii) *Rifling of a Gun Barrel*—The screw thread in the gun barrel causes the shell to twist as it passes through the barrel. This ensures that the shell maintains its direction of flight.

348   THE ELEMENTS OF NAVIGATION AND NAUTICAL ASTRONOMY

(iii) *Spinning Top*—The axis of the top remains fixed in the vertical so long as it is spinning rapidly.

(iv) *The Earth*—Because it spins, the Earth tends to maintain its plane of spin and, therefore, the North Pole of the Earth tends to point to a fixed position on the celestial sphere. This position is the Celestial Pole.

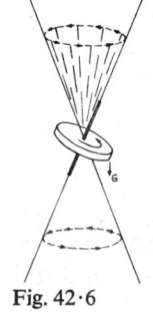

Examples of Precession

(i) *The Spinning Top*—Because of the friction at the pivot of the top, the top tilts when its speed of rotation is insufficient to overcome this friction. As soon as this happens, the force of gravity tries to capsize it, but the effect of the force of gravity does not act at the point where it would act were the wheel not spinning: it acts 90° in advance of this position in the direction of spin of the wheel. The axis of the top, therefore, describes a conical movement the direction of which is the same as that of the spin of the wheel. This is illustrated in fig. 42·6.

Fig. 42·6

(ii) *Precession of the Earth's Spin Axis*—The precession of the Earth's polar axis is due to a combination of the following contributory causes:

(*a*) The Earth's axial rotation.

(*b*) The Earth's orbital motion.

(*c*) The Solar attraction.

(*d*) The Earth's oblate shape.

(*e*) The obliquity of the ecliptic.

The attraction of the Sun on the Earth is such that it tends to force the Earth's axis perpendicular into the plane of its orbit around the Sun. This is due to the greater amount of matter in the Earth's equatorial plane than in its polar plane. Because the Earth rotates this force acts so that the extension of the Earth's polar axis on the celestial sphere sweeps out a conical path on the celestial sphere. This has a radius of 23½° and is centred at the pole of the ecliptic. The direction of this precessional motion is opposite to that of the Earth's axial spin.

The effect of the Earth's axial precession is to cause the equinoctial points to revolve in the ecliptic with a motion known as the Precession of the Equinoxes.

The First Points of Aries and Libra revolve in the ecliptic at an average rate of 50″ per year, taking 26,000 years to make the complete circuit. An effect of this is that the declination and S.H.A. of all fixed stars change gradually.

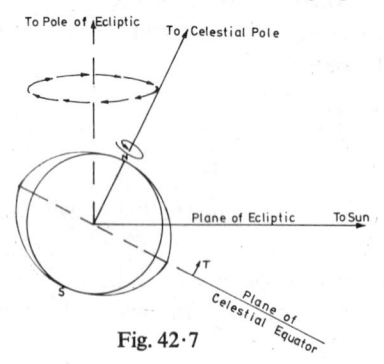

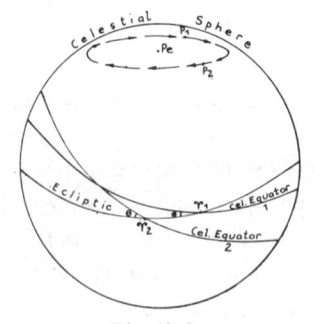

Fig. 42·7

Fig. 42·8

THE GYROSCOPIC COMPASS

Fig. 42·7 illustrates the precession of the Earth's axis. T represents the torque exerted by the Sun endeavouring to force the plane of the Earth's equator into the plane of the ecliptic.

Fig. 42·8 illustrates the precession of the equinoxes. When the celestial pole is at P_1, the equinoctial cuts the ecliptic at ♈. As the celestial pole precesses from P_1 to P_2, the First Point of Aries, or the Spring Equinox, precesses from ♈$_1$ to ♈$_2$.

Another important effect of the precession of the equinoxes is that Polaris, which is less than 1° from the celestial pole at the present time will, in 13,000 years, be about 46° from the celestial pole.

3. Tilt and Drift

The direction which a gyro axle endeavours to maintain is fixed with reference to space. Because of the Earth's rotation fixed directions in space change relative to the horizontal and vertical planes of an observer. All celestial bodies which are fixed in space appear to revolve around the Earth once in the time taken for the Earth to make one rotation on its axis.

Were the axle of a free gyro pointed to a fixed star, the axle would endeavour to maintain rigidly this direction because of the gyroscopic inertia of the gyroscope. It is for this reason that the property of gyroscopic inertia is sometimes referred to as Rigidity in Space.

Relative to the horizon and meridian of a terrestrial observer a star continually changes its altitude and azimuth. A gyro axle tends to do likewise. The component motions of a gyro axle correspond to changes in altitude and azimuth of a fixed star to which the axle is pointing. The angle of inclination of a gyro axle to the horizontal plane corresponds to altitude: this angle is called Tilt. The angle which the vertical plane through the gyro axle makes with the plane of the meridian corresponds to azimuth: this angle is called Drift.

Suppose a gyro wheel is set spinning on the equator with its axle horizontal and lying in a direction at right angles to the meridian. The effect of gyroscopic inertia is to cause that end of the axle which is pointing due East, to change its tilt upwards at the rate of 15° per hour. The gyro axle is simply illustrating the Earth's spin and, in one sidereal day it would perform a complete rotation about the horizontal N./S. axis.

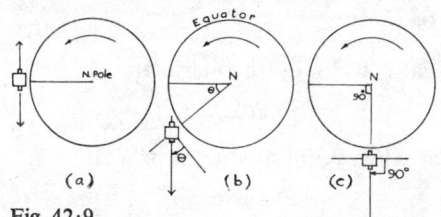

Fig. 42·9

The gyro in diagram (a) of fig. 42·9 has its axle horizontal. After the Earth has spun through an angle θ the gyro axle will have changed its tilt from 0° to θ°. This is illustrated in diagram (b) of fig. 42·9. After the Earth has spun through 90°, as illustrated in diagram (c) of fig. 42·9, the axle will have changed its tilt from 0° to 90°.

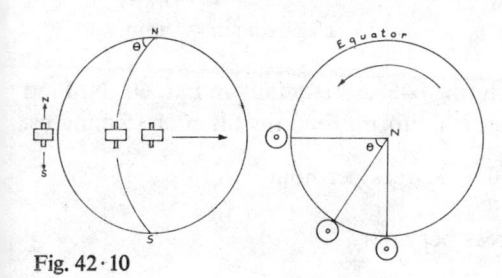

Had the gyro axle been set horizontal and lying in the plane of the meridian it would have remained on the meridian. In this case the gyro axle would be pointing to the celestial pole.

Fig. 42·10 illustrates the gyro axle of a gyro set horizontal on the plane of the meridian at the equator, pointing continually to the celestial pole. In other words the tilt of the axle remains 0° continually.

Fig. 42·10

Imagine a gyro axle set horizontally at the north pole of the Earth as illustrated in fig. 42·11. The gyro axle is at right angles to the Earth's axis of rotation and, because of gyroscopic inertia, it will remain so. The gyro axle will, therefore, remain horizontal, and its tilt will always be 0°. The gyro axle will, however, appear to rotate clockwise looking down on it. In so doing it is manifesting the Earth's real rotation in an anticlockwise direction. The apparent turntable movement of a gyro axle is analogous to change in azimuth of a fixed star: it is therefore, Change of the Drift of the gyro axle.

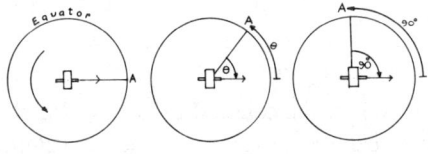

Fig. 42·11

At either pole change of drift takes place at the rate of 15° per hour.

At either pole, provided that the gyro axle is not in the vertical direction, the apparent movement of the gyro axle is one of change of drift only.

At the equator, provided that the gyro axle is not in the plane of the meridian, the apparent movement of the horizontal gyro axle is change of tilt only.

In any position other than at either pole or at the equator, a horizontal gyro axle will change its tilt and drift simultaneously.

4. Tilting and Drifting

The rate at which the tilt of a gyro axle changes is called Tilting. The rate at which a gyro axle changes its drift is called Drifting.

$$\text{Tilting} = \text{rate of change of Tilt}$$
$$\text{Drifting} = \text{rate of change of Drift}$$

Tilting of a horizontal gyro axle at the equator when on the meridian
$$= 0° \text{ per hr.}$$

Tilting of a horizontal gyro axle at the equator when pointing due East/West
$$= 15° \text{ per hr.}$$

Therefore:

Tilting of a horizontal gyro axle at the equator $= 15 \, . \, \sin Az.$ per hr.
Tilting of a horizontal gyro axle at either pole $= 0°$ per hr.

Therefore:

Tilting of a horizontal gyro axle in Lat. θ Az. $\varphi = 15 \cos \theta \sin \varphi$ °/hr.
$$= 15 \cos \theta \sin \varphi \text{ '/min.}$$

Example 42·1—A gyro is spinning with its axle pointing 045° horizontally in Lat. 60° N. Find the change of tilt in 15 minutes assuming the change is uniform. Find the tilt after 15 minutes.

$$\text{Tilting} = 15 \, . \, \sin 45° \cos 60° \text{ degrees per hour}$$
$$= 15 \cdot \frac{1}{\sqrt{2}} \cdot \frac{1}{2} \text{ degrees per hour.}$$

THE GYROSCOPIC COMPASS

Now, Change in Tilt

in any interval t = Rate of change of tilt $\times t$

Therefore:

Change of tilt in 15 minutes $= 15 \cdot \dfrac{1}{\sqrt{2}} \cdot \dfrac{1}{2} \cdot \dfrac{1}{4}$ degrees.

$= 1° \ 19\frac{1}{2}'$

Initial Tilt $= 0° \ 00'$

Final Tilt $= 1° \ 19\frac{1}{2}'$ End pointing eastwards inclined upwards.

Answer—Change in tilt $= 1° \ 19\frac{1}{2}'$

New Tilt $= 1° \ 19\frac{1}{2}'$ East end up.

The Rate of change of Drift of a gyro axle is called Drifting.

Drifting of a horizontal gyro axle in Lat. $0° = 0°$ per hr.

Drifting of a horizontal gyro axle in Lat. $90° = 15°$ per hr.

Therefore:

Drifting of a horizontal gyro axle in Lat. $\theta = 15 \cdot \sin \theta \ °/\text{hr}$.

$= 15 \cdot \sin \theta \ '/\text{min}$.

Example 42·2—A gyro axle is set spinning with its axle horizontal and pointing 045° in Latitude 60° N. Assuming that the rate of change of drift is uniform, find the azimuth of the gyro axle after 15 minutes.

Drifting $= 15 \cdot \sin \text{Lat.} \ °/\text{hr}$.

Ch. of Drift in $\frac{1}{4}$ hr. $= 15 \cdot \sin 60° \cdot \frac{1}{4}$

$= 15 \cdot \dfrac{\sqrt{3}}{2} \cdot \dfrac{1}{4}$ degrees

$= 3° \ 25'$ End pointing eastwards
changes its drift towards the
right in the northern hemisphere

Initial Azimuth $= 045°$

Final Azimuth $= 048° \ 25'$

Answer—Azimuth $= 048° \ 25'$.

5. The Gravity Controlled Gyroscope

The mechanical gyroscope described above is sometimes described as a Free Gyroscope. To make a free gyroscope into a Meridian-Seeking Gyroscope, it is necessary to control the apparent movements of the gyroscope axle due to the motion of the Earth on her axis. The Earth's force of gravity is used to effect this control. The harnessing of the Earth's force of gravity to control the apparent movement of the free gyro due to the Earth's rotation is achieved by fitting a device known as a Gravity Control.

There are several forms of gravity control. In the Anschutz and the Sperry Mark 20 compasses the gravity control consists of a mass fitted to the casing in which the gyro wheel is housed. This mass, sometimes known as a Bail Weight, moves out of the vertical when the gyro axle moves out of the horizontal, the force of gravity acting on it thereby providing the torque which precesses the axle. When the axle lies in the plane of the meridian, the tendency to change tilt is zero so that, provided the axle is horizontal, it remains on the meridian.

Another type of gravity control is the Liquid Gravity Control.

6. The Liquid Gravity Control

Fig. 42·12 illustrates a liquid gravity control fitted to a mechanical gyroscope. The gravity control, which consists of two pots interconnected by a small bore tube lying in the vertical plane through the gyro axle, is secured to the base of the rotor case.

When the axle is horizontal the liquid in the system is such that there is an equal amount in each pot. When the axle is inclined to the horizontal the system is also inclined and more liquid is to be found in the lower than in the higher pot. When this is so, the force of gravity acting on the excess liquid in the lower pot provides a torque which causes the rotor axle to precess. The direction of precession is dependent upon the direction of the spin of the rotor.

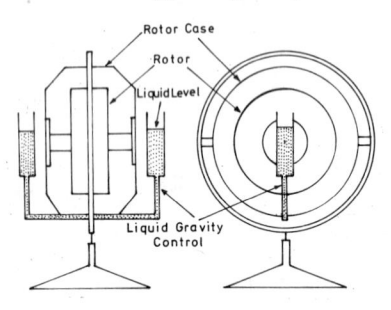

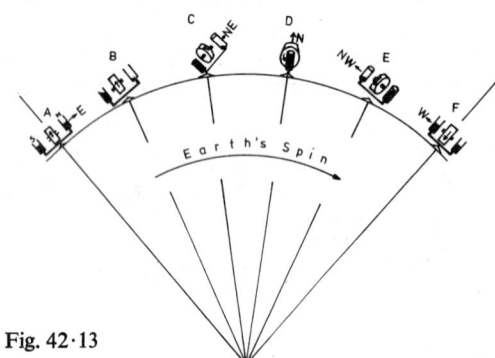

Fig. 42·12

Fig. 42·13 represents six successive positions of a liquid gravity controlled gyroscope, as it is carried around the Earth's axis due to the rotation of the Earth. The gyro wheel is supposed to be spinning anticlockwise when viewed from the South end, marked S in the diagrams, and the positions are supposed to be viewed from South.

At position A the gyro axle is horizontal and at right angles to the meridian. The end marked N (for North) is pointing due east. When the gyro has been carried around to position B, the north end has inclined upwards and the liquid has flowed to the south pot. Torque due to gravity causes the north end of the gyro axle to precess towards the northwards. Note that the wheel is spinning anticlockwise when viewed from the south. At position D the axle is on the meridian, but the north end of the axle is inclined up with maximum tilt. At position E, the north end has precessed to the west of the meridian and the north end has begun to decline causing the liquid to level itself again. At position F the liquid is level and the north end is now pointing due west. The other end now begins to incline upwards, and the north end then precesses towards the meridian, with the axle inclining south end upwards.

Fig. 42·13

7. The Movement of the Axle of a Gravity Controlled Gyroscope

(a) *At the Equator*—Suppose a gyro axle is set horizontal with one end, say the north end, pointing N. θ E. At the equator the gyro axle tends to change its tilt but not its azimuth or drift. Were the movement of the north end of the axle projected onto a vertical screen lying in the east–west plane, the path traced out would be an ellipse. The diameter of the east–west axis of this ellipse is 2 . θ, and that of the vertical axis is dependent upon the magnitude of the gravity control. The vertical axis of the elliptical path traced out by the axle of an undamped gyro compass is never more than a couple of degrees.

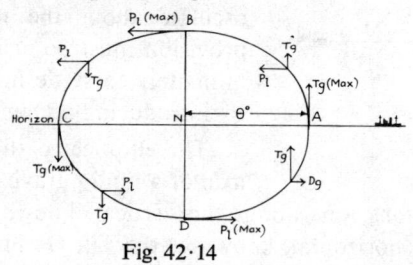

Fig. 42·14

Referring to fig. 42·14, the north end of the gyro axle is pointing to position *A*. Because of gyroscopic inertia, the north end will incline upwards. The rate of change of tilt is maximum at *A* and is represented by *Tg*. The liquid in the gravity control flows to the south pot and, as a consequence, the north end of the axle precesses towards north. Note that the direction of spin of the gyro wheel is clockwise when viewed from the north end.

As the azimuth decreases, tilting (rate of change of tilt) decreases, but the angle of tilt increases, until it reaches a maximum when the axle is on the meridian. At point *B*, the precession due to the liquid, which is represented by P_1 is maximum. The north end of the axle crosses the meridian and begins to decline, because the south end is now east of the meridian and it will, therefore, incline upwards. The liquid now flows back to the north pot, the precession due to the liquid decreases and tilting increases until the north end of the axle is pointing N. θ W., when the axle will again be horizontal. At *C* the south end of the axle, which continues to incline upwards results in the liquid flowing to the north pot. Precession of the north end is now towards the meridian again, and it reaches the meridian with maximum tilt north end downwards. Excess of liquid in the north pot causes the north end to continue precessing towards the east, and it will point N. θ E. when the axle is again restored to the horizontal.

(*b*) *In Any North Latitude*—Suppose the axle is horizontal and the north end is pointing N. θ E. Because of gyroscopic inertia the north end inclines upwards and changes its azimuth towards the eastwards. As the axle increases its tilt liquid flows from the north to the south pot and a torque is introduced which tends to cause the north end to precess towards the meridian. The north end does move towards the meridian after the instant at which the precession due to the liquid exceeds drifting due to the Earth's rotation, which latter is represented by *Dg* in fig. 42·15.

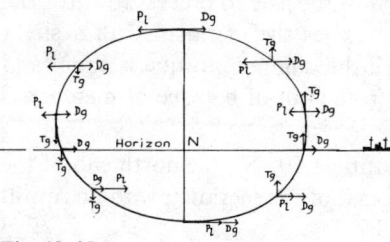

Fig. 42·15

Fig. 42·15 illustrates the path traced out by the north end of a liquid gravity controlled gyro in a northern Latitude. The centre of the elliptical path is vertically above the north point of the horizon.

(*c*) *In Any South Latitude*—In any south Latitude, where the change of azimuth or drift of the north end of a gyro axle is the opposite direction to that in the northern hemisphere, the path traced out by the north end of a liquid gravity controlled gyro is an ellipse, the centre of which is vertically below the north point of the horizon. This is illustrated in fig. 42·16 (overleaf).

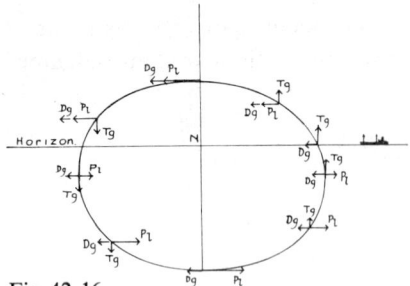

8. Damping

The gravity controlled gyro described above is useless as a compass: the axle would perpetually oscillate about the meridian. In a gyro compass provision must be made to cause the oscillating axle ultimately to settle in the meridian. This is achieved by introducing a Damping Torque.

Fig. 42·16

The elliptical path traced out by each end of the axle of a liquid gravity controlled gyro is made up of two component oscillations, one about the vertical, known as the Azimuthal Oscillation; and the other about the horizontal, known as the Tilt Oscillation.

In the Sperry Mark 14 compass a slight torque about the vertical axis of the gyro wheel is introduced. This torque causes precession which damps the tilt oscillations. In the Brown compass a torque about the horizontal axis causes precession which damps the azimuthal oscillations.

9. The Natural Errors of the Gyro Compass

A gyro compass on board ship may be affected by three Natural Errors. These are called: Latitude Error; Rolling Error; and Latitude/Course/Speed Error.

Latitude Error—This error arises when the damping torque acts about the vertical axis of the gyro wheel. Because of this, in every Latitude except the equator, the axle of such a gyro settles off the meridian with a slight tilt.

In the northern hemisphere the north end of the gyro axle settles east of the meridian with an up tilt. In southern Latitudes the north end of the axle settles west of the meridian with a slight down tilt. The horizontal angle between the vertical plane through the gyro axle and the true meridian due to this is known as Latitude Error.

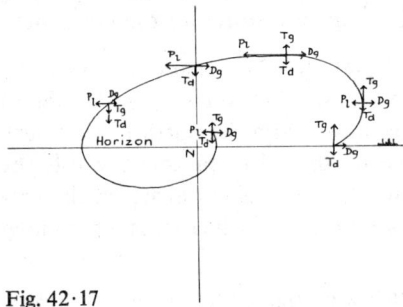

Fig. 42·17 represents the path traced out by the projection of the north end of the gyro axle of a Sperry Mark 14 compass in a northern Latitude. In the diagram, *d* represents the effect of the damping torque.

If the north end of the gyro wheel is set horizontal and east of the meridian, tilting due to gravity causes the mercury in the control system to flow to the south pot. A torque due to the out-of-balance of mercury causes the north end of the axle to precess towards the

Fig. 42·17

meridian. Damping takes effect such that the north end reaches the meridian with a slight down tilt. The axle ultimately settles in a position at which the damping torque is equal and opposite to tilting due to gravity, and the precession due to the out-of-balance of mercury is equal and opposite to the drifting due to gravity.

Latitude error increases as the Latitude increases. In Latitude 50° N., the north end of the axle of a Sperry Mark 14 compass settles about $1\frac{1}{2}°$ to the east of the meridian with an up tilt of about 2' of arc.

Because Latitude error is due to the method of damping it is sometimes known as Damping Error.

THE GYROSCOPIC COMPASS

Rolling Error—When a ship rolls or pitches, the gyro is subjected to athwartship and fore-and-aft accelerations which may cause error in two ways. The first is due to the asymmetry of the gyro, the mass of which in the east–west plane, that is in the plane of the wheel, is greater than that in the north–south plane.

Any mass which is caused to swing tends to align itself such that the plane of greatest moment of inertia lies in the plane of the swing. To compensate rolling error due to this cause, the gyro is fitted with Compensator Weights in north–south alignment so that the moment of inertia of the gyro in the plane of the axle is equal to that in the plane of the wheel.

Rolling error may also be caused in some compasses through the surging of the liquid in the gravity control when the ship rolls or pitches. When the ship rolls or pitches the liquid in the control system is subjected to accelerations resulting, in general, in an out-of-balance of liquid in the system. This provides a torque which causes error.

It may be shown that rolling error is maximum on intercardinal headings and nil on cardinal headings.

Latitude/Course/Speed Error—This error is sometimes referred to as Steaming Error. The tilting of a gyro axle is due to the Earth's rotation only when the gyro is placed on a stationary ship. Tilting, as we have seen, is maximum where the Earth's circumferential speed is maximum, that is at the equator.

The tilting of the gyro axle of a compass on board a moving ship is due to the resultant of the Earth's circumferential speed and the speed of the ship over the Earth's surface. In all cases, the gyro axle settles at right angles to the direction in which it is being carried through space.

On a ship moving due east or due west tilting is increased or decreased respectively, but the settling position will tend to be on the meridian. If however the gyro is carried in any direction other than east or west, the settling position will tend to lie off the meridian.

The axle of the gyro tends to settle in the plane of the Virtual Meridian, this lying at right angles to the resultant motion of the gyro through space. The horizontal angle between the vertical planes of the virtual and true meridians is called Latitude/Course/Speed Error.

Latitude/Course/Speed error is usually very small because the ship's speed over the ground is small relative to the Earth's circumferential speed.

Latitude/Course/Speed error depends not only upon the speed of the ship but also upon her course and the Latitude. It is the northerly or southerly component of the ship's speed which is responsible for this error. The error varies from zero when the course angle is 90°, to a maximum when the course angle is zero. It varies, therefore, as the cosine of the course angle. The Earth's circumferential speed varies with Latitude, it being 900 knots at the equator and zero at the pole. The speed, therefore, varies as the cosine of the Latitude, the Latitude/Course/Speed error being least in Latitude 0° increasing as the Latitude increases.

Fig. 42·18 shows that Latitude/Course/Speed error may be computed for any combination of Latitude, course and speed.

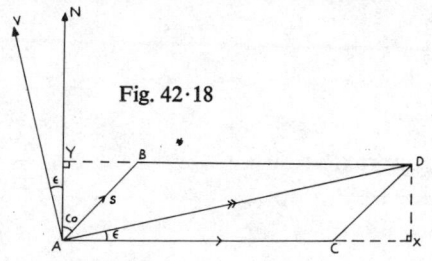

Fig. 42·18

In fig. 42·18:

AN represents the true meridian.

AB represents the ship's velocity, where s is the ship's speed.

AC represents the Earth's circumferential velocity.

By the parallelogram law, AD represents the velocity of the gyro through space. VA represents the virtual meridian which lies at right angles to AD.

$$NAV = DAC = \text{Latitude/Course/Speed Error} = \varepsilon$$

$$\tan \varepsilon = \frac{DX}{AX}$$

$$= \frac{AY}{AC + BY}$$

$$= \frac{s \cdot \cos co}{900 \cdot \cos \text{Lat.} + s \cdot \sin co}$$

Had the course been westerly with the same course angle:

$$\tan \varepsilon = \frac{s \cdot \cos co}{900 \cdot \cos \text{Lat.} - s \cdot \sin co}$$

In general:

$$\tan \varepsilon = \frac{s \cdot \cos co}{900 \cos \text{Lat.} \pm s \cdot \sin co}$$

Because the error ε is a small angle:

$$\varepsilon \text{ in radians} = \frac{s \cdot \cos co}{900 \cdot \cos \text{Lat.} \pm s \cdot \sin co}$$

Now $s \cdot \sin co$ is a small quantity and it may normally be ignored without introducing material error.

Therefore:

$$\varepsilon \text{ in radians} = \frac{s \cdot \cos co}{900 \cdot \cos \text{Lat.}}$$

$$\varepsilon \text{ in degrees} = \frac{s \cdot \cos co}{900 \cdot \cos \text{Lat.}} \cdot \frac{180}{\pi}$$

$$\varepsilon^\circ = \frac{s \cdot \cos co}{5\pi \cdot \cos \text{Lat.}}$$

For northerly courses error is West and gyro reads too high.

For southerly courses error is East and gyro reads too low.

Example 42·3—Find the Latitude/Course/Speed error in 60° N. when steaming at 30 knots due South.

$$\text{Speed Error} = \frac{s \cdot \cos co}{5\pi \cdot \cos \text{Lat.}}$$

$$= \frac{30}{5\pi \cdot \tfrac{1}{2}}$$

$$= \frac{12}{\pi}$$

$$= 4° \text{ W. approx.}$$

Answer—Error = 4° W. or gyro reads too LOW.

10. Ballistic Deflection

The gyro compass is pendulously supported, so that when the ship accelerates, that is to say when she changes course and/or speed, the force causing the acceleration acts at the point of support of the gyro. The inertia of the gyro causes it to move out of the vertical plane, but the gyroscopic inertia of the wheel causes the plane of the spin to remain the same. The liquid in a gravity control that may be fitted to the compass, being mobile, surges from the pot in the direction of the acceleration, to the other pot. The excess of liquid in the pot lying in the direction opposite to that of the acceleration, provides a torque which causes the axle to precess. The amount of precession consequent upon an acceleration is known as Ballistic Deflection.

Fig. 42·19(a) illustrates the gyro compass having a liquid control in a ship travelling due north at uniform velocity. Fig. 42·19(b) illustrates the same compass when the ship increases her speed towards the north.

When a ship accelerates the angle between the vertical planes of the virtual and true meridians changes and, therefore, the settling position of the gyro axle changes an amount equal to the change in Latitude/Course/Speed error consequent upon the acceleration. Referring to example 42·3, if the ship altered course to due south at 30 knots, the Latitude/Course/Speed error would change to the extent of 8°, that is from 4° W. to 4° E. This means that the virtual meridian would change its direction by the same amount, that is 8°. Now it is important that the gyro axle is always in the vertical plane of the virtual meridian. Therefore, when a ship accelerates, the gyro axle must be made to follow the virtual meridian. This is achieved by making the liquid gravity control of such a magnitude that the ballistic deflection is exactly equal to the change in course and/or speed error. The magnitude of the control influences the Period of the compass, this being the time taken for the axle to swing through one oscillation about the meridian before it settles on the meridian. The gravity control has the correct magnitude—the magnitude being influenced by the distance between the pots; the diameters of the pots and the interconnecting tube; and the density of the liquid used—when the period of the undamped gyro is about 85 minutes

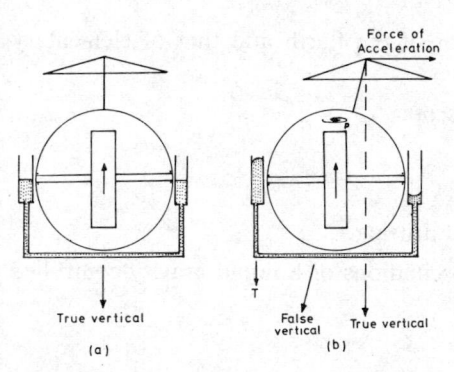

Fig. 42·19

11. Ballistic Tilt

In the Sperry Mark 14 compass the liquid control is secured to the rotor case slightly to the east of the vertical through the centre of the wheel. This provides for damping. When the ship accelerates the surge of the liquid in the gravity control causes, not only a torque about the horizontal axis which causes precession about the vertical plane, but also a small torque about the vertical axis which causes precession about the horizontal plane. This latter precession is known as Ballistic Tilt. The ballistic tilt is very much smaller than the ballistic deflection for any given acceleration, because the offset of the point of attachment of the liquid control to the gyro case is very small.

Ballistic tilt causes the axle to wander slightly after an alteration of course and/or speed, because of the inclination of the axle. This wandering is referred to as Ballistic Tilt Effect.

Exercises on Chapter 42

1. Describe a mechanical gyroscope and the two properties of all spinning bodies for which the mechanical gyroscope may be used to demonstrate.
2. Give examples of gyroscopic inertia.
3. Explain the principle of the conservation of angular momentum, and thence explain why the axle of a spinning body tends to maintain a fixed direction in space.
4. Explain why a gyro axle precesses when a torque is applied at right angles to the axle at one of its ends.
5. State a rule for ascertaining the direction of precession for any given torque and direction of wheel spin.
6. Using your knowledge of the shape and motions of the Earth and that of elementary mechanics, explain the precession of the equinoxes.
7. Distinguish between a Free and Controlled gyroscope.
8. What is the function of a gravity control?
9. Describe a simple form of liquid gravity control.
10. Distinguish between tilt and tilting, and drift and drifting.
11. What is meant by damping? Explain how the oscillations of a liquid gravity controlled gyroscope are damped.
12. State the natural errors of a gyro compass.
13. Discuss Latitude error and its cause.
14. Discuss rolling error and its causes and compensation.
15. What is Steaming error? Derive a formula for finding steaming error for any combination of Latitude, course, and speed.
16. Define Virtual meridian. Explain how a gyro compass maintains the virtual meridian when the ship on which the compass is fitted, changes her course and/or speed.
17. Describe ballistic deflection and ballistic tilt.
18. Compare the usefulness of a magnetic compass with that of a gyroscopic compass.

CHAPTER 43

THE SEXTANT AND CHRONOMETER

1. Description of Sextant

The sextant is an instrument used for measuring the angle between two distant objects. It may be used in Coastal Navigation for measuring the horizontal angle between two conspicuous shore objects, or for measuring the vertical angle of a lighthouse or peak.

In the practice of Nautical Astronomy the sextant is used for measuring altitudes of celestial bodies. The angle, in this case, is the angle at the observer's eye between the direction of the celestial body and that of the visible horizon measured in the plane of the vertical circle on which the body lies.

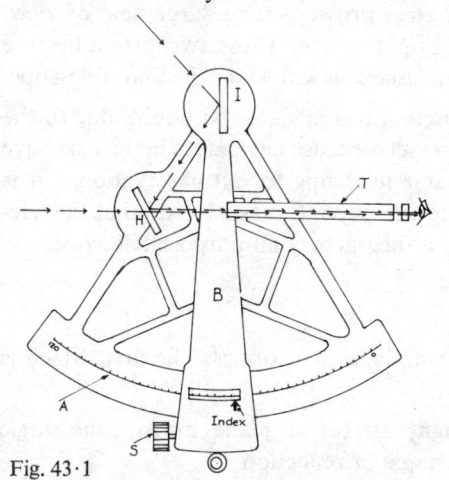

Fig. 43·1

The sextant consists of a Metal Frame on which is fitted a Graduated Arc denoted by A in fig. 43·1. At the centre of the circle of which the arc forms part is pivoted a bar B on which is a Pointer or Index to facilitate the reading on the arc. Fitted perpendicularly to the Index Bar is a metal frame in which is housed the Index Mirror indicated by I in fig. 43·1. Another metal frame houses the Horizon Glass indicated by H. The horizon glass is half silvered: the half which is nearer to the Plane of the Sextant, which is defined as the plane on which the arc lies, is silvered, the other half being clear glass. Both the index mirror and the horizon glass may be adjusted so that they may be set perpendicularly to the plane of the sextant.

An Adjustable Collar is sometimes fitted to the sextant frame. The purpose of this is to house a telescope indicated by T in fig. 43·1. The collar is adjusted so that the axis of the telescope lies parallel to the plane of the sextant. The axis of the telescope passes through the horizon glass, and the telescope may on some instruments be adjusted so that the amount of the silvered portion of the horizon glass visible through the telescope may be altered to suit the observer and the observation. The device which facilitates this adjustment is known as the Rising Piece. Most modern sextants are not provided with this device.

The index bar may be clamped to any position along the arc of the sextant. For fine adjustment of the position of the index bar a Tangent Screw indicated by S in fig. 43·1 is used; or, alternatively, the sextant may be fitted with a Micrometer.

The sextant is provided with several tinted glass Shades of different intensities, which are used to reduce the brilliance of the Sun, and sometimes the horizon, when observing the Sun's altitude. The shades at the index mirror are known as the Index Shades, and those at the horizon glass, the Horizon Shades.

2. The Sextant Telescopes

A good sextant outfit is usually provided with two telescopes. The telescope of higher magnification is an Astronomical or Inverting telescope, so called because the image appears inverted. The inverting telescope should be used for all observations of the Sun when the horizon is clear and the vessel reasonably steady. The high magnifying power of this telescope ensures an accurate grazing contact of the reflected image of the Sun's limb with the direct image of the horizon. The inverting telescope should also be used for measuring vertical angles of peaks and lighthouses, because, in this type of observation, great accuracy of the measured angle is essential to ensure a good position line.

The magnifying power of a telescope has an effect on the brightness of the image. The lower the magnifying power of a telescope, for a given object glass, the brighter is the image and the larger will be the field of view. Daylight observations, using the inverting telescope are not affected by the comparatively low degree of brilliance of the objects observed, nor by the small field of view. The inverting telescope, however, is useless for twilight observations of stars, hence the necessity for providing a second telescope.

The Erecting Telescope provided in the sextant outfit is bell-shaped. It has a large object glass and a low magnifying power. The large object glass provides for a large field of view, and the low magnifying power provides for a very bright image. These two properties are necessary for star observations, for which reason this telescope is known as the Star Telescope.

Many navigators use the star telescope for all observations. This is, no doubt, due to the slight difficulty in using the inverting telescope. And, in fact, because navigators in the past have tended to ignore the inverting telescope—using the star telescope for all observations—it is rare for a modern sextant outfit to include an inverting telescope. It is fair to add that the telescope provided in modern sextant outfits is generally a satisfactory all-purpose telescope.

3. The Principle of the Sextant

The principle of the sextant is based on the two simple laws of optics. The first, which is illustrated in fig. 43·2, is:

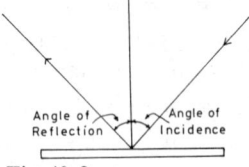

"When a ray of light strikes a plane mirror, the angle incidence is equal to the angle of reflection."

Fig. 43·2

The second law of optics is:

"The incident ray, the normal, and the reflected ray all lie in the same plane."

It follows that if a ray of light is doubly reflected by two plane mirrors, the angle between the first incident ray and the second reflected ray is twice the angle between the mirrors. This is illustrated in fig. 43·3.

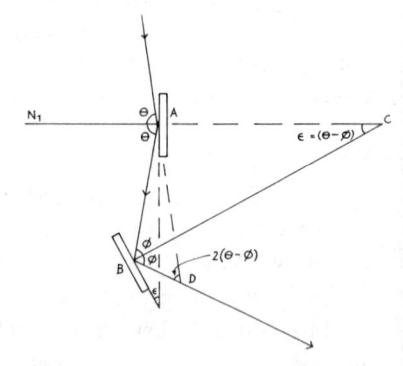

Fig. 43·3

THE SEXTANT AND CHRONOMETER

Let A and B in fig. 43·3 be two mirrors. Let the first angle of incidence be θ, and the second angle of reflection be φ.

Angle between the mirrors = Angle between their normals
$$= \varepsilon$$

In triangle ABC: $\qquad ACB = (\theta - \varphi)$
In triangle ABD: $\qquad ADB = 2\theta - 2\varphi$
$$= 2(\theta - \varphi)$$
Therefore: $\qquad ADB = 2 \cdot ACB$

Hence the angle between the first incident ray and second reflected ray is twice the angle between the mirrors.

Fig. 43·4 serves to illustrate that when a plane mirror is rotated through any given angle the angle which the reflected ray turns is equal to twice the angle through which the mirror has been rotated.

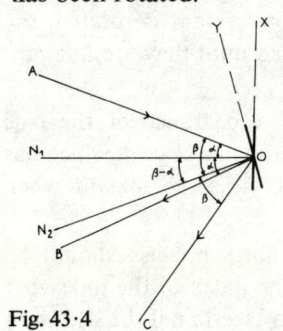

Suppose the plane mirror illustrated in fig. 43·4 is rotated through an angle XOY.

Let the first angle of incidence be α

Let the second angle of incidence be β

Angle through which mirror rotates $= N_1ON_2$
$$= (\beta - \alpha)$$

Angle through which reflected ray rotates $= BOX$
$$= 2\beta - 2\alpha$$
$$= 2(\beta - \alpha)$$

Fig. 43·4

Therefore, the angle through which the reflected ray rotates is twice the angle through which the mirror rotates.

For this reason the arc of a sextant is graduated to indicate twice the angle through which the index bar—and hence the index mirror—rotates from the zero mark on the scale. Although the sextant arc extends to only about 60°, the instrument may be used to measure angles up to about 120°.

Angles measured with the sextant may be determined with an accuracy of 10″ of arc. This is achieved by fitting either a Vernier or Micrometer to the index bar. Most modern sextants are fitted with the popular micrometer.

4. The Errors and Adjustments of the Sextant

The errors of a sextant are classified under two headings:

(1) Adjustable Errors.
(2) Non-adjustable Errors.

There are four adjustable errors: These are error of perpendicularity; side error; index error; and collimation error.

Error of Perpendicularity is due to the index mirror not being perpendicular to the plane of the sextant. To ascertain if a sextant has error of perpendicularity, hold the sextant face upwards with the arc away. Move the index bar along the arc until the true arc and the reflected arc from the index mirror may be seen simultaneously. If the true and reflected images of the arc are in alignment the sextant is free from this error. If they are not, the adjusting screw at the back of the frame, which holds the mirror, must be turned until they are in line so as to adjust the sextant for this error.

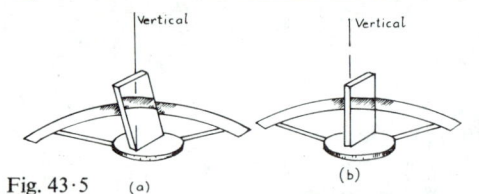

Fig. 43·5(a) indicates the appearance of the true and reflected images of the arc in a sextant which has error of perpendicularity.

Fig. 43·5(b) depicts the same sextant when the error has been removed.

Fig. 43·5 (a)   (b)

If the horizon glass is not perpendicular to the plane of the sextant, Side Error exists. To detect the presence of this error, the index bar should be adjusted until the true image of the horizon, as seen in the unsilvered part of the horizon glass, is in alignment with the reflected image as seen in the silvered part of the horizon glass. If the true and reflected images remain in alignment when the sextant is rotated about the line of sight, the horizon glass is perpendicular to the plane of the instrument and the sextant is free from side error. If the images are not in alignment when the sextant is rotated, the adjusting screw at the back of the horizon glass frame must be turned until they are in alignment, so as to adjust the Side Error.

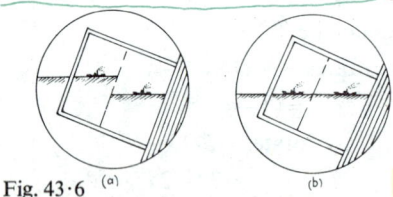

Fig. 43·6(a) illustrates the appearance of the true and reflected images of the horizon in a sextant which has side error. Fig. 43·6(b) shows the same sextant when side error has been removed.

Fig. 43·6 (a)   (b)

The index mirror and the horizon glass should be parallel to one another when the index of the index bar is at the zero mark of the scale: if they are not, Index Error exists. To ascertain if the sextant is free from index error, the sextant is held vertically, and the index bar adjusted until the true and reflected images of the horizon appear in alignment. When this is so, the index on the index bar coincides with the zero mark on the scale. If the index is not at the zero mark, the reading is referred to as the Index Error of the sextant.

Index error may be removed by means of the adjusting screw at the back of the horizon glass frame. Note that there are two screws at the back of the horizon glass frame: one at the top of the frame and the other at the bottom. The top one is for adjusting the horizon glass so as to remove side error. The lower screw is used to eliminate index error.

An alternative method for ascertaining index error is by means of a star. If the true and reflected images of a star are in exact coincidence when the index bar is at zero, no index error exists. When using this method the reflected image of the star is observed at the very edge of the mirror of the horizon glass; that is to say on the line which divides the silvered from the unsilvered part of the horizon glass.

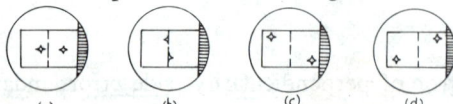

Fig. 43·7 illustrates the process of removing index error and side error using a star.

Fig. 43·7 (a)   (b)   (c)   (d)

When adjusting a sextant it will be found that an adjustment for side error affects the adjustment for index error, and *vice versa*. Therefore, it is the correct practice to make these two adjustments simultaneously, by first removing the initial side error and the initial index error. When this has been done half the remaining side error is removed and half the remaining index error is removed—and so on until both errors are eliminated.

Using the star method for adjusting side and index error, if side error but no index error exists, the true and reflected images of the star appear as in diagram (*a*) of fig. 43·7.

If index error but no side error exists, the images appear as in diagram (*b*).

If both errors exist, the images appear as in diagrams (*c*) or (*d*).

Collimation Error exists when the axis of the telescope is not parallel to the plane of the sextant. Collimation error may be detected by the method outlined as follows, if the sextant outfit includes an inverting telescope.

Ship the inverting telescope and set the eyepiece so that one pair of cross wires is parallel to the plane of the sextant. Select two stars which are at least 90° apart, and adjust the index bar until the true image of one and the reflected image of the other are in perfect contact on one of the cross wires. Tilt the sextant until the images are on the other cross wire. If they close or separate, the telescope is not parallel to the plane of the sextant. It should be made so by adjusting the two screws of the collar of the telescope.

In many modern sextants the telescope collar is permanently set and no adjustment is possible. It is important when using the sextant to ensure that the observed objects occupy the centre of the field of view of the telescope. If they do not collimation error will result even if the telescope axis is correctly set.

Most modern sextants are manufactured by highly skilled craftsmen, who have at their disposal tools of the highest precision. Modern sextants, therefore, leave the hands of the makers in a state as near to perfection as is possible. However, it is not unlikely that many old sextants do possess certain errors which cannot be adjusted. Most important of these non-adjustable errors is Centring Error. This error is due to the axis of the index bar not coinciding with the centre of the circle of which the arc forms part. It is very likely that an old well-used sextant possesses centring error on account of the wear of the bush on which the index bar pivots.

Other non-adjustable errors are:

(1) Shade Error, due to the surfaces of the coloured shades not being parallel to each other.
(2) Prismatic Error, due to the surfaces of the index mirror or horizon glass not being parallel to each other.
(3) Graduation Error, due to inaccuracy in the dividing of the vernier or micrometer and/or the arc.

5. Using the Sextant

(i) *Observing Horizontal Angles*—The star telescope is used for horizontal angle observations. The sextant is held horizontally face up. The index bar is set to the zero mark on the scale. The right-hand object is observed. As the index bar is rotated to higher readings the sextant is turned about a vertical axis, so that the reflected image of the right-hand object is kept in the silvered portion of the horizon glass. When the left-hand object is seen through the unsilvered part of the horizon glass the index bar is clamped and the tangent screw or micrometer is used to make an accurate contact between the reflected image of the right-hand object and the direct image of the left-hand object.

If the right-hand object is indistinct, or if it is a flashing light, and the left-hand object is a fixed light, it is better to hold the sextant face downwards, and bring the reflected image of the left-hand object into coincidence with the direct image of the right-hand object.

(ii) *Observing Vertical Angles*—The inverting telescope should be used for observing vertical angles. It is a good plan to observe the vertical angle both On and Off the Arc, using the Arc of Excess to measure the angle off the arc. The arc of excess is the part of the graduated arc which lies to the right of the zero mark on the scale. The average of the two readings should then be taken as the vertical angle.

(iii) *Observing a Star's Altitude*—The star telescope is shipped and focused and the sextant held vertically with the index set at zero. The star is observed, and the true and reflected images distinguished. The index bar is now rotated and the sextant tilted about a horizontal axis simultaneously, so that the reflected image of the star is kept in view in the silvered part of the horizon glass, until the horizon is seen through the unsilvered part. The index bar is then clamped. To ensure that the arc of a vertical circle is being measured, the sextant is rocked gently to and fro, so that the reflected image of the star appears to describe an arc of a circle. The tangent screw is adjusted so that this arc just grazes the horizon, as the sextant is rocked.

If the horizon is indistinct it is better to hold the sextant vertically with the arc uppermost, and to bring the reflected image of the horizon up to the true image of the star. When this has been accomplished, the sextant is held right way up and the tangent screw used for the accurate measurement.

(iv) *Observing the Sun's Altitude*—To observe the Sun, the inverting telescope is shipped and focused. Appropriate glass shades are turned down into position, and the sextant held with its plane vertically. The observer faces the direction of the Sun, and the index bar is gently rotated to and fro in the vicinity of the arc which reads the approximate altitude of the Sun until the reflected image of the Sun is observed through the silvered portion of the horizon glass, at the same time as the direct image of the horizon is visible through the unsilvered portion. The tangent screw is then used to make the fine adjustment.

6. Care of the Sextant

At all times the sextant should be handled with the greatest of care. The slightest knock may derange the adjustments or permanently damage the instrument.

The silvering of the mirrors may be impaired if moisture is allowed to remain on the mirrors so that after using the sextant, the surfaces of the index mirror and the horizon glass should be wiped carefully and lightly with a piece of soft chamois, which is kept specially for the purpose. The working parts of the sextant should occasionally be lightly smeared with high grade lubricating oil to reduce the rate of wear of working parts.

It may be thought that a modern sextant remains in perfect adjustment for an indefinite period. Experience shows that this is not so. Navigators are advised, therefore, to check the adjustments of their sextants frequently. Rather than meddle with the adjusting screws to the extent that the threads wear causing the screws to become loose, it is better to measure the extent of the error and to apply it to the sextant reading.

7. The Chronometer and its Care

A chronometer is an accurate timekeeper suitable for keeping time, for nautical astronomical purposes, on a lively vessel at sea and which is capable of maintaining a uniform and small rate of gaining or losing.

We have seen in Chapter 29 that time and Longitude are intimately related, and that the essence of Longitude-finding at sea by nautical astronomical methods is the comparing of L.M.T. with G.M.T. The L.M.T. is found from astronomical observation and the chronometer provides G.M.T. if its error is known.

In recent years precision chronometers, capable of maintaining an accuracy of 0·1 seconds per week, have become available: these employing quartz crystal oscillators.

A quartz chronometer provides a digital read-out and operates from the mains supply. Should this fail a built-in battery circuit automatically takes over. The instrument requires minimal maintenance and attention. The traditional chronometer, on the other hand, is a mechanical time-piece which depends on the energy of a wound mainspring which, through a train of gears, is transmitted to the hands which register the time on a dial of the usual clock-face style. The important feature of this type of chronometer is the provision of compensation for changes in temperature—without which the going of the time-piece would be so erratic as to render the instrument entirely unfit for nautical astronomical purposes.

A mechanical chronometer is usually designed to run for about 56 hours without re-winding. Such an instrument is known as a Two-Day Chronometer. To ensure a systematic routine for winding, so as to ensure that the daily rate is kept as steady as possible, the chronometer should be wound at the same time each day, and preferably by the same officer.

The chronometer is mounted on gymbals and housed in a locker in the chartroom which should be dust- and draught-free, and insulated to offset the effects on the chronometer of temperature changes. To wind the chronometer the instrument is turned on its side, within its gymbals, and the metal guard covering the keyhole slid back. The key is then inserted and the chronometer wound until the key butts—this normally requiring about seven half-turns. The winding key, known as a Tipsy Key, is designed so that it cannot be turned in the wrong direction. The tipsy key is also used for resetting the hands of the chronometer should this become necessary; and, in no circumstances, should the hands be altered except by means of the tipsy key. A small dial on the face of the chronometer indicates the state of winding.

A chronometer should be cleaned and re-oiled at intervals of about two years. This should be done by the manufacturer or by an approved instrument-maker. When it is necessary to transport a chronometer the balance wheel should be wedged by means of two thin cork wedges cut for the purpose. To do this it is necessary to remove the instrument from its box and to lift out the working mechanism from its brass case. This requires the greatest of care: the glass front is first unscrewed and the brass case is inverted, the key being used if necessary to ease the mechanism from the case.

8. Use of Chronometer

In general an altitude observation must be timed by the chronometer. Ideally an assistant should be employed to record the time of a sight; although, with practice at counting seconds, the observer himself is able to time his own sights with accuracy.

When recording a chronometer time the three hands should be read in descending order of the rapidity of their motions. This means that the seconds should be read first; then the minutes; and finally the hours. It is customary to record the chart-room clock time as well, so that a future check may be made if necessary.

The error of the chronometer which, together with its daily rate, is recorded in the Chronometer Journal kept by the navigating officer, is to be applied to the chronometer reading to obtain the corresponding G.M.T.

The error of the chronometer should be checked frequently—at least once daily when the vessel is at sea—by means of Radio Time Signals, full particulars of which are to be found in the *Admiralty List of Radio Signals*, Volume 5.

Exercises on Chapter 43

1. Describe the angles which a sextant is designed to measure.
2. Describe the construction of a sextant.
3. Discuss sextant telescopes and their uses.
4. Define: Rising Piece; Plane of the Sextant.
5. State the laws of optics on which the principle of the sextant is based.
6. Prove that when a plane mirror is rotated the angle through which a reflected ray turns is twice the angle through which the mirror is rotated.
7. Explain why the sextant arc, which is an arc of 60°, is graduated from 0° to 120°.
8. Enumerate the errors of the sextant.
9. Explain carefully how you would detect and eliminate Side Error.
10. Explain how you would detect and eliminate Error of Perpendicularity.
11. Explain why Side Error and Index Error are eliminated simultaneously. Explain how you would eliminate these errors using a star.
12. What is Collimation Error: How would you detect collimation error in a sextant, and how would you remove it?
13. Describe the non-adjustable errors of a sextant.
14. Explain how a sextant is used for measuring the altitude of a star.
15. Explain how a Sun-sight is made with a sextant. Why should an inverting telescope be used for Sun observations: assuming that one is available?
16. Write an essay on the Care of Sextants.
17. Describe, in detail, the process of timing a sight.
18. Discuss the chronometer and the care it should receive by the navigating officer.
19. Explain how a chronometer should be prepared for transport from a vessel to the shore.
20. Examine the *Admiralty List of Radio Signals*, Volume 5, and describe the "English" and the "ONOGO" systems of radio time signal transmissions.

CHAPTER 44

SOUNDING INSTRUMENTS AND LOGS

1. The Lead Line and Mechanical Sounding Machine

When navigating in coastal waters in thick weather, a knowledge of the depth of water under the keel is often of great value in affording the means of checking a vessel's position. The earliest method of ascertaining the depth of water was by means of the Lead Line. The Hand Lead Line was used for measuring shallow water depths of up to about 20 fathoms, and the Deep Sea Lead Line was used for measuring depths of up to about 100 fathoms. The cumbersome method of sounding by means of a lead line has been superseded by the Kelvin Sounding Machine and the Kelvin Sounding Tube; and, more recently, by the Echo-sounder.

2. The Sounding Machine and Sounding Tube

The sounding machine consists of a metal frame in which is housed a drum which is controlled by a suitable brake. Around the drum is wound about 300 fathoms of fine gauge galvanized steel wire. A brass case, in which is fitted the sounding tube, is secured to the outboard end of the sounding wire. To the end of the brass case attachment is secured a 28 lb. sinker. The drum is fitted with crank handles or an electric motor so that the wire may be wound in after a cast has been taken. The Kelvin sounding tube is a glass tube of small bore sealed at one end. The inside of the tube is coated with a chemical composition which changes colour when acted upon by salt water. The sounding tube is placed in the brass container with its open end downwards. The sounding wire is led through a suitable fairlead, and the sinker and container hung over the ship's side. When ready to cast the brake of the sounding machine is released, whereupon the sinker carries the sounding tube to the sea bed. With increasing depth the water pressure causes the air within the sounding tube to be compressed into an increasingly smaller volume. The sea water which enters the tube discolours the chemical coating. The depth of water corresponding to the length of discoloration is obtained from a graduated boxwood scale.

The principle of the sounding tube is based upon Boyle's Law, which is:

> "The volume of a given mass of gas at constant temperature varies directly as the pressure."

In other words, the product of the volume and the pressure of a given mass of gas at a uniform temperature is a constant amount.

A useful rule, the proof of which illustrates Boyle's Law as it applies to the sounding tube, is stated and derived as follows:

> The air pressure at sea level is equal to the pressure of a head of sea water about 33 feet or 5·5 fathoms.

Suppose the length of discoloration of a sounding tube is W (wet), and the length of the remaining part D (dry). Let the sounding be S fathoms.

Then: by Boyle's Law:
$$\text{Pressure} \times \text{Volume} = \text{constant}$$
so that:
$$A \cdot D(5 \cdot 5 + S) = A(W + D) \cdot 5 \cdot 5$$
where A is the cross sectional area of the sounding tube.

Therefore:
$$5 \cdot 5 \cdot D + D \cdot S = 5 \cdot 5 \cdot W + 5 \cdot 5 \cdot D$$

From which:
$$S = \frac{W}{D} \cdot 5 \cdot 5$$

Example 44·1—A sounding tube 20 inches long is discoloured for 16 inches. Find the approximate depth to which it had descended.

$$\text{Depth} = \frac{16}{4} \times 5 \cdot 5 \text{ fathoms}$$

$$= 22 \text{ fathoms}$$

Answer—Depth = 22 fathoms.

Example 44·2—A sounding tube is found to be discoloured for exactly half its length. Find the depth to which it had descended.

$$\text{Depth} = \frac{W}{D} \times 5 \cdot 5$$

$$= \frac{W}{W} \times 5 \cdot 5 \quad \text{because } W = D$$

$$= 5 \cdot 5 \text{ fathoms}$$

Answer—Depth = 5·5 fathoms.

Other types of sounding tube appeared on the market after the advent of Lord Kelvin's chemical tube. Notable amongst these is the Wigzell Tube, which is a plastic tube sealed at one end, and having a cap fitted with a non-return valve at the other end. When the tube is lowered into the sea, water is driven into the tube owing to the increase of water pressure. As the tube is being hove up, the valve prevents the entrapped water from escaping from the tube. The length of the entrapped water column is read against a suitable boxwood scale on which depths are marked.

3. The Echo-sounder

The principle of echo sounding is simple, although the sounding instrument is a very complex and delicate piece of equipment. If the time interval between the instant of transmission of a sound pulse and the receipt of the echo pulse from the sea-bed is measured, the depth of water may be found by a simple calculation, the speed of sound in sea water being known. Although the speed of sound in sea water varies with the temperature and salinity of the water, the average speed of 800 fathoms per second is sufficiently accurate for sounding purposes in almost all cases.

The echo-sounder consists of three basic parts, these being a Transmitter, a Receiver, and a Recorder. An electrical impulse causes a pulse of sound energy of high frequency, to be sent out from the transmitter which is fitted to the bottom of the hull of the vessel. The sound energy is reflected from the sea-bed and is returned to the vessel where it is received, as a low energy pulse, by the receiver. The receiver, on receipt of the echo pulse, sends a faint electrical impulse to a valve amplifier, which converts the weak signal into a comparatively strong electrical current.

Fitted in the recorder is a roll of sensitised paper saturated with an iodine solution. The paper is drawn by means of an electric motor, across the Platen, which is merely a metal plate. A stylus is caused to sweep across the paper starting its traverse each time a pulse of sound energy is transmitted. The incoming impulse causes an electric current to pass through the paper from the stylus to the platen. The current passing through the recording paper causes the paper to be marked, through electrolytic action, at the point occupied by the stylus at the instant the incoming pulse is received. The paper moves across a graduated scale against which the depth of water may be read.

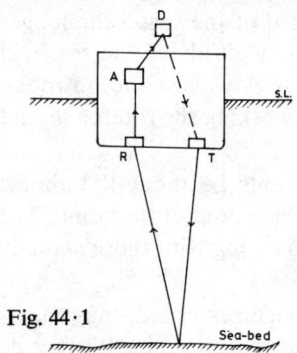

Fig. 44·1 illustrates diagrammatically the essential parts of an echo-sounder. T represents the transmitter, R the receiver, A the valve amplifier, and D the recorder.

Fig. 44·1

4. Logs

The earliest device for measuring the speed of a vessel through the water consisted of a piece of wood in the form of a quadrant, weighted on its curved edge so that it floated more or less vertically. This was secured, by means of a three-legged bridle, to the Log-Line. Made fast to the log-line at equally-spaced intervals were pieces of knotted cord. The knotted cord nearest to the Log-Ship—the name given to the quadrant of wood—had one knot; the next, two knots; the next three, and so on. The log-line was wound around a wooden reel.

When measuring the speed of the vessel the log-reel, log-line, and log-ship, were assembled aft. The log-ship was hove into the wake and the log-line allowed to run freely from the reel as the vessel sailed away from the log-ship. After a certain amount of line had been rendered—sufficient for the log-ship to have cleared the confused water of the wake—the hand in charge of the operation called out "turn", whereupon a sand glass would be turned. When the sand had run out—usually in 14 or 28 seconds—the order "hold" was given, and the log-line held. The vessel's speed through the water was then found from the knotted cord nearest to the hand of the seaman who held the line.

The distance between successive knotted cords and the running time of the glass were proportional, respectively, to the length of a nautical mile and an hour. Thus, if say four knotted cords passed over the stern during the running time of the glass, the speed of the vessel would have been four knots.

It is from the old-fashioned knotted log-line that we get the name Knot, which denotes the navigational unit of speed. From the wooden log-ship we get the name Log which, nowadays, is used to describe any device that measures speed or distance through the water.

The mechanical log commonly in use at the present time registers not speed but distance travelled through the water since the log-clock was set to zero. The log-clock dial is usually graduated from 0 to 100 with secondary dials graduated from 0·0 to 10·0 and from 100 to 1000 miles.

Fastened to the log-clock, which fits into a shoe on the taffrail or at the end of a log-boom, is the log-line at the end of which is a brass rotator in the form of a screw.

The inboard end of the log-line is fitted with a wheel which, acting as a governor, ensures the smooth running of the clockwork mechanism of the log-clock.

To stream the log, the inboard end is first clipped onto the log-clock, and the rotator is clipped onto the outboard end. The rotator is lowered into the water, care being taken to prevent it knocking the vessel's hull and damaging the fins of the rotator, and allowed to stream astern until the line is taut.

To house or ship the log, the inboard end is disconnected from the governor and, as the rotator is hove in, the inboard end is allowed to run out over the stern. This is necessary in order that the turns in the line due to the rotator turning as the log is hove in, will come out; and the coiling of the line, starting at the rotator end, is thus facilitated.

The accuracy of a mechanical log is affected by the length of the log-line and the distance of the taffrail from the sea surface. In general, the greater the speed of the vessel the longer should be the log-line. A line too short, especially if the taffrail is high above the sea, will cause the rotator to lie so near to the surface that it will be affected by waves; and, in rough weather, it will sometimes be dragged out of the water. For correct working the rotator should be well below the sea surface.

As a rough guide, for a speed of about 10 knots the log-line should be about 40 fathoms long; for 15 knots about 60 fathoms; and for 20 knots it should be about 80 fathoms. The correct length is best found by trial and error, comparing distances by log with those actually made through the water as found from observations.

The Taffrail Log, as the mechanical log described above is sometimes called, may incorporate a small dynamo, and the registering mechanism, in this case, conveniently may be fitted in the chart-room.

The Pitometer- and Chernikeeff-logs are sophisticated devices for measuring the ship's speed through the water.

The pitometer log depends for its action on a specially designed Pitot Tube which projects below the hull of the vessel. The orifice of the tube points ahead, and the pressure of water which enters the tube, and which thrusts on a float which is geared to the registering mechanism, is a function of the speed of the ship through the water.

The Chernikeeff log consists of a small propeller which projects below the hull of the vessel, the speed of rotation of which is a function of the vessel's speed.

Exercises on Chapter 44

1. Describe a sounding machine, and explain clearly how a cast is made using a sounding machine.
2. Describe a sounding tube and the principle on which it is based.
3. Derive a simple formula, in terms of the lengths of the discoloured and un-discoloured lengths of a Kelvin sounding tube, for finding the depth to which the tube descended in fathoms.
4. Show that if the discoloured length of a Kelvin sounding tube is a quarter of the length of the tube, the tube had descended to about 11 feet.
5. Describe an echo-sounder of the type fitted on your ship.
6. Explain carefully how a taffrail log should be streamed and housed.

CHAPTER 45

RADIO DIRECTION FINDING

1. The Simple Direction Finder

The simplest form of radio direction finder comprises a Loop Aerial, an Amplifying Unit, and a Headphone. The loop aerial consists of one or more turns of wire mounted in such a way that it may be rotated about a vertical axis. Radio signals from a transmitting station may be received by the loop aerial. The ends of the aerial are connected to the headphone by way of the amplifying unit.

The strength of the signal received is dependent upon the angle between the directions of the transmitting station and the plane of the loop aerial. When the plane of the loop aerial lies in the direction of the transmitting station the e.m.f. induced in the aerial, by the electromagnetic energy which emanates from the transmitter, is maximum, and the signal strength is greatest. When the plane of the aerial lies at right angles to the direction of the transmitter, no e.m.f. is induced in the loop aerial and, consequently, the signal strength is zero.

The signal strength varies as the cosine of the angle which the plane of the loop aerial makes with the direction of the transmitter.

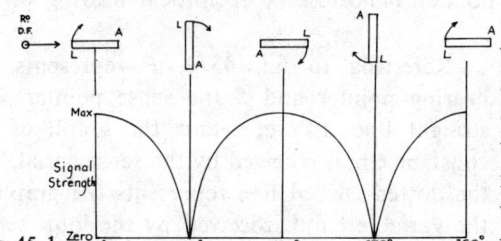

Fig. 45·1 serves to illustrate graphically the relationship between the signal strength and the angle contained between the plane of the loop aerial and the direction of the transmitter. Imagine the loop aerial, represented in plan by *LA*, to be rotated clockwise as indicated in fig. 45·1. It will be noticed that the maximum signal strength remains more or less the same for an appreciable angle in the vicinity of 0° and 180° on the scale; and that the zero strength is approached very sharply when the angle between the plane of the loop aerial and the direction of the transmitter is 90° and 270°. The angle which the plane of the loop aerial makes with the vessel's fore-and-aft line is indicated by a Bearing Pointer which can be rotated within a graduated dial.

Because it is easier to discriminate the zero signal than the maximum signal, the bearing pointer is fixed at right angles to the directions indicated by the plane of the loop aerial, such that when the plane of the aerial is at right angles to the direction of the transmitter the signal strength is a minimum or zero. In other words, when the signal strength is zero the bearing pointer indicates the direction of the transmitter or its reciprocal direction.

In order to resolve the ambiguity—the 180°-Ambiguity, as it is called—a Sense-Finding Unit is fitted to the direction finder.

The sense-finding unit is provided with a vertical aerial which is separately connected to the amplifying unit, as is the loop aerial. The strength of the signal received by the sense aerial is adjusted so that it is the same as that received by the loop aerial when the latter lies

372 THE ELEMENTS OF NAVIGATION AND NAUTICAL ASTRONOMY

in the direction of the transmitter. In other words, the strength of the signal received by the sense aerial is equal to the maximum received by the loop aerial.

Fitted to the graduated dial, in addition to the bearing pointer, is a Sense Pointer which is set at right angles to the bearing pointer.

When the sense aerial switch is made, the sense pointer indicates the direction of the transmitter (or its reciprocal direction) when the plane of the loop aerial is in alignment with the direction of the transmitter. If the sense pointer is then turned through 180° the strength of the signal, although a maximum, will have a different value from the maximum signal received when the sense pointer was set to its original direction.

The sense pointer is fitted relative to the bearing pointer such that when the former indicates the weaker of the two maximum signals, it also indicates the direction of the transmitter.

The principle of sensing is as follows: Depending upon whether the transmitter lies in a certain direction or the opposite direction, the e.m.f. induced in the loop aerial is altered in phase by 180°. The phase of the e.m.f. induced in the vertical aerial, on the other hand, is not affected by the direction of the transmitter. When the sense aerial switch is made the signals from both loop and vertical aerials are received. The e.m.f. in the loop aerial is a maximum when the plane of the aerial lies in the direction of the transmitter. When the loop aerial is rotated 180° from this direction, the strength is again maximum but the phase is different and it may be described as negative maximum.

Thus, if the signals from the loop aerial and the sense aerial are received simultaneously they will oppose each other when the sense pointer indicates the bearing of the transmitter, and reinforce each other when the sense pointer indicates the reciprocal bearing of the transmitter.

Referring to fig. 45·2, P represents the bearing pointer and S the sense pointer. The straight line AX represents the graph of the constant e.m.f. received by the sense aerial, and the dotted curved line represents the graph of the variable e.m.f. received by the loop aerial. The combined effect produces the curve $ABCDE$.

The simultaneous reception of the signals received by the loop and sense aerials produces a single minimum and a single maximum for 360° rotation of the loop aerial. The single minimum occurs at a position on the graduated dial which is the bearing of the transmitter indicated by the sense pointer. The single maximum occurs at the opposite position.

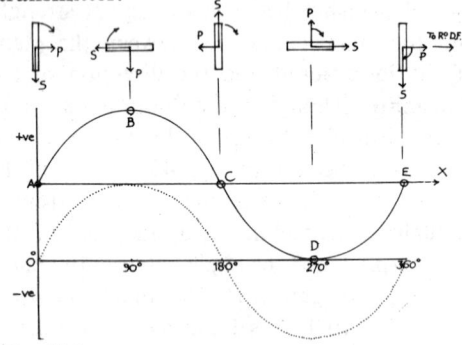

Fig. 45·2

2. The Bellini-Tosi Direction Finder

The Bellini-Tosi direction finder consists of a fixed "cross-loop" aerial instead of a rotating single loop aerial as in the simple direction finder described in Paragraph 1. The ends of the loop aerials are led to two coils which are mounted so that their planes are at right angles to each other in an instrument known as a Goniometer. Within the two crossed coils of the goniometer is another coil—a small coil known as a Search Coil. The search coil may be rotated, and the e.m.f. induced in it is dependent upon the e.m.f.s induced in the crossed coils of the goniometer; and these e.m.f.s, in turn, depend upon the direction of the transmitter relative to the planes of the two cross-loop aerials.

The bearing and sense pointers are fitted to the spindle which carries the search coil, and the method of ascertaining the bearing of a transmitter is similar to the method used with the simple direction finder.

3. Errors in Radio Direction Finding

(i) *Night Effect*—Hitherto, it has been assumed that the radio energy received by the aerial of a direction finder has travelled from the transmitter along a path which coincides with the great circle arc connecting transmitter and receiver. The reception, however, is sometimes due, at least in part, to energy that has been reflected from ionized layers high up in the Earth's atmosphere. This radiation, not being horizontal, induces an e.m.f. in the loop different from what it would be had the radiation been direct- or Ground Radiation. The resulting error is said to be due to Night Effect.

During the hours of darkness, the indirect radiation affects reception of radio energy to a greater extent than during the daylight. This is especially the case when the distance between transmitter and receiver is great; for, in this circumstance, ground signals are weak and sky-signals, as radiations reflected from ionized layers are called, become important for reception.

During the dark hours the reliable range for radio direction finding is reckoned to be within about 25 miles. At and near the times of sunrise and sunset, when the ionized layers of the atmosphere are particularly agitated, reliable bearings cannot be obtained.

(ii) *Land Effect*—The direction of a ray of radio energy over the Earth is influenced by the nature of the surface over which it travels. A radio ray which travels partly over land and partly over sea may be refracted at the coast. This may result in an error in an observed radio bearing which is said to be due to Land Effect. Before using a radio bearing for the purpose of fixing a vessel it is prudent to ascertain, either from the chart or from the *Admiralty List of Radio Signals*, if land effect is likely to be present.

(iii) *Quadrantal Error*—Electro-magnetic energy emanating from a transmitter may, on arriving at a vessel, induce currents in the metal parts of the vessel and these, in turn, may radiate energy which may be received by the direction finder aerials, and which may result in error in an observed radio bearing. The effect of this tends to be least when the transmitter is dead ahead, astern or on either beam. The effect is greatest for transmitters which lie 45° on the low or quarter, for this reason the error is known as Quadrantal Error.

The direction finder should be calibrated from simultaneous radio and visual bearings, and a table or curve of errors prepared, from which the correction to an observed bearing may be lifted as occasion demands.

(iv) *Half Convergency*—Radio energy tends to travel from transmitter to receiver along a great circle arc joining transmitter and receiver. The observed bearing of a transmitter is, therefore, a great circle bearing. Before laying down a position line from a radio observation, it is necessary to apply a correction to the great circle bearing to obtain a rhumb-line bearing. The correction is equal to the Half-convergency of the meridians at the transmitting and receiving stations. Convergency is discussed in detail in Chapter 23.

Exercises on Chapter 45

1. Describe a simple radio direction finder.

2. Explain sensing, and indicate how the 180°-ambiguity in radio direction finding is resolved.

3. Explain the Bellini-Tosi direction finder.

4. Discuss the errors in Radio Direction Finding due to Night Effect and Land Effect.

5. Describe quadrantal error in radio direction finding.

6. Explain how a radio direction finder is calibrated.

CHAPTER 46

HYPERBOLIC NAVIGATION

1. Introduction

A number of navigational instruments employing radio techniques have, since about 1940, come into general use. By means of this type of instrument a navigator may fix his vessel at the intersection of two position lines which are the projections on the chart of spherical hyperbolae on the Earth. The techniques in which these instruments are employed form a branch of navigation which has become known as Hyperbolic Navigation.

2. The Hyperbola

A hyperbola is one of the conic sections; but, for our purposes, we shall define it in terms of one of its important geometrical properties. It is a curve in a plane such that the difference between the distances of any point on it from two fixed points, is a constant amount.

Fig. 46·1 illustrates a typical hyperbola. The two fixed points denoted in fig. 46·1 by A and B, are known as the focal points, or foci, of the hyperbola XY.

At all points P on the hyperbola XY the difference between AP and BP is constant. Thus:

$$AP_1 - BP_1 = AP_2 - BP_2 = AP_3 - BP_3$$

It is for this reason that we may define a hyperbola as a locus such that the difference between the distances from any point on it to each of two fixed points, called the foci of the hyperbola, is constant.

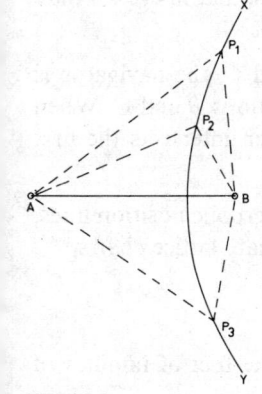

Fig. 46·1

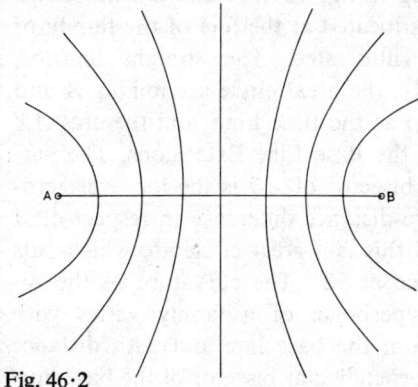

Fig. 46·2 illustrates a series of hyperbolae all having the same focal points A and B. Such a series is known as a Family of Con-focal Hyperbolae.

Spherical hyperbolae are loci on the surface of a sphere such that the difference between the great circle distances from any point on such a spherical hyperbola to two fixed points on the sphere is constant. Families of terrestrial spherical hyperbolae, when projected on a chart, give rise to a complicated network of lines. Such a chart is called a Lattice Chart, and such charts are important parts of any system of hyperbolic navigation.

Fig. 46·2

3. Principles of Hyperbolic Navigation

Hyperbolic Navigation is based on the accurate measurement of the difference in times taken by signals transmitted from each of two fixed radio stations to reach an observer. If the velocity of radio energy is assumed to be constant, it follows that distances travelled are proportional to travel times. Thus, a hyperbolic navigation system may be dependent upon the accurate measurement of differences of "Distance" instead of differences of "Time". For this reason some hyperbolic navigation systems are known as "Distance-Difference Systems", and others are known as "Time-Difference Systems".

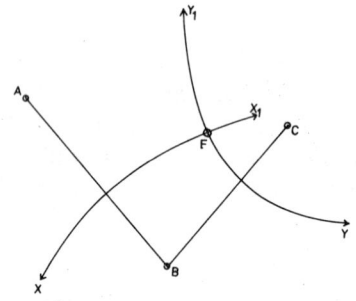

Fig. 46·3

Referring to fig. 46·3: let us suppose that an observer on board a vessel received radio signals, respectively, from radio stations A and B, from which he could measure the intervals of time taken for the radio energy to travel from A and from B, to his vessel. Knowing that radio energy travels at the rate of 300×10^6 metres per second, he could translate the difference of time intervals into a corresponding distance difference and, accordingly, plot the hyperbola XX_1, which has for its foci the positions of the respective radio stations A and B. The hyperbola XX_1 is a locus of constant distance-difference relative to A and B; so that, having plotted it on the chart, the navigator is able to say with confidence that his vessel may be fixed on XX_1, which line therefore, is a position line.

By repeating the procedure, but this time using radio stations B and C, the navigator is able to determine a second hyperbola YY_1, which has for its foci the stations B and C. When plotted on the chart this provides him with a second position line which intersects the first at F, which is a fix obtained from two hyperbolic position lines.

In practice, of course, it is not necessary for the navigator to plot hyperbolic position lines. This tedious task is rendered unnecessary by the availability of appropriate lattice charts.

4. The Nature of Hyperbolic Position Lines

In a hyperbolic navigation system the transmitters are located at the foci of families of hyperbolae.

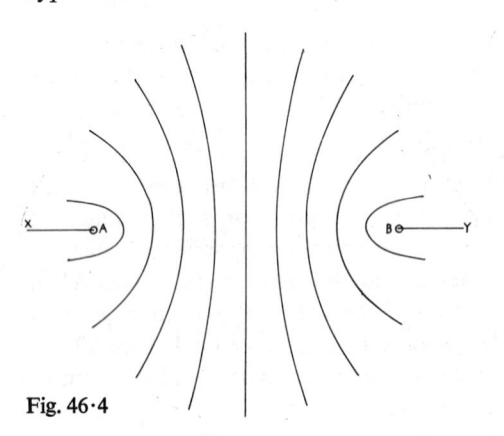

Fig. 46·4

Referring to fig. 46·4, A and B denote two transmitters located at the foci of the family of hyperbolae illustrated. The straight line or, more strictly, the great circle arc joining A and B, is known as the Base Line, and the arcs AX and BY as the Base Line Extensions. The perpendicular bisector of AB is the locus of zero-time or zero-distance difference in respect of A and B, and this is a great circle arc which cuts the base line at $90°$. The curvature of the remaining hyperbolae of a family varies with distance from the base line and with distance from the perpendicular bisector of the base line.

Accuracy of a hyperbolic position line is greatest along the base line where the members of the family of hyperbolae are most closely spaced. As any two adjacent hyperbolae separate more and more as distance from the base line increases, position line accuracy falls off. But the rate of decrease of accuracy is greatest along the base line extension: and, for this reason, in many hyperbolic systems the areas near the base line extensions are of no use for navigational purposes.

Any given member of a family of hyperbolae, other than the base line and the base line extensions, has its greatest curvature at the point where it intersects the base line. As distance from the base line increases the curvature diminishes until a point is reached at which the hyperbola may, for practical purposes, be considered to coincide with the great circle arc. The length of the base line determines the distance from the base line at which the hyperbola and great circle arc are considered to be coincident. In some systems the transmitters are closely spaced and these systems are more "directional" than "hyperbolic". In others the base line may be of many hundreds of miles, and nowhere may the hyperbola be considered to blend with a great circle arc.

5. Consol

Consol is a long-range medium-frequency hyperbolic system in which the transmitters are closely spaced; so that it is, essentially, a directional navigation system in which a navigator may determine the great circle bearing of the Consol station with a relatively high degree of accuracy for ranges of 1000 miles or more in favourable conditions. The principle of Consol is illustrated in fig. 46·5.

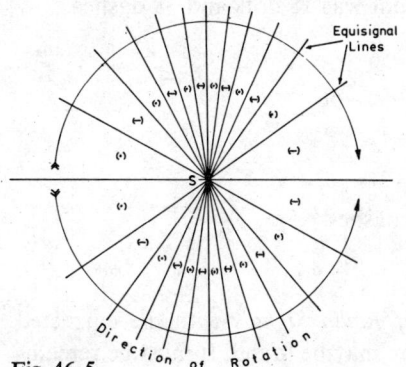

Fig. 46·5

At the Consol transmitting station, denoted by S in fig. 46·5, three aerials radiate energy which produce a "dot and dash" system of audible signals which may be received by means of an ordinary radio receiver on board. By counting the numbers of dots and dashes during a transmission cycle, the bearing of the Consol station may be found from a suitable table or from a Consol lattice chart. The intersecting lines illustrated in fig. 46·5 are the boundaries of sectors within which dots or dashes are received at the beginnings of the transmission cycles. On the boundary lines themselves, a continuous signal is heard. The continuous note is known as the Equisignal, and the boundary is known as the Equisignal Line.

The transmission cycle consists of signals each of 60 dots and dashes. The dots and dashes have the same period, and the cycle is usually completed in one minute. During the transmission of the 60 dots and dashes, the pattern of the dot and dash system is made to rotate at a uniform rate such that at the end of the cycle each equisignal line has swung through one sector. This means that if a vessel is located in a sector in which dots are heard at the beginning of the cycle, dashes will be heard during the latter part of the cycle, and *vice versa*.

At the end of the cycle of transmission the station sends a coded identification signal, during which the equisignal lines are brought back to their original positions ready for the next cycle to commence.

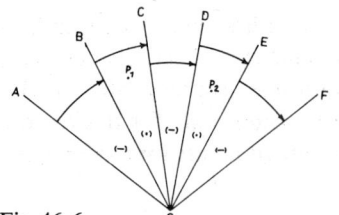

Consider an observer at P_1 in fig. 46·6. At the beginning of the transmission cycle he will receive dash signals but, because of the rotation of the pattern, the signal will presently change to dots. The relative numbers of dots and dashes received during the cycle will assist him in establishing his position relative to the equisignal line BS.

Fig. 46·6

It is evident that ambiguity may arise; an observer cannot determine if he is within the sector BSC or the sector DSE from the dot and dash count alone, the count being the same at P_2 as it is at P_1. His estimated position may assist him in resolving the ambiguity that may exist; but, failing this, the radio direction finder may be used to obtain a rough bearing of the station. The coded identification signal, which is transmitted between successive dot and dash cycles, will assist in this process.

The advantages of Consol are; first, simplicity of operation; and, second, no special receiving apparatus is needed on board. At the present time Consol stations provide facilities for position finding in the North Atlantic Ocean and the adjacent seas.

The operation of Consol is performed by tuning a radio receiver to the frequency of the Consol station and counting the number of dots and dashes in the keying cycle. The total number of dot and dash characters should be 60; but, because the change from dots to dashes, or from dashes to dots, is masked by the width of the equisignal, the number counted is normally less than 60. The count, therefore, has to be corrected for the lost characters. This involves subtracting the total count from 60 and adding half the difference to each of the dot and dash counts.

Example 46·1—Find the correct count if the observed count was 18 dots and 38 dashes.

$$\text{Total Count} = 18 + 38 = 56$$
$$\text{Number of lost characters} = 60 - 56$$
$$= 4$$
$$\text{Correction necessary} = 2$$
$$\text{Corrected Count: 20 dots, 40 dashes.}$$

Answer—20 dots, 40 dashes.

By entering a table provided in the *Admiralty List of Radio Signals* with the corrected count, the true bearing of the vessel from the Consol station may be found. It must be remembered that the true bearing found from a Consol table is a great circle bearing and it may be necessary to apply a half-convergency correction to it before laying it down on a chart as a position line.

On a Consol lattice chart, the lines of the lattice are marked with the count numbers so that a position line is readily found. There will be ambiguity, as explained earlier, unless the bearing of the station is known to an accuracy of about 10°.

6. Loran

The name "Loran" is derived from **Lo**ng **Ra**nge **N**avigation. It is a long-wave hyperbolic system by means of which an observer may determine his vessel's position by measuring the time interval—using suitable equipment to do so—between the instants of receipt of synchronized signals from each of two Loran transmitting stations.

Loran stations are usually located many hundreds of miles apart, groups of stations forming Loran Chains. Within the coverage area, which at present is most of the North Atlantic and North Pacific Oceans, signals may be received from at least two pairs of stations of a chain. Three Loran stations working as two pairs, are sufficient to give two hyperbolic position lines which, if they intersect at a good angle of cut, give a reliable fix of relatively low accuracy.

Long-wave energy of the type used in Loran is reflected from the ionized regions of the atmosphere, and such reflected energy forms the so-called "sky-waves". The ionized region of the atmosphere rises to higher levels during night-time, and this has the effect of increasing the range of sky-wave reception during the hours of darkness.

In general, ground-wave reception is possible for receivers within about 700 miles of the transmitter, but sky-wave reception, especially at night, increases the range to about double this distance.

Signals received at a Loran receiver on board a vessel activate an indicator in the form of a Cathode Ray Tube. On the face of the C.R.T., "traces" of the signals from each of a pair of Loran transmitters have to be matched, after which the required time-difference is obtained. Families of Loran hyperbolic position lines are over-printed on the appropriate navigation chart, and such a Loran Chart facilitates fixing from Loran observations.

Loran is a useful aid, especially for aircraft flying over oceans, in which an approximate position is sufficient for the navigator's purpose. The accuracy of a Loran fix is related to the manner in which the radiated energy travels between transmitter and receiver, and on whether ground- or sky-waves are used for measuring the required time differences. It also depends upon the degree of skill of the observer in identifying the signals as they appear on the C.R.T. indicator.

7. Decca Navigator

The Decca Navigation System employs a chain of transmitters formed by a Master and three Slave stations, the Slave stations being designated Red, Green and Purple, respectively. The distance between a Master station, which is located centrally in the chain, from a Slave station is about 70 miles, and the system operates on medium frequency, continuous wave (C.W.) transmissions.

Each station of a chain transmits C.W.s at a specified frequency which is related harmonically to the frequencies used by the other stations. The four frequencies used are in the ratio 5, 6, 8 and 9.

On board the vessel is fitted a multi-channel receiver tuned to receive the four frequencies used by the stations of the chain. The Master frequency is combined with each of the frequencies of the Slaves, in turn, to form a Comparison Frequency which is the L.C.M. of the Master and appropriate Slave frequencies.

The C.W. transmissions of Master and each Slave, in turn, are phase-locked, so that the signals which combine to make the Comparison Frequency have the same phase on lines on the Earth's surface which are spherical hyperbolae having the Master and Slave stations at their foci.

Decca Lattice Charts provide the projections of the three families of hyperbolae appropriate to a given chain; the families of hyperbolae being coloured Red, Green and Purple, for each of the Master and Slave combinations.

The exact phase relationship between the C.W. signals from Master and Slave is obtained by means of a phase-measuring instrument known as a Decometer. There are three Decometers in the indicating unit, these being coloured Red, Green and Purple, and which give details of the particular Red, Green and Purple hyperbolae on which the vessel is located.

In moving across a Decca coverage area from one point to another during which the phase difference of the signals from Master and Slave changes from $0°$ to $360°$, the vessel is said to traverse a Decca Lane. A Decca Lane is merely the space between two hyperbolic position lines on both of which the phase difference of the signals is $0°$; or, in other words, the signals are "in phase".

The $360°$-phase difference resulting from traversing one lane is divided into hundredths of a lane-width, and the corresponding fraction of a lane-width from the boundary of a lane, together with the lane "number" and "zone", are indicated on each of the three Decometers. It is an easy matter, therefore, to transfer the Decometer readings to the Decca Lattice Chart, and hence to fix a vessel.

The lane-width on the base line depends upon the comparison frequency used, but its maximum value is under 2000 feet. It follows, therefore, that a Decometer reading given to the nearest hundredth of a lane-width permits the navigator to fix his vessel to a theoretical accuracy of better than 20 feet on the base line. Of course the lane-width varies according to position in the coverage area, and propagation and other errors may affect the fix; but there can be no doubt that a Decca fix is highly accurate, and that the Decca Navigation System is a remarkable radio aid to coastal navigation.

The reliable maximum range of Decca is about 250 miles from the Master station of a given chain. Details of propagation and other errors are promulgated by the Decca Navigation Company in the form of Data Sheets which are issued periodically as needed. The *Admiralty List of Radio Signals* also gives information about Decca and other hyperbolic navigation systems.

A large number of Decca Chains have been established in many parts of the globe where shipping activity warrants the use of an accurate position fixing system.

8. Omega

The Omega Navigation System is a very powerful long-wave hyperbolic system which, like the Decca System, is based on phase comparison of C.W. transmissions from each of a pair of Omega Stations.

The base line in the Omega system is in the order of thousands of miles (compared with tens of miles in the Decca System), and not more than eight suitably-located transmitters will, when the system is fully operational, provide world-wide coverage.

Omega Tables are available from which a hyperbolic position line may be laid down on a navigational chart. Omega lattice charts have, however, been provided for certain ocean areas.

The lane-width on the base line in the Omega system is about 15 miles, and fix accuracy is about a half a mile at a range of about 5000 miles during the daytime. But, because of the uncertainties of sky-wave reception, the accuracy is somewhat less at night.

Exercises on Chapter 46

1. Define a hyperbola in terms of its important geometrical property of value to position line navigation.
2. Describe carefully the relationship between the curvature of hyperbolae of a given family, and the distance between the foci of the hyperbolae.
3. Describe a lattice chart of the type used in hyperbolic navigation.
4. Describe the principles of Consol, and explain how to correct a Consol Count.
5. Describe how to resolve ambiguity of sector when using Consol.
6. Explain why Consol is described as being a "Directional" hyperbolic system.
7. Describe the principles of Loran.
8. What errors may affect the position obtained from a Loran observation?
9. Describe the Decca Navigation System.
10. Compare the Loran System of Navigation with that of Decca.
11. What is meant by "sky-wave reception"?
12. Discuss the Omega System of Navigation.

CHAPTER 47

RADAR NAVIGATION

1. Principles of Radar

Radar is an instrument by which the bearing and range of a distant object may be found provided that the object is within the so-called Radar Horizon. The name Radar is derived from **Ra**dio **D**irection **a**nd **R**ange.

The principle of radar is similar to that of echo-sounding. In radar, pulses of radio energy of very high frequency—known as radar frequency—are transmitted, and corresponding echoes received, just as in echo-sounding. If the speed at which a radar pulse travels is known, and the interval between the instants of transmission of the pulse and the receipt of its echo can be measured, the range of the object responsible for returning the echo is immediately determined. Moreover, as in echo-sounding, radar pulses are transmitted through the atmosphere in a narrow beam so that the direction of the object, as well as its range, is determined.

Radar energy, like light, is refracted as it passes through the Earth's atmosphere; but, because of the difference of frequencies of radar energy and light, the Radar Horizon has a range of about 15% more than that of the Visible Horizon. The range R of the Radar Horizon, for an aerial height of H feet above sea level is given by the formula:

$$R = 1 \cdot 22\sqrt{H}$$

The radar equipment comprises a transmitter which generates the pulses or signals at a rate, known as the Pulse Repetition Frequency (P.R.F.), in the order of about 1000 per second. The signals are passed to a specially-designed aerial from which they are transmitted horizontally, and the same aerial is used to receive the echoes of signals. The interval between the transmission of a pulse and the receipt of its corresponding echo is timed, and the range of the object responsible for the echo is indicated in the form of a light-spot on the face of a cathode ray tube which forms the display.

The normal display is in a form known as a P.P.I., or Plan Position Indicator, on which bearings, as well as ranges, are indicated.

The principal use of radar is as an anti-collision aid. But radar is also capable of providing navigational information of particular value when coasting in thick weather when visual observations are not possible.

2. The Use of Primary Radar in Navigation

The term Primary Radar applies to the radar equipment of a vessel by means of which ranges and bearings of objects such as other vessels, land, ice-bergs, may be found.

Provided that an identifiable object appears on a radar display, a single observation gives the object's range and bearing; and hence the vessel may be fixed on the navigational chart at the intersection of a position line (obtained from the observed bearing) and a position circle (obtained from the observed range). But the bearing discrimination of radar is not as good as that of visual observations, although the range discrimination is excellent. For this reason, when using radar as a navigational aid, a vessel is best fixed by radar ranges of two suitably-placed identifiable objects. In other words, a fix by cross-ranges, or two position circles, is to be preferred to a cross-bearing fix from radar observations.

The radar response of different objects varies considerably, and a good deal of skill and experience is necessary if radar is to be used to best advantage for navigating coastwise. A variety of Chart Comparison Units are available by means of which the radar display may be matched with the charted information.

The use of primary radar in navigation is hampered by the presence on the display of so-called unwanted, or false, echoes. These may arise from returns from wave fronts—a form of unwanted echo known as sea-clutter which is troublesome in rough seas; multiple and indirect reflections; and echoes due to rain.

On the normal P.P.I. display the range of an object which appears as a light-spot is proportional to the distance of the light-spot from the centre of the display. The navigator has the choice of a "North-up" or "Head-up" display. In the former true bearings of objects are obtained, whereas with the head-up display relative bearings are obtained. The navigator may also choose between a "Relative" or "True-Motion" display. In the relative display the centre of the display denotes the observer's position at all times, whereas in the true-motion display the centre of the display denotes a given geographical position, and on this display light-spots representing all moving objects detectable by radar, as well as the observer's own vessel, move in their real directions across the display at speeds proportional to those of the objects they represent.

Provision is often made for offsetting the centre of a P.P.I. display to allow an extended period of observation of distant objects lying in particular directions.

3. Radar Beacons

The term Secondary Radar applies to radar equipment located ashore or on a light vessel, which is triggered by pulses from the primary radar with which a vessel in the vicinity is equipped. Secondary radar equipment is usually in the form of a Radar Beacon known as a Racon.

A Racon, a name derived from the term **Ra**dar Bea**con**, consists of a transponder which transmits a signal only when the beacon has been interrogated by the original transmitted signal made by the vessel's radar. The re-transmitted signal made by the beacon is in the form of a coded group which manifests itself on the display by a line of dots and/or dashes radially aligned in the direction of the beacon from the vessel. Not only is direction of the racon obtained but so also is its range.

Another type of radar beacon is known as Ramark, a name derived from the term **Ra**dar **Mark**. This form of beacon transmits continuously in all directions. The bearing of a ramark is indicated on the P.P.I. of a vessel in the vicinity as a radial line which indicates the bearing of the radar beacon.

Ramarks and racons are still in the experimental stage of their development for marine use.

Exercises on Chapter 47

1. Explain the principle of radar and state the more important parts of a radar equipment.
2. Distinguish between primary and secondary radar.
3. Discuss the use of radar for navigating coastwise when the visibility is poor.
4. Explain why cross-ranges are to be preferred to cross-bearings when fixing by radar.
5. Describe the unwanted echoes that may hamper the use of radar as an aid to navigation.
6. What is meant by Radar Horizon? Explain why the range of the radar horizon exceeds that of the visible horizon.
7. Describe a Ramark and a Racon.
8. Discuss the advantages and disadvantages of
    (*a*) True Motion and Relative Motion Display.
    (*b*) North-up and Head-up Displays.

CHAPTER 48

NAVIGATIONAL SATELLITES AND INERTIAL NAVIGATION

1. Introduction

It is interesting to reflect in the closing chapter of this book that the essential processes in navigation are finding and setting course, and determining the distance to travel to reach one's destination. The basic instruments of navigation are, therefore, the compass and the log. Had these instruments been capable of providing the navigator with completely reliable information, systems of position-finding at sea (except, perhaps, for those related to finding position in respect of the depth of water under a vessel's keel) would have been unnecessary. In other words, had D.R. Navigation been perfect Nautical Astronomy and the whole range of navigational equipment based on modern technology need never have been invented.

Of course, in the earliest days of ocean navigation, the mariner realized the insufficiency of his compass and log (and even of his chart, as well); and he was quick to enlist the support of land-based scholars who have, down the ages, devoted untiring attention to the improvement of navigational instruments and techniques. The importance of safe voyaging, especially at times when maritime trade has grown rapidly, has resulted in numerous of the world's leading philosophers, including men of the intellectual stature of Flamsteed, Newton, Halley, Lalande, Euler, Mayer and Lord Kelvin, to name but few, giving serious attention to the improvement of navigation and nautical astronomy. This applies particularly to the present when the most advanced technology finds its application in the field of navigation.

The concluding chapter of this book is devoted to navigational satellites and inertial navigation. These navigational systems will be but briefly described. However, it is appropriate that mention be made of the remarkable techniques that may, in the near future, be commonly used by many readers of this book.

Implicit in the following descriptions is a concept of which every practising navigator ought to be aware. This is the concept of "perpetual change", which applies to Nature in general and to human activities—ideas, science and technology—in particular.

Navigators of the past often were censured for their staunch and dogged resistance to change. The accusations, in large measure, were justified; for many a navigator, refusing to accept new ideas, clung to archaic methods long after these had outlived their usefulness. Young navigators of the present, however, cannot be but keenly aware of the rapid advances currently being made in science and technology. These men (and some women too) realize, most surely, that their noble art is continually in a state of evolution.

2. Navigational Satellites

Soon after the first artificial Earth satellite was launched in 1957, physicists of the Applied Physics Laboratory of Johns Hopkins University, who had set up equipment for receiving radio signals from such satellites, were struck by the change in frequency of the signals, familiarly known as the Doppler Shift, which results when there is a change of relative motion between transmitter and receiver.

The Doppler Shift, Δf, is given by the formula:

$$\Delta f = -\frac{f}{c} \cdot r$$

where f is the transmitted frequency; c the velocity of radio energy, viz. 300×10^6 metres per second; and r is the rate of change of distance between transmitter and receiver. If f and c are known, a measure of the Doppler Shift (Δf) is equivalent to that of the rate at which the range of the transmitter is changing.

Because the motion of an artificial satellite is completely predictable—excepting small errors due to atmospheric drag and those due to imprecise knowledge of the Earth's gravitational field—it is possible to determine the details of the orbit of the satellite; that is to say, the "parameters" of the orbit, from observations of the Doppler Shift.

It was realized that if it is possible to determine the orbit of an artificial satellite from Doppler Shift observations, Doppler information received on board a vessel could be used to fix the vessel's position. It was this realization that led to the satellite navigational system now known as TRANSIT.

Transit employs a number of artificial satellites—six at present—which circle the Earth in polar orbits at a distance from the Earth's surface of about 600 nautical miles. This means that the period of each satellite is about 108 minutes, and that a stationary observer on Earth passes under each satellite orbit twice each 24 hours. Because an observer is in radio line of sight of each satellite at least twice during each passage of a satellite over or near his position, it follows that there are at least four opportunities each day when a vessel may be fixed from each satellite in the system. With six satellites this gives an average of one opportunity each hour.

The satellite accommodates a stable oscillator which controls the frequency of a transmitter. Error in position finding arising from unknown refraction of the radio energy at the Earth's ionosphere, is eliminated by arranging for the satellite to transmit signals at two different frequencies. In the absence of shipboard equipment capable of receiving both frequencies the position finding accuracy is about half a mile.

If the position of an artificial satellite could be predicted accurately for a long period of time it would be possible to provide the navigator with a satellite ephemeris, similar to that of the Moon as given in the *Nautical Almanac*, which would permit the navigator to know the satellite's position in the sky, and hence its geographical position, for any given G.M.T. But predictions sufficiently accurate for navigational purposes are not possible for more than a few days. The positions of the satellites, therefore, which must be known if they are to provide position-finding capacity, are provided by the satellites themselves in the following manner.

Each satellite is equipped with a magnetic memory capable of storing its own ephemeral data for 12 hours or so. The ephemeral data is provided by a ground station, called an Injection Station, which transmits the data to the satellite each time the satellite makes a passage at or near the station. Before the 12-hour ephemeris is transmitted for storage in the satellite, the satellite's memory of the previous 12-hour ephemeris is erased.

The ephemeral data is determined through the agency of a number of tracking stations at which Doppler data is measured and transmitted to a Computing Centre which analyses the data and computes up-dated ephemerides. These are transmitted by teletype to the injection station for transmission to the satellites as they come into line of sight.

Details of the satellite's position, which the navigator receives direct from the satellite, together with an observed Doppler Shift as the satellite passes the observer's position, enable the navigator to fix his vessel.

The Doppler Shift is ascertained by comparing the frequency of the signal received from the satellite with a signal of an appropriate frequency generated by a Local Oscillator. It is measured by counting the number of cycles of a beat frequency over an extended time interval of 2 minutes, during which the satellite transmits the appropriate ephemeral data from its memory. Since the Doppler Shift is a measure of slanting range difference, each range difference is associated with a hyperboloid (a three-dimensional figure swept out by rotating a hyperbola about an axis on which its foci lie) having the satellite at one of its focal points. Such a hyperboloid intersects the surface of the Earth (assumed spherical) along a spherical hyperbola. Two observations of the ephemeral data and Doppler Shift provide two spherical hyperbolae at the intersection of which the observer is located.

The calculations necessary to find a vessel's position from satellite observations is facilitated by means of a purpose-designed computer. This is fed with ephemeral data received from the satellite; the measured Doppler Shift; the latitude and longitude of the vessel by estimation; the vessel's course and speed; and the height of the satellite above a reference surface corresponding to the Earth's surface.

The Transit System gives world-wide coverage at all times of the year and at all times of the day and night. It provides position finding to a high order of accuracy for navigation, this being 0·1 miles when the vessel is equipped with a receiver capable of handling both frequencies at which the satellite transmits.

The most significant source of error arises from incorrect knowledge of the vessel's course and speed, especially the meridianal component. To reduce this error means are provided whereby inputs are made to the computer from the gyro-compass and an electro-magnetic or other sophisticated type of log.

3. Inertial Navigation

The system of navigation designated "inertial" is based on Newton's First Law of Motion in which it is stated that every body tends to maintain its present velocity and that it will do so if no resultant force acts on it. Velocity, which is a vector quantity, changes whenever the speed or direction of a moving body changes; and the rate at which velocity changes is known as acceleration. Newton's Second Law relates the force, f, which, acting on a body of mass m, causes it to accelerate at acceleration a. If appropriate units of mass, force and acceleration, are used Newton's Second Law is expressed as:

$$f = ma$$

It is possible to detect or "sense" an acceleration of any moving body by means of a pendulum which, when fitted to an accelerating body on the Earth's surface, takes up a "false" vertical; and the angle which the false vertical makes with the true vertical is a measure of the acceleration of the body at the instant.

From the well-known equations of motion, viz: $v = at$, and $s = vt$, it is clear that, by integrating acceleration a with respect to time t, it is possible to find velocity v; and that by integrating velocity v with respect to time t, distance s is determined.

A device which is capable of sensing accelerations is known as an Accelerometer, and such devices are basic to inertial navigation.

Inertial navigation is, essentially, a sophisticated Dead Reckoning System in which the motion of a vessel, given its initial velocity, is sensed, without compass or log, so that the vessel's position relative to its starting point is at all times known.

Two accelerometers are needed in an inertial system suitable for surface vehicles, each to measure horizontal accelerations in mutually perpendicular planes. The accelerometers are fitted to a device known as a Stable Platform. This has three planes of freedom and it maintains a fixed orientation in space irrespective of the motion of the vessel. The platform is stabilized by means of gyroscopes which sense the rotation of the platform relative to space.

The accuracy of an inertial navigation system is closely related to the degree of precision of the gyroscopes and accelerometers which are vital to its performance; and it was not until relatively recently that the state of technology made it possible to manufacture these devices to the necessary precision required.

An important feature of inertial navigation for surface vessels is the so-called Schuler tuning* by which the platform on which the accelerometers are mounted is maintained in a horizontal plane. The merest angle of tilt of the platform out of the horizontal results in substantial error. Error in an inertial navigation system, like that of D.R. by compass and log, tends to be proportional to time, so that the importance of precision gyroscopes is paramount.

In addition to the accelerometers and gyroscopes, the third requirement of an inertial navigational system is a computer designed to deal with the complex integration problems associated with the accelerations of the vessel.

Errors in inertial navigation systems are primarily due to gyro drift; and such errors, which, for a gyro drift of 1° per hour amounts to about 6 miles per hour, are cumulative and increase with time. It becomes necessary, therefore, to update the inertial system by other position finding methods.

The inertial system used for surface vessels is known as SINS, which stands for **Ship's Inertial Navigation System**. Its important feature is that it is a self-contained system which functions independently of weather conditions which can hamper nautical astronomy; and of radio energy which may suffer from interference—man-made as well as natural. For this latter reason SINS is of importance for naval vessels, particularly submarines. At present its high cost, coupled with the fact that alternative navigational systems are available, does not warrant the use of the inertial navigation system in commercial vessels. But who can foresee the future?

---

*It is Schuler tuning that is used in gyro compasses to ensure that change in course and speed error is equal to ballistic deflection.

Extracts from
Admiralty Tide Tables
and
The Nautical Almanac

TABLE Ia

MULTIPLICATION TABLE for use with Tables I and II

TABLE Ia (cont.)
MULTIPLICATION TABLE for use with Tables I and II

THE ELEMENTS OF NAVIGATION AND NAUTICAL ASTRONOMY

CARDIFF
MEAN SPRING AND NEAP CURVES

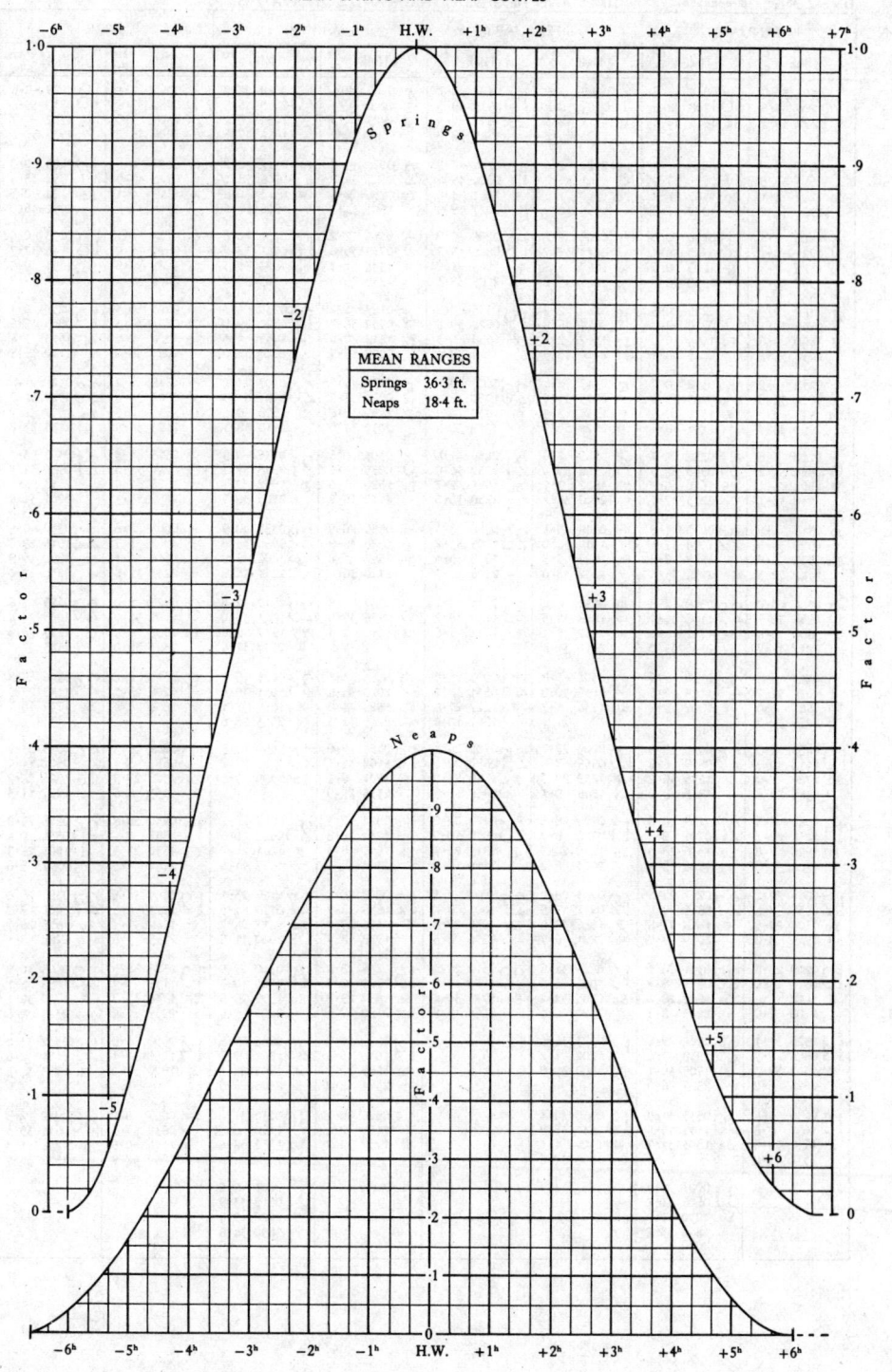

MEAN RANGES	
Springs	36·3 ft.
Neaps	18·4 ft.

WALES — CARDIFF
Lat. 51° 27' N.   Long. 3° 09' W.

TIME ZONE: Greenwich.

TIMES AND HEIGHTS OF HIGH AND LOW WATERS

JANUARY		FEBRUARY		MARCH		APRIL	
TIME / Ht.Ft.	TIME / Ht.Ft.	TIME / Ht.Ft.	TIME / Ht.Ft.	TIME / Ht.Ft.	TIME / Ht.Ft.	TIME / Ht.Ft.	TIME / Ht.Ft.
1 F 0603 33.7 / 1206 3.8 / 1822 33.7	**16** Sa 0543 35.5 / 1208 2.8 / 1814 36.5	**1** M 0048 3.6 / 0707 34.4 / 1312 2.6 / 1923 34.2	**16** Tu 0127 0.7 / 0719 38.3 / 1357 -0.3 / 1945 38.7	**1** M 0605 32.1 / 1210 4.3 / 1823 32.5	**16** Tu 0019 1.7 / 0618 36.8 / 1252 0.4 / 1844 37.5	**1** Th 0045 2.1 / 0650 35.6 / 1309 1.0 / 1906 36.2	**16** F 0129 -0.6 / 0717 38.4 / 1352 -0.7 / 1937 38.2
2 Sa 0029 3.7 / 0643 34.5 / 1247 3.1 / 1901 34.2	**17** Su 0036 1.9 / 0638 37.2 / 1309 1.4 / 1907 37.8	**2** Tu 0128 2.6 / 0741 35.2 / 1350 1.9 / 1956 34.9	**17** W 0214 -0.4 / 0804 39.4 / 1443 -1.0 / 2027 39.3	**2** Tu 0029 3.8 / 0644 34.2 / 1254 2.5 / 1901 34.3	**17** W 0112 0.2 / 0704 38.1 / 1342 -0.6 / 1926 38.6	**2** F 0126 0.9 / 0724 36.9 / 1349 0.1 / 1941 37.5	**17** Sa 0203 -0.8 / 0754 38.4 / 1424 -0.5 / 2012 38.0
3 Su 0105 3.1 / 0721 35.1 / 1325 2.5 / 1937 34.4	**18** M 0134 0.9 / 0729 38.4 / 1405 0.5 / 1958 38.6	**3** W 0205 2.2 / 0813 35.7 / 1427 1.9 / 2027 35.2	**18** Th 0256 -0.8 / 0845 39.7 / 1523 -1.0 / 2109 39.1	**3** W 0111 2.2 / 0719 35.6 / 1334 1.3 / 1935 35.8	**18** Th 0156 -0.8 / 0744 39.2 / 1423 -1.3 / 2005 39.0	**3** Sa 0206 0.1 / 0800 38.0 / 1429 -0.5 / 2016 38.2	**18** Su 0235 -0.6 / 0825 37.7 / 1451 0.3 / 2043 37.2
4 M 0140 2.9 / 0755 35.1 / 1400 2.6 / 2010 34.2	**19** Tu 0225 0.3 / 0818 39.2 / 1454 0.0 / 2045 38.9	**4** Th 0240 2.2 / 0844 35.8 / 1502 2.1 / 2100 35.3	**19** F 0335 -0.8 / 0924 39.4 / 1559 -0.3 / 2147 38.1	**4** Th 0149 1.2 / 0752 36.7 / 1412 0.6 / 2008 36.7	**19** F 0235 -1.3 / 0820 39.5 / 1457 -1.3 / 2041 39.0	**4** Su 0243 -0.3 / 0833 38.3 / 1504 -0.4 / 2049 38.0	**19** M 0259 0.1 / 0855 36.6 / 1513 1.4 / 2110 35.9
5 Tu 0212 3.2 / 0821 34.8 / 1433 3.3 / 2040 33.7	**20** W 0312 0.2 / 0903 39.2 / 1539 0.3 / 2131 38.5	**5** F 0313 2.4 / 0915 35.6 / 1535 2.5 / 2129 35.0	**20** Sa 0408 0.0 / 1000 38.1 / 1630 0.9 / 2219 36.4	**5** F 0226 0.7 / 0823 37.2 / 1450 0.4 / 2039 37.0	**20** Sa 0307 -1.2 / 0857 39.0 / 1528 -0.6 / 2115 38.1	**5** M 0318 0.2 / 0907 37.6 / 1535 0.6 / 2123 36.9	**20** Tu 0321 1.4 / 0920 34.8 / 1527 2.9 / 2135 33.8
6 W 0242 3.8 / 0851 34.2 / 1504 4.1 / 2110 33.3	**21** Th 0354 0.7 / 0948 38.6 / 1620 0.9 / 2213 37.1	**6** Sa 0345 2.8 / 0944 35.0 / 1602 3.1 / 2200 34.2	**21** Su 0436 1.4 / 1032 36.0 / 1651 2.7 / 2250 34.0	**6** Sa 0303 0.6 / 0855 37.3 / 1524 0.6 / 2112 36.5	**21** Su 0336 -0.5 / 0928 37.7 / 1552 0.7 / 2145 36.5	**6** Tu 0345 1.3 / 0941 36.0 / 1558 2.0 / 2156 35.1	**21** W 0335 3.3 / 0946 32.3 / 1538 4.9 / 2201 31.4
7 Th 0313 4.5 / 0925 33.5 / 1534 4.9 / 2142 32.5	**22** F 0435 1.5 / 1028 37.1 / 1659 2.2 / 2254 35.3	**7** Su 0408 3.4 / 1014 34.0 / 1627 3.9 / 2232 33.0	**22** M 0454 3.3 / 1103 33.2 / 1708 5.1 / 2324 31.2	**7** Su 0335 0.9 / 0925 36.9 / 1552 1.2 / 2140 36.0	**22** M 0358 0.9 / 0956 35.8 / 1606 2.4 / 2211 34.2	**7** W 0403 2.8 / 1015 33.7 / 1618 4.1 / 2233 32.6	**22** Th 0350 5.4 / 1016 29.8 / 1554 6.9 / 2234 29.0
8 F 0344 5.0 / 0958 32.7 / 1606 5.4 / 2217 31.7	**23** Sa 0508 2.8 / 1108 35.1 / 1732 4.0 / 2335 33.0	**8** M 0431 4.4 / 1048 32.5 / 1652 5.1 / 2310 31.4	**23** Tu 0514 5.7 / 1138 30.2 / 1732 7.6	**8** M 0357 1.8 / 0956 35.5 / 1611 2.3 / 2212 34.4	**23** Tu 0412 2.8 / 1021 33.0 / 1617 4.6 / 2237 31.4	**8** Th 0429 5.0 / 1101 31.0 / 1653 6.4 / 2323 30.0	**23** F 0415 7.6 / 1055 27.6 / 1627 8.8 / 2323 27.0
9 Sa 0417 5.6 / 1034 31.7 / 1641 6.0 / 2255 30.8	**24** Su 0540 4.6 / 1148 32.6 / 1804 5.9	**9** Tu 0502 5.8 / 1136 30.8 / 1734 6.7	**24** W 0003 28.5 / 0551 8.2 / 1227 27.4 / 1817 10.0	**9** Tu 0416 3.1 / 1028 33.5 / 1628 4.0 / 2245 32.3	**24** W 0425 5.2 / 1051 30.0 / 1634 7.1 / 2313 28.8	**9** F 0517 7.4 / 1208 28.7 / 1804 8.4	**24** Sa 0506 9.8 / 1157 25.5 / 1737 10.7
10 Su 0453 6.5 / 1118 30.7 / 1724 7.0 / 2346 29.8	**25** M 0017 30.6 / 0619 6.6 / 1235 30.2 / 1846 8.1	**10** W 0005 29.7 / 0559 7.6 / 1239 29.3 / 1846 8.1	**25** Th 0102 26.4 / 0655 10.4 / 1340 25.8 / 1939 11.3	**10** W 0438 5.0 / 1108 31.1 / 1701 6.3 / 2335 29.9	**25** Th 0450 7.9 / 1135 27.2 / 1709 9.5	**10** Sa 0042 28.3 / 0657 8.7 / 1348 27.9 / 1953 8.6	**25** Su 0036 25.5 / 0641 11.1 / 1326 24.9 / 1927 11.3
11 M 0545 7.5 / 1214 29.7 / 1826 7.8	**26** Tu 0109 28.6 / 0707 8.4 / 1336 28.5 / 1946 9.4	**11** Th 0121 28.8 / 0728 8.7 / 1410 28.9 / 2023 8.2	**26** F 0228 23.3 / 0830 10.9 / 1513 25.9 / 2115 10.8	**11** Th 0522 7.3 / 1211 28.8 / 1810 8.3	**26** F 0004 26.4 / 0548 10.3 / 1241 25.2 / 1827 11.4	**11** Su 0230 28.8 / 0857 7.4 / 1530 29.9 / 2141 6.5	**26** M 0215 25.8 / 0834 10.4 / 1503 26.3 / 2112 9.8
12 Tu 0051 29.2 / 0655 8.4 / 1327 29.5 / 1944 8.0	**27** W 0217 27.6 / 0815 9.2 / 1448 27.8 / 2057 9.7	**12** F 0253 29.4 / 0912 7.9 / 1546 30.6 / 2201 6.7	**27** Sa 0400 27.1 / 1003 9.4 / 1638 27.8 / 2236 8.5	**12** F 0050 28.3 / 0656 9.4 / 1350 27.9 / 1958 8.9	**27** Sa 0126 25.1 / 0730 11.5 / 1417 24.7 / 2020 11.5	**12** M 0403 31.5 / 1027 4.4 / 1644 33.1 / 2255 3.6	**27** Tu 0344 28.0 / 1000 7.9 / 1616 28.7 / 2228 7.1
13 W 0209 29.6 / 0818 8.0 / 1448 30.4 / 2108 7.0	**28** Th 0331 28.1 / 1604 28.6 / 2208 8.5	**13** Sa 0423 31.6 / 1044 5.5 / 1705 33.2 / 2320 4.4	**28** Su 0512 29.7 / 1120 6.8 / 1739 30.4 / 2341 5.9	**13** Sa 0236 28.4 / 0900 9.7 / 1537 29.7 / 2150 7.1	**28** Su 0310 25.9 / 0920 10.3 / 1555 26.6 / 2200 9.4	**13** Tu 0508 34.5 / 1136 1.4 / 1738 35.7 / 2359 1.4	**28** W 0445 30.9 / 1059 5.1 / 1710 32.1 / 2323 4.6
14 Th 0328 31.1 / 0946 6.7 / 1607 32.3 / 2224 5.2	**29** F 0441 30.1 / 1038 7.4 / 1709 30.1 / 2224 4.2	**14** Su 0533 34.2 / 1202 3.2 / 1808 35.8		**14** Su 0415 28.5 / 1038 5.3 / 1659 32.9 / 2308 6.5	**29** M 0434 28.5 / 1041 7.5 / 1703 29.6 / 2308 6.5	**14** W 0558 36.8 / 1230 0.1 / 1820 37.2	**29** Th 0532 33.4 / 1151 2.8 / 1753 34.5
15 F 0439 33.4 / 1059 4.8 / 1711 34.6 / 2334 3.4	**30** Sa 0539 31.5 / 1139 5.5 / 1803 31.7	**15** M 0030 2.1 / 0631 36.8 / 1304 1.1 / 1858 37.6		**15** M 0526 34.2 / 1154 2.4 / 1757 35.7	**30** Tu 0532 31.4 / 1140 4.7 / 1752 32.4	**15** Th 0047 0.2 / 0639 37.8 / 1314 -0.6 / 1859 37.9	**30** F 0012 2.5 / 0614 35.5 / 1234 1.3 / 1832 36.4
	31 Su 0003 4.9 / 0626 33.2 / 1230 3.9 / 1846 33.2				**31** W 0003 4.0 / 0615 33.9 / 1228 2.4 / 1830 34.6		

TABLE V

TIDAL LEVELS AT STANDARD PORTS
(with data concerning predictions, etc.)

Standard Port	L.A.T. (a)	M.L.W.S. (b)	M.L.W.N. (c)	M.L. (d)	M.H.W.N. (c)	M.H.W.S. (b)	H.A.T. (a)	Observations	Constants	Predictions	Method of Predicting (f)	Years of Observations (g)
Devonport	−1.7	+0.5	+ 5.4	+ 8.6	+12.5	+16.1	+17.9	Hyd.	L.T.I.	L.T.I.	H.	1961–62
Portland	−1.2	+0.3	+ 2.4	+ 3.5	+ 4.5	+ 7.0	+ 7.8	Hyd.	L.T.I.	L.T.I.	H.C.	1961–62
Southampton	0.0	+1.5	+ 5.8	+ 8.6	+12.3	+14.9	+16.2	H.A.	L.T.I.	L.T.I.	H.C.	1924
Portsmouth	−0.1	+2.0	+ 5.8	+ 8.9	+12.5	+15.4	+16.9	Hyd.	L.T.I.	L.T.I.	H.C.	1961–62
Shoreham	−2.1	+0.3	+ 4.2	+ 9.3	+14.2	+18.4	+20.0	H.A.	L.T.I.	L.T.I.	H.	1959
Dover	0.0	+2.5	+ 6.5	+12.1	+17.5	+21.9	+23.6	H.A.	L.T.I.	L.T.I.	H.C.	1960
Sheerness	−1.2	+0.9	+ 3.8	+ 9.3	+14.7	+17.8	+19.2	Hyd.	—	Hyd.	Diff.	—
Chatham	−1.8	+0.3	+ 3.4	+ 9.4	+15.3	+18.6	+20.3	Hyd.	—	Hyd.	Diff.	—
London Bridge	−1.0	+0.8	+ 3.5	+11.3	+18.6	+22.5	+24.0	H.A.	L.T.I.	L.T.I.	H.C.	—
Harwich	−0.3	+1.2	+ 3.2	+ 6.9	+10.4	+12.8	+13.7	H.A.	L.T.I.	L.T.I.	H.C.	1931–32
Lowestoft	−0.9	+0.7	+ 2.5	+ 4.0	+ 5.9	+ 7.1	+ 8.3	H.A.	L.T.I.	L.T.I.	H.	1959
Immingham	−1.6	+1.5	+ 6.5	+11.6	+16.8	+21.6	+24.5	H.A.	L.T.I.	L.T.I.	H.	1955–56
Hull	−2.4	+0.8	+ 6.1	+11.5	+17.1	+22.1	+25.0	H.A.	L.T.I.	L.T.I.	H.C.	1960
River Tees	−1.1	+1.5	+ 5.5	+ 9.1	+12.9	+16.7	+18.6	H.A.	L.T.I.	L.T.I.	H.	1948
River Tyne	−1.9	+0.5	+ 4.3	+ 7.7	+11.3	+14.8	+16.9	H.A.	L.T.I.	L.T.I.	H.	1947
Leith	−0.4	+2.0	+ 6.4	+10.1	+14.3	+17.7	+19.5	H.A.	L.T.I.	L.T.I.	H.C.	1955
Rosyth	−1.2	+1.3	+ 5.8	+ 9.7	+14.0	+17.7	+19.7	Hyd.	L.T.I.	L.T.I.	H.C.	1945, 1947
Aberdeen	−1.7	+0.4	+ 3.7	+ 6.5	+ 9.5	+12.3	+14.0	H.A.	L.T.I.	L.T.I.	H.	1930–31
Stromness	−1.6	+0.4	+ 3.5	+ 5.4	+ 7.5	+10.3	+12.3	Hyd.	L.T.I.	L.T.I.	H.C.	1910–12
Stornoway	−1.5	+0.7	+ 5.0	+ 7.6	+10.5	+14.1	+16.1	Hyd.	L.T.I.	L.T.I.	H.	1929, 1959
Oban	−0.5	+0.9	+ 4.3	+ 6.1	+ 8.1	+11.3	+12.4	Hyd.	Rob.	L.T.I.	H.	1910–11
Greenock	−0.8	+0.7	+ 2.8	+ 5.8	+ 9.0	+10.8	+12.6	H.A.	L.T.I.	L.T.I.	H.C.	1948
Glasgow	−2.7	−0.2	+ 2.9	+ 6.7	+10.8	+13.3	+15.0	H.A.	L.T.I.	L.T.I.	H.C.	—
Liverpool	−2.0	+1.5	+ 7.8	+15.3	+22.8	+29.0	+32.1	H.A.	L.T.I.	L.T.I.	H.C.	8 years
Holyhead	−1.8	+0.5	+ 4.7	+ 8.7	+12.8	+16.7	+19.4	H.A.	L.T.I.	L.T.I.	H.	1959–60
Milford Haven	−0.2	+2.2	+ 8.1	+12.6	+17.2	+23.0	+26.1	Hyd.	L.T.I.	L.T.I.	H.	1953–54
Swansea	−2.2	+1.5	+ 8.4	+15.2	+21.7	+29.1	+32.3	H.A.	—	—	—	—
Cardiff	−2.8	+1.3	+ 9.6	+19.1	+28.0	+37.6	+41.1	H.A.	—	L.T.I.	Diff.	1956
Port of Bristol (Avonmouth)	−1.1	+1.9	+10.4	+21.5	+31.6	+42.2	+46.4	H.A.	L.T.I.	L.T.I.	H.C.	10 years
Dublin	−0.7	+1.0	+ 4.0	+ 7.0	+10.4	+12.7	+14.5	H.A.	L.T.I.	L.T.I.	H.	1948–49
Belfast	0.0	+1.4	+ 3.5	+ 6.5	+ 9.7	+11.4	+12.8	H.A.	L.T.I.	L.T.I.	H.	1947–48
Londonderry	−0.7	+0.5	+ 2.6	+ 4.3	+ 6.0	+ 8.2	+ 9.6	H.A.	L.T.I.	L.T.I.	H.	1935–36
Galway	−0.5	+1.2	+ 5.2	+ 8.3	+11.7	+15.3	+17.0	H.A.	Hyd.	L.T.I.	H.	1959
Cobh	−1.7	+0.1	+ 2.7	+ 6.0	+ 9.2	+12.0	+13.7	Hyd.	L.T.I.	L.T.I.	H.	1906
Reykjavik	−1.6	+0.4	+ 4.2	+ 6.7	+ 9.4	+12.9	+14.5	I.	L.T.I.	L.T.I.	H.	1951
Port of K'yem	+0.3	+1.2	+ 2.1	+ 3.6	+ 5.3	+ 6.1	+ 6.7	R.	R.	L.T.I.	H.	1910
Yekaterininskaya	−0.3	+1.7	+ 4.3	+ 7.0	+ 9.7	+12.1	+13.5	R.	G.	G.	H.	1906–07
Narvik	+0.2	+1.6	+ 4.0	+ 6.0	+ 7.8	+10.6	+11.9	—	—	Nor.	—	—
Bergen	−0.2	+0.6	+ 1.7	+ 2.6	+ 3.5	+ 4.6	+ 5.3	—	—	Nor.	—	—
Esbjerg	−1.2	−0.4	+ 0.8	+ 2.5	+ 4.5	+ 5.1	+ 6.1	D.	D.	L.T.I.	H.	—
Helgoland	−1.1	−0.1	+ 1.2	+ 4.3	+ 7.5	+ 8.6	+ 9.5	—	—	G.	—	—
Cuxhaven	−1.3	−0.4	+ 0.8	+ 4.9	+ 8.9	+10.2	+11.2	—	—	G.	—	—
Wilhelmshaven	−1.5	−0.3	+ 1.5	+ 6.4	+11.4	+12.9	+14.0	—	—	G.	—	—
Hook of Holland	+0.3	+1.0	+ 1.3	+ 3.7	+ 5.6	+ 6.7	+ 7.7	—	—	N.	—	—
Flushing	+0.1	+1.2	+ 3.3	+ 8.2	+12.7	+15.6	+16.5	—	—	N.	—	—
Antwerp	−1.1	+0.1	+ 1.6	+ 8.6	+14.8	+17.7	+19.1	B.	L.T.I.	L.T.I.	H.C.	1952
Dunkerque	+0.8	+2.0	+ 4.8	+10.3	+15.5	+19.0	+20.2	—	—	F.	—	—
Calais	+0.8	+2.3	+ 6.0	+12.4	+18.4	+22.7	+24.0	—	—	F.	—	—
Boulogne	+0.8	+2.8	+ 8.6	+16.0	+23.4	+29.1	+31.2	—	—	F.	—	—
Dieppe	+0.7	+2.0	+ 7.8	+15.8	+23.3	+30.0	+32.8	—	—	F.	—	—
Le Havre	+1.1	+3.3	+ 8.9	+14.8	+20.9	+25.5	+27.2	—	—	F.	—	—
Cherbourg	+1.5	+2.7	+ 7.7	+11.6	+15.7	+20.3	+23.0	—	—	F.	—	—
St. Helier	−1.0	+3.2	+12.5	+19.2	+25.7	+35.3	+39.4	H.A.	L.T.I.	L.T.I.	H.	1952
Brest	+1.8	+4.4	+ 9.9	+14.4	+18.9	+24.3	+27.1	—	—	F.	—	—
Cordouan	+0.5	+2.1	+ 5.9	+ 9.1	+12.3	+16.2	+18.5	—	—	—	—	—
Lisbon	+0.4	+1.6	+ 4.6	+ 7.1	+ 9.8	+12.5	+14.0	—	—	P.	—	—
Gibraltar	−0.1	+0.1	+ 0.9	+ 1.6	+ 2.3	+ 3.1	+ 3.4	Hyd.	L.T.I.	L.T.I.	H.	1961–62
Venezia (Venice)	−0.3	+0.4	+ 1.3	+ 1.7	+ 2.1	+ 2.8	+ 3.3	—	—	It.	—	—

The above levels are referred to *CHART DATUM*, which is the same as the zero of the tidal predictions in all cases. For notes (*a*) to (*g*), see page xxvi.

A2 ALTITUDE CORRECTION TABLES 10°–90°—SUN, STARS, PLANETS

OCT.–MAR. SUN APR.–SEPT.			STARS AND PLANETS		DIP			
App. Alt.	Lower Limb / Upper Limb	App. Alt.	Lower Limb / Upper Limb	App. Alt.	Corrⁿ	App. Alt.	Additional Corrⁿ	Ht. of Eye / Corrⁿ	Ht. of Eye / Corrⁿ
° ′	′	° ′	′	° ′	′	° ′	′	ft.	ft.
9 34	+10·8 −22·7	9 39	+10·6 −22·4	9 56	−5·3	**1958**		1·1 −1·1	44 −6·5
9 45	+10·9 −22·6	9 51	+10·7 −22·3	10 08	−5·2	**VENUS**		1·4 −1·2	45 −6·6
9 56	+11·0 −22·5	10 03	+10·8 −22·2	10 20	−5·1			1·6 −1·3	47 −6·7
10 08	+11·1 −22·4	10 15	+10·9 −22·1	10 33	−5·0	Jan. 1–Jan. 10		1·9 −1·4	48 −6·8
10 21	+11·2 −22·3	10 27	+11·0 −22·0	10 46	−4·9			2·2 −1·5	49 −6·9
10 34	+11·3 −22·2	10 40	+11·1 −21·9	11 00	−4·8	°	′	2·5 −1·6	51 −7·0
10 47	+11·4 −22·1	10 54	+11·2 −21·8	11 14	−4·7	0	+0·5	2·8 −1·7	52 −7·1
11 01	+11·5 −22·0	11 08	+11·3 −21·7	11 29	−4·6	6	+0·6	3·2 −1·8	54 −7·2
11 15	+11·6 −21·9	11 23	+11·4 −21·6	11 45	−4·5	20	+0·7	3·6 −1·9	55 −7·3
11 30	+11·7 −21·8	11 38	+11·5 −21·5	12 01	−4·4	31		4·0 −2·0	57 −7·4
11 46	+11·8 −21·7	11 54	+11·6 −21·4	12 18	−4·3	Jan. 11–Feb. 14		4·4 −2·1	58 −7·5
12 02	+11·9 −21·6	12 10	+11·7 −21·3	12 35	−4·2	°	′	4·9 −2·2	60 −7·6
12 19	+12·0 −21·5	12 28	+11·8 −21·2	12 54	−4·1	0	+0·6	5·3 −2·3	62 −7·7
12 37	+12·1 −21·4	12 46	+11·9 −21·1	13 13	−4·0	4	+0·7	5·8 −2·4	63 −7·8
12 55	+12·2 −21·3	13 05	+12·0 −21·0	13 33	−3·9	12	+0·8	6·3 −2·5	65 −7·9
13 14	+12·3 −21·2	13 24	+12·1 −20·9	13 54	−3·8	22		6·9 −2·6	67 −8·0
13 35	+12·4 −21·1	13 45	+12·2 −20·8	14 16	−3·7	Feb. 15–Feb. 21		7·4 −2·7	68 −8·1
13 56	+12·5 −21·0	14 07	+12·3 −20·7	14 40	−3·6	°	′	8·0 −2·8	70 −8·2
14 18	+12·6 −20·9	14 30	+12·4 −20·6	15 04	−3·5	0	+0·5	8·6 −2·9	72 −8·3
14 42	+12·7 −20·8	14 54	+12·5 −20·5	15 30	−3·4	6	+0·6	9·2 −3·0	74 −8·4
15 06	+12·8 −20·7	15 19	+12·6 −20·4	15 57	−3·3	20	+0·7	9·8 −3·1	75 −8·5
15 32	+12·9 −20·6	15 46	+12·7 −20·3	16 26	−3·2	31		10·5 −3·2	77 −8·6
15 59	+13·0 −20·5	16 14	+12·8 −20·2	16 56	−3·1	Feb. 22–Mar. 9		11·2 −3·3	79 −8·7
16 28	+13·1 −20·4	16 44	+12·9 −20·1	17 28	−3·0	°	′	11·9 −3·4	81 −8·8
16 59	+13·2 −20·3	17 15	+13·0 −20·0	18 02	−2·9	0	+0·4	12·6 −3·5	83 −8·9
17 32	+13·3 −20·2	17 48	+13·1 −19·9	18 38	−2·8	11	+0·5	13·3 −3·6	85 −9·0
18 06	+13·4 −20·1	18 24	+13·2 −19·8	19 17	−2·7	41		14·1 −3·7	87 −9·1
18 42	+13·5 −20·0	19 01	+13·3 −19·7	19 58	−2·6	Mar. 10–Apr. 4		14·9 −3·8	88 −9·2
19 21	+13·6 −19·9	19 42	+13·4 −19·6	20 42	−2·5	°	′	15·7 −3·9	90 −9·3
20 03	+13·7 −19·8	20 25	+13·5 −19·5	21 28	−2·4	0	+0·3	16·5 −4·0	92 −9·4
20 48	+13·8 −19·7	21 11	+13·6 −19·4	22 19	−2·3	46		17·4 −4·1	94 −9·5
21 35	+13·9 −19·6	22 00	+13·7 −19·3	23 13	−2·2	Apr. 5–May 19		18·3 −4·2	96 −9·6
22 26	+14·0 −19·5	22 54	+13·8 −19·2	24 11	−2·1	°	′	19·1 −4·3	98 −9·7
23 22	+14·1 −19·4	23 51	+13·9 −19·1	25 14	−2·0	0	+0·2	20·1 −4·4	101 −9·8
24 21	+14·2 −19·3	24 53	+14·0 −19·0	26 22	−1·9	47		21·0 −4·5	103 −9·9
25 26	+14·3 −19·2	26 00	+14·1 −18·9	27 36	−1·8	May 20–Dec. 31		22·0 −4·6	105 −10·0
26 36	+14·4 −19·1	27 13	+14·2 −18·8	28 56	−1·7	°	′	22·9 −4·7	107 −10·1
27 52	+14·5 −19·0	28 33	+14·3 −18·7	30 24	−1·6	0	+0·1	23·9 −4·8	109 −10·2
29 15	+14·6 −18·9	30 00	+14·4 −18·6	32 00	−1·5	42		24·9 −4·9	111 −10·3
30 46	+14·7 −18·8	31 35	+14·5 −18·5	33 45	−1·4	**MARS**		26·0 −5·0	113 −10·4
32 26	+14·8 −18·7	33 20	+14·6 −18·4	35 40	−1·3	Jan. 1–Sept. 3		27·1 −5·1	116 −10·5
34 17	+14·9 −18·6	35 17	+14·7 −18·3	37 48	−1·2	°	′	28·1 −5·2	118 −10·6
36 20	+15·0 −18·5	37 26	+14·8 −18·2	40 08	−1·1	0	+0·1	29·2 −5·3	120 −10·7
38 36	+15·1 −18·4	39 50	+14·9 −18·1	42 44	−1·0	60		30·4 −5·4	122 −10·8
41 08	+15·2 −18·3	42 31	+15·0 −18·0	45 36	−0·9	Sept. 4–Dec. 31		31·5 −5·5	125 −10·9
43 59	+15·3 −18·2	45 31	+15·1 −17·9	48 47	−0·8	°	′	32·7 −5·6	127 −11·0
47 10	+15·4 −18·1	48 55	+15·2 −17·8	52 18	−0·7	0	+0·1	33·9 −5·7	129 −11·1
50 46	+15·5 −18·0	52 44	+15·3 −17·7	56 11	−0·6	60		35·1 −5·8	132 −11·2
54 49	+15·6 −17·9	57 02	+15·4 −17·6	60 28	−0·5			36·3 −5·9	134 −11·3
59 23	+15·7 −17·8	61 51	+15·5 −17·5	65 08	−0·4			37·6 −6·0	136 −11·4
64 30	+15·8 −17·7	67 17	+15·6 −17·4	70 11	−0·3	°	′	38·9 −6·1	139 −11·5
70 12	+15·9 −17·6	73 16	+15·7 −17·3	75 34	−0·2	34	+0·3	40·1 −6·2	141 −11·6
76 26	+16·0 −17·5	79 43	+15·8 −17·2	81 13	−0·1	60	+0·2	41·5 −6·3	144 −11·7
83 05	+16·1 −17·4	86 32	+15·9 −17·1	87 03	0·0	80	+0·1	42·8 −6·4	146 −11·8
90 00		90 00		90 00				44·2	149

App. Alt. = Apparent altitude = Sextant altitude corrected for index error and dip.

THE ELEMENTS OF NAVIGATION AND NAUTICAL ASTRONOMY

ALTITUDE CORRECTION TABLES 0°–10°—SUN, STARS, PLANETS A3

App. Alt.	OCT.–MAR. SUN Lower Limb	Upper Limb	APR.–SEPT. Lower Limb	Upper Limb	STARS PLANETS	App. Alt.	OCT.–MAR. SUN Lower Limb	Upper Limb	APR.–SEPT. Lower Limb	Upper Limb	STARS PLANETS
° ′	′	′	′	′	′	° ′	′	′	′	′	′
0 00	−18·2	−51·7	−18·4	−51·4	−34·5	3 30	+3·3	−30·2	+3·1	−29·9	−13·0
03	17·5	51·0	17·8	50·8	33·8	35	3·6	29·9	3·3	29·7	12·7
06	16·9	50·4	17·1	50·1	33·2	40	3·8	29·7	3·5	29·5	12·5
09	16·3	49·8	16·5	49·5	32·6	45	4·0	29·5	3·7	29·3	12·3
12	15·7	49·2	15·9	48·9	32·0	50	4·2	29·3	3·9	29·1	12·1
15	15·1	48·6	15·3	48·3	31·4	3 55	4·4	29·1	4·1	28·9	11·9
0 18	−14·5	−48·0	−14·8	−47·8	−30·8	4 00	+4·5	−29·0	+4·3	−28·7	−11·8
21	14·0	47·5	14·2	47·2	30·3	05	4·7	28·8	4·5	28·5	11·6
24	13·5	47·0	13·7	46·7	29·8	10	4·9	28·6	4·6	28·4	11·4
27	12·9	46·4	13·2	46·2	29·2	15	5·1	28·4	4·8	28·2	11·2
30	12·4	45·9	12·7	45·7	28·7	20	5·2	28·3	5·0	28·0	11·1
33	11·9	45·4	12·2	45·2	28·2	25	5·4	28·1	5·1	27·9	10·9
0 36	−11·5	−45·0	−11·7	−44·7	−27·8	4 30	+5·6	−27·9	+5·3	−27·7	−10·7
39	11·0	44·5	11·2	44·2	27·3	35	5·7	27·8	5·5	27·5	10·6
42	10·5	44·0	10·8	43·8	26·8	40	5·9	27·6	5·6	27·4	10·4
45	10·1	43·6	10·3	43·3	26·4	45	6·0	27·5	5·8	27·2	10·3
48	9·6	43·1	9·9	42·9	25·9	50	6·2	27·3	5·9	27·1	10·1
51	9·2	42·7	9·5	42·5	25·5	4 55	6·3	27·2	6·0	27·0	10·0
0 54	−8·8	−42·3	−9·1	−42·1	−25·1	5 00	+6·4	−27·1	+6·2	−26·8	−9·9
0 57	8·4	41·9	8·7	41·7	24·7	05	6·6	26·9	6·3	26·7	9·7
1 00	8·0	41·5	8·3	41·3	24·3	10	6·7	26·8	6·4	26·6	9·6
03	7·7	41·2	7·9	40·9	24·0	15	6·8	26·7	6·6	26·4	9·5
06	7·3	40·8	7·5	40·5	23·6	20	6·9	26·6	6·7	26·3	9·4
09	6·9	40·4	7·2	40·2	23·2	25	7·1	26·4	6·8	26·2	9·2
1 12	−6·6	−40·1	−6·8	−39·8	−22·9	5 30	+7·2	−26·3	+6·9	−26·1	−9·1
15	6·2	39·7	6·5	39·5	22·5	35	7·3	26·2	7·0	26·0	9·0
18	5·9	39·4	6·2	39·2	22·2	40	7·4	26·1	7·2	25·8	8·9
21	5·6	39·1	5·8	38·8	21·9	45	7·5	26·0	7·3	25·7	8·8
24	5·3	38·8	5·5	38·5	21·6	50	7·6	25·9	7·4	25·6	8·7
27	4·9	38·4	5·2	38·2	21·2	5 55	7·7	25·8	7·5	25·5	8·6
1 30	−4·6	−38·1	−4·9	−37·9	−20·9	6 00	+7·8	−25·7	+7·6	−25·4	−8·5
35	4·2	37·7	4·4	37·4	20·5	10	8·0	25·5	7·8	25·2	8·3
40	3·7	37·2	4·0	37·0	20·0	20	8·2	25·3	8·0	25·0	8·1
45	3·2	36·7	3·5	36·5	19·5	30	8·4	25·1	8·1	24·9	7·9
50	2·8	36·3	3·1	36·1	19·1	40	8·6	24·9	8·3	24·7	7·7
1 55	2·4	35·9	2·6	35·6	18·7	6 50	8·7	24·8	8·5	24·5	7·6
2 00	−2·0	−35·5	−2·2	−35·2	−18·3	7 00	+8·9	−24·6	+8·6	−24·4	−7·4
05	1·6	35·1	1·8	34·8	17·9	10	9·1	24·4	8·8	24·2	7·2
10	1·2	34·7	1·5	34·5	17·5	20	9·2	24·3	9·0	24·0	7·1
15	0·9	34·4	1·1	34·1	17·2	30	9·3	24·2	9·1	23·9	7·0
20	0·5	34·0	0·8	33·8	16·8	40	9·5	24·0	9·2	23·8	6·8
25	−0·2	33·7	0·4	33·4	16·5	7 50	9·6	23·9	9·4	23·6	6·7
2 30	+0·2	−33·3	−0·1	−33·1	−16·1	8 00	+9·7	−23·8	+9·5	−23·5	−6·6
35	0·5	33·0	+0·2	32·8	15·8	10	9·9	23·6	9·6	23·4	6·4
40	0·8	32·7	0·5	32·5	15·5	20	10·0	23·5	9·7	23·3	6·3
45	1·1	32·4	0·8	32·2	15·2	30	10·1	23·4	9·8	23·2	6·2
50	1·4	32·1	1·1	31·9	14·9	40	10·2	23·3	10·0	23·0	6·1
2 55	1·6	31·9	1·4	31·6	14·7	8 50	10·3	23·2	10·1	22·9	6·0
3 00	+1·9	−31·6	+1·7	−31·3	−14·4	9 00	+10·4	−23·1	+10·2	−22·8	−5·9
05	2·2	31·3	1·9	31·1	14·1	10	10·5	23·0	10·3	22·7	5·8
10	2·4	31·1	2·1	30·9	13·9	20	10·6	22·9	10·4	22·6	5·7
15	2·6	30·9	2·4	30·6	13·7	30	10·7	22·8	10·5	22·5	5·6
20	2·9	30·6	2·6	30·4	13·4	40	10·8	22·7	10·6	22·4	5·5
25	3·1	30·4	2·9	30·1	13·2	9 50	10·9	22·6	10·6	22·4	5·4
3 30	+3·3	−30·2	+3·1	−29·9	−13·0	10 00	+11·0	−22·5	+10·7	−22·3	−5·3

Additional corrections for temperature and pressure are given on the following page.
For bubble sextant observations ignore dip and use the star corrections for Sun, planets, and stars.

A4 ALTITUDE CORRECTION TABLES—ADDITIONAL CORRECTIONS
ADDITIONAL REFRACTION CORRECTIONS FOR NON-STANDARD CONDITIONS

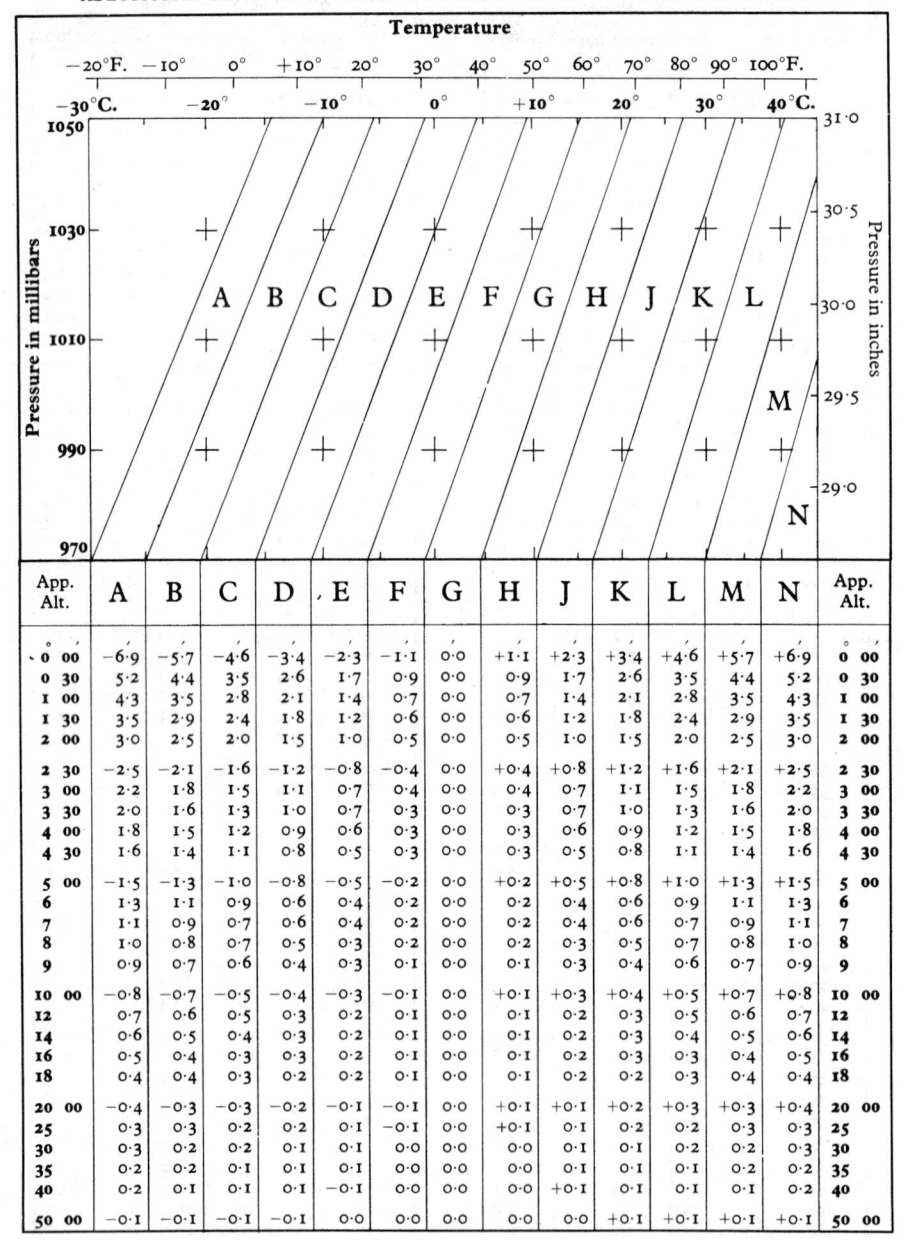

App. Alt.	A	B	C	D	E	F	G	H	J	K	L	M	N	App. Alt.
° ′	′	′	′	′	′	′	′	′	′	′	′	′	′	° ′
0 00	−6·9	−5·7	−4·6	−3·4	−2·3	−1·1	0·0	+1·1	+2·3	+3·4	+4·6	+5·7	+6·9	0 00
0 30	5·2	4·4	3·5	2·6	1·7	0·9	0·0	0·9	1·7	2·6	3·5	4·4	5·2	0 30
1 00	4·3	3·5	2·8	2·1	1·4	0·7	0·0	0·7	1·4	2·1	2·8	3·5	4·3	1 00
1 30	3·5	2·9	2·4	1·8	1·2	0·6	0·0	0·6	1·2	1·8	2·4	2·9	3·5	1 30
2 00	3·0	2·5	2·0	1·5	1·0	0·5	0·0	0·5	1·0	1·5	2·0	2·5	3·0	2 00
2 30	−2·5	−2·1	−1·6	−1·2	−0·8	−0·4	0·0	+0·4	+0·8	+1·2	+1·6	+2·1	+2·5	2 30
3 00	2·2	1·8	1·5	1·1	0·7	0·4	0·0	0·4	0·7	1·1	1·5	1·8	2·2	3 00
3 30	2·0	1·6	1·3	1·0	0·7	0·3	0·0	0·3	0·7	1·0	1·3	1·6	2·0	3 30
4 00	1·8	1·5	1·2	0·9	0·6	0·3	0·0	0·3	0·6	0·9	1·2	1·5	1·8	4 00
4 30	1·6	1·4	1·1	0·8	0·5	0·3	0·0	0·3	0·5	0·8	1·1	1·4	1·6	4 30
5 00	−1·5	−1·3	−1·0	−0·8	−0·5	−0·2	0·0	+0·2	+0·5	+0·8	+1·0	+1·3	+1·5	5 00
6	1·3	1·1	0·9	0·6	0·4	0·2	0·0	0·2	0·4	0·6	0·9	1·1	1·3	6
7	1·1	0·9	0·7	0·6	0·4	0·2	0·0	0·2	0·4	0·6	0·7	0·9	1·1	7
8	1·0	0·8	0·7	0·5	0·3	0·2	0·0	0·2	0·3	0·5	0·7	0·8	1·0	8
9	0·9	0·7	0·6	0·4	0·3	0·1	0·0	0·1	0·3	0·4	0·6	0·7	0·9	9
10 00	−0·8	−0·7	−0·5	−0·4	−0·3	−0·1	0·0	+0·1	+0·3	+0·4	+0·5	+0·7	+0·8	10 00
12	0·7	0·6	0·5	0·3	0·2	0·1	0·0	0·1	0·2	0·3	0·5	0·6	0·7	12
14	0·6	0·5	0·4	0·3	0·2	0·1	0·0	0·1	0·2	0·3	0·4	0·5	0·6	14
16	0·5	0·4	0·3	0·3	0·2	0·1	0·0	0·1	0·2	0·3	0·3	0·4	0·5	16
18	0·4	0·4	0·3	0·2	0·2	0·1	0·0	0·1	0·2	0·2	0·3	0·4	0·4	18
20 00	−0·4	−0·3	−0·3	−0·2	−0·1	−0·1	0·0	+0·1	+0·1	+0·2	+0·3	+0·3	+0·4	20 00
25	0·3	0·3	0·2	0·2	0·1	−0·1	0·0	+0·1	0·1	0·2	0·2	0·3	0·3	25
30	0·3	0·2	0·2	0·1	0·1	0·0	0·0	0·0	0·1	0·1	0·2	0·2	0·3	30
35	0·2	0·2	0·1	0·1	0·1	0·0	0·0	0·0	0·1	0·1	0·1	0·2	0·2	35
40	0·2	0·1	0·1	0·1	−0·1	0·0	0·0	0·0	+0·1	0·1	0·1	0·1	0·2	40
50 00	−0·1	−0·1	−0·1	−0·1	0·0	0·0	0·0	0·0	0·0	+0·1	+0·1	+0·1	+0·1	50 00

The graph is entered with arguments temperature and pressure to find a zone letter; using as arguments this zone letter and apparent altitude (sextant altitude corrected for dip), a correction is taken from the table. This correction is to be applied to the sextant altitude in addition to the corrections for standard conditions (for the Sun, stars and planets from page A2 and for the Moon from pages xxxiv and xxxv).

THE ELEMENTS OF NAVIGATION AND NAUTICAL ASTRONOMY

JUNE 12, 13, 14 (THURS., FRI., SAT.)

G.M.T.	ARIES G.H.A.	VENUS −3.5 G.H.A.	Dec.	MARS +0.5 G.H.A.	Dec.	JUPITER −1.8 G.H.A.	Dec.	SATURN +0.2 G.H.A.	Dec.	STARS Name	S.H.A.	Dec.
d h	° ′	° ′	° ′	° ′	° ′	° ′	° ′	° ′	° ′		° ′	° ′
12 00	259 48.8	219 05.8	N13 35.1	255 51.9	S 0 37.5	59 07.1	S 7 14.2	357 49.3	S21 48.8	Acamar	315 50.1	S 40 28.2
01	274 51.3	234 05.4	35.9	270 52.7	36.8	74 09.6	14.1	12 52.0	48.7	Achernar	335 57.9	S 57 26.6
02	289 53.8	249 05.0	36.8	285 53.5	36.1	89 12.1	14.1	27 54.6	48.7	Acrux	173 55.3	S 62 52.5
03	304 56.2	264 04.7	·· 37.7	300 54.4	·· 35.4	104 14.6	·· 14.1	42 57.3	·· 48.7	Adhara	255 45.4	S 28 55.1
04	319 58.7	279 04.3	38.6	315 55.2	34.8	119 17.1	14.1	58 00.0	48.7	Aldebaran	291 37.2	N 16 25.5
05	335 01.2	294 03.9	39.5	330 56.0	34.1	134 19.7	14.1	73 02.6	48.7			
06	350 03.6	309 03.5	N13 40.4	345 56.8	S 0 33.4	149 22.2	S 7 14.1	88 05.3	S21 48.7	Alioth	166 56.8	N 56 11.3
07	5 06.1	324 03.2	41.3	0 57.7	32.7	164 24.7	14.1	103 08.0	48.7	Alkaid	153 31.2	N 49 31.4
T 08	20 08.6	339 02.8	42.2	15 58.5	32.0	179 27.2	14.1	118 10.6	48.7	Al Na'ir	28 35.5	S 47 09.5
H 09	35 11.0	354 02.4	·· 43.1	30 59.3	·· 31.4	194 29.7	·· 14.1	133 13.3	·· 48.7	Alnilam	276 28.7	S 1 13.8
U 10	50 13.5	9 02.0	44.0	46 00.1	30.7	209 32.2	14.1	148 15.9	48.7	Alphard	218 36.9	S 8 28.9
R 11	65 15.9	24 01.6	44.9	61 01.0	30.0	224 34.8	14.1	163 18.6	48.7			
S 12	80 18.4	39 01.3	N13 45.8	76 01.8	S 0 29.3	239 37.3	S 7 14.0	178 21.3	S21 48.7	Alphecca	126 45.7	N 26 51.4
D 13	95 20.9	54 00.9	46.7	91 02.6	28.6	254 39.8	14.0	193 23.9	48.7	Alpheratz	358 26.3	N 28 51.5
A 14	110 23.3	69 00.5	47.6	106 03.4	27.9	269 42.3	14.0	208 26.6	48.7	Altair	62 48.3	N 8 45.5
Y 15	125 25.8	84 00.1	·· 48.5	121 04.3	·· 27.3	284 44.8	·· 14.0	223 29.3	·· 48.6	Ankaa	353 56.6	S 42 31.6
16	140 28.3	98 59.7	49.4	136 05.1	26.6	299 47.3	14.0	238 31.9	48.6	Antares	113 16.6	S 26 20.4
17	155 30.7	113 59.4	50.2	151 05.9	25.9	314 49.8	14.0	253 34.6	48.6			
18	170 33.2	128 59.0	N13 51.1	166 06.8	S 0 25.2	329 52.4	S 7 14.0	268 37.2	S21 48.6	Arcturus	146 33.2	N 19 24.0
19	185 35.7	143 58.6	52.0	181 07.6	24.5	344 54.9	14.0	283 39.9	48.6	Atria	108 55.0	S 68 57.2
20	200 38.1	158 58.2	52.9	196 08.4	23.9	359 57.4	14.0	298 42.6	48.6	Avior	234 35.4	S 59 22.9
21	215 40.6	173 57.8	·· 53.8	211 09.2	·· 23.2	14 59.9	·· 14.0	313 45.2	·· 48.6	Bellatrix	279 16.7	N 6 18.7
22	230 43.1	188 57.4	54.7	226 10.1	22.5	30 02.4	14.0	328 47.9	48.6	Betelgeuse	271 46.4	N 7 23.9
23	245 45.5	203 57.1	55.6	241 10.9	21.8	45 04.9	14.0	343 50.6	48.6			
13 00	260 48.0	218 56.7	N13 56.5	256 11.7	S 0 21.2	60 07.4	S 7 14.0	358 53.2	S21 48.6	Canopus	264 15.0	S 52 40.6
01	275 50.4	233 56.3	57.4	271 12.5	20.5	75 09.9	14.0	13 55.9	48.6	Capella	281 36.0	N 45 57.3
02	290 52.9	248 55.9	58.2	286 13.4	19.8	90 12.4	13.9	28 58.5	48.6	Deneb	49 59.3	N 45 07.8
03	305 55.4	263 55.5	13 59.1	301 14.2	·· 19.1	105 15.0	·· 13.9	44 01.2	·· 48.6	Denebola	183 15.8	N 14 48.3
04	320 57.8	278 55.1	14 00.0	316 15.0	18.4	120 17.5	13.9	59 03.9	48.6	Diphda	349 37.5	S 18 12.8
05	336 00.3	293 54.7	00.9	331 15.9	17.8	135 20.0	13.9	74 06.5	48.5			
06	351 02.8	308 54.3	N14 01.8	346 16.7	S 0 17.1	150 22.5	S 7 13.9	89 09.2	S21 48.5	Dubhe	194 42.5	N 61 58.7
07	6 05.2	323 54.0	02.7	1 17.5	16.4	165 25.0	13.9	104 11.9	48.5	Elnath	279 05.3	N 28 34.3
08	21 07.7	338 53.6	03.5	16 18.3	15.7	180 27.5	13.9	119 14.5	48.5	Eltanin	91 04.8	N 51 29.7
F 09	36 10.2	353 53.2	·· 04.4	31 19.2	·· 15.0	195 30.0	·· 13.9	134 17.2	·· 48.5	Enif	34 27.5	N 9 41.1
R 10	51 12.6	8 52.8	05.3	46 20.0	14.4	210 32.5	13.9	149 19.8	48.5	Fomalhaut	16 09.5	S 29 50.3
I 11	66 15.1	23 52.4	06.2	61 20.8	13.7	225 35.0	13.9	164 22.5	48.5			
D 12	81 17.6	38 52.0	N14 07.1	76 21.7	S 0 13.0	240 37.5	S 7 13.9	179 25.2	S21 48.5	Gacrux	172 46.8	S 56 53.2
A 13	96 20.0	53 51.6	08.0	91 22.5	12.3	255 40.0	13.9	194 27.8	48.5	Gienah	176 34.8	S 17 18.8
Y 14	111 22.5	68 51.2	08.8	106 23.3	11.6	270 42.5	13.9	209 30.5	48.5	Hadar	149 46.1	S 60 10.6
15	126 24.9	83 50.8	·· 09.7	121 24.1	·· 11.0	285 45.1	·· 13.9	224 33.2	·· 48.5	Hamal	328 47.7	N 23 15.9
16	141 27.4	98 50.4	10.6	136 25.0	10.3	300 47.6	13.8	239 35.8	48.5	Kaus Aust.	84 38.2	S 34 24.2
17	156 29.9	113 50.0	11.5	151 25.8	09.6	315 50.1	13.9	254 38.5	48.5			
18	171 32.3	128 49.6	N14 12.4	166 26.6	S 0 08.9	330 52.6	S 7 13.9	269 41.1	S21 48.5	Kochab	137 17.4	N 74 19.8
19	186 34.8	143 49.3	13.2	181 27.5	08.3	345 55.1	13.9	284 43.8	48.4	Markab	14 19.5	N 14 58.9
20	201 37.3	158 48.9	14.1	196 28.3	07.6	0 57.6	13.9	299 46.5	48.4	Menkar	314 58.6	N 3 55.6
21	216 39.7	173 48.5	·· 15.0	211 29.1	·· 06.9	16 00.1	·· 13.8	314 49.1	·· 48.4	Menkent	148 56.1	S 36 10.1
22	231 42.2	188 48.1	15.9	226 30.0	06.2	31 02.6	13.8	329 51.8	48.4	Miaplacidus	221 48.9	S 69 33.2
23	246 44.7	203 47.7	16.8	241 30.8	05.6	46 05.1	13.8	344 54.5	48.4			
14 00	261 47.1	218 47.3	N14 17.6	256 31.6	S 0 04.9	61 07.6	S 7 13.8	359 57.1	S21 48.4	Mirfak	309 40.0	N 49 42.7
01	276 49.6	233 46.9	18.5	271 32.4	04.2	76 10.1	13.8	14 59.8	48.4	Nunki	76 49.2	S 26 20.8
02	291 52.0	248 46.5	19.4	286 33.3	03.5	91 12.6	13.8	30 02.4	48.4	Peacock	54 23.9	S 56 51.9
03	306 54.5	263 46.1	·· 20.3	301 34.1	·· 02.8	106 15.1	·· 13.8	45 05.1	·· 48.4	Pollux	244 18.6	N 28 07.6
04	321 57.0	278 45.7	21.1	316 34.9	02.2	121 17.6	13.8	60 07.8	48.4	Procyon	245 43.3	N 5 19.8
05	336 59.4	293 45.3	22.0	331 35.8	01.5	136 20.1	13.8	75 10.4	48.4			
06	352 01.9	308 44.9	N14 22.9	346 36.6	S 0 00.8	151 22.6	S 7 13.8	90 13.1	S21 48.4	Rasalhague	96 44.5	N 12 35.4
07	7 04.4	323 44.5	23.7	1 37.4	S 0 00.1	166 25.1	13.8	105 15.8	48.4	Regulus	208 27.6	N 12 10.2
S 08	22 06.8	338 44.1	24.6	16 38.3	N 0 00.5	181 27.6	13.8	120 18.4	48.3	Rigel	281 52.1	S 8 15.0
A 09	37 09.3	353 43.7	·· 25.5	31 39.1	·· 01.2	196 30.1	·· 13.8	135 21.1	·· 48.3	Rigil Kent.	140 47.7	S 60 40.1
T 10	52 11.8	8 43.3	26.4	46 39.9	01.9	211 32.6	13.8	150 23.7	48.3	Sabik	102 59.6	S 15 40.4
U 11	67 14.2	23 42.9	27.2	61 40.8	02.6	226 35.1	13.8	165 26.4	48.3			
R 12	82 16.7	38 42.5	N14 28.1	76 41.6	N 0 03.3	241 37.6	S 7 13.8	180 29.1	S21 48.3	Schedar	350 27.8	N 56 18.3
D 13	97 19.2	53 42.1	29.0	91 42.4	03.9	256 40.1	13.8	195 31.7	48.3	Shaula	97 17.6	S 37 04.4
A 14	112 21.6	68 41.7	29.8	106 43.3	04.6	271 42.6	13.8	210 34.4	48.3	Sirius	259 10.5	S 16 39.7
Y 15	127 24.1	83 41.3	·· 30.7	121 44.1	·· 05.3	286 45.1	·· 13.8	225 37.1	·· 48.3	Spica	159 14.7	S 10 56.8
16	142 26.5	98 40.9	31.6	136 44.9	06.0	301 47.6	13.8	240 39.7	48.3	Suhail	223 23.1	S 43 16.2
17	157 29.0	113 40.5	32.4	151 45.8	06.6	316 50.1	13.8	255 42.4	48.3			
18	172 31.5	128 40.0	N14 33.3	166 46.6	N 0 07.3	331 52.6	S 7 13.8	270 45.1	S21 48.3	Vega	81 06.5	N 38 44.7
19	187 33.9	143 39.6	34.2	181 47.4	08.0	346 55.1	13.8	285 47.7	48.3	Zuben'ubi	137 50.9	S 15 52.2
20	202 36.4	158 39.2	35.0	196 48.3	08.7	1 57.6	13.8	300 50.4	48.3		S.H.A.	Mer. Pass.
21	217 38.9	173 38.8	·· 35.9	211 49.1	·· 09.3	17 00.1	·· 13.8	315 53.0	·· 48.2		° ′	h m
22	232 41.3	188 38.4	36.8	226 49.9	10.0	32 02.6	13.8	330 55.7	48.2	Venus	318 08.7	9 25
23	247 43.8	203 38.0	37.6	241 50.7	10.7	47 05.1	13.8	345 58.4	48.2	Mars	355 23.7	6 55
Mer. Pass. 6 35.7		v −0.4	d 0.9	v 0.8	d 0.7	v 2.5	d 0.0	v 2.7	d 0.0	Jupiter	159 19.4	19 56
										Saturn	98 05.2	0 04

400 THE ELEMENTS OF NAVIGATION AND NAUTICAL ASTRONOMY

JUNE 12, 13, 14 (THURS., FRI., SAT.)

G.M.T.	SUN G.H.A.	Dec.	MOON G.H.A.	v	Dec.	d	H.P.	Lat.	Twilight Naut.	Civil	Sun-rise	Moonrise 12	13	14	15
d h	° ′	° ′	° ′	′	° ′	′	′	°	h m	h m	h m	h m	h m	h m	h m
12 00	180 07·1	N23 06·2	241 41·3	14·8	N 8 40·2	8·8	54·3	N 72	□	□	□	[00 03 / 23 54]	23 44	23 26	□
01	195 06·9	06·4	256 15·1	14·7	8 49·0	8·7	54·3	N 70	□	□	□	00 15	00 14	00 15	00 18
02	210 06·8	06·6	270 48·8	14·7	8 57·7	8·6	54·3	68	□	□	□	00 26	00 30	00 38	00 50
03	225 06·7	·· 06·8	285 22·5	14·6	9 06·3	8·7	54·3	66	□	□	□	00 34	00 43	00 56	01 10
04	240 06·5	06·9	299 56·1	14·7	9 15·0	8·6	54·3	64	////	////	01 36	00 42	00 54	01 10	01 33
05	255 06·4	07·1	314 29·8	14·6	9 23·6	8·5	54·3	62	////	////	02 12	00 48	01 03	01 23	01 48
06	270 06·3	N23 07·3	329 03·4	14·5	N 9 32·1	8·6	54·3	60	////	00 57	02 37	00 53	01 11	01 33	02 01
07	285 06·2	07·4	343 36·9	14·6	9 40·7	8·4	54·3	N 58	////	01 43	02 57	00 58	01 18	01 42	02 12
T 08	300 06·0	07·6	358 10·5	14·4	9 49·1	8·5	54·3	56	////	02 12	03 14	01 03	01 25	01 51	02 22
H 09	315 05·9	·· 07·8	12 43·9	14·5	9 57·6	8·4	54·3	54	00 52	02 34	03 28	01 06	01 30	01 58	02 31
U 10	330 05·8	07·9	27 17·4	14·4	10 06·0	8·4	54·3	52	01 35	02 51	03 40	01 10	01 35	02 04	02 38
R 11	345 05·6	08·1	41 50·8	14·4	10 14·4	8·3	54·3	50	02 02	03 06	03 50	01 13	01 40	02 10	02 45
S 12	0 05·5	N23 08·2	56 24·2	14·4	N10 22·7	8·3	54·3	45	02 46	03 35	04 13	01 20	01 50	02 23	03 00
D 13	15 05·4	08·4	70 57·6	14·3	10 31·0	8·3	54·4	N 40	03 16	03 58	04 30	01 26	01 58	02 33	03 13
A 14	30 05·3	·· 08·6	85 30·9	14·2	10 39·3	8·2	54·4	35	03 39	04 16	04 45	01 31	02 05	02 42	03 23
Y 15	45 05·1	·· 08·7	100 04·1	14·3	10 47·5	8·1	54·4	30	03 58	04 31	04 58	01 36	02 11	02 50	03 32
16	60 05·0	08·9	114 37·4	14·2	10 55·6	8·2	54·4	20	04 26	04 55	05 20	01 44	02 22	03 04	03 48
17	75 04·9	09·0	129 10·6	14·1	11 03·8	8·0	54·4	N 10	04 49	05 16	05 39	01 51	02 32	03 16	04 02
18	90 04·7	N23 09·2	143 43·7	14·1	N11 11·8	8·1	54·4	0	05 08	05 34	05 56	01 57	02 41	03 27	04 15
19	105 04·6	09·4	158 16·8	14·1	11 19·9	8·0	54·4	S 10	05 24	05 51	06 13	02 04	02 50	03 39	04 28
20	120 04·5	09·5	172 49·9	14·0	11 27·9	7·9	54·4	20	05 40	06 08	06 32	02 11	03 00	03 51	04 43
21	135 04·4	·· 09·7	187 22·9	14·0	11 35·8	7·9	54·4	30	05 57	06 26	06 53	02 19	03 11	04 05	04 59
22	150 04·2	09·8	201 55·9	14·0	11 43·7	7·9	54·4	35	06 05	06 37	07 05	02 23	03 18	04 13	05 08
23	165 04·1	10·0	216 28·9	13·9	11 51·6	7·8	54·5	40	06 15	06 48	07 19	02 29	03 25	04 22	05 19
								45	06 25	07 02	07 36	02 35	03 34	04 33	05 32
13 00	180 04·0	N23 10·1	231 01·8	13·9	N11 59·4	7·7	54·5	S 50	06 36	07 18	07 56	02 42	03 44	04 46	05 47
01	195 03·8	10·3	245 34·7	13·8	12 07·1	7·8	54·5	52	06 41	07 25	08 06	02 46	03 49	04 52	05 55
02	210 03·7	10·4	260 07·5	13·8	12 14·9	7·6	54·5	54	06 47	07 33	08 17	02 50	03 55	04 59	06 03
03	225 03·6	·· 10·6	274 40·3	13·7	12 22·5	7·6	54·5	56	06 53	07 42	08 29	02 54	04 01	05 07	06 12
04	240 03·4	10·7	289 13·0	13·7	12 30·1	7·6	54·5	58	07 00	07 52	08 44	02 59	04 07	05 16	06 22
05	255 03·3	10·9	303 45·7	13·6	12 37·7	7·5	54·5	S 60	07 07	08 04	09 01	03 04	04 15	05 26	06 34
06	270 03·2	N23 11·0	318 18·3	13·6	N12 45·2	7·5	54·5								
07	285 03·1	11·2	332 50·9	13·6	12 52·7	7·4	54·5	Lat.	Sun-set	Twilight Civil	Naut.	Moonset 12	13	14	15
08	300 02·9	11·3	347 23·5	13·5	13 00·1	7·3	54·6								
F 09	315 02·8	·· 11·5	1 56·0	13·4	13 07·4	7·3	54·6	°	h m	h m	h m	h m	h m	h m	h m
R 10	330 02·7	11·6	16 28·4	13·4	13 14·7	7·3	54·6	N 72	□	□	□	16 49	18 34	20 29	□
I 11	345 02·5	11·8	31 00·8	13·4	13 22·0	7·2	54·6	N 70	□	□	□	16 30	18 04	19 39	21 12
D 12	0 02·4	N23 11·9	45 33·2	13·3	N13 29·2	7·1	54·6	68	□	□	□	16 15	17 41	19 07	20 28
A 13	15 02·3	12·1	60 05·5	13·3	13 36·3	7·1	54·6	66	□	□	□	16 03	17 24	18 44	19 58
Y 14	30 02·1	12·2	74 37·8	13·2	13 43·4	7·0	54·6	64	22 25	////	////	15 54	17 10	18 25	19 36
15	45 02·0	·· 12·3	89 10·0	13·2	13 50·4	7·0	54·7	62	21 49	////	////	15 45	16 59	18 10	19 18
16	60 01·9	12·5	103 42·2	13·1	13 57·4	6·9	54·7	60	21 23	23 04	////	15 38	16 49	17 58	19 04
17	75 01·7	12·6	118 14·3	13·1	14 04·3	6·9	54·7								
18	90 01·6	N23 12·8	132 46·4	13·0	N14 11·2	6·8	54·7	N 58	21 03	22 17	////	15 31	16 40	17 47	18 51
19	105 01·5	12·9	147 18·4	13·0	14 18·0	6·7	54·7	56	20 46	21 48	////	15 26	16 32	17 38	18 40
20	120 01·4	13·1	161 50·4	12·9	14 24·7	6·7	54·7	54	20 32	21 26	23 09	15 21	16 26	17 29	18 31
21	135 01·2	·· 13·2	176 22·3	12·9	14 31·4	6·6	54·7	52	20 20	21 09	22 26	15 16	16 20	17 22	18 22
22	150 01·1	13·3	190 54·2	12·8	14 38·0	6·5	54·8	50	20 10	20 54	21 59	15 12	16 14	17 15	18 15
23	165 01·0	13·5	205 26·0	12·8	14 44·5	6·5	54·8	45	19 47	20 25	21 14	15 03	16 02	17 01	17 59
14 00	180 00·8	N23 13·6	219 57·8	12·7	N14 51·0	6·4	54·8	N 40	19 29	20 02	20 44	14 56	15 53	16 49	17 46
01	195 00·7	13·7	234 29·5	12·6	14 57·4	6·4	54·8	35	19 15	19 44	20 21	14 50	15 44	16 39	17 34
02	210 00·6	13·9	249 01·1	12·7	15 03·8	6·3	54·8	30	19 02	19 29	20 02	14 44	15 37	16 31	17 25
03	225 00·4	·· 14·0	263 32·8	12·5	15 10·1	6·2	54·8	20	18 40	19 04	19 33	14 34	15 24	16 15	17 08
04	240 00·3	14·1	278 04·3	12·5	15 16·3	6·2	54·9	N 10	18 21	18 44	19 11	14 26	15 12	16 02	16 53
05	255 00·2	14·3	292 35·8	12·5	15 22·5	6·1	54·9	0	18 03	18 26	18 52	14 18	15 03	15 50	16 39
06	270 00·0	N23 14·4	307 07·3	12·4	N15 28·6	6·0	54·9	S 10	17 46	18 09	18 35	14 10	14 53	15 38	16 25
07	284 59·9	14·5	321 38·7	12·4	15 34·6	6·0	54·9	20	17 28	17 52	18 19	14 02	14 42	15 24	16 11
S 08	299 59·8	14·7	336 10·1	12·3	15 40·6	5·8	54·9	30	17 07	17 33	18 03	13 52	14 29	15 10	15 54
A 09	314 59·6	·· 14·8	350 41·4	12·2	15 46·4	5·9	54·9	35	16 55	17 23	17 54	13 47	14 22	15 01	15 44
T 10	329 59·5	14·9	5 12·6	12·2	15 52·3	5·7	55·0	40	16 41	17 11	17 45	13 41	14 14	14 51	15 33
U 11	344 59·4	15·0	19 43·8	12·2	15 58·0	5·7	55·0	45	16 24	16 58	17 35	13 33	14 04	14 39	15 20
R 12	359 59·2	N23 15·2	34 15·0	12·1	N16 03·7	5·6	55·0	S 50	16 04	16 42	17 23	13 25	13 53	14 25	15 04
D 13	14 59·1	15·3	48 46·1	12·0	16 09·3	5·5	55·0	52	15 54	16 35	17 18	13 21	13 48	14 19	14 56
A 14	29 59·0	15·4	63 17·1	12·0	16 14·8	5·5	55·0	54	15 43	16 26	17 12	13 16	13 42	14 12	14 48
Y 15	44 58·9	·· 15·5	77 48·1	12·0	16 20·3	5·4	55·0	56	15 30	16 17	17 06	13 12	13 35	14 04	14 39
16	59 58·7	15·7	92 19·1	11·9	16 25·7	5·3	55·1	58	15 16	16 07	17 00	13 06	13 28	13 55	14 28
17	74 58·6	15·8	106 50·0	11·8	16 31·0	5·2	55·1	S 60	14 59	15 56	16 52	13 00	13 20	13 44	14 16
18	89 58·5	N23 15·9	121 20·8	11·8	N16 36·2	5·2	55·1			SUN			MOON		
19	104 58·3	16·0	135 51·6	11·7	16 41·4	5·1	55·1	Day	Eqn. of Time		Mer.	Mer. Pass.		Age	Phase
20	119 58·2	16·2	150 22·3	11·7	16 46·5	5·0	55·1		00ʰ	12ʰ	Pass.	Upper	Lower		
21	134 58·1	·· 16·3	164 53·0	11·6	16 51·5	4·9	55·2		m s	m s	h m	h m	h m	d	
22	149 57·9	16·4	179 23·6	11·6	16 56·4	4·9	55·2	12	00 28	00 22	12 00	08 07	20 30	25	
23	164 57·8	16·5	193 54·2	11·5	17 01·3	4·8	55·2	13	00 16	00 10	12 00	08 52	21 15	26	●
	S.D. 15·8	d 0·1	S.D. 14·8		14·9		15·0	14	00 04	[00 03]	12 00	09 38	22 02	27	

JUNE 15, 16, 17 (SUN., MON., TUES.)

G.M.T.	ARIES G.H.A.	VENUS −3.5 G.H.A. / Dec.	MARS +0.4 G.H.A. / Dec.	JUPITER −1.8 G.H.A. / Dec.	SATURN +0.2 G.H.A. / Dec.	STARS Name / S.H.A. / Dec.
d h	° ′	° ′ / ° ′	° ′ / ° ′	° ′ / ° ′	° ′ / ° ′	° ′ / ° ′
15 00	262 46.3	218 37.6 N14 38.5	256 51.6 N 0 11.4	62 07.6 S 7 13.8	1 01.0 S21 48.2	Acamar 315 50.1 S 40 28.1
01	277 48.7	233 37.2 · · 39.4	271 52.4 · · 12.0	77 10.1 · · 13.8	16 03.7 · · 48.2	Achernar 335 57.9 S 57 26.6
02	292 51.2	248 36.8 · · 40.2	286 53.2 · · 12.7	92 12.6 · · 13.8	31 06.4 · · 48.2	Acrux 173 55.3 S 62 52.5
03	307 53.7	263 36.4 · · 41.1	301 54.1 · · 13.4	107 15.1 · · 13.8	46 09.0 · · 48.2	Adhara 255 45.4 S 28 55.1
04	322 56.1	278 36.0 · · 42.0	316 54.9 · · 14.1	122 17.6 · · 13.8	61 11.7 · · 48.2	Aldebaran 291 37.2 N 16 25.5
05	337 58.6	293 35.6 · · 42.8	331 55.8 · · 14.7	137 20.1 · · 13.8	76 14.3 · · 48.2	
06	353 01.0	308 35.1 N14 43.7	346 56.6 N 0 15.4	152 22.6 S 7 13.8	91 17.0 S21 48.2	Alioth 166 56.8 N 56 11.3
07	8 03.5	323 34.7 · · 44.5	1 57.4 · · 16.1	167 25.1 · · 13.8	106 19.7 · · 48.2	Alkaid 153 31.2 N 49 31.4
08	23 06.0	338 34.3 · · 45.4	16 58.3 · · 16.8	182 27.6 · · 13.8	121 22.3 · · 48.2	Al Na'ir 28 35.4 S 47 09.4
S 09	38 08.4	353 33.9 · · 46.3	31 59.1 · · 17.4	197 30.1 · · 13.8	136 25.0 · · 48.2	Alnilam 276 28.7 S 1 13.8
U 10	53 10.9	8 33.5 · · 47.1	46 59.9 · · 18.1	212 32.6 · · 13.8	151 27.7 · · 48.1	Alphard 218 36.9 S 8 28.8
N 11	68 13.4	23 33.1 · · 48.0	62 00.8 · · 18.8	227 35.1 · · 13.8	166 30.3 · · 48.1	
D 12	83 15.8	38 32.7 N14 48.8	77 01.6 N 0 19.5	242 37.6 S 7 13.8	181 33.0 S21 48.1	Alphecca 126 45.7 N 26 51.4
A 13	98 18.3	53 32.3 · · 49.7	92 02.4 · · 20.1	257 40.1 · · 13.8	196 35.6 · · 48.1	Alpheratz 358 26.3 N 28 51.5
Y 14	113 20.8	68 31.8 · · 50.5	107 03.3 · · 20.8	272 42.6 · · 13.8	211 38.3 · · 48.1	Altair 62 48.3 N 8 45.6
15	128 23.2	83 31.4 · · 51.4	122 04.1 · · 21.5	287 45.1 · · 13.8	226 41.0 · · 48.1	Ankaa 353 56.6 S 42 31.6
16	143 25.7	98 31.0 · · 52.3	137 04.9 · · 22.2	302 47.5 · · 13.8	241 43.6 · · 48.1	Antares 113 16.6 S 26 20.4
17	158 28.1	113 30.6 · · 53.1	152 05.8 · · 22.8	317 50.0 · · 13.8	256 46.3 · · 48.1	
18	173 30.6	128 30.2 N14 54.0	167 06.6 N 0 23.5	332 52.5 S 7 13.8	271 49.0 S21 48.1	Arcturus 146 33.2 N 19 24.0
19	188 33.1	143 29.8 · · 54.8	182 07.4 · · 24.2	347 55.0 · · 13.8	286 51.6 · · 48.1	Atria 108 55.0 S 68 57.2
20	203 35.5	158 29.3 · · 55.7	197 08.3 · · 24.9	2 57.5 · · 13.8	301 54.3 · · 48.1	Avior 234 35.4 S 59 22.9
21	218 38.0	173 28.9 · · 56.5	212 09.1 · · 25.5	18 00.0 · · 13.8	316 56.9 · · 48.1	Bellatrix 279 16.7 N 6 18.7
22	233 40.5	188 28.5 · · 57.4	227 09.9 · · 26.2	33 02.5 · · 13.8	331 59.6 · · 48.1	Betelgeuse 271 46.4 N 7 23.9
23	248 42.9	203 28.1 · · 58.2	242 10.8 · · 26.9	48 05.0 · · 13.8	347 02.3 · · 48.1	
16 00	263 45.4	218 27.7 N14 59.1	257 11.6 N 0 27.6	63 07.5 S 7 13.8	2 04.9 S21 48.0	Canopus 264 15.0 S 52 40.5
01	278 47.9	233 27.2 · · 14 59.9	272 12.5 · · 28.2	78 10.0 · · 13.8	17 07.6 · · 48.0	Capella 281 36.0 N 45 57.3
02	293 50.3	248 26.8 · · 15 00.8	287 13.3 · · 28.9	93 12.5 · · 13.8	32 10.3 · · 48.0	Deneb 49 59.3 N 45 07.8
03	308 52.8	263 26.4 · · 01.6	302 14.1 · · 29.6	108 14.9 · · 13.8	47 12.9 · · 48.0	Denebola 183 15.8 N 14 48.3
04	323 55.3	278 26.0 · · 02.5	317 15.0 · · 30.3	123 17.4 · · 13.8	62 15.6 · · 48.0	Diphda 349 37.5 S 18 12.7
05	338 57.7	293 25.6 · · 03.3	332 15.8 · · 30.9	138 19.9 · · 13.8	77 18.2 · · 48.0	
06	354 00.2	308 25.1 N15 04.2	347 16.6 N 0 31.6	153 22.4 S 7 13.8	92 20.9 S21 48.0	Dubhe 194 42.5 N 61 58.7
07	9 02.6	323 24.7 · · 05.0	2 17.5 · · 32.3	168 24.9 · · 13.8	107 23.6 · · 48.0	Elnath 279 05.3 N 28 34.3
08	24 05.1	338 24.3 · · 05.9	17 18.3 · · 33.0	183 27.4 · · 13.8	122 26.2 · · 48.0	Eltanin 91 04.8 N 51 29.7
M 09	39 07.6	353 23.9 · · 06.7	32 19.1 · · 33.6	198 29.9 · · 13.8	137 28.9 · · 48.0	Enif 34 27.5 N 9 41.1
O 10	54 10.0	8 23.4 · · 07.6	47 20.0 · · 34.3	213 32.4 · · 13.8	152 31.6 · · 48.0	Fomalhaut 16 09.5 S 29 50.3
N 11	69 12.5	23 23.0 · · 08.4	62 20.8 · · 35.0	228 34.8 · · 13.8	167 34.2 · · 48.0	
D 12	84 15.0	38 22.6 N15 09.3	77 21.7 N 0 35.7	243 37.3 S 7 13.8	182 36.9 S21 48.0	Gacrux 172 46.8 S 56 53.2
A 13	99 17.4	53 22.2 · · 10.1	92 22.5 · · 36.3	258 39.8 · · 13.8	197 39.5 · · 48.0	Gienah 176 34.8 S 17 18.8
Y 14	114 19.9	68 21.7 · · 11.0	107 23.3 · · 37.0	273 42.3 · · 13.8	212 42.2 · · 47.9	Hadar 149 46.1 S 60 10.6
15	129 22.4	83 21.3 · · 11.8	122 24.2 · · 37.7	288 44.8 · · 13.8	227 44.9 · · 47.9	Hamal 328 47.7 N 23 15.9
16	144 24.8	98 20.9 · · 12.6	137 25.0 · · 38.3	303 47.3 · · 13.8	242 47.5 · · 47.9	Kaus Aust. 84 38.2 S 34 24.2
17	159 27.3	113 20.5 · · 13.5	152 25.8 · · 39.0	318 49.8 · · 13.8	257 50.2 · · 47.9	
18	174 29.8	128 20.0 N15 14.3	167 26.7 N 0 39.7	333 52.2 S 7 13.8	272 52.9 S21 47.9	Kochab 137 17.4 N 74 19.8
19	189 32.2	143 19.6 · · 15.2	182 27.5 · · 40.4	348 54.7 · · 13.8	287 55.5 · · 47.9	Markab 14 19.5 N 14 58.9
20	204 34.7	158 19.2 · · 16.0	197 28.4 · · 41.0	3 57.2 · · 13.8	302 58.2 · · 47.9	Menkar 314 58.6 N 3 55.6
21	219 37.1	173 18.7 · · 16.9	212 29.2 · · 41.7	18 59.7 · · 13.8	318 00.8 · · 47.9	Menkent 148 56.1 S 36 10.1
22	234 39.6	188 18.3 · · 17.7	227 30.0 · · 42.4	34 02.2 · · 13.8	333 03.5 · · 47.9	Miaplacidus 221 49.0 S 69 33.1
23	249 42.1	203 17.9 · · 18.5	242 30.9 · · 43.1	49 04.7 · · 13.9	348 06.2 · · 47.9	
17 00	264 44.5	218 17.5 N15 19.4	257 31.7 N 0 43.7	64 07.2 S 7 13.9	3 08.8 S21 47.9	Mirfak 309 40.0 N 49 42.7
01	279 47.0	233 17.0 · · 20.2	272 32.6 · · 44.4	79 09.6 · · 13.9	18 11.5 · · 47.9	Nunki 76 49.2 S 26 20.8
02	294 49.5	248 16.6 · · 21.1	287 33.4 · · 45.1	94 12.1 · · 13.9	33 14.2 · · 47.9	Peacock 54 23.9 S 56 51.9
03	309 51.9	263 16.2 · · 21.9	302 34.2 · · 45.7	109 14.6 · · 13.9	48 16.8 · · 47.9	Pollux 244 18.6 N 28 07.6
04	324 54.4	278 15.7 · · 22.7	317 35.1 · · 46.4	124 17.1 · · 13.9	63 19.5 · · 47.8	Procyon 245 43.3 N 5 19.8
05	339 56.9	293 15.3 · · 23.6	332 35.9 · · 47.1	139 19.6 · · 13.9	78 22.1 · · 47.8	
06	354 59.3	308 14.9 N15 24.4	347 36.8 N 0 47.8	154 22.0 S 7 13.9	93 24.8 S21 47.8	Rasalhague 96 44.5 N 12 35.5
07	10 01.8	323 14.4 · · 25.2	2 37.6 · · 48.4	169 24.5 · · 13.9	108 27.6 · · 47.8	Regulus 208 27.6 N 12 10.2
08	25 04.3	338 14.0 · · 26.1	17 38.4 · · 49.1	184 27.0 · · 13.9	123 30.1 · · 47.8	Rigel 281 52.1 S 8 15.0
T 09	40 06.7	353 13.6 · · 26.9	32 39.3 · · 49.8	199 29.5 · · 13.9	138 32.8 · · 47.8	Rigil Kent. 140 47.7 S 60 40.1
U 10	55 09.2	8 13.1 · · 27.7	47 40.1 · · 50.4	214 32.0 · · 13.9	153 35.5 · · 47.8	Sabik 102 59.6 S 15 40.4
E 11	70 11.6	23 12.7 · · 28.6	62 41.0 · · 51.1	229 34.4 · · 13.9	168 38.1 · · 47.8	
S 12	85 14.1	38 12.2 N15 29.4	77 41.8 N 0 51.8	244 36.9 S 7 13.9	183 40.8 S21 47.8	Schedar 350 27.8 N 56 18.3
D 13	100 16.6	53 11.8 · · 30.2	92 42.6 · · 52.5	259 39.4 · · 13.9	198 43.4 · · 47.8	Shaula 97 17.6 S 37 04.4
A 14	115 19.0	68 11.4 · · 31.1	107 43.5 · · 53.1	274 41.9 · · 13.9	213 46.1 · · 47.8	Sirius 259 10.5 S 16 39.7
Y 15	130 21.5	83 10.9 · · 31.9	122 44.3 · · 53.8	289 44.4 · · 13.9	228 48.8 · · 47.8	Spica 159 14.7 S 10 56.8
16	145 24.0	98 10.5 · · 32.7	137 45.2 · · 54.5	304 46.8 · · 13.9	243 51.4 · · 47.8	Suhail 223 23.1 S 43 16.2
17	160 26.4	113 10.1 · · 33.6	152 46.0 · · 55.1	319 49.3 · · 14.0	258 54.1 · · 47.8	
18	175 28.9	128 09.6 N15 34.4	167 46.8 N 0 55.8	334 51.8 S 7 14.0	273 56.8 S21 47.7	Vega 81 06.5 N 38 44.7
19	190 31.4	143 09.2 · · 35.2	182 47.7 · · 56.5	349 54.3 · · 14.0	288 59.4 · · 47.7	Zuben'ubi 137 50.9 S 15 52.2
20	205 33.8	158 08.7 · · 36.0	197 48.5 · · 57.2	4 56.8 · · 14.0	304 02.1 · · 47.7	S.H.A. / Mer. Pass.
21	220 36.3	173 08.3 · · 36.9	212 49.4 · · 57.8	19 59.2 · · 14.0	319 04.7 · · 47.7	° ′ / h m
22	235 38.8	188 07.8 · · 37.7	227 50.2 · · 58.5	35 01.7 · · 14.0	334 07.4 · · 47.7	Venus 314 42.3 9 26
23	250 41.2	203 07.4 · · 38.5	242 51.0 · · 59.2	50 04.2 · · 14.0	349 10.1 · · 47.7	Mars 353 26.2 6 51
	h m					Jupiter 159 22.1 19 44
Mer. Pass.	6 23.9	v −0.4 d 0.8	v 0.8 d 0.7	v 2.5 d 0.0	v 2.7 d 0.0	Saturn 98 19.5 23 47

JUNE 15, 16, 17 (SUN., MON., TUES.)

G.M.T.	SUN G.H.A.	Dec.	MOON G.H.A.	v	Dec.	d	H.P.	Lat.	Twilight Naut.	Civil	Sun-rise	Moonrise 15	16	17	18
d h	° ′	° ′	° ′	′	° ′	′	′	°	h m	h m	h m	h m	h m	h m	h m
15 00	179 57·7	N23 16·7	208 24·7	11·5	N17 06·1	4·7	55·2	N 72	▭	▭	▭	▭	▭	▭	▭
01	194 57·5	16·8	222 55·2	11·4	17 10·8	4·6	55·2	70	▭	▭	▭	00 18	00 28	00 59	02 11
02	209 57·4	16·9	237 25·6	11·4	17 15·4	4·5	55·3	68	▭	▭	▭	00 50	01 13	01 53	02 57
03	224 57·3 ··	17·0	251 56·0	11·3	17 19·9	4·5	55·3	66	▭	▭	▭	01 14	01 42	02 26	03 28
04	239 57·1	17·1	266 26·3	11·2	17 24·4	4·4	55·3	64	////	////	01 33	01 33	02 05	02 50	03 50
05	254 57·0	17·2	280 56·5	11·3	17 28·8	4·3	55·3	62	////	////	02 10	01 48	02 23	03 09	04 09
								60	////	00 52	02 36	02 01	02 37	03 25	04 23
06	269 56·9	N23 17·4	295 26·8	11·1	N17 33·1	4·2	55·3	N 58	////	01 41	02 56	02 12	02 50	03 38	04 36
07	284 56·7	17·5	309 56·9	11·1	17 37·3	4·1	55·4	56	////	02 11	03 13	02 22	03 01	03 49	04 47
08	299 56·6	17·6	324 27·0	11·1	17 41·4	4·0	55·4	54	00 48	02 33	03 27	02 31	03 11	03 59	04 57
S 09	314 56·5 ··	17·7	338 57·1	11·0	17 45·4	4·0	55·4	52	01 33	02 51	03 39	02 38	03 19	04 08	05 05
U 10	329 56·3	17·8	353 27·1	11·0	17 49·4	3·8	55·4	50	02 00	03 06	03 50	02 45	03 27	04 16	05 13
N 11	344 56·2	17·9	7 57·1	10·9	17 53·2	3·8	55·4	45	02 46	03 35	04 13	03 00	03 43	04 33	05 29
D 12	359 56·1	N23 18·0	22 27·0	10·9	N17 57·0	3·7	55·5	N 40	03 16	03 58	04 30	03 13	03 57	04 47	05 43
A 13	14 55·9	18·1	36 56·9	10·8	18 00·7	3·6	55·5	35	03 39	04 16	04 45	03 23	04 09	04 59	05 54
Y 14	29 55·8	18·2	51 26·7	10·7	18 04·3	3·5	55·5	30	03 58	04 31	04 58	03 32	04 19	05 10	06 04
15	44 55·7 ··	18·4	65 56·4	10·8	18 07·8	3·5	55·5	20	04 27	04 56	05 20	03 48	04 36	05 27	06 22
16	59 55·5	18·5	80 26·2	10·6	18 11·3	3·3	55·5	N 10	04 49	05 16	05 39	04 02	04 51	05 43	06 36
17	74 55·4	18·6	94 55·8	10·6	18 14·6	3·2	55·6	0	05 08	05 34	05 57	04 15	05 05	05 58	06 51
18	89 55·3	N23 18·7	109 25·4	10·6	N18 17·8	3·2	55·6	S 10	05 25	05 51	06 14	04 28	05 20	06 12	07 05
19	104 55·1	18·8	123 55·0	10·6	18 21·0	3·1	55·6	20	05 41	06 09	06 33	04 43	05 35	06 28	07 19
20	119 55·0	18·9	138 24·6	10·4	18 24·1	2·9	55·6	30	05 58	06 27	06 54	04 59	05 53	06 46	07 37
21	134 54·9 ··	19·0	152 54·0	10·5	18 27·0	2·9	55·7	35	06 06	06 38	07 06	05 08	06 03	06 56	07 47
22	149 54·7	19·1	167 23·5	10·4	18 29·9	2·8	55·7	40	06 16	06 50	07 20	05 19	06 14	07 08	07 58
23	164 54·6	19·2	181 52·9	10·3	18 32·7	2·7	55·7	45	06 26	07 03	07 37	05 32	06 28	07 22	08 11
16 00	179 54·5	N23 19·3	196 22·2	10·3	N18 35·4	2·6	55·7	S 50	06 38	07 19	07 58	05 47	06 45	07 39	08 28
01	194 54·3	19·4	210 51·5	10·3	18 38·0	2·5	55·7	52	06 43	07 27	08 08	05 55	06 53	07 47	08 35
02	209 54·2	19·5	225 20·8	10·2	18 40·5	2·4	55·8	54	06 48	07 35	08 19	06 03	07 02	07 56	08 44
03	224 54·1 ··	19·6	239 50·0	10·2	18 42·9	2·3	55·8	56	06 55	07 44	08 31	06 12	07 12	08 07	08 53
04	239 53·9	19·7	254 19·2	10·1	18 45·2	2·3	55·8	58	07 01	07 54	08 46	06 22	07 23	08 18	09 04
05	254 53·8	19·8	268 48·3	10·1	18 47·5	2·1	55·8	S 60	07 09	08 06	09 03	06 34	07 37	08 32	09 17
06	269 53·7	N23 19·9	283 17·4	10·0	N18 49·6	2·0	55·9								
07	284 53·5	20·0	297 46·4	10·0	18 51·6	2·0	55·9		Sun-	Twilight		Moonset			
08	299 53·4	20·1	312 15·4	10·0	18 53·6	1·8	55·9	Lat.	set	Civil	Naut.	15	16	17	18
M 09	314 53·2 ··	20·2	326 44·4	9·9	18 55·4	1·7	55·9								
O 10	329 53·1	20·3	341 13·3	9·9	18 57·1	1·7	55·9	°	h m	h m	h m	h m	h m	h m	h m
N 11	344 53·0	20·4	355 42·2	9·9	18 58·8	1·5	56·0	N 72	▭	▭	▭	▭	▭	▭	▭
D 12	359 52·8	N23 20·5	10 11·1	9·8	N19 00·3	1·4	56·0	N 70	▭	▭	▭	21 12	22 28	23 07	23 20
A 13	14 52·7	20·6	24 39·9	9·7	19 01·7	1·4	56·0	68	▭	▭	▭	20 28	21 34	22 20	22 46
Y 14	29 52·6	20·7	39 08·6	9·8	19 03·1	1·2	56·0	66	▭	▭	▭	19 58	21 01	21 49	22 21
15	44 52·4 ··	20·7	53 37·4	9·7	19 04·3	1·1	56·1	64	22 29	////	////	19 36	20 37	21 26	22 02
16	59 52·3	20·8	68 06·1	9·7	19 05·4	1·0	56·1	62	21 51	////	////	19 18	20 18	21 08	21 46
17	74 52·2	20·9	82 34·8	9·6	19 06·4	1·0	56·1	60	21 25	23 10	////	19 04	20 03	20 53	21 33
18	89 52·0	N23 21·0	97 03·4	9·6	N19 07·4	0·8	56·1	N 58	21 05	22 20	////	18 51	19 49	20 40	21 22
19	104 51·9	21·1	111 32·0	9·5	19 08·2	0·7	56·1	56	20 48	21 51	////	18 40	19 38	20 29	21 12
20	119 51·8	21·2	126 00·5	9·6	19 08·9	0·6	56·2	54	20 34	21 29	23 14	18 31	19 28	20 19	21 03
21	134 51·6 ··	21·3	140 29·1	9·5	19 09·6	0·5	56·2	52	20 22	21 11	22 29	18 22	19 19	20 10	20 55
22	149 51·5	21·4	154 57·6	9·4	19 10·1	0·4	56·2	50	20 11	20 56	22 01	18 15	19 11	20 02	20 48
23	164 51·4	21·4	169 26·0	9·5	19 10·5	0·3	56·2	45	19 49	20 26	21 16	17 59	18 54	19 46	20 33
17 00	179 51·2	N23 21·5	183 54·5	9·4	N19 10·8	0·2	56·3	N 40	19 31	20 03	20 45	17 46	18 40	19 32	20 20
01	194 51·1	21·6	198 22·9	9·4	19 11·0	0·1	56·3	35	19 16	19 45	20 22	17 34	18 28	19 20	20 09
02	209 51·0	21·7	212 51·3	9·3	19 11·1	0·0	56·3	30	19 03	19 30	20 03	17 25	18 18	19 10	20 00
03	224 50·8 ··	21·8	227 19·6	9·3	19 11·1	0·1	56·3	20	18 41	19 05	19 34	17 08	18 00	18 53	19 44
04	239 50·7	21·9	241 47·9	9·3	19 11·0	0·2	56·3	N 10	18 22	18 45	19 12	16 53	17 45	18 37	19 29
05	254 50·5	21·9	256 16·2	9·3	19 10·8	0·3	56·4	0	18 04	18 27	18 53	16 39	17 30	18 23	19 16
06	269 50·4	N23 22·0	270 44·5	9·3	N19 10·5	0·4	56·4	S 10	17 47	18 10	18 36	16 25	17 16	18 08	19 03
07	284 50·3	22·1	285 12·8	9·2	19 10·1	0·6	56·4	20	17 28	17 52	18 20	16 11	17 00	17 53	18 48
08	299 50·1	22·2	299 41·0	9·2	19 09·5	0·6	56·4	30	17 07	17 34	18 05	15 54	16 42	17 35	18 32
T 09	314 50·0 ··	22·3	314 09·2	9·2	19 08·9	0·7	56·5	35	16 55	17 23	17 55	15 44	16 32	17 25	18 22
U 10	329 49·9	22·3	328 37·4	9·1	19 08·2	0·9	56·5	40	16 41	17 11	17 45	15 33	16 20	17 13	18 11
E 11	344 49·7	22·4	343 05·5	9·2	19 07·3	0·9	56·5	45	16 24	16 58	17 35	15 20	16 06	16 59	17 58
S 12	359 49·6	N23 22·5	357 33·7	9·1	N19 06·4	1·1	56·5	S 50	16 03	16 42	17 23	15 04	15 49	16 42	17 42
D 13	14 49·5	22·6	12 01·8	9·1	19 05·3	1·1	56·6	52	15 53	16 34	17 18	14 56	15 41	16 34	17 35
A 14	29 49·3	22·6	26 29·9	9·0	19 04·2	1·3	56·6	54	15 42	16 26	17 12	14 48	15 32	16 25	17 26
Y 15	44 49·2 ··	22·7	40 57·9	9·1	19 02·9	1·4	56·6	56	15 30	16 17	17 06	14 39	15 22	16 15	17 17
16	59 49·1	22·8	55 26·0	9·0	19 01·5	1·5	56·6	58	15 15	16 07	17 00	14 28	15 10	16 03	17 07
17	74 48·9	22·9	69 54·0	9·1	19 00·0	1·6	56·6	S 60	14 58	15 55	16 52	14 16	14 57	15 50	16 54
18	89 48·8	N23 22·9	84 22·1	9·0	N18 58·4	1·7	56·7		SUN			MOON			
19	104 48·6	23·0	98 50·1	9·0	18 56·7	1·8	56·7	Day	Eqn. of Time		Mer. Pass.	Mer. Pass.		Age	Phase
20	119 48·5	23·1	113 18·1	8·9	18 54·9	1·9	56·7		00h	12h		Upper	Lower		
21	134 48·4 ··	23·2	127 46·0	9·0	18 53·0	2·0	56·7		m s	m s	h m	h m	h m	d	
22	149 48·2	23·2	142 14·0	9·0	18 51·0	2·1	56·8	15	00 09	00 15	12 00	10 27	22 52	28	●
23	164 48·1	23·3	156 42·0	8·9	18 49·2	2·3	56·8	16	00 22	00 28	12 00	11 18	23 44	29	
	S.D. 15·8	d 0·1	S.D. 15·1		15·3		15·4	17	00 35	00 41	12 01	12 10	24 37	00	

THE ELEMENTS OF NAVIGATION AND NAUTICAL ASTRONOMY 403

SEPTEMBER 22, 23, 24 (MON., TUES., WED.)

G.M.T. d h	ARIES G.H.A. ° '	VENUS −3.4 G.H.A. ° ' / Dec. ° '	MARS −1.0 G.H.A. ° ' / Dec. ° '	JUPITER −1.3 G.H.A. ° ' / Dec. ° '	SATURN +0.7 G.H.A. ° ' / Dec. ° '	STARS Name / S.H.A. ° ' / Dec. ° '
22 00	0 21.0	193 12.3 N 7 01.0	301 54.4 N18 19.2	149 25.8 S11 29.9	101 23.1 S21 54.0	Acamar 315 49.3 S 40 28.0
01	15 23.5	208 11.9 6 59.8	316 56.3 19.4	164 27.8 30.1	116 25.4 54.0	Achernar 335 56.9 S 57 26.6
02	30 25.9	223 11.5 58.6	331 58.1 19.5	179 29.8 30.2	131 27.8 54.0	Acrux 173 55.9 S 62 52.2
03	45 28.4	238 11.0 ·· 57.4	347 00.0 ·· 19.7	194 31.8 ·· 30.4	146 30.1 ·· 54.0	Adhara 255 44.9 S 28 54.8
04	60 30.8	253 10.6 56.3	2 01.8 19.9	209 33.8 30.6	161 32.5 54.0	Aldebaran 291 36.5 N 16 25.6
05	75 33.3	268 10.2 55.1	17 03.7 20.1	224 35.8 30.8	176 34.8 54.1	
06	90 35.8	283 09.8 N 6 53.9	32 05.6 N18 20.3	239 37.8 S11 30.9	191 37.1 S21 54.1	Alioth 166 57.4 N 56 11.1
07	105 38.2	298 09.3 52.7	47 07.4 20.4	254 39.8 31.1	206 39.5 54.1	Alkaid 153 31.8 N 49 31.3
08	120 40.7	313 08.9 51.6	62 09.3 20.6	269 41.7 31.3	221 41.8 54.1	Al Na'ir 28 35.0 S 47 09.6
M 09	135 43.2	328 08.5 ·· 50.4	77 11.1 ·· 20.8	284 43.7 ·· 31.4	236 44.2 ·· 54.1	Alnilam 276 28.1 S 1 13.6
O 10	150 45.6	343 08.1 49.2	92 13.0 21.0	299 45.7 31.6	251 46.5 54.1	Alphard 218 36.7 S 8 28.7
N 11	165 48.1	358 07.6 48.0	107 14.9 21.2	314 47.7 31.8	266 48.8 54.2	
D 12	180 50.6	13 07.2 N 6 46.8	122 16.7 N18 21.3	329 49.7 S11 32.0	281 51.2 S21 54.2	Alphecca 126 46.1 N 26 51.5
A 13	195 53.0	28 06.8 45.7	137 18.6 21.5	344 51.7 32.1	296 53.5 54.2	Alpheratz 358 25.7 N 28 51.9
Y 14	210 55.5	43 06.4 44.5	152 20.5 21.7	359 53.7 32.3	311 55.9 54.2	Altair 62 48.2 N 8 45.8
15	225 57.9	58 05.9 ·· 43.3	167 22.3 ·· 21.9	14 55.7 ·· 32.5	326 58.2 ·· 54.2	Ankaa 353 55.9 S 42 31.6
16	241 00.4	73 05.5 42.1	182 24.2 22.0	29 57.7 32.7	342 00.5 54.2	Antares 113 16.9 S 26 20.4
17	256 02.9	88 05.1 40.9	197 26.1 22.2	44 59.7 32.8	357 02.9 54.2	
18	271 05.3	103 04.7 N 6 39.7	212 27.9 N18 22.4	60 01.7 S11 33.0	12 05.2 S21 54.3	Arcturus 146 33.6 N 19 24.0
19	286 07.8	118 04.2 38.6	227 29.8 22.6	75 03.6 33.2	27 07.5 54.3	Atria 108 55.8 S 68 57.4
20	301 10.3	133 03.8 37.4	242 31.7 22.8	90 05.6 33.4	42 09.9 54.3	Avior 234 35.2 S 59 22.4
21	316 12.7	148 03.4 ·· 36.2	257 33.6 ·· 22.9	105 07.6 ·· 33.5	57 12.2 ·· 54.3	Bellatrix 279 16.1 N 6 18.8
22	331 15.2	163 03.0 35.0	272 35.4 23.1	120 09.6 33.7	72 14.6 54.3	Betelgeuse 271 45.8 N 7 24.0
23	346 17.7	178 02.5 33.8	287 37.3 23.3	135 11.6 33.9	87 16.9 54.3	
23 00	1 20.1	193 02.1 N 6 32.6	302 39.2 N18 23.5	150 13.6 S11 34.1	102 19.2 S21 54.4	Canopus 264 14.5 S 52 40.2
01	16 22.6	208 01.7 31.5	317 41.1 23.6	165 15.6 34.2	117 21.6 54.4	Capella 281 35.2 N 45 57.2
02	31 25.1	223 01.3 30.3	332 43.0 23.8	180 17.6 34.4	132 23.9 54.4	Deneb 49 59.2 N 45 08.3
03	46 27.5	238 00.8 ·· 29.1	347 44.9 ·· 24.0	195 19.6 ·· 34.6	147 26.3 ·· 54.4	Denebola 183 15.9 N 14 48.2
04	61 30.0	253 00.4 27.9	2 46.7 24.2	210 21.6 34.8	162 28.6 54.4	Diphda 349 36.8 S 18 12.6
05	76 32.4	268 00.0 26.7	17 48.6 24.3	225 23.5 34.9	177 30.9 54.4	
06	91 34.9	282 59.6 N 6 25.5	32 50.5 N18 24.5	240 25.5 S11 35.1	192 33.3 S21 54.5	Dubhe 194 42.8 N 61 58.4
07	106 37.4	297 59.1 24.3	47 52.4 24.7	255 27.5 35.3	207 35.6 54.5	Elnath 279 04.6 N 28 34.3
08	121 39.8	312 58.7 23.1	62 54.3 24.9	270 29.5 35.4	222 37.9 54.5	Eltanin 91 05.3 N 51 30.4
T 09	136 42.3	327 58.3 ·· 22.0	77 56.2 ·· 25.0	285 31.5 ·· 35.6	237 40.3 ·· 54.5	Enif 34 27.3 N 9 41.4
U 10	151 44.8	342 57.9 20.8	92 58.1 25.2	300 33.5 35.8	252 42.6 54.5	Fomalhaut 16 09.0 S 29 50.3
E 11	166 47.2	357 57.5 19.6	108 00.0 25.4	315 35.5 36.0	267 44.9 54.5	
S 12	181 49.7	12 57.0 N 6 18.4	123 01.9 N18 25.5	330 37.5 S11 36.1	282 47.3 S21 54.6	Gacrux 172 47.3 S 56 52.9
D 13	196 52.2	27 56.6 17.2	138 03.8 25.7	345 39.4 36.3	297 49.6 54.6	Gienah 176 35.0 S 17 18.7
A 14	211 54.6	42 56.2 16.0	153 05.7 25.9	0 41.4 36.5	312 51.9 54.6	Hadar 149 46.8 S 60 10.5
Y 15	226 57.1	57 55.8 ·· 14.8	168 07.5 ·· 26.1	15 43.4 ·· 36.7	327 54.3 ·· 54.6	Hamal 328 46.9 N 23 16.1
16	241 59.5	72 55.4 13.6	183 09.4 26.2	30 45.4 36.8	342 56.6 54.6	Kaus Aust. 84 38.3 S 34 24.3
17	257 02.0	87 54.9 12.5	198 11.3 26.4	45 47.4 37.0	357 59.0 54.6	
18	272 04.5	102 54.5 N 6 11.3	213 13.3 N18 26.6	60 49.4 S11 37.2	13 01.3 S21 54.7	Kochab 137 19.2 N 74 19.7
19	287 06.9	117 54.1 10.1	228 15.2 26.8	75 51.4 37.4	28 03.6 54.7	Markab 14 19.0 N 14 59.2
20	302 09.4	132 53.7 08.9	243 17.1 26.9	90 53.4 37.5	43 06.0 54.7	Menkar 314 57.9 N 3 55.8
21	317 11.9	147 53.3 ·· 07.7	258 19.0 ·· 27.1	105 55.3 ·· 37.7	58 08.3 ·· 54.7	Menkent 148 56.4 S 36 10.0
22	332 14.3	162 52.8 06.5	273 20.9 27.3	120 57.3 37.9	73 10.6 54.7	Miaplacidus 221 49.0 S 69 32.7
23	347 16.8	177 52.4 05.3	288 22.8 27.4	135 59.3 38.1	88 13.0 54.7	
24 00	2 19.3	192 52.0 N 6 04.1	303 24.7 N18 27.6	151 01.3 S11 38.2	103 15.3 S21 54.8	Mirfak 309 39.0 N 49 42.8
01	17 21.7	207 51.6 02.9	318 26.6 27.8	166 03.3 38.4	118 17.6 54.8	Nunki 76 49.2 S 26 20.9
02	32 24.2	222 51.2 01.7	333 28.5 28.0	181 05.3 38.6	133 20.0 54.8	Peacock 54 23.7 S 56 52.2
03	47 26.7	237 50.7 6 00.5	348 30.4 ·· 28.1	196 07.3 ·· 38.8	148 22.3 ·· 54.8	Pollux 244 18.2 N 28 07.5
04	62 29.1	252 50.3 5 59.3	3 32.3 28.3	211 09.3 38.9	163 24.6 54.8	Procyon 245 42.9 N 5 19.9
05	77 31.6	267 49.9 58.1	18 34.3 28.5	226 11.2 39.1	178 27.0 54.8	
06	92 34.0	282 49.5 N 5 57.0	33 36.2 N18 28.6	241 13.2 S11 39.3	193 29.3 S21 54.9	Rasalhague 96 44.7 N 12 35.6
W 07	107 36.5	297 49.1 55.8	48 38.1 28.8	256 15.2 39.5	208 31.6 54.9	Regulus 208 27.6 N 12 10.2
E 08	122 39.0	312 48.7 54.6	63 40.0 29.0	271 17.2 39.6	223 34.0 54.9	Rigel 281 51.6 S 8 14.8
D 09	137 41.4	327 48.2 ·· 53.4	78 41.9 ·· 29.1	286 19.2 ·· 39.8	238 36.3 ·· 54.9	Rigil Kent. 140 48.4 S 60 40.0
N 10	152 43.9	342 47.8 52.2	93 43.9 29.3	301 21.2 40.0	253 38.6 54.9	Sabik 102 59.8 S 15 40.4
E 11	167 46.4	357 47.4 51.0	108 45.8 29.5	316 23.2 40.2	268 41.0 54.9	
S 12	182 48.8	12 47.0 N 5 49.8	123 47.7 N18 29.6	331 25.1 S11 40.3	283 43.3 S21 55.0	Schedar 350 26.9 N 56 18.7
D 13	197 51.3	27 46.6 48.6	138 49.6 29.8	346 27.1 40.5	298 45.6 55.0	Shaula 97 17.9 S 37 04.5
A 14	212 53.8	42 46.2 47.4	153 51.6 30.0	1 29.1 40.7	313 48.0 55.0	Sirius 259 10.1 S 16 39.4
Y 15	227 56.2	57 45.7 ·· 46.2	168 53.5 ·· 30.2	16 31.1 ·· 40.9	328 50.3 ·· 55.0	Spica 159 14.9 S 10 56.7
16	242 58.7	72 45.3 45.0	183 55.4 30.3	31 33.1 41.0	343 52.6 55.0	Suhail 223 23.0 S 43 15.8
17	258 01.2	87 44.9 43.8	198 57.3 30.5	46 35.1 41.2	358 54.9 55.0	
18	273 03.6	102 44.5 N 5 42.6	213 59.3 N18 30.7	61 37.1 S11 41.4	13 57.3 S21 55.1	Vega 81 06.8 N 38 45.1
19	288 06.1	117 44.1 41.4	229 01.2 30.8	76 39.0 41.6	28 59.6 55.1	Zuben'ubi 137 51.2 S 15 52.2
20	303 08.5	132 43.7 40.2	244 03.1 31.0	91 41.0 41.7	44 01.9 55.1	S.H.A. / Mer. Pass. ° ' / h m
21	318 11.0	147 43.2 ·· 39.0	259 05.1 ·· 31.2	106 43.0 ·· 41.9	59 04.3 ·· 55.1	Venus 191 42.0 / 11 08
22	333 13.5	162 42.8 37.8	274 07.0 31.3	121 45.0 42.1	74 06.6 55.1	Mars 301 19.1 / 3 49
23	348 15.9	177 42.4 36.6	289 09.0 31.5	136 47.0 42.3	89 08.9 55.1	Jupiter 148 53.5 / 13 57
Mer. Pass. 23 50.7		v −0.4 d 1.2	v 1.9 d 0.2	v 2.0 d 0.2	v 2.3 d 0.0	Saturn 100 59.1 / 17 08

SEPTEMBER 22, 23, 24 (MON., TUES., WED.)

G.M.T.	SUN G.H.A.	Dec.	MOON G.H.A.	v	Dec.	d	H.P.	Lat.	Twilight Naut.	Civil	Sun-rise	Moonrise 22	23	24	25
d h	° ′	° ′	° ′	′	° ′	′	′	°	h m	h m	h m	h m	h m	h m	h m
22 00	181 44.5 N	0 36.2	68 05.7 10.2		S16 33.5	5.0	56.6	N 72	03 01	04 32	05 41	17 46	17 37	17 29	17 23
01	196 44.7	35.2	82 34.9 10.2		16 28.5	5.1	56.6	N 70	03 21	04 40	05 42	17 13	17 16	17 16	17 16
02	211 44.9	34.2	97 04.1 10.3		16 23.4	5.1	56.6	68	03 36	04 47	05 43	16 49	16 59	17 05	17 10
03	226 45.2 ··	33.3	111 33.4 10.4		16 18.3	5.3	56.5	66	03 48	04 52	05 44	16 31	16 46	16 56	17 05
04	241 45.4	32.3	126 02.8 10.4		16 13.0	5.3	56.5	64	03 58	04 57	05 44	16 16	16 34	16 49	17 01
05	256 45.6	31.3	140 32.2 10.4		16 07.7	5.4	56.5	62	04 07	05 01	05 45	16 03	16 25	16 42	16 57
								60	04 14	05 04	05 45	15 52	16 17	16 37	16 54
06	271 45.8 N	0 30.3	155 01.6 10.6		S16 02.3	5.4	56.5	N 58	04 20	05 07	05 46	15 43	16 09	16 32	16 51
07	286 46.0	29.4	169 31.2 10.5		15 56.9	5.6	56.4	56	04 25	05 09	05 46	15 35	16 03	16 27	16 48
08	301 46.3	28.4	184 00.7 10.7		15 51.3	5.6	56.4	54	04 29	05 11	05 46	15 27	15 57	16 23	16 46
M 09	316 46.5 ··	27.4	198 30.4 10.7		15 45.7	5.7	56.4	52	04 33	05 13	05 47	15 21	15 52	16 19	16 44
O 10	331 46.7	26.4	213 00.1 10.8		15 40.0	5.8	56.4	50	04 37	05 15	05 47	15 15	15 47	16 16	16 42
N 11	346 46.9	25.5	227 29.9 10.8		15 34.2	5.9	56.3	45	04 44	05 18	05 48	15 02	15 37	16 09	16 38
D 12	1 47.1 N	0 24.5	241 59.7 10.9		S15 28.3	5.9	56.3	N 40	04 49	05 21	05 48	14 51	15 29	16 03	16 34
A 13	16 47.4	23.5	256 29.6 10.9		15 22.4	6.0	56.3	35	04 53	05 23	05 48	14 42	15 21	15 58	16 31
Y 14	31 47.6	22.6	270 59.5 11.0		15 16.4	6.1	56.3	30	04 57	05 25	05 48	14 34	15 15	15 53	16 28
15	46 47.8 ··	21.6	285 29.5 11.1		15 10.3	6.2	56.2	20	05 01	05 27	05 49	14 20	15 04	15 45	16 24
16	61 48.0	20.6	299 59.6 11.1		15 04.1	6.2	56.2	N 10	05 04	05 28	05 49	14 08	14 54	15 38	16 20
17	76 48.2	19.6	314 29.7 11.1		14 57.9	6.3	56.2	0	05 05	05 29	05 49	13 56	14 45	15 31	16 16
18	91 48.5 N	0 18.7	328 59.8 11.3		S14 51.6	6.4	56.2	S 10	05 04	05 28	05 49	13 45	14 36	15 25	16 12
19	106 48.7	17.7	343 30.1 11.3		14 45.2	6.4	56.1	20	05 02	05 27	05 49	13 33	14 26	15 17	16 08
20	121 48.9	16.7	358 00.4 11.3		14 38.8	6.5	56.1	30	04 57	05 25	05 49	13 19	14 15	15 09	16 03
21	136 49.1 ··	15.7	12 30.7 11.4		14 32.3	6.6	56.1	35	04 54	05 24	05 49	13 11	14 08	15 05	16 00
22	151 49.3	14.8	27 01.1 11.5		14 25.7	6.6	56.1	40	04 50	05 22	05 49	13 01	14 01	14 59	15 57
23	166 49.6	13.8	41 31.6 11.5		14 19.1	6.7	56.0	45	04 45	05 19	05 48	12 51	13 52	14 53	15 54
23 00	181 49.8 N	0 12.8	56 02.1 11.6		S14 12.4	6.8	56.0	S 50	04 38	05 16	05 48	12 37	13 41	14 46	15 49
01	196 50.0	11.8	70 32.7 11.6		14 05.6	6.8	56.0	52	04 34	05 14	05 48	12 31	13 37	14 42	15 47
02	211 50.2	10.9	85 03.3 11.7		13 58.8	6.9	56.0	54	04 31	05 12	05 47	12 25	13 31	14 39	15 45
03	226 50.4 ··	09.9	99 34.0 11.7		13 51.9	7.0	55.9	56	04 26	05 10	05 47	12 17	13 25	14 34	15 43
04	241 50.7	08.9	114 04.7 11.8		13 44.9	7.0	55.9	58	04 21	05 08	05 47	12 08	13 19	14 30	15 40
05	256 50.9	08.0	128 35.5 11.9		13 37.9	7.1	55.9	S 60	04 15	05 05	05 47	11 59	13 11	14 24	15 37

								Lat.	Sun-set	Twilight Civil	Naut.	Moonset 22	23	24	25
06	271 51.1 N	0 07.0	143 06.4 11.9		S13 30.8	7.1	55.9								
07	286 51.3	06.0	157 37.3 11.9		13 23.7	7.2	55.8	°	h m	h m	h m	h m	h m	h m	h m
T 08	301 51.5	05.0	172 08.2 12.1		13 16.5	7.3	55.8	N 72	18 02	19 10	20 39	22 43	24 32	00 32	02 14
U 09	316 51.7 ··	04.1	186 39.3 12.0		13 09.2	7.3	55.8	N 70	18 01	19 03	20 21	23 15	24 51	00 51	02 25
E 10	331 52.0	03.1	201 10.3 12.2		13 01.9	7.4	55.8	68	18 01	18 56	20 06	23 38	25 07	01 07	02 35
S 11	346 52.2	02.1	215 41.5 12.1		12 54.5	7.4	55.8	66	18 00	18 51	19 54	23 55	25 19	01 19	02 42
D 12	1 52.4 N	0 01.1	230 12.6 12.3		S12 47.1	7.5	55.7	64	17 59	18 47	19 45	24 10	00 10	01 30	02 49
A 13	16 52.6 N	0 00.2	244 43.9 12.3		12 39.6	7.5	55.7	62	17 59	18 43	19 37	24 22	00 22	01 38	02 54
Y 14	31 52.8 S	0 00.8	259 15.2 12.3		12 32.1	7.6	55.7	60	17 58	18 40	19 30	24 32	00 32	01 46	02 59
15	46 53.1 ··	01.8	273 46.5 12.4		12 24.5	7.7	55.7	N 58	17 58	18 38	19 24	24 41	00 41	01 52	03 03
16	61 53.3	02.8	288 17.9 12.4		12 16.8	7.7	55.6	56	17 58	18 35	19 19	24 49	00 49	01 58	03 07
17	76 53.5	03.7	302 49.3 12.5		12 09.1	7.7	55.6	54	17 58	18 33	19 15	24 56	00 56	02 03	03 10
18	91 53.7 S	0 04.7	317 20.8 12.6		S12 01.4	7.8	55.6	52	17 57	18 31	19 11	25 02	01 02	02 08	03 14
19	106 53.9	05.7	331 52.4 12.5		11 53.6	7.9	55.6	50	17 57	18 29	19 07	00 04	01 08	02 12	03 16
20	121 54.2	06.6	346 23.9 12.7		11 45.7	7.9	55.6	45	17 57	18 26	19 00	00 18	01 20	02 21	03 22
21	136 54.4 ··	07.6	0 55.6 12.7		11 37.8	8.0	55.5								
22	151 54.6	08.6	15 27.3 12.7		11 29.8	8.0	55.5	N 40	17 56	18 23	18 55	00 30	01 30	02 29	03 27
23	166 54.8	09.6	29 59.0 12.8		11 21.8	8.0	55.5	35	17 56	18 21	18 51	00 41	01 38	02 35	03 32
24 00	181 55.0 S	0 10.5	44 30.8 12.9		S11 13.8	8.1	55.5	30	17 56	18 20	18 48	00 49	01 46	02 41	03 35
01	196 55.2	11.5	59 02.7 12.9		11 05.7	8.2	55.5	20	17 56	18 18	18 43	01 05	01 58	02 51	03 42
02	211 55.5	12.5	73 34.6 12.9		10 57.5	8.2	55.4	N 10	17 56	18 17	18 41	01 18	02 10	02 59	03 47
03	226 55.7 ··	13.5	88 06.5 13.0		10 49.3	8.2	55.4	0	17 56	18 16	18 40	01 30	02 20	03 07	03 53
04	241 55.9	14.4	102 38.5 13.0		10 41.1	8.3	55.4	S 10	17 56	18 17	18 41	01 43	02 30	03 15	03 58
05	256 56.1	15.4	117 10.5 13.1		10 32.8	8.3	55.4	20	17 56	18 18	18 44	01 56	02 41	03 24	04 04
06	271 56.3 S	0 16.4	131 42.6 13.1		S10 24.5	8.4	55.4	30	17 56	18 20	18 48	02 11	02 54	03 33	04 10
W 07	286 56.6	17.4	146 14.7 13.2		10 16.1	8.4	55.3	35	17 57	18 22	18 52	02 19	03 01	03 39	04 14
E 08	301 56.8	18.3	160 46.9 13.2		10 07.7	8.4	55.3	40	17 57	18 24	18 56	02 29	03 09	03 45	04 18
D 09	316 57.0 ··	19.3	175 19.1 13.3		9 59.3	8.5	55.3	45	17 58	18 27	19 01	02 41	03 19	03 52	04 22
N 10	331 57.2	20.3	189 51.4 13.3		9 50.8	8.5	55.3	S 50	17 58	18 31	19 08	02 54	03 30	04 01	04 28
E 11	346 57.4	21.3	204 23.7 13.3		9 42.3	8.6	55.3	52	17 58	18 32	19 12	03 01	03 35	04 05	04 31
S 12	1 57.6 S	0 22.2	218 56.0 13.4		S 9 33.7	8.6	55.2	54	17 59	18 34	19 16	03 08	03 41	04 09	04 33
D 13	16 57.9	23.2	233 28.4 13.5		9 25.1	8.6	55.2	56	17 59	18 36	19 21	03 16	03 48	04 14	04 36
A 14	31 58.1	24.2	248 00.8 13.5		9 16.5	8.7	55.2	58	17 59	18 39	19 26	03 25	03 55	04 19	04 40
Y 15	46 58.3 ··	25.1	262 33.3 13.5		9 07.8	8.7	55.2	S 60	18 00	18 41	19 32	03 35	04 03	04 25	04 44
16	61 58.5	26.1	277 05.8 13.6		8 59.1	8.8	55.2								
17	76 58.7	27.1	291 38.4 13.6		8 50.3	8.8	55.1		SUN			MOON			
18	91 59.0 S	0 28.1	306 11.0 13.6		S 8 41.5	8.8	55.1	Day	Eqn. of Time 00h	12h	Mer. Pass.	Mer. Pass. Upper	Lower	Age	Phase
19	106 59.2	29.0	320 43.6 13.7		8 32.7	8.8	55.1		m s	m s	h m	h m	h m	d	
20	121 59.4	30.0	335 16.3 13.7		8 23.9	8.9	55.1	22	06 58	07 08	11 53	20 08	07 43	09	◯
21	136 59.6 ··	31.0	349 49.0 13.8		8 15.0	8.9	55.1	23	07 19	07 29	11 53	20 56	08 32	10	
22	151 59.8	32.0	4 21.8 13.8		8 06.1	9.0	55.1	24	07 40	07 50	11 52	21 42	09 19	11	
23	167 00.0	32.9	18 54.6 13.8		7 57.1	8.9	55.0								
	S.D. 16.0	d 1.0	S.D. 15.3		15.2		15.1								

THE ELEMENTS OF NAVIGATION AND NAUTICAL ASTRONOMY

DEC. 30, 31, 1959 JAN. 1 (TUES., WED., THURS.)

G.M.T.	ARIES G.H.A.	VENUS −3.4 G.H.A. / Dec.	MARS −0.6 G.H.A. / Dec.	JUPITER −1.4 G.H.A. / Dec.	SATURN +0.7 G.H.A. / Dec.	STARS Name	S.H.A.	Dec.
d h	° ′	° ′ ° ′	° ′ ° ′	° ′ ° ′	° ′ ° ′		° ′	° ′
30 00	97 55.7	166 37.2 S23 08.1	53 46.2 N18 43.8	226 27.9 S17 46.0	188 45.4 S22 29.1	Acamar	315 49.2	S 40 28.4
01	112 58.2	181 36.3 07.7	68 48.4 44.0	241 29.9 46.1	203 47.5 29.1	Achernar	335 57.1	S 57 27.0
02	128 00.6	196 35.3 07.3	83 50.6 44.1	256 31.9 46.2	218 49.7 29.1	Acrux	173 55.0	S 62 52.1
03	143 03.1	211 34.4 ·· 06.9	98 52.7 ·· 44.2	271 33.9 ·· 46.4	233 51.8 ·· 29.1	Adhara	255 44.3	S 28 55.1
04	158 05.6	226 33.5 06.5	113 54.9 44.3	286 35.9 46.5	248 54.0 29.1	Aldebaran	291 36.1	N 16 25.6
05	173 08.0	241 32.6 06.1	128 57.1 44.4	301 37.9 46.6	263 56.1 29.1			
06	188 10.5	256 31.6 S23 05.8	143 59.2 N18 44.5	316 39.9 S17 46.7	278 58.3 S22 29.1	Alioth	166 56.7	N 56 10.6
07	203 13.0	271 30.7 05.4	159 01.4 44.7	331 41.9 46.8	294 00.4 29.1	Alkaid	153 31.4	N 49 30.8
T 08	218 15.4	286 29.8 05.0	174 03.6 44.8	346 43.9 46.9	309 02.5 29.1	Al Na'ir	28 35.5	S 47 09.7
U 09	233 17.9	301 28.9 ·· 04.6	189 05.7 ·· 44.9	1 45.9 ·· 47.0	324 04.7 ·· 29.1	Alnilam	276 27.6	S 1 13.8
E 10	248 20.4	316 28.0 04.2	204 07.9 45.0	16 47.8 47.1	339 06.8 29.1	Alphard	218 36.1	S 8 28.9
S 11	263 22.8	331 27.0 03.8	219 10.0 45.1	31 49.8 47.3	354 09.0 29.1			
D 12	278 25.3	346 26.1 S23 03.4	234 12.2 N18 45.3	46 51.8 S17 47.4	9 11.1 S22 29.1	Alphecca	126 46.0	N 26 51.1
A 13	293 27.7	1 25.2 03.0	249 14.4 45.4	61 53.8 47.5	24 13.3 29.1	Alpheratz	358 25.9	N 28 52.0
Y 14	308 30.2	16 24.3 02.6	264 16.5 45.5	76 55.8 47.6	39 15.4 29.1	Altair	62 48.5	N 8 45.7
15	323 32.7	31 23.4 ·· 02.2	279 18.7 ·· 45.6	91 57.8 ·· 47.7	54 17.6 ·· 29.1	Ankaa	353 56.2	S 42 31.9
16	338 35.1	46 22.5 01.8	294 20.8 45.7	106 59.8 47.8	69 19.7 29.1	Antares	113 16.8	S 26 20.4
17	353 37.6	61 21.5 01.4	309 23.0 45.9	122 01.8 47.9	84 21.9 29.1			
18	8 40.1	76 20.6 S23 01.0	324 25.1 N18 46.0	137 03.8 S17 48.0	99 24.0 S22 29.1	Arcturus	146 33.3	N 19 23.6
19	23 42.5	91 19.7 00.6	339 27.3 46.1	152 05.8 48.2	114 26.2 29.1	Atria	108 55.9	S 68 57.0
20	38 45.0	106 18.8 23 00.2	354 29.4 46.2	167 07.8 48.3	129 28.3 29.1	Avior	234 34.1	S 59 22.6
21	53 47.5	121 17.9 22 59.7	9 31.6 ·· 46.3	182 09.8 ·· 48.4	144 30.5 ·· 29.1	Bellatrix	279 15.6	N 6 18.7
22	68 49.9	136 17.0 59.3	24 33.7 46.5	197 11.7 48.5	159 32.6 29.1	Betelgeuse	271 45.3	N 7 23.9
23	83 52.4	151 16.0 58.9	39 35.9 46.6	212 13.7 48.6	174 34.8 29.1			
31 00	98 54.9	166 15.1 S22 58.5	54 38.0 N18 46.7	227 15.7 S17 48.7	189 36.9 S22 29.2	Canopus	264 13.8	S 52 40.5
01	113 57.3	181 14.2 58.1	69 40.1 46.8	242 17.7 48.8	204 39.0 29.2	Capella	281 34.5	N 45 57.4
02	128 59.8	196 13.3 57.7	84 42.3 46.9	257 19.7 48.9	219 41.2 29.2	Deneb	49 59.8	N 45 08.2
03	144 02.2	211 12.4 ·· 57.3	99 44.4 ·· 47.1	272 21.7 ·· 49.1	234 43.3 ·· 29.2	Denebola	183 15.3	N 14 47.9
04	159 04.7	226 11.5 56.8	114 46.6 47.2	287 23.7 49.2	249 45.5 29.2	Diphda	349 37.0	S 18 12.8
05	174 07.2	241 10.6 56.4	129 48.7 47.3	302 25.7 49.3	264 47.6 29.2			
06	189 09.6	256 09.6 S22 56.0	144 50.8 N18 47.4	317 27.7 S17 49.4	279 49.8 S22 29.2	Dubhe	194 41.7	N 61 58.0
07	204 12.1	271 08.7 55.6	159 53.0 47.6	332 29.7 49.5	294 51.9 29.2	Elnath	279 04.0	N 28 34.4
W 08	219 14.6	286 07.8 55.2	174 55.1 47.7	347 31.7 49.6	309 54.1 29.2	Eltanin	91 05.7	N 51 29.7
E 09	234 17.0	301 06.9 ·· 54.7	189 57.2 ·· 47.8	2 33.7 ·· 49.7	324 56.2 ·· 29.2	Enif	34 27.6	N 9 41.3
D 10	249 19.5	316 06.0 54.3	204 59.4 47.9	17 35.7 49.8	339 58.4 29.2	Fomalhaut	16 09.3	S 29 50.5
N 11	264 22.0	331 05.1 53.9	220 01.5 48.1	32 37.7 50.0	355 00.5 29.2			
E 12	279 24.4	346 04.2 S22 53.5	235 03.6 N18 48.2	47 39.7 S17 50.1	10 02.7 S22 29.2	Gacrux	172 46.5	S 56 52.8
S 13	294 26.9	1 03.3 53.0	250 05.8 48.3	62 41.6 50.2	25 04.8 29.2	Gienah	176 34.4	S 17 18.8
D 14	309 29.4	16 02.3 52.6	265 07.9 48.4	77 43.6 50.3	40 07.0 29.2	Hadar	149 46.2	S 60 10.3
A 15	324 31.8	31 01.4 ·· 52.2	280 10.0 ·· 48.6	92 45.6 ·· 50.4	55 09.1 ·· 29.2	Hamal	328 46.9	N 23 16.2
Y 16	339 34.3	46 00.5 51.7	295 12.1 48.7	107 47.6 50.5	70 11.3 29.2	Kaus Aust.	84 38.5	S 34 24.2
17	354 36.7	60 59.6 51.3	310 14.3 48.8	122 49.6 50.6	85 13.4 29.2			
18	9 39.2	75 58.7 S22 50.9	325 16.4 N18 48.9	137 51.6 S17 50.7	100 15.6 S22 29.2	Kochab	137 19.1	N 74 19.2
19	24 41.7	90 57.8 50.4	340 18.5 49.1	152 53.6 50.8	115 17.7 29.2	Markab	14 19.3	N 14 59.2
20	39 44.1	105 56.9 50.0	355 20.6 49.2	167 55.6 51.0	130 19.8 29.2	Menkar	314 57.7	N 3 55.7
21	54 46.6	120 56.0 ·· 49.6	10 22.7 ·· 49.3	182 57.6 ·· 51.1	145 22.0 ·· 29.2	Menkent	148 56.0	S 36 09.9
22	69 49.1	135 55.1 49.1	25 24.9 49.5	197 59.6 51.2	160 24.1 29.2	Miaplacidus	221 47.5	S 69 32.8
23	84 51.5	150 54.2 48.7	40 27.0 49.6	213 01.6 51.3	175 26.3 29.2			
1 00	99 54.0	165 53.3 S22 48.2	55 29.1 N18 49.7	228 03.6 S17 51.4	190 28.4 S22 29.2	Mirfak	309 38.6	N 49 43.1
01	114 56.5	180 52.4 47.8	70 31.2 49.8	243 05.6 51.5	205 30.6 29.2	Nunki	76 49.4	S 26 20.8
02	129 58.9	195 51.4 47.3	85 33.3 50.0	258 07.6 51.6	220 32.7 29.2	Peacock	54 24.3	S 56 52.1
03	145 01.4	210 50.5 ·· 46.9	100 35.4 ·· 50.1	273 09.6 ·· 51.7	235 34.9 ·· 29.2	Pollux	244 17.4	N 28 07.4
04	160 03.9	225 49.6 46.5	115 37.5 50.2	288 11.6 51.8	250 37.0 29.2	Procyon	245 42.3	N 5 19.7
05	175 06.3	240 48.7 46.0	130 39.7 50.4	303 13.6 51.9	265 39.2 29.2			
06	190 08.8	255 47.8 S22 45.6	145 41.8 N18 50.5	318 15.6 S17 52.1	280 41.3 S22 29.3	Rasalhague	96 44.8	N 12 35.4
07	205 11.2	270 46.9 45.1	160 43.9 50.6	333 17.6 52.2	295 43.5 29.3	Regulus	208 26.9	N 12 09.9
T 08	220 13.7	285 46.0 44.7	175 46.0 50.7	348 19.6 52.3	310 45.6 29.3	Rigel	281 51.1	S 8 15.0
H 09	235 16.2	300 45.1 ·· 44.2	190 48.1 ·· 50.9	3 21.6 ·· 52.4	325 47.8 ·· 29.3	Rigil Kent.	140 48.0	S 60 39.7
U 10	250 18.6	315 44.2 43.7	205 50.2 51.0	18 23.6 52.5	340 49.9 29.3	Sabik	102 59.9	S 15 40.4
R 11	265 21.1	330 43.3 43.3	220 52.3 51.1	33 25.6 52.6	355 52.1 29.3			
S 12	280 23.6	345 42.4 S22 42.8	235 54.4 N18 51.3	48 27.6 S17 52.7	10 54.2 S22 29.3	Schedar	350 27.2	N 56 19.0
D 13	295 26.0	0 41.5 42.4	250 56.5 51.4	63 29.6 52.8	25 56.4 29.3	Shaula	97 18.0	S 37 04.4
A 14	310 28.5	15 40.6 41.9	265 58.6 51.5	78 31.5 52.9	40 58.5 29.3	Sirius	259 09.5	S 16 39.7
Y 15	325 31.0	30 39.7 ·· 41.5	281 00.7 ·· 51.7	93 33.5 ·· 53.0	56 00.7 ·· 29.3	Spica	159 14.5	S 10 56.8
16	340 33.4	45 38.8 41.0	296 02.8 51.8	108 35.5 53.2	71 02.8 29.3	Suhail	223 22.2	S 43 16.0
17	355 35.9	60 37.9 40.5	311 04.9 51.9	123 37.5 53.3	86 05.0 29.3			
18	10 38.3	75 37.0 S22 40.1	326 07.0 N18 52.1	138 39.5 S17 53.4	101 07.1 S22 29.3	Vega	81 07.1	N 38 44.8
19	25 40.8	90 36.1 39.6	341 09.1 52.2	153 41.5 53.5	116 09.3 29.3	Zuben'ubi	137 51.0	S 15 52.2
20	40 43.3	105 35.2 39.1	356 11.2 52.3	168 43.5 53.6	131 11.4 29.3		S.H.A.	Mer. Pass.
21	55 45.7	120 34.3 ·· 38.7	11 13.2 ·· 52.5	183 45.5 ·· 53.7	146 13.6 ·· 29.3		° ′	h m
22	70 48.2	135 33.4 38.2	26 15.3 52.6	198 47.5 53.8	161 15.7 29.3	Venus	67 20.2	12 56
23	85 50.7	150 32.5 37.7	41 17.4 52.7	213 49.5 53.9	176 17.9 29.3	Mars	315 43.1	20 19
Mer. Pass.	h m 17 21.5	v −0.9 d 0.4	v 2.1 d 0.1	v 2.0 d 0.1	v 2.1 d 0.0	Jupiter Saturn	128 20.8 90 42.0	8 50 11 20

DEC. 30, 31, 1959 JAN. 1 (TUES., WED., THURS.)

G.M.T.	SUN G.H.A.	Dec.	MOON G.H.A.	v	Dec.	d	H.P.	Lat.	Twilight Naut.	Civil	Sun-rise	Moonrise 30	31	1	2
d h	° '	° '	° '	'	° '	'	'	°	h m	h m	h m	h m	h m	h m	h m
30 00	179 27.3	S 23 12.8	312 58.5 10.8	N 9 39.0	9.3	57.6	N 72	08 25	10 46	▓	20 32	22 22	24 14	00 14
01	194 27.0	12.7	327 28.3 10.8	9 29.7	9.4	57.6	N 70	08 06	09 51	▓	20 41	22 25	24 09	00 09
02	209 26.7	12.5	341 58.1 10.9	9 20.3	9.5	57.6	68	07 50	09 18	▓	20 49	22 26	24 05	00 05
03	224 26.4	12.4	356 28.0 10.8	9 10.8	9.5	57.6	66	07 38	08 54	10 30	20 55	22 28	24 02	00 02
04	239 26.1	12.2	10 57.8 10.9	9 01.3	9.5	57.6	64	07 27	08 34	09 51	21 00	22 29	23 59	25 31
05	254 25.8	12.1	25 27.7 10.8	8 51.8	9.6	57.7	62	07 17	08 19	09 23	21 04	22 30	23 57	25 25
06	269 25.5	S 23 11.9	39 57.5 10.9	N 8 42.2	9.7	57.7	60	07 09	08 06	09 03	21 08	22 31	23 55	25 20
07	284 25.2	11.7	54 27.4 10.9	8 32.5	9.7	57.7	N 58	07 02	07 54	08 46	21 12	22 32	23 53	25 16
T 08	299 24.9	11.6	68 57.3 10.9	8 22.8	9.8	57.7	56	06 55	07 44	08 32	21 15	22 33	23 52	25 12
U 09	314 24.6	11.4	83 27.2 10.9	8 13.0	9.8	57.7	54	06 49	07 36	08 19	21 18	22 33	23 50	25 08
E 10	329 24.3	11.3	97 57.1 10.9	8 03.2	9.9	57.7	52	06 44	07 28	08 08	21 20	22 34	23 49	25 05
S 11	344 24.0	11.1	112 27.0 10.9	7 53.3	9.9	57.8	50	06 39	07 20	07 59	21 22	22 35	23 48	25 02
D 12	359 23.7	S 23 11.0	126 56.9 10.9	N 7 43.4	9.9	57.8	45	06 27	07 05	07 38	21 27	22 36	23 45	24 56
A 13	14 23.4	10.8	141 26.8 11.0	7 33.5	10.0	57.8	N 40	06 17	06 51	07 22	21 31	22 37	23 43	24 51
Y 14	29 23.1	10.6	155 56.8 10.9	7 23.5	10.1	57.8	35	06 08	06 40	07 08	21 35	22 38	23 42	24 46
15	44 22.8	10.5	170 26.7 10.9	7 13.4	10.1	57.8	30	05 59	06 29	06 55	21 38	22 39	23 40	24 42
16	59 22.5	10.3	184 56.6 11.0	7 03.3	10.1	57.9	20	05 43	06 11	06 35	21 43	22 40	23 37	24 36
17	74 22.2	10.2	199 26.6 10.9	6 53.2	10.2	57.9	N 10	05 27	05 54	06 16	21 48	22 41	23 35	24 30
								0	05 11	05 37	05 59	21 53	22 42	23 33	24 24
18	89 21.9	S 23 10.0	213 56.5 11.0	N 6 43.0	10.2	57.9	S 10	04 52	05 19	05 42	21 57	22 44	23 31	24 19
19	104 21.6	09.8	228 26.5 10.9	6 32.8	10.2	57.9	20	04 29	04 59	05 23	22 02	22 45	23 28	24 13
20	119 21.3	09.7	242 56.4 11.0	6 22.6	10.3	57.9	30	04 01	04 34	05 01	22 07	22 46	23 26	24 07
21	134 21.0	09.5	257 26.4 11.0	6 12.3	10.4	58.0	35	03 42	04 19	04 48	22 10	22 47	23 24	24 03
22	149 20.7	09.3	271 56.4 10.9	6 01.9	10.3	58.0	40	03 19	04 01	04 34	22 14	22 48	23 23	23 59
23	164 20.4	09.2	286 26.3 11.0	5 51.6	10.4	58.0	45	02 49	03 39	04 16	22 18	22 49	23 21	23 54
31 00	179 20.1	S 23 09.0	300 56.3 11.0	N 5 41.2	10.5	58.0	S 50	02 05	03 10	03 54	22 22	22 51	23 19	23 48
01	194 19.8	08.8	315 26.3 10.9	5 30.7	10.5	58.0	52	01 38	02 55	03 43	22 25	22 51	23 17	23 45
02	209 19.5	08.7	329 56.2 11.0	5 20.2	10.5	58.0	54	00 56	02 37	03 31	22 27	22 52	23 16	23 42
03	224 19.2	08.5	344 26.2 11.0	5 09.7	10.5	58.1	56	////	02 16	03 17	22 30	22 53	23 15	23 39
04	239 18.9	08.3	358 56.2 11.0	4 59.2	10.6	58.1	58	////	01 47	03 01	22 33	22 53	23 14	23 36
05	254 18.6	08.1	13 26.2 10.9	4 48.6	10.6	58.1	S 60	////	01 01	02 41	22 36	22 54	23 12	23 32
06	269 18.3	S 23 08.0	27 56.1 11.0	N 4 38.0	10.7	58.1								
W 07	284 18.0	07.8	42 26.1 10.9	4 27.3	10.6	58.1	Lat.	Sun-set	Twilight Civil	Naut.	Moonset 30	31	1	2
E 08	299 17.7	07.6	56 56.0 11.0	4 16.7	10.7	58.2								
D 09	314 17.4	07.5	71 26.0 10.9	4 06.0	10.8	58.2	°	h m	h m	h m	h m	h m	h m	h m
N 10	329 17.1	07.3	85 56.0 10.9	3 55.2	10.7	58.2	N 72	▓	13 19	15 41	11 09	11 03	10 57	10 50
E 11	344 16.8	07.1	100 25.9 11.0	3 44.5	10.8	58.2	N 70	▓	14 14	16 00	10 58	10 58	10 58	10 57
S 12	359 16.5	S 23 06.9	114 55.9 10.9	N 3 33.7	10.8	58.2	68	▓	14 47	16 15	10 49	10 54	10 59	11 03
D 13	14 16.2	06.7	129 25.8 10.9	3 22.9	10.8	58.2	66	13 35	15 12	16 28	10 41	10 51	10 59	11 08
A 14	29 15.9	06.6	143 55.7 11.0	3 12.1	10.9	58.3	64	14 15	15 31	16 39	10 35	10 48	11 00	11 12
Y 15	44 15.6	06.4	158 25.7 10.9	3 01.2	10.8	58.3	62	14 42	15 46	16 48	10 29	10 45	11 00	11 16
16	59 15.3	06.2	172 55.6 10.9	2 50.4	10.9	58.3	60	15 03	16 00	16 56	10 24	10 43	11 01	11 19
17	74 15.0	06.0	187 25.5 10.9	2 39.5	11.0	58.3								
18	89 14.7	S 23 05.8	201 55.4 10.9	N 2 28.5	10.9	58.4	N 58	15 20	16 11	17 03	10 20	10 41	11 01	11 22
19	104 14.4	05.7	216 25.3 10.9	2 17.6	10.9	58.4	56	15 34	16 21	17 10	10 16	10 39	11 02	11 25
20	119 14.1	05.5	230 55.2 10.9	2 06.7	11.0	58.4	54	15 47	16 30	17 16	10 12	10 38	11 02	11 27
21	134 13.8	05.3	245 25.1 10.8	1 55.7	11.0	58.4	52	15 58	16 39	17 22	10 09	10 36	11 02	11 29
22	149 13.5	05.1	259 54.9 10.9	1 44.7	11.0	58.4	50	16 07	16 46	17 27	10 06	10 35	11 03	11 31
23	164 13.2	04.9	274 24.8 10.9	1 33.7	11.0	58.4	45	16 28	17 01	17 39	10 00	10 32	11 03	11 35
1 00	179 12.9	S 23 04.7	288 54.6 10.9	N 1 22.7	11.0	58.4	N 40	16 44	17 15	17 49	09 55	10 29	11 04	11 39
01	194 12.6	04.5	303 24.5 10.8	1 11.7	11.1	58.5	35	16 58	17 26	17 58	09 50	10 27	11 04	11 42
02	209 12.3	04.4	317 54.3 10.8	1 00.6	11.1	58.5	30	17 10	17 37	18 06	09 46	10 25	11 05	11 44
03	224 12.0	04.2	332 24.1 10.8	0 49.6	11.1	58.5	20	17 31	17 55	18 23	09 39	10 22	11 05	11 49
04	239 11.7	04.0	346 53.9 10.7	0 38.5	11.1	58.5	N 10	17 50	18 12	18 39	09 33	10 19	11 06	11 53
05	254 11.4	03.8	1 23.6 10.8	0 27.4	11.1	58.5	0	18 07	18 29	18 55	09 27	10 16	11 06	11 57
06	269 11.1	S 23 03.6	15 53.4 10.7	N 0 16.3	11.0	58.6	S 10	18 24	18 47	19 14	09 21	10 13	11 07	12 01
07	284 10.8	03.4	30 23.1 10.7	N 0 05.3	11.1	58.6	20	18 43	19 07	19 36	09 14	10 10	11 07	12 05
T 08	299 10.5	03.2	44 52.8 10.7	S 0 05.8	11.1	58.6	30	19 04	19 32	20 05	09 07	10 07	11 08	12 09
H 09	314 10.2	03.0	59 22.5 10.7	0 16.9	11.1	58.6	35	19 17	19 47	20 23	09 02	10 05	11 08	12 12
U 10	329 09.9	02.8	73 52.2 10.7	0 28.0	11.1	58.6	40	19 32	20 05	20 46	08 58	10 02	11 08	12 15
R 11	344 09.6	02.6	88 21.9 10.6	0 39.1	11.2	58.6	45	19 50	20 27	21 16	08 52	10 00	11 09	12 19
S 12	359 09.3	S 23 02.4	102 51.5 10.6	S 0 50.3	11.1	58.6	S 50	20 12	20 56	22 00	08 45	09 56	11 09	12 23
D 13	14 09.0	02.2	117 21.1 10.6	1 01.4	11.1	58.7	52	20 22	21 11	22 26	08 42	09 55	11 10	12 25
A 14	29 08.7	02.0	131 50.7 10.6	1 12.5	11.1	58.7	54	20 34	21 28	23 07	08 38	09 53	11 10	12 27
Y 15	44 08.4	01.8	146 20.3 10.6	1 23.6	11.1	58.7	56	20 48	21 49	////	08 34	09 51	11 10	12 29
16	59 08.1	01.6	160 49.9 10.5	1 34.7	11.1	58.7	58	21 05	22 18	////	08 30	09 49	11 10	12 32
17	74 07.8	01.4	175 19.4 10.5	1 45.8	11.1	58.7	S 60	21 24	23 03	////	08 25	09 47	11 11	12 35
18	89 07.5	S 23 01.2	189 48.9 10.5	S 1 56.9	11.1	58.8		SUN			MOON			
19	104 07.3	01.0	204 18.4 10.5	2 08.0	11.1	58.8	Day	Eqn. of Time 00h	12h	Mer. Pass.	Mer. Pass. Upper	Lower	Age	Phase
20	119 07.0	00.8	218 47.9 10.4	2 19.1	11.1	58.8								
21	134 06.7	00.6	233 17.3 10.4	2 30.2	11.1	58.8		m s	m s	h m	h m	h m	d	
22	149 06.4	00.4	247 46.7 10.4	2 41.3	11.0	58.8	30	02 10	02 25	12 02	03 15	15 39	20	
23	164 06.1	00.2	262 16.1 10.3	2 52.3	11.1	58.8	31	02 39	02 53	12 03	04 04	16 29	21	
	S.D. 16.3	d 0.2	S.D. 15.7		15.9		16.0	1	03 08	03 22	12 03	04 54	17 19	22	◐

THE ELEMENTS OF NAVIGATION AND NAUTICAL ASTRONOMY 407

36ᵐ   INCREMENTS AND CORRECTIONS   **37ᵐ**

36ᵐ	SUN PLANETS	ARIES	MOON	v or d Corrⁿ	v or d Corrⁿ	v or d Corrⁿ	37ᵐ	SUN PLANETS	ARIES	MOON	v or d Corrⁿ	v or d Corrⁿ	v or d Corrⁿ
s	° ′	° ′	° ′	′ ′	′ ′	′ ′	s	° ′	° ′	° ′	′ ′	′ ′	′ ′
00	9 00·0	9 01·5	8 35·4	0·0 0·0	6·0 3·7	12·0 7·3	00	9 15·0	9 16·5	8 49·7	0·0 0·0	6·0 3·8	12·0 7·5
01	9 00·3	9 01·7	8 35·6	0·1 0·1	6·1 3·7	12·1 7·4	01	9 15·3	9 16·8	8 50·0	0·1 0·1	6·1 3·8	12·1 7·6
02	9 00·5	9 02·0	8 35·9	0·2 0·1	6·2 3·8	12·2 7·4	02	9 15·5	9 17·0	8 50·2	0·2 0·1	6·2 3·9	12·2 7·6
03	9 00·8	9 02·2	8 36·1	0·3 0·2	6·3 3·8	12·3 7·5	03	9 15·8	9 17·3	8 50·4	0·3 0·2	6·3 3·9	12·3 7·7
04	9 01·0	9 02·5	8 36·4	0·4 0·2	6·4 3·9	12·4 7·5	04	9 16·0	9 17·5	8 50·7	0·4 0·3	6·4 4·0	12·4 7·8
05	9 01·3	9 02·7	8 36·6	0·5 0·3	6·5 4·0	12·5 7·6	05	9 16·3	9 17·8	8 50·9	0·5 0·3	6·5 4·1	12·5 7·8
06	9 01·5	9 03·0	8 36·8	0·6 0·4	6·6 4·0	12·6 7·7	06	9 16·5	9 18·0	8 51·1	0·6 0·4	6·6 4·1	12·6 7·9
07	9 01·8	9 03·2	8 37·1	0·7 0·4	6·7 4·1	12·7 7·7	07	9 16·8	9 18·3	8 51·4	0·7 0·4	6·7 4·2	12·7 7·9
08	9 02·0	9 03·5	8 37·3	0·8 0·5	6·8 4·1	12·8 7·8	08	9 17·0	9 18·5	8 51·6	0·8 0·5	6·8 4·3	12·8 8·0
09	9 02·3	9 03·7	8 37·5	0·9 0·5	6·9 4·2	12·9 7·8	09	9 17·3	9 18·8	8 51·9	0·9 0·6	6·9 4·3	12·9 8·1
10	9 02·5	9 04·0	8 37·8	1·0 0·6	7·0 4·3	13·0 7·9	10	9 17·5	9 19·0	8 52·1	1·0 0·6	7·0 4·4	13·0 8·1
11	9 02·8	9 04·2	8 38·0	1·1 0·7	7·1 4·3	13·1 8·0	11	9 17·8	9 19·3	8 52·3	1·1 0·7	7·1 4·4	13·1 8·2
12	9 03·0	9 04·5	8 38·3	1·2 0·7	7·2 4·4	13·2 8·0	12	9 18·0	9 19·5	8 52·6	1·2 0·8	7·2 4·5	13·2 8·3
13	9 03·3	9 04·7	8 38·5	1·3 0·8	7·3 4·4	13·3 8·1	13	9 18·3	9 19·8	8 52·8	1·3 0·8	7·3 4·6	13·3 8·3
14	9 03·5	9 05·0	8 38·7	1·4 0·9	7·4 4·5	13·4 8·2	14	9 18·5	9 20·0	8 53·1	1·4 0·9	7·4 4·6	13·4 8·4
15	9 03·8	9 05·2	8 39·0	1·5 0·9	7·5 4·6	13·5 8·2	15	9 18·8	9 20·3	8 53·3	1·5 0·9	7·5 4·7	13·5 8·4
16	9 04·0	9 05·5	8 39·2	1·6 1·0	7·6 4·6	13·6 8·3	16	9 19·0	9 20·5	8 53·5	1·6 1·0	7·6 4·8	13·6 8·5
17	9 04·3	9 05·7	8 39·5	1·7 1·0	7·7 4·7	13·7 8·3	17	9 19·3	9 20·8	8 53·8	1·7 1·1	7·7 4·8	13·7 8·6
18	9 04·5	9 06·0	8 39·7	1·8 1·1	7·8 4·7	13·8 8·4	18	9 19·5	9 21·0	8 54·0	1·8 1·1	7·8 4·9	13·8 8·6
19	9 04·8	9 06·2	8 39·9	1·9 1·2	7·9 4·8	13·9 8·5	19	9 19·8	9 21·3	8 54·3	1·9 1·2	7·9 4·9	13·9 8·7
20	9 05·0	9 06·5	8 40·2	2·0 1·2	8·0 4·9	14·0 8·5	20	9 20·0	9 21·5	8 54·5	2·0 1·3	8·0 5·0	14·0 8·8
21	9 05·3	9 06·7	8 40·4	2·1 1·3	8·1 4·9	14·1 8·6	21	9 20·3	9 21·8	8 54·7	2·1 1·3	8·1 5·1	14·1 8·8
22	9 05·5	9 07·0	8 40·6	2·2 1·3	8·2 5·0	14·2 8·6	22	9 20·5	9 22·0	8 55·0	2·2 1·4	8·2 5·1	14·2 8·9
23	9 05·8	9 07·2	8 40·9	2·3 1·4	8·3 5·0	14·3 8·7	23	9 20·8	9 22·3	8 55·2	2·3 1·4	8·3 5·2	14·3 8·9
24	9 06·0	9 07·5	8 41·1	2·4 1·5	8·4 5·1	14·4 8·8	24	9 21·0	9 22·5	8 55·4	2·4 1·5	8·4 5·3	14·4 9·0
25	9 06·3	9 07·7	8 41·4	2·5 1·5	8·5 5·2	14·5 8·8	25	9 21·3	9 22·8	8 55·7	2·5 1·6	8·5 5·3	14·5 9·1
26	9 06·5	9 08·0	8 41·6	2·6 1·6	8·6 5·2	14·6 8·9	26	9 21·5	9 23·0	8 55·9	2·6 1·6	8·6 5·4	14·6 9·1
27	9 06·8	9 08·2	8 41·8	2·7 1·6	8·7 5·3	14·7 8·9	27	9 21·8	9 23·3	8 56·2	2·7 1·7	8·7 5·4	14·7 9·2
28	9 07·0	9 08·5	8 42·1	2·8 1·7	8·8 5·4	14·8 9·0	28	9 22·0	9 23·5	8 56·4	2·8 1·8	8·8 5·5	14·8 9·3
29	9 07·3	9 08·7	8 42·3	2·9 1·8	8·9 5·4	14·9 9·1	29	9 22·3	9 23·8	8 56·6	2·9 1·8	8·9 5·6	14·9 9·3
30	9 07·5	9 09·0	8 42·6	3·0 1·8	9·0 5·5	15·0 9·1	30	9 22·5	9 24·0	8 56·9	3·0 1·9	9·0 5·6	15·0 9·4
31	9 07·8	9 09·2	8 42·8	3·1 1·9	9·1 5·5	15·1 9·2	31	9 22·8	9 24·3	8 57·1	3·1 1·9	9·1 5·7	15·1 9·4
32	9 08·0	9 09·5	8 43·0	3·2 1·9	9·2 5·6	15·2 9·2	32	9 23·0	9 24·5	8 57·4	3·2 2·0	9·2 5·8	15·2 9·5
33	9 08·3	9 09·8	8 43·3	3·3 2·0	9·3 5·7	15·3 9·3	33	9 23·3	9 24·8	8 57·6	3·3 2·1	9·3 5·8	15·3 9·6
34	9 08·5	9 10·0	8 43·5	3·4 2·1	9·4 5·7	15·4 9·4	34	9 23·5	9 25·0	8 57·8	3·4 2·1	9·4 5·9	15·4 9·6
35	9 08·8	9 10·3	8 43·8	3·5 2·1	9·5 5·8	15·5 9·4	35	9 23·8	9 25·3	8 58·1	3·5 2·2	9·5 5·9	15·5 9·7
36	9 09·0	9 10·5	8 44·0	3·6 2·2	9·6 5·8	15·6 9·5	36	9 24·0	9 25·5	8 58·3	3·6 2·3	9·6 6·0	15·6 9·8
37	9 09·3	9 10·8	8 44·2	3·7 2·3	9·7 5·9	15·7 9·6	37	9 24·3	9 25·8	8 58·5	3·7 2·3	9·7 6·1	15·7 9·8
38	9 09·5	9 11·0	8 44·5	3·8 2·3	9·8 6·0	15·8 9·6	38	9 24·5	9 26·0	8 58·8	3·8 2·4	9·8 6·1	15·8 9·9
39	9 09·8	9 11·3	8 44·7	3·9 2·4	9·9 6·0	15·9 9·7	39	9 24·8	9 26·3	8 59·0	3·9 2·4	9·9 6·2	15·9 9·9
40	9 10·0	9 11·5	8 44·9	4·0 2·4	10·0 6·1	16·0 9·7	40	9 25·0	9 26·5	8 59·3	4·0 2·5	10·0 6·3	16·0 10·0
41	9 10·3	9 11·8	8 45·2	4·1 2·5	10·1 6·1	16·1 9·8	41	9 25·3	9 26·8	8 59·5	4·1 2·6	10·1 6·3	16·1 10·1
42	9 10·5	9 12·0	8 45·4	4·2 2·6	10·2 6·2	16·2 9·9	42	9 25·5	9 27·0	8 59·7	4·2 2·6	10·2 6·4	16·2 10·1
43	9 10·8	9 12·3	8 45·7	4·3 2·6	10·3 6·3	16·3 9·9	43	9 25·8	9 27·3	9 00·0	4·3 2·7	10·3 6·4	16·3 10·2
44	9 11·0	9 12·5	8 45·9	4·4 2·7	10·4 6·3	16·4 10·0	44	9 26·0	9 27·5	9 00·2	4·4 2·8	10·4 6·5	16·4 10·3
45	9 11·3	9 12·8	8 46·1	4·5 2·7	10·5 6·4	16·5 10·0	45	9 26·3	9 27·8	9 00·5	4·5 2·8	10·5 6·6	16·5 10·3
46	9 11·5	9 13·0	8 46·4	4·6 2·8	10·6 6·4	16·6 10·1	46	9 26·5	9 28·1	9 00·7	4·6 2·9	10·6 6·6	16·6 10·4
47	9 11·8	9 13·3	8 46·6	4·7 2·9	10·7 6·5	16·7 10·2	47	9 26·8	9 28·3	9 00·9	4·7 2·9	10·7 6·7	16·7 10·4
48	9 12·0	9 13·5	8 46·9	4·8 2·9	10·8 6·6	16·8 10·2	48	9 27·0	9 28·6	9 01·2	4·8 3·0	10·8 6·8	16·8 10·5
49	9 12·3	9 13·8	8 47·1	4·9 3·0	10·9 6·6	16·9 10·3	49	9 27·3	9 28·8	9 01·4	4·9 3·1	10·9 6·8	16·9 10·6
50	9 12·5	9 14·0	8 47·3	5·0 3·0	11·0 6·7	17·0 10·3	50	9 27·5	9 29·1	9 01·6	5·0 3·1	11·0 6·9	17·0 10·6
51	9 12·8	9 14·3	8 47·6	5·1 3·1	11·1 6·8	17·1 10·4	51	9 27·8	9 29·3	9 01·9	5·1 3·2	11·1 6·9	17·1 10·7
52	9 13·0	9 14·5	8 47·8	5·2 3·2	11·2 6·8	17·2 10·5	52	9 28·0	9 29·6	9 02·1	5·2 3·3	11·2 7·0	17·2 10·8
53	9 13·3	9 14·8	8 48·0	5·3 3·2	11·3 6·9	17·3 10·5	53	9 28·3	9 29·8	9 02·4	5·3 3·3	11·3 7·1	17·3 10·8
54	9 13·5	9 15·0	8 48·3	5·4 3·3	11·4 6·9	17·4 10·6	54	9 28·5	9 30·1	9 02·6	5·4 3·4	11·4 7·1	17·4 10·9
55	9 13·8	9 15·3	8 48·5	5·5 3·3	11·5 7·0	17·5 10·6	55	9 28·8	9 30·3	9 02·8	5·5 3·4	11·5 7·2	17·5 10·9
56	9 14·0	9 15·5	8 48·8	5·6 3·4	11·6 7·1	17·6 10·7	56	9 29·0	9 30·6	9 03·1	5·6 3·5	11·6 7·3	17·6 11·0
57	9 14·3	9 15·8	8 49·0	5·7 3·5	11·7 7·1	17·7 10·8	57	9 29·3	9 30·8	9 03·3	5·7 3·6	11·7 7·3	17·7 11·1
58	9 14·5	9 16·0	8 49·2	5·8 3·5	11·8 7·2	17·8 10·8	58	9 29·5	9 31·1	9 03·6	5·8 3·6	11·8 7·4	17·8 11·1
59	9 14·8	9 16·3	8 49·5	5·9 3·6	11·9 7·2	17·9 10·9	59	9 29·8	9 31·3	9 03·8	5·9 3·7	11·9 7·4	17·9 11·2
60	9 15·0	9 16·5	8 49·7	6·0 3·7	12·0 7·3	18·0 11·0	60	9 30·0	9 31·6	9 04·0	6·0 3·8	12·0 7·5	18·0 11·3

INCREMENTS AND CORRECTIONS

38m

38s	SUN PLANETS	ARIES	MOON	v or d Corrn	v or d Corrn	v or d Corrn
	° ′	° ′	° ′	′ ′	′ ′	′ ′
00	9 30·0	9 31·6	9 04·0	0·0 0·0	6·0 3·9	12·0 7·7
01	9 30·3	9 31·8	9 04·3	0·1 0·1	6·1 3·9	12·1 7·8
02	9 30·5	9 32·1	9 04·5	0·2 0·1	6·2 4·0	12·2 7·8
03	9 30·8	9 32·3	9 04·7	0·3 0·2	6·3 4·0	12·3 7·9
04	9 31·0	9 32·6	9 05·0	0·4 0·3	6·4 4·1	12·4 8·0
05	9 31·3	9 32·8	9 05·2	0·5 0·3	6·5 4·2	12·5 8·0
06	9 31·5	9 33·1	9 05·5	0·6 0·4	6·6 4·2	12·6 8·1
07	9 31·8	9 33·3	9 05·7	0·7 0·4	6·7 4·3	12·7 8·1
08	9 32·0	9 33·6	9 05·9	0·8 0·5	6·8 4·4	12·8 8·2
09	9 32·3	9 33·8	9 06·2	0·9 0·6	6·9 4·4	12·9 8·3
10	9 32·5	9 34·1	9 06·4	1·0 0·6	7·0 4·5	13·0 8·3
11	9 32·8	9 34·3	9 06·7	1·1 0·7	7·1 4·6	13·1 8·4
12	9 33·0	9 34·6	9 06·9	1·2 0·8	7·2 4·6	13·2 8·5
13	9 33·3	9 34·8	9 07·1	1·3 0·8	7·3 4·7	13·3 8·5
14	9 33·5	9 35·1	9 07·4	1·4 0·9	7·4 4·7	13·4 8·6
15	9 33·8	9 35·3	9 07·6	1·5 1·0	7·5 4·8	13·5 8·7
16	9 34·0	9 35·6	9 07·9	1·6 1·0	7·6 4·9	13·6 8·7
17	9 34·3	9 35·8	9 08·1	1·7 1·1	7·7 4·9	13·7 8·8
18	9 34·5	9 36·1	9 08·3	1·8 1·2	7·8 5·0	13·8 8·9
19	9 34·8	9 36·3	9 08·6	1·9 1·2	7·9 5·1	13·9 8·9
20	9 35·0	9 36·6	9 08·8	2·0 1·3	8·0 5·1	14·0 9·0
21	9 35·3	9 36·8	9 09·0	2·1 1·3	8·1 5·2	14·1 9·0
22	9 35·5	9 37·1	9 09·3	2·2 1·4	8·2 5·3	14·2 9·1
23	9 35·8	9 37·3	9 09·5	2·3 1·5	8·3 5·3	14·3 9·2
24	9 36·0	9 37·6	9 09·8	2·4 1·5	8·4 5·4	14·4 9·2
25	9 36·3	9 37·8	9 10·0	2·5 1·6	8·5 5·5	14·5 9·3
26	9 36·5	9 38·1	9 10·2	2·6 1·7	8·6 5·5	14·6 9·4
27	9 36·8	9 38·3	9 10·5	2·7 1·7	8·7 5·6	14·7 9·4
28	9 37·0	9 38·6	9 10·7	2·8 1·8	8·8 5·6	14·8 9·5
29	9 37·3	9 38·8	9 11·0	2·9 1·9	8·9 5·7	14·9 9·6
30	9 37·5	9 39·1	9 11·2	3·0 1·9	9·0 5·8	15·0 9·6
31	9 37·8	9 39·3	9 11·4	3·1 2·0	9·1 5·8	15·1 9·7
32	9 38·0	9 39·6	9 11·7	3·2 2·1	9·2 5·9	15·2 9·8
33	9 38·3	9 39·8	9 11·9	3·3 2·1	9·3 6·0	15·3 9·8
34	9 38·5	9 40·1	9 12·1	3·4 2·2	9·4 6·0	15·4 9·9
35	9 38·8	9 40·3	9 12·4	3·5 2·2	9·5 6·1	15·5 9·9
36	9 39·0	9 40·6	9 12·6	3·6 2·3	9·6 6·2	15·6 10·0
37	9 39·3	9 40·8	9 12·9	3·7 2·4	9·7 6·2	15·7 10·1
38	9 39·5	9 41·1	9 13·1	3·8 2·4	9·8 6·3	15·8 10·1
39	9 39·8	9 41·3	9 13·3	3·9 2·5	9·9 6·4	15·9 10·2
40	9 40·0	9 41·6	9 13·6	4·0 2·6	10·0 6·4	16·0 10·3
41	9 40·3	9 41·8	9 13·8	4·1 2·6	10·1 6·5	16·1 10·3
42	9 40·5	9 42·1	9 14·1	4·2 2·7	10·2 6·5	16·2 10·4
43	9 40·8	9 42·3	9 14·3	4·3 2·8	10·3 6·6	16·3 10·5
44	9 41·0	9 42·6	9 14·5	4·4 2·8	10·4 6·7	16·4 10·5
45	9 41·3	9 42·8	9 14·8	4·5 2·9	10·5 6·7	16·5 10·6
46	9 41·5	9 43·1	9 15·0	4·6 3·0	10·6 6·8	16·6 10·7
47	9 41·8	9 43·3	9 15·2	4·7 3·0	10·7 6·9	16·7 10·7
48	9 42·0	9 43·6	9 15·5	4·8 3·1	10·8 6·9	16·8 10·8
49	9 42·3	9 43·8	9 15·7	4·9 3·1	10·9 7·0	16·9 10·8
50	9 42·5	9 44·1	9 16·0	5·0 3·2	11·0 7·1	17·0 10·9
51	9 42·8	9 44·3	9 16·2	5·1 3·3	11·1 7·1	17·1 11·0
52	9 43·0	9 44·6	9 16·4	5·2 3·3	11·2 7·2	17·2 11·0
53	9 43·3	9 44·8	9 16·7	5·3 3·4	11·3 7·3	17·3 11·1
54	9 43·5	9 45·1	9 16·9	5·4 3·5	11·4 7·3	17·4 11·2
55	9 43·8	9 45·3	9 17·2	5·5 3·5	11·5 7·4	17·5 11·2
56	9 44·0	9 45·6	9 17·4	5·6 3·6	11·6 7·4	17·6 11·3
57	9 44·3	9 45·8	9 17·6	5·7 3·7	11·7 7·5	17·7 11·4
58	9 44·5	9 46·1	9 17·9	5·8 3·7	11·8 7·6	17·8 11·4
59	9 44·8	9 46·4	9 18·1	5·9 3·8	11·9 7·6	17·9 11·5
60	9 45·0	9 46·6	9 18·4	6·0 3·9	12·0 7·7	18·0 11·6

39m

39s	SUN PLANETS	ARIES	MOON	v or d Corrn	v or d Corrn	v or d Corrn
	° ′	° ′	° ′	′ ′	′ ′	′ ′
00	9 45·0	9 46·6	9 18·4	0·0 0·0	6·0 4·0	12·0 7·9
01	9 45·3	9 46·9	9 18·6	0·1 0·1	6·1 4·0	12·1 8·0
02	9 45·5	9 47·1	9 18·8	0·2 0·1	6·2 4·1	12·2 8·0
03	9 45·8	9 47·4	9 19·1	0·3 0·2	6·3 4·1	12·3 8·1
04	9 46·0	9 47·6	9 19·3	0·4 0·3	6·4 4·2	12·4 8·2
05	9 46·3	9 47·9	9 19·5	0·5 0·3	6·5 4·3	12·5 8·2
06	9 46·5	9 48·1	9 19·8	0·6 0·4	6·6 4·3	12·6 8·3
07	9 46·8	9 48·4	9 20·0	0·7 0·5	6·7 4·4	12·7 8·4
08	9 47·0	9 48·6	9 20·3	0·8 0·5	6·8 4·5	12·8 8·4
09	9 47·3	9 48·9	9 20·5	0·9 0·6	6·9 4·5	12·9 8·5
10	9 47·5	9 49·1	9 20·7	1·0 0·7	7·0 4·6	13·0 8·6
11	9 47·8	9 49·4	9 21·0	1·1 0·7	7·1 4·7	13·1 8·6
12	9 48·0	9 49·6	9 21·2	1·2 0·8	7·2 4·7	13·2 8·7
13	9 48·3	9 49·9	9 21·5	1·3 0·9	7·3 4·8	13·3 8·8
14	9 48·5	9 50·1	9 21·7	1·4 0·9	7·4 4·9	13·4 8·8
15	9 48·8	9 50·4	9 21·9	1·5 1·0	7·5 4·9	13·5 8·9
16	9 49·0	9 50·6	9 22·2	1·6 1·1	7·6 5·0	13·6 9·0
17	9 49·3	9 50·9	9 22·4	1·7 1·1	7·7 5·1	13·7 9·0
18	9 49·5	9 51·1	9 22·6	1·8 1·2	7·8 5·1	13·8 9·1
19	9 49·8	9 51·4	9 22·9	1·9 1·3	7·9 5·2	13·9 9·2
20	9 50·0	9 51·6	9 23·1	2·0 1·3	8·0 5·3	14·0 9·2
21	9 50·3	9 51·9	9 23·4	2·1 1·4	8·1 5·3	14·1 9·3
22	9 50·5	9 52·1	9 23·6	2·2 1·4	8·2 5·4	14·2 9·3
23	9 50·8	9 52·4	9 23·8	2·3 1·5	8·3 5·5	14·3 9·4
24	9 51·0	9 52·6	9 24·1	2·4 1·6	8·4 5·5	14·4 9·5
25	9 51·3	9 52·9	9 24·3	2·5 1·6	8·5 5·6	14·5 9·5
26	9 51·5	9 53·1	9 24·6	2·6 1·7	8·6 5·7	14·6 9·6
27	9 51·8	9 53·4	9 24·8	2·7 1·8	8·7 5·7	14·7 9·7
28	9 52·0	9 53·6	9 25·0	2·8 1·8	8·8 5·8	14·8 9·7
29	9 52·3	9 53·9	9 25·3	2·9 1·9	8·9 5·9	14·9 9·8
30	9 52·5	9 54·1	9 25·5	3·0 2·0	9·0 5·9	15·0 9·9
31	9 52·8	9 54·4	9 25·7	3·1 2·0	9·1 6·0	15·1 9·9
32	9 53·0	9 54·6	9 26·0	3·2 2·1	9·2 6·1	15·2 10·0
33	9 53·3	9 54·9	9 26·2	3·3 2·2	9·3 6·1	15·3 10·1
34	9 53·5	9 55·1	9 26·5	3·4 2·2	9·4 6·2	15·4 10·1
35	9 53·8	9 55·4	9 26·7	3·5 2·3	9·5 6·3	15·5 10·2
36	9 54·0	9 55·6	9 26·9	3·6 2·4	9·6 6·3	15·6 10·3
37	9 54·3	9 55·9	9 27·2	3·7 2·4	9·7 6·4	15·7 10·3
38	9 54·5	9 56·1	9 27·4	3·8 2·5	9·8 6·5	15·8 10·4
39	9 54·8	9 56·4	9 27·7	3·9 2·6	9·9 6·5	15·9 10·5
40	9 55·0	9 56·6	9 27·9	4·0 2·6	10·0 6·6	16·0 10·5
41	9 55·3	9 56·9	9 28·1	4·1 2·7	10·1 6·6	16·1 10·6
42	9 55·5	9 57·1	9 28·4	4·2 2·8	10·2 6·7	16·2 10·7
43	9 55·8	9 57·4	9 28·6	4·3 2·8	10·3 6·8	16·3 10·7
44	9 56·0	9 57·6	9 28·8	4·4 2·9	10·4 6·8	16·4 10·8
45	9 56·3	9 57·9	9 29·1	4·5 3·0	10·5 6·9	16·5 10·9
46	9 56·5	9 58·1	9 29·3	4·6 3·0	10·6 7·0	16·6 10·9
47	9 56·8	9 58·4	9 29·6	4·7 3·1	10·7 7·0	16·7 11·0
48	9 57·0	9 58·6	9 29·8	4·8 3·2	10·8 7·1	16·8 11·1
49	9 57·3	9 58·9	9 30·0	4·9 3·2	10·9 7·2	16·9 11·1
50	9 57·5	9 59·1	9 30·3	5·0 3·3	11·0 7·2	17·0 11·2
51	9 57·8	9 59·4	9 30·5	5·1 3·4	11·1 7·3	17·1 11·3
52	9 58·0	9 59·6	9 30·8	5·2 3·4	11·2 7·4	17·2 11·3
53	9 58·3	9 59·9	9 31·0	5·3 3·5	11·3 7·4	17·3 11·4
54	9 58·5	10 00·1	9 31·2	5·4 3·6	11·4 7·5	17·4 11·5
55	9 58·8	10 00·4	9 31·5	5·5 3·6	11·5 7·6	17·5 11·5
56	9 59·0	10 00·6	9 31·7	5·6 3·7	11·6 7·6	17·6 11·6
57	9 59·3	10 00·9	9 32·0	5·7 3·8	11·7 7·7	17·7 11·7
58	9 59·5	10 01·1	9 32·2	5·8 3·8	11·8 7·8	17·8 11·7
59	9 59·8	10 01·4	9 32·4	5·9 3·9	11·9 7·8	17·9 11·8
60	10 00·0	10 01·6	9 32·7	6·0 4·0	12·0 7·9	18·0 11·9

THE ELEMENTS OF NAVIGATION AND NAUTICAL ASTRONOMY 409

INCREMENTS AND CORRECTIONS

40ᵐ

40	SUN PLANETS	ARIES	MOON	v or d / Corrⁿ	v or d / Corrⁿ	v or d / Corrⁿ
s	° ′	° ′	° ′	′ ′	′ ′	′ ′
00	10 00·0	10 01·6	9 32·7	0·0 0·0	6·0 4·1	12·0 8·1
01	10 00·3	10 01·9	9 32·9	0·1 0·1	6·1 4·1	12·1 8·2
02	10 00·5	10 02·1	9 33·1	0·2 0·1	6·2 4·2	12·2 8·2
03	10 00·8	10 02·4	9 33·4	0·3 0·2	6·3 4·3	12·3 8·3
04	10 01·0	10 02·6	9 33·6	0·4 0·3	6·4 4·3	12·4 8·4
05	10 01·3	10 02·9	9 33·9	0·5 0·3	6·5 4·4	12·5 8·4
06	10 01·5	10 03·1	9 34·1	0·6 0·4	6·6 4·5	12·6 8·5
07	10 01·8	10 03·4	9 34·3	0·7 0·5	6·7 4·5	12·7 8·6
08	10 02·0	10 03·6	9 34·6	0·8 0·5	6·8 4·6	12·8 8·6
09	10 02·3	10 03·9	9 34·8	0·9 0·6	6·9 4·7	12·9 8·7
10	10 02·5	10 04·1	9 35·1	1·0 0·7	7·0 4·7	13·0 8·8
11	10 02·8	10 04·4	9 35·3	1·1 0·7	7·1 4·8	13·1 8·8
12	10 03·0	10 04·7	9 35·5	1·2 0·8	7·2 4·9	13·2 8·9
13	10 03·3	10 04·9	9 35·8	1·3 0·9	7·3 4·9	13·3 9·0
14	10 03·5	10 05·2	9 36·0	1·4 0·9	7·4 5·0	13·4 9·0
15	10 03·8	10 05·4	9 36·2	1·5 1·0	7·5 5·1	13·5 9·1
16	10 04·0	10 05·7	9 36·5	1·6 1·1	7·6 5·1	13·6 9·2
17	10 04·3	10 05·9	9 36·7	1·7 1·1	7·7 5·2	13·7 9·2
18	10 04·5	10 06·2	9 37·0	1·8 1·2	7·8 5·3	13·8 9·3
19	10 04·8	10 06·4	9 37·2	1·9 1·3	7·9 5·3	13·9 9·4
20	10 05·0	10 06·7	9 37·4	2·0 1·4	8·0 5·4	14·0 9·5
21	10 05·3	10 06·9	9 37·7	2·1 1·4	8·1 5·5	14·1 9·5
22	10 05·5	10 07·2	9 37·9	2·2 1·5	8·2 5·5	14·2 9·6
23	10 05·8	10 07·4	9 38·2	2·3 1·6	8·3 5·6	14·3 9·7
24	10 06·0	10 07·7	9 38·4	2·4 1·6	8·4 5·7	14·4 9·7
25	10 06·3	10 07·9	9 38·6	2·5 1·7	8·5 5·7	14·5 9·8
26	10 06·5	10 08·2	9 38·9	2·6 1·8	8·6 5·8	14·6 9·9
27	10 06·8	10 08·4	9 39·1	2·7 1·8	8·7 5·9	14·7 9·9
28	10 07·0	10 08·7	9 39·3	2·8 1·9	8·8 5·9	14·8 10·0
29	10 07·3	10 08·9	9 39·6	2·9 2·0	8·9 6·0	14·9 10·1
30	10 07·5	10 09·2	9 39·8	3·0 2·0	9·0 6·1	15·0 10·1
31	10 07·8	10 09·4	9 40·1	3·1 2·1	9·1 6·1	15·1 10·2
32	10 08·0	10 09·7	9 40·3	3·2 2·2	9·2 6·2	15·2 10·3
33	10 08·3	10 09·9	9 40·5	3·3 2·2	9·3 6·3	15·3 10·3
34	10 08·5	10 10·2	9 40·8	3·4 2·3	9·4 6·3	15·4 10·4
35	10 08·8	10 10·4	9 41·0	3·5 2·4	9·5 6·4	15·5 10·5
36	10 09·0	10 10·7	9 41·3	3·6 2·4	9·6 6·5	15·6 10·5
37	10 09·3	10 10·9	9 41·5	3·7 2·5	9·7 6·5	15·7 10·6
38	10 09·5	10 11·2	9 41·7	3·8 2·6	9·8 6·6	15·8 10·7
39	10 09·8	10 11·4	9 42·0	3·9 2·6	9·9 6·7	15·9 10·7
40	10 10·0	10 11·7	9 42·2	4·0 2·7	10·0 6·8	16·0 10·8
41	10 10·3	10 11·9	9 42·4	4·1 2·8	10·1 6·8	16·1 10·9
42	10 10·5	10 12·2	9 42·7	4·2 2·8	10·2 6·9	16·2 10·9
43	10 10·8	10 12·4	9 42·9	4·3 2·9	10·3 7·0	16·3 11·0
44	10 11·0	10 12·7	9 43·2	4·4 3·0	10·4 7·0	16·4 11·1
45	10 11·3	10 12·9	9 43·4	4·5 3·0	10·5 7·1	16·5 11·1
46	10 11·5	10 13·2	9 43·6	4·6 3·1	10·6 7·2	16·6 11·2
47	10 11·8	10 13·4	9 43·9	4·7 3·2	10·7 7·2	16·7 11·3
48	10 12·0	10 13·7	9 44·1	4·8 3·2	10·8 7·3	16·8 11·3
49	10 12·3	10 13·9	9 44·4	4·9 3·3	10·9 7·4	16·9 11·4
50	10 12·5	10 14·2	9 44·6	5·0 3·4	11·0 7·4	17·0 11·5
51	10 12·8	10 14·4	9 44·8	5·1 3·4	11·1 7·5	17·1 11·5
52	10 13·0	10 14·7	9 45·1	5·2 3·5	11·2 7·6	17·2 11·6
53	10 13·3	10 14·9	9 45·3	5·3 3·6	11·3 7·6	17·3 11·7
54	10 13·5	10 15·2	9 45·6	5·4 3·6	11·4 7·7	17·4 11·7
55	10 13·8	10 15·4	9 45·8	5·5 3·7	11·5 7·8	17·5 11·8
56	10 14·0	10 15·7	9 46·0	5·6 3·8	11·6 7·8	17·6 11·9
57	10 14·3	10 15·9	9 46·3	5·7 3·8	11·7 7·9	17·7 11·9
58	10 14·5	10 16·2	9 46·5	5·8 3·9	11·8 8·0	17·8 12·0
59	10 14·8	10 16·4	9 46·7	5·9 4·0	11·9 8·0	17·9 12·1
60	10 15·0	10 16·7	9 47·0	6·0 4·1	12·0 8·1	18·0 12·2

41ᵐ

41	SUN PLANETS	ARIES	MOON	v or d / Corrⁿ	v or d / Corrⁿ	v or d / Corrⁿ
s	° ′	° ′	° ′	′ ′	′ ′	′ ′
00	10 15·0	10 16·7	9 47·0	0·0 0·0	6·0 4·2	12·0 8·3
01	10 15·3	10 16·9	9 47·2	0·1 0·1	6·1 4·2	12·1 8·4
02	10 15·5	10 17·2	9 47·5	0·2 0·1	6·2 4·3	12·2 8·4
03	10 15·8	10 17·4	9 47·7	0·3 0·2	6·3 4·4	12·3 8·5
04	10 16·0	10 17·7	9 47·9	0·4 0·3	6·4 4·4	12·4 8·6
05	10 16·3	10 17·9	9 48·2	0·5 0·3	6·5 4·5	12·5 8·6
06	10 16·5	10 18·2	9 48·4	0·6 0·4	6·6 4·6	12·6 8·7
07	10 16·8	10 18·4	9 48·7	0·7 0·5	6·7 4·6	12·7 8·8
08	10 17·0	10 18·7	9 48·9	0·8 0·6	6·8 4·7	12·8 8·9
09	10 17·3	10 18·9	9 49·1	0·9 0·6	6·9 4·8	12·9 8·9
10	10 17·5	10 19·2	9 49·4	1·0 0·7	7·0 4·8	13·0 9·0
11	10 17·8	10 19·4	9 49·6	1·1 0·8	7·1 4·9	13·1 9·1
12	10 18·0	10 19·7	9 49·8	1·2 0·8	7·2 5·0	13·2 9·1
13	10 18·3	10 19·9	9 50·1	1·3 0·9	7·3 5·0	13·3 9·2
14	10 18·5	10 20·2	9 50·3	1·4 1·0	7·4 5·1	13·4 9·3
15	10 18·8	10 20·4	9 50·6	1·5 1·0	7·5 5·2	13·5 9·3
16	10 19·0	10 20·7	9 50·8	1·6 1·1	7·6 5·3	13·6 9·4
17	10 19·3	10 20·9	9 51·0	1·7 1·2	7·7 5·3	13·7 9·5
18	10 19·5	10 21·2	9 51·3	1·8 1·2	7·8 5·4	13·8 9·5
19	10 19·8	10 21·4	9 51·5	1·9 1·3	7·9 5·5	13·9 9·6
20	10 20·0	10 21·7	9 51·8	2·0 1·4	8·0 5·5	14·0 9·7
21	10 20·3	10 21·9	9 52·0	2·1 1·5	8·1 5·6	14·1 9·8
22	10 20·5	10 22·2	9 52·2	2·2 1·5	8·2 5·7	14·2 9·8
23	10 20·8	10 22·4	9 52·5	2·3 1·6	8·3 5·7	14·3 9·9
24	10 21·0	10 22·7	9 52·7	2·4 1·7	8·4 5·8	14·4 10·0
25	10 21·3	10 23·0	9 52·9	2·5 1·7	8·5 5·9	14·5 10·0
26	10 21·5	10 23·2	9 53·2	2·6 1·8	8·6 5·9	14·6 10·1
27	10 21·8	10 23·5	9 53·4	2·7 1·9	8·7 6·0	14·7 10·2
28	10 22·0	10 23·7	9 53·7	2·8 1·9	8·8 6·1	14·8 10·2
29	10 22·3	10 24·0	9 53·9	2·9 2·0	8·9 6·2	14·9 10·3
30	10 22·5	10 24·2	9 54·1	3·0 2·1	9·0 6·2	15·0 10·4
31	10 22·8	10 24·5	9 54·4	3·1 2·1	9·1 6·3	15·1 10·4
32	10 23·0	10 24·7	9 54·6	3·2 2·2	9·2 6·4	15·2 10·5
33	10 23·3	10 25·0	9 54·9	3·3 2·3	9·3 6·4	15·3 10·6
34	10 23·5	10 25·2	9 55·1	3·4 2·4	9·4 6·5	15·4 10·7
35	10 23·8	10 25·5	9 55·3	3·5 2·4	9·5 6·6	15·5 10·7
36	10 24·0	10 25·7	9 55·6	3·6 2·5	9·6 6·6	15·6 10·8
37	10 24·3	10 26·0	9 55·8	3·7 2·6	9·7 6·7	15·7 10·9
38	10 24·5	10 26·2	9 56·1	3·8 2·6	9·8 6·8	15·8 10·9
39	10 24·8	10 26·5	9 56·3	3·9 2·7	9·9 6·8	15·9 11·0
40	10 25·0	10 26·7	9 56·5	4·0 2·8	10·0 6·9	16·0 11·1
41	10 25·3	10 27·0	9 56·8	4·1 2·8	10·1 7·0	16·1 11·1
42	10 25·5	10 27·2	9 57·0	4·2 2·9	10·2 7·1	16·2 11·2
43	10 25·8	10 27·5	9 57·2	4·3 3·0	10·3 7·1	16·3 11·3
44	10 26·0	10 27·7	9 57·5	4·4 3·0	10·4 7·2	16·4 11·3
45	10 26·3	10 28·0	9 57·7	4·5 3·1	10·5 7·3	16·5 11·4
46	10 26·5	10 28·2	9 58·0	4·6 3·2	10·6 7·3	16·6 11·5
47	10 26·8	10 28·5	9 58·2	4·7 3·3	10·7 7·4	16·7 11·6
48	10 27·0	10 28·7	9 58·4	4·8 3·3	10·8 7·5	16·8 11·6
49	10 27·3	10 29·0	9 58·7	4·9 3·4	10·9 7·5	16·9 11·7
50	10 27·5	10 29·2	9 58·9	5·0 3·5	11·0 7·6	17·0 11·8
51	10 27·8	10 29·5	9 59·2	5·1 3·5	11·1 7·7	17·1 11·8
52	10 28·0	10 29·7	9 59·4	5·2 3·6	11·2 7·7	17·2 11·9
53	10 28·3	10 30·0	9 59·6	5·3 3·7	11·3 7·8	17·3 12·0
54	10 28·5	10 30·2	9 59·9	5·4 3·7	11·4 7·9	17·4 12·0
55	10 28·8	10 30·5	10 00·1	5·5 3·8	11·5 8·0	17·5 12·1
56	10 29·0	10 30·7	10 00·3	5·6 3·9	11·6 8·0	17·6 12·2
57	10 29·3	10 31·0	10 00·6	5·7 3·9	11·7 8·1	17·7 12·2
58	10 29·5	10 31·2	10 00·8	5·8 4·0	11·8 8·2	17·8 12·3
59	10 29·8	10 31·5	10 01·1	5·9 4·1	11·9 8·2	17·9 12·4
60	10 30·0	10 31·7	10 01·3	6·0 4·2	12·0 8·3	18·0 12·5

THE ELEMENTS OF NAVIGATION AND NAUTICAL ASTRONOMY

42ᵐ INCREMENTS AND CORRECTIONS 43ᵐ

42ᵐ	SUN PLANETS	ARIES	MOON	v or d / Corrⁿ	v or d / Corrⁿ	v or d / Corrⁿ	43ᵐ	SUN PLANETS	ARIES	MOON	v or d / Corrⁿ	v or d / Corrⁿ	v or d / Corrⁿ
s	° ′	° ′	° ′	′ ′	′ ′	′ ′	s	° ′	° ′	° ′	′ ′	′ ′	′ ′
00	10 30·0	10 31·7	10 01·3	0·0 0·0	6·0 4·3	12·0 8·5	00	10 45·0	10 46·8	10 15·6	0·0 0·0	6·0 4·4	12·0 8·7
01	10 30·3	10 32·0	10 01·5	0·1 0·1	6·1 4·3	12·1 8·6	01	10 45·3	10 47·0	10 15·9	0·1 0·1	6·1 4·4	12·1 8·8
02	10 30·5	10 32·2	10 01·8	0·2 0·1	6·2 4·4	12·2 8·6	02	10 45·5	10 47·3	10 16·1	0·2 0·1	6·2 4·5	12·2 8·8
03	10 30·8	10 32·5	10 02·0	0·3 0·2	6·3 4·5	12·3 8·7	03	10 45·8	10 47·5	10 16·3	0·3 0·2	6·3 4·6	12·3 8·9
04	10 31·0	10 32·7	10 02·3	0·4 0·3	6·4 4·5	12·4 8·8	04	10 46·0	10 47·8	10 16·6	0·4 0·3	6·4 4·6	12·4 9·0
05	10 31·3	10 33·0	10 02·5	0·5 0·4	6·5 4·6	12·5 8·9	05	10 46·3	10 48·0	10 16·8	0·5 0·4	6·5 4·7	12·5 9·1
06	10 31·5	10 33·2	10 02·7	0·6 0·4	6·6 4·7	12·6 8·9	06	10 46·5	10 48·3	10 17·0	0·6 0·4	6·6 4·8	12·6 9·1
07	10 31·8	10 33·5	10 03·0	0·7 0·5	6·7 4·7	12·7 9·0	07	10 46·8	10 48·5	10 17·3	0·7 0·5	6·7 4·9	12·7 9·2
08	10 32·0	10 33·7	10 03·2	0·8 0·6	6·8 4·8	12·8 9·1	08	10 47·0	10 48·8	10 17·5	0·8 0·6	6·8 4·9	12·8 9·3
09	10 32·3	10 34·0	10 03·4	0·9 0·6	6·9 4·9	12·9 9·1	09	10 47·3	10 49·0	10 17·8	0·9 0·7	6·9 5·0	12·9 9·4
10	10 32·5	10 34·2	10 03·7	1·0 0·7	7·0 5·0	13·0 9·2	10	10 47·5	10 49·3	10 18·0	1·0 0·7	7·0 5·1	13·0 9·4
11	10 32·8	10 34·5	10 03·9	1·1 0·8	7·1 5·0	13·1 9·3	11	10 47·8	10 49·5	10 18·2	1·1 0·8	7·1 5·1	13·1 9·5
12	10 33·0	10 34·7	10 04·2	1·2 0·9	7·2 5·1	13·2 9·4	12	10 48·0	10 49·8	10 18·5	1·2 0·9	7·2 5·2	13·2 9·6
13	10 33·3	10 35·0	10 04·4	1·3 0·9	7·3 5·2	13·3 9·4	13	10 48·3	10 50·0	10 18·7	1·3 0·9	7·3 5·3	13·3 9·6
14	10 33·5	10 35·2	10 04·6	1·4 1·0	7·4 5·2	13·4 9·5	14	10 48·5	10 50·3	10 19·0	1·4 1·0	7·4 5·4	13·4 9·7
15	10 33·8	10 35·5	10 04·9	1·5 1·1	7·5 5·3	13·5 9·6	15	10 48·8	10 50·5	10 19·2	1·5 1·1	7·5 5·4	13·5 9·8
16	10 34·0	10 35·7	10 05·1	1·6 1·1	7·6 5·4	13·6 9·6	16	10 49·0	10 50·8	10 19·4	1·6 1·2	7·6 5·5	13·6 9·9
17	10 34·3	10 36·0	10 05·4	1·7 1·2	7·7 5·5	13·7 9·7	17	10 49·3	10 51·0	10 19·7	1·7 1·2	7·7 5·6	13·7 9·9
18	10 34·5	10 36·2	10 05·6	1·8 1·3	7·8 5·5	13·8 9·8	18	10 49·5	10 51·3	10 19·9	1·8 1·3	7·8 5·7	13·8 10·0
19	10 34·8	10 36·5	10 05·8	1·9 1·3	7·9 5·6	13·9 9·8	19	10 49·8	10 51·5	10 20·2	1·9 1·4	7·9 5·7	13·9 10·1
20	10 35·0	10 36·7	10 06·1	2·0 1·4	8·0 5·7	14·0 9·9	20	10 50·0	10 51·8	10 20·4	2·0 1·5	8·0 5·8	14·0 10·2
21	10 35·3	10 37·0	10 06·3	2·1 1·5	8·1 5·7	14·1 10·0	21	10 50·3	10 52·0	10 20·6	2·1 1·5	8·1 5·9	14·1 10·2
22	10 35·5	10 37·2	10 06·5	2·2 1·6	8·2 5·8	14·2 10·1	22	10 50·5	10 52·3	10 20·9	2·2 1·6	8·2 5·9	14·2 10·3
23	10 35·8	10 37·5	10 06·8	2·3 1·6	8·3 5·9	14·3 10·1	23	10 50·8	10 52·5	10 21·1	2·3 1·7	8·3 6·0	14·3 10·4
24	10 36·0	10 37·7	10 07·0	2·4 1·7	8·4 6·0	14·4 10·2	24	10 51·0	10 52·8	10 21·3	2·4 1·7	8·4 6·1	14·4 10·4
25	10 36·3	10 38·0	10 07·3	2·5 1·8	8·5 6·0	14·5 10·3	25	10 51·3	10 53·0	10 21·6	2·5 1·8	8·5 6·2	14·5 10·5
26	10 36·5	10 38·2	10 07·5	2·6 1·8	8·6 6·1	14·6 10·3	26	10 51·5	10 53·3	10 21·8	2·6 1·9	8·6 6·2	14·6 10·6
27	10 36·8	10 38·5	10 07·7	2·7 1·9	8·7 6·2	14·7 10·4	27	10 51·8	10 53·5	10 22·1	2·7 2·0	8·7 6·3	14·7 10·7
28	10 37·0	10 38·7	10 08·0	2·8 2·0	8·8 6·2	14·8 10·5	28	10 52·0	10 53·8	10 22·3	2·8 2·0	8·8 6·4	14·8 10·7
29	10 37·3	10 39·0	10 08·2	2·9 2·1	8·9 6·3	14·9 10·6	29	10 52·3	10 54·0	10 22·5	2·9 2·1	8·9 6·5	14·9 10·8
30	10 37·5	10 39·2	10 08·5	3·0 2·1	9·0 6·4	15·0 10·6	30	10 52·5	10 54·3	10 22·8	3·0 2·2	9·0 6·5	15·0 10·9
31	10 37·8	10 39·5	10 08·7	3·1 2·2	9·1 6·4	15·1 10·7	31	10 52·8	10 54·5	10 23·0	3·1 2·2	9·1 6·6	15·1 10·9
32	10 38·0	10 39·7	10 08·9	3·2 2·3	9·2 6·5	15·2 10·8	32	10 53·0	10 54·8	10 23·3	3·2 2·3	9·2 6·7	15·2 11·0
33	10 38·3	10 40·0	10 09·2	3·3 2·3	9·3 6·6	15·3 10·8	33	10 53·3	10 55·0	10 23·5	3·3 2·4	9·3 6·7	15·3 11·1
34	10 38·5	10 40·2	10 09·4	3·4 2·4	9·4 6·7	15·4 10·9	34	10 53·5	10 55·3	10 23·7	3·4 2·5	9·4 6·8	15·4 11·2
35	10 38·8	10 40·5	10 09·7	3·5 2·5	9·5 6·7	15·5 11·0	35	10 53·8	10 55·5	10 24·0	3·5 2·5	9·5 6·9	15·5 11·2
36	10 39·0	10 40·7	10 09·9	3·6 2·6	9·6 6·8	15·6 11·1	36	10 54·0	10 55·8	10 24·2	3·6 2·6	9·6 7·0	15·6 11·3
37	10 39·3	10 41·0	10 10·1	3·7 2·6	9·7 6·9	15·7 11·1	37	10 54·3	10 56·0	10 24·4	3·7 2·7	9·7 7·0	15·7 11·4
38	10 39·5	10 41·3	10 10·4	3·8 2·7	9·8 6·9	15·8 11·2	38	10 54·5	10 56·3	10 24·7	3·8 2·8	9·8 7·1	15·8 11·5
39	10 39·8	10 41·5	10 10·6	3·9 2·8	9·9 7·0	15·9 11·3	39	10 54·8	10 56·5	10 24·9	3·9 2·8	9·9 7·2	15·9 11·5
40	10 40·0	10 41·8	10 10·8	4·0 2·8	10·0 7·1	16·0 11·3	40	10 55·0	10 56·8	10 25·2	4·0 2·9	10·0 7·3	16·0 11·6
41	10 40·3	10 42·0	10 11·1	4·1 2·9	10·1 7·2	16·1 11·4	41	10 55·3	10 57·0	10 25·4	4·1 3·0	10·1 7·3	16·1 11·7
42	10 40·5	10 42·3	10 11·3	4·2 3·0	10·2 7·2	16·2 11·5	42	10 55·5	10 57·3	10 25·6	4·2 3·0	10·2 7·4	16·2 11·7
43	10 40·8	10 42·5	10 11·6	4·3 3·0	10·3 7·3	16·3 11·5	43	10 55·8	10 57·5	10 25·9	4·3 3·1	10·3 7·5	16·3 11·8
44	10 41·0	10 42·8	10 11·8	4·4 3·1	10·4 7·4	16·4 11·6	44	10 56·0	10 57·8	10 26·1	4·4 3·2	10·4 7·5	16·4 11·9
45	10 41·3	10 43·0	10 12·0	4·5 3·2	10·5 7·4	16·5 11·7	45	10 56·3	10 58·0	10 26·4	4·5 3·3	10·5 7·6	16·5 12·0
46	10 41·5	10 43·3	10 12·3	4·6 3·3	10·6 7·5	16·6 11·8	46	10 56·5	10 58·3	10 26·6	4·6 3·3	10·6 7·7	16·6 12·0
47	10 41·8	10 43·5	10 12·5	4·7 3·3	10·7 7·6	16·7 11·8	47	10 56·8	10 58·5	10 26·8	4·7 3·4	10·7 7·8	16·7 12·1
48	10 42·0	10 43·8	10 12·8	4·8 3·4	10·8 7·7	16·8 11·9	48	10 57·0	10 58·8	10 27·1	4·8 3·5	10·8 7·8	16·8 12·2
49	10 42·3	10 44·0	10 13·0	4·9 3·5	10·9 7·7	16·9 12·0	49	10 57·3	10 59·0	10 27·3	4·9 3·6	10·9 7·9	16·9 12·3
50	10 42·5	10 44·3	10 13·2	5·0 3·5	11·0 7·8	17·0 12·0	50	10 57·5	10 59·3	10 27·5	5·0 3·6	11·0 8·0	17·0 12·3
51	10 42·8	10 44·5	10 13·5	5·1 3·6	11·1 7·9	17·1 12·1	51	10 57·8	10 59·5	10 27·8	5·1 3·7	11·1 8·0	17·1 12·4
52	10 43·0	10 44·8	10 13·7	5·2 3·7	11·2 7·9	17·2 12·2	52	10 58·0	10 59·8	10 28·0	5·2 3·8	11·2 8·1	17·2 12·5
53	10 43·3	10 45·0	10 13·9	5·3 3·8	11·3 8·0	17·3 12·3	53	10 58·3	11 00·1	10 28·3	5·3 3·8	11·3 8·2	17·3 12·5
54	10 43·5	10 45·3	10 14·2	5·4 3·8	11·4 8·1	17·4 12·3	54	10 58·5	11 00·3	10 28·5	5·4 3·9	11·4 8·3	17·4 12·6
55	10 43·8	10 45·5	10 14·4	5·5 3·9	11·5 8·1	17·5 12·4	55	10 58·8	11 00·6	10 28·7	5·5 4·0	11·5 8·3	17·5 12·7
56	10 44·0	10 45·8	10 14·7	5·6 4·0	11·6 8·2	17·6 12·5	56	10 59·0	11 00·8	10 29·0	5·6 4·1	11·6 8·4	17·6 12·8
57	10 44·3	10 46·0	10 14·9	5·7 4·0	11·7 8·3	17·7 12·5	57	10 59·3	11 01·1	10 29·2	5·7 4·1	11·7 8·5	17·7 12·8
58	10 44·5	10 46·3	10 15·1	5·8 4·1	11·8 8·4	17·8 12·6	58	10 59·5	11 01·3	10 29·5	5·8 4·2	11·8 8·6	17·8 12·9
59	10 44·8	10 46·5	10 15·4	5·9 4·2	11·9 8·4	17·9 12·7	59	10 59·8	11 01·6	10 29·7	5·9 4·3	11·9 8·6	17·9 13·0
60	10 45·0	10 46·8	10 15·6	6·0 4·3	12·0 8·5	18·0 12·8	60	11 00·0	11 01·8	10 29·9	6·0 4·4	12·0 8·7	18·0 13·1

THE ELEMENTS OF NAVIGATION AND NAUTICAL ASTRONOMY 411

STARS, JANUARY—JUNE

Mag.	Name and No.			S.H.A.						Declination					
				JAN.	FEB.	MAR.	APR.	MAY	JUNE		JAN.	FEB.	MAR.	APR.	MAY	JUNE
1·6	α	Geminorum		247	00·6	00·6	00·7	00·8	00·9	01·0	N. 31	58·8	58·8	58·8	58·9	58·8	58·8
3·3	σ	Puppis*		248	00·9	01·0	01·1	01·3	01·5	01·6	S. 43	13·1	13·3	13·4	13·4	13·3	13·2
3·1	β	Canis Minoris		248	46·4	46·4	46·5	46·6	46·7	46·7	N. 8	22·3	22·3	22·3	22·3	22·3	22·3
2·4	η	Canis Majoris		249	23·0	23·0	23·1	23·3	23·4	23·5	S. 29	13·4	13·5	13·6	13·6	13·6	13·5
2·7	π	Puppis*		251	04·6	04·6	04·7	04·9	05·1	05·1	S. 37	01·5	01·6	01·7	01·7	01·7	01·6
2·0	δ	Canis Majoris		253	19·2	19·3	19·4	19·5	19·7	19·7	S. 26	19·7	19·9	19·9	19·9	19·9	19·8
3·1	o	Canis Majoris		254	40·4	40·5	40·6	40·7	40·8	40·9	S. 23	46·5	46·6	46·6	46·6	46·6	46·5
1·6	ε	Canis Majoris	19	255	44·9	44·9	45·0	45·2	45·3	45·4	S. 28	55·1	55·2	55·2	55·2	55·2	55·1
2·8	τ	Puppis*		257	46·0	46·1	46·3	46·5	46·8	46·8	S. 50	34·1	34·2	34·3	34·3	34·2	34·1
−1·6	α	Canis Majoris	18	259	10·1	10·1	10·2	10·4	10·5	10·5	S. 16	39·7	39·7	39·8	39·8	39·7	39·7
1·9	γ	Geminorum		261	10·2	10·2	10·3	10·4	10·5	10·5	N. 16	26·0	26·0	26·0	26·0	26·0	26·0
−0·9	α	Carinæ*	17	264	14·1	14·3	14·5	14·7	15·0	15·0	S. 52	40·6	40·7	40·8	40·8	40·7	40·5
2·0	β	Canis Majoris		264	46·8	46·8	47·0	47·1	47·2	47·2	S. 17	56·3	56·3	56·4	56·4	56·3	56·2
2·7	θ	Aurigæ		270	46·5	46·6	46·7	46·9	47·0	47·0	N. 37	12·6	12·6	12·7	12·6	12·6	12·6
2·1	β	Aurigæ		270	52·6	52·7	52·8	53·0	53·1	53·1	N. 44	56·7	56·7	56·7	56·7	56·7	56·6
Var.‡	α	Orionis	16	271	46·0	46·1	46·2	46·3	46·4	46·4	N. 7	23·9	23·9	23·9	23·9	23·9	23·9
2·2	κ	Orionis		273	33·1	33·1	33·3	33·4	33·5	33·5	S. 9	41·2	41·3	41·3	41·3	41·2	41·1
1·9	ζ	Orionis		275	20·0	20·0	20·1	20·3	20·3	20·3	S. 1	58·0	58·0	58·0	58·0	58·0	57·9
2·8	α	Columbæ		275	27·6	27·7	27·9	28·1	28·2	28·2	S. 34	06·0	06·1	06·1	06·1	06·0	05·9
3·0	ζ	Tauri		276	12·5	12·5	12·7	12·8	12·9	12·8	N. 21	07·0	07·0	07·0	07·0	07·0	07·0
1·8	ε	Orionis	15	276	28·3	28·4	28·5	28·6	28·7	28·7	S. 1	13·8	13·9	13·9	13·9	13·8	13·8
2·9	ι	Orionis		276	38·9	39·0	39·1	39·2	39·3	39·2	S. 5	56·3	56·4	56·4	56·4	56·4	56·3
2·7	α	Leporis		277	16·4	16·5	16·6	16·7	16·8	16·8	S. 17	51·3	51·4	51·4	51·4	51·3	51·2
2·5	δ	Orionis		277	31·6	31·7	31·8	32·0	32·0	32·0	S. 0	19·9	19·9	20·0	19·9	19·9	19·8
3·0	β	Leporis		278	22·9	23·0	23·1	23·2	23·3	23·3	S. 20	47·7	47·8	47·8	47·8	47·7	47·6
1·8	β	Tauri	14	279	04·9	05·0	05·1	05·3	05·3	05·3	N. 28	34·4	34·4	34·4	34·3	34·3	34·3
1·7	γ	Orionis	13	279	16·4	16·4	16·6	16·7	16·7	16·7	N. 6	18·6	18·6	18·6	18·6	18·6	18·7
0·2	α	Aurigæ	12	281	35·6	35·7	35·9	36·0	36·1	36·0	N. 45	57·4	57·5	57·5	57·5	57·4	57·3
0·3	β	Orionis	11	281	51·8	51·9	52·0	52·1	52·2	52·1	S. 8	15·1	15·2	15·2	15·2	15·1	15·0
2·9	β	Eridani		283	32·8	32·9	33·0	33·1	33·2	33·1	S. 5	08·5	08·6	08·6	08·6	08·5	08·5
2·9	ι	Aurigæ †		286	25·6	25·7	25·8	26·0	26·0	25·9	N. 33	06·1	06·1	06·1	06·0	06·0	06·0
1·1	α	Tauri	10	291	36·9	37·0	37·1	37·2	37·3	37·2	N. 16	25·5	25·5	25·5	25·5	25·5	25·5
3·2	γ	Eridani		300	58·6	58·7	58·9	59·0	59·0	58·9	S. 13	37·8	37·8	37·8	37·8	37·7	37·6
3·0	ε	Persei		301	14·0	14·2	14·3	14·4	14·4	14·3	N. 39	53·5	53·5	53·5	53·4	53·4	53·3
2·9	ζ	Persei		302	07·2	07·3	07·5	07·6	07·6	07·4	N. 31	45·7	45·7	45·6	45·6	45·6	45·5
3·0	η	Tauri		303	44·8	44·9	45·0	45·1	45·1	45·0	N. 23	58·6	58·6	58·6	58·5	58·5	58·5
1·9	α	Persei	9	309	39·7	39·9	40·1	40·2	40·1	40·0	N. 49	43·0	42·9	42·9	42·8	42·7	42·7
Var.§	β	Persei †		313	38·1	38·2	38·4	38·5	38·4	38·3	N. 40	47·9	47·8	47·8	47·7	47·6	47·6
2·8	α	Ceti	8	314	58·5	58·6	58·7	58·7	58·7	58·6	N. 3	55·5	55·5	55·5	55·5	55·5	55·6
3·1	θ	Eridani	7	315	49·8	49·9	50·1	50·2	50·2	50·1	S. 40	28·6	28·6	28·6	28·5	28·3	28·2
3·1	β	Trianguli		328	14·1	14·2	14·3	14·4	14·3	14·1	N. 34	47·5	47·5	47·4	47·4	47·3	47·3
2·2	α	Arietis	6	328	47·7	47·8	47·9	47·9	47·8	47·6	N. 23	16·0	15·9	15·9	15·9	15·8	15·9
2·2	γ	Andromedæ		329	39·9	40·1	40·2	40·2	40·1	39·9	N. 42	07·9	07·9	07·8	07·7	07·7	07·7
3·0	α	Hydri		330	38·3	38·6	38·8	38·9	38·9	38·6	S. 61	46·7	46·7	46·6	46·4	46·2	46·1
2·7	β	Arietis		331	55·0	55·1	55·2	55·2	55·1	54·9	N. 20	36·3	36·3	36·2	36·2	36·2	36·2
2·1	α	Ursæ Minoris †		331	14·7	23·9	30·8	33·7	30·5	23·1	N. 89	04·4	04·4	04·3	04·2	04·0	03·9
0·6	α	Eridani	5	335	57·8	58·0	58·2	58·2	58·1	57·9	N. 57	27·3	27·2	27·1	26·9	26·8	26·6
2·8	δ	Cassiopeiæ		339	13·8	14·1	14·2	14·2	14·1	13·7	N. 60	01·4	01·3	01·2	01·1	01·0	00·9
2·4	β	Andromedæ		343	09·2	09·3	09·3	09·3	09·2	09·0	N. 35	24·1	24·0	24·0	23·9	23·9	23·9
Var. ‖	γ	Cassiopeiæ		346	27·3	27·6	27·7	27·7	27·4	27·1	N. 60	29·7	29·6	29·5	29·4	29·3	29·3
2·2	β	Ceti	4	349	37·7	37·8	37·8	37·8	37·7	37·5	S. 18	13·1	13·1	13·0	13·0	12·9	12·7
2·5	α	Cassiopeiæ	3	350	28·1	28·3	28·4	28·3	28·1	27·8	N. 56	18·7	18·7	18·6	18·4	18·4	18·4
2·4	α	Phœnicis	2	353	56·9	57·0	57·0	57·0	56·8	56·6	S. 42	32·2	32·2	32·1	31·9	31·8	31·6
2·9	β	Hydri †		354	07·2	07·7	08·0	07·9	07·5	06·8	S. 77	29·7	29·6	29·4	29·2	29·0	28·9
2·9	γ	Pegasi		357	13·8	13·9	13·9	13·9	13·7	13·5	N. 14	57·1	57·1	57·0	57·0	57·0	57·1
2·4	β	Cassiopeiæ		358	16·0	16·2	16·2	16·1	15·9	15·5	N. 58	55·4	55·3	55·2	55·1	55·0	55·0
2·2	α	Andromedæ	1	358	26·8	26·8	26·8	26·7	26·6	26·3	N. 28	51·7	51·6	51·6	51·5	51·5	51·6

* Formerly Argus ‡ 0·1—1·2 § 2·3—3·5 ‖ Irregular variable; 1955 mag. 2·8
† Not suitable for use with H.O. 214 (H.D. 486)

STARS   JULY—DECEMBER

Mag.	Name and No.			S.H.A.						Declination					
				JULY	AUG.	SEPT.	OCT.	NOV.	DEC.	JULY	AUG.	SEPT.	OCT.	NOV.	DEC.
1·6		Castor		246 60·9	60·8	60·6	60·3	60·1	59·9	N. 31 58·8	58·7	58·7	58·7	58·6	58·6
3·3	σ	Puppis*		248 01·6	01·5	01·3	01·1	00·8	00·6	S. 43 13·1	13·0	12·9	12·9	12·9	13·1
3·1	β	Canis Minoris		248 46·7	46·6	46·4	46·2	45·9	45·8	N. 8 22·4	22·4	22·4	22·4	22·3	22·3
2·4	η	Canis Majoris		249 23·4	23·3	23·2	22·9	22·7	22·5	S. 29 13·4	13·3	13·2	13·2	13·3	13·4
2·7	π	Puppis*		251 05·1	05·0	04·8	04·6	04·3	04·2	S. 37 01·4	01·3	01·2	01·2	01·3	01·4
2·0		Wezen		253 19·7	19·6	19·4	19·1	18·9	18·8	S. 26 19·7	19·6	19·5	19·5	19·6	19·7
3·1	o	Canis Majoris		254 40·9	40·7	40·5	40·3	40·1	39·9	S. 23 46·4	46·3	46·2	46·2	46·3	46·4
1·6		Adhara	19	255 45·3	45·2	45·0	44·8	44·5	44·4	S. 28 55·0	54·8	54·8	54·8	54·9	55·0
2·8	τ	Puppis*		257 46·8	46·7	46·4	46·1	45·9	45·7	S. 50 34·0	33·8	33·7	33·7	33·8	34·0
−1·6		Sirius	18	259 10·4	10·3	10·1	09·9	09·7	09·5	S. 16 39·6	39·5	39·4	39·4	39·5	39·6
1·9		Alhena		261 10·4	10·3	10·1	09·9	09·6	09·5	N. 16 26·1	26·1	26·1	26·1	26·0	26·0
−0·9		Canopus	17	264 15·0	14·8	14·5	14·3	14·0	13·8	S. 52 40·4	40·2	40·2	40·2	40·3	40·5
2·0		Mirzam		264 47·1	47·0	46·8	46·6	46·4	46·2	S. 17 56·1	56·0	56·0	56·0	56·1	56·2
2·7	θ	Aurigæ		270 46·8	46·6	46·3	46·1	45·8	45·7	N. 37 12·5	12·5	12·5	12·5	12·5	12·5
2·1		Menkalinan		270 52·9	52·7	52·4	52·1	51·9	51·7	N. 44 56·5	56·5	56·5	56·5	56·5	56·6
Var.‡		Betelgeuse	16	271 46·3	46·1	45·9	45·7	45·5	45·4	N. 7 24·0	24·0	24·0	24·0	24·0	23·9
2·2	κ	Orionis		273 33·4	33·2	33·0	32·8	32·6	32·5	S. 9 41·0	41·0	40·9	41·0	41·0	41·1
1·9		Alnitak		275 20·2	20·0	19·8	19·6	19·4	19·3	S. 1 57·9	57·8	57·8	57·8	57·8	57·9
2·8		Phact		275 28·1	27·9	27·7	27·4	27·2	27·1	S. 34 05·7	05·6	05·6	05·6	05·7	05·8
3·0	ζ	Tauri		276 12·7	12·5	12·3	12·0	11·8	11·7	N. 21 07·0	07·0	07·0	07·0	07·0	07·0
1·8		Alnilam	15	276 28·5	28·4	28·2	27·9	27·8	27·6	S. 1 13·7	13·6	13·6	13·6	13·7	13·7
2·9	ι	Orionis		276 39·1	38·9	38·7	38·5	38·4	38·2	S. 5 56·2	56·1	56·1	56·1	56·2	56·3
2·7	α	Leporis		277 16·7	16·5	16·3	16·1	15·9	15·8	S. 17 51·1	51·0	50·9	51·0	51·1	51·2
2·5	δ	Orionis		277 31·9	31·7	31·5	31·3	31·1	31·0	S. 0 19·8	19·7	19·7	19·7	19·8	19·8
3·0	β	Leporis		278 23·2	23·0	22·8	22·6	22·4	22·3	S. 20 47·5	47·4	47·4	47·4	47·5	47·6
1·8		Elnath	14	279 05·1	04·9	04·7	04·4	04·2	04·1	N. 28 34·3	34·3	34·3	34·3	34·3	34·4
1·7		Bellatrix	13	279 16·6	16·4	16·2	16·0	15·8	15·7	N. 6 18·7	18·8	18·8	18·8	18·8	18·7
0·2		Capella	12	281 35·8	35·6	35·3	35·0	34·7	34·6	N. 45 57·3	57·3	57·3	57·3	57·3	57·4
0·3		Rigel	11	281 52·0	51·8	51·6	51·4	51·2	51·1	S. 8 14·9	14·9	14·8	14·8	14·9	15·0
2·9	β	Eridani		283 33·0	32·8	32·6	32·4	32·3	32·2	S. 5 08·4	08·3	08·3	08·3	08·3	08·4
2·9	ι	Aurigæ †		286 25·8	25·5	25·3	25·0	24·9	24·7	N. 33 06·0	06·0	06·0	06·0	06·0	06·1
1·1		Aldebaran	10	291 37·0	36·8	36·6	36·4	36·2	36·1	N. 16 25·5	25·6	25·6	25·6	25·6	25·6
3·2	γ	Eridani		300 58·7	58·5	58·3	58·1	58·0	57·9	S. 13 37·5	37·4	37·4	37·4	37·5	37·6
3·0	ε	Persei		301 14·1	13·8	13·5	13·3	13·1	13·1	N. 39 53·3	53·3	53·4	53·5	53·5	53·6
2·9	ζ	Persei		302 07·2	07·0	06·7	06·5	06·4	06·3	N. 31 45·5	45·6	45·6	45·7	45·7	45·8
3·0		Alcyone		303 44·8	44·6	44·3	44·1	44·0	44·0	N. 23 58·6	58·6	58·7	58·7	58·8	58·8
1·9		Mirfak	9	309 39·7	39·4	39·1	38·8	38·7	38·6	N. 49 42·7	42·7	42·8	42·9	43·0	43·1
Var. §		Algol †		313 38·0	37·7	37·5	37·3	37·2	37·1	N. 40 47·7	47·7	47·8	47·8	47·9	48·0
2·8		Menkar	8	314 58·4	58·1	57·9	57·8	57·7	57·7	N. 3 55·7	55·7	55·8	55·8	55·8	55·7
3·1		Acamar	7	315 49·8	49·6	49·3	49·2	49·1	49·1	S. 40 28·0	28·0	28·0	28·1	28·2	28·3
3·1	β	Trianguli		328 13·8	13·6	13·3	13·2	13·2	13·2	N. 34 47·4	47·5	47·5	47·6	47·7	47·8
2·2		Hamal	6	328 47·4	47·2	47·0	46·9	46·8	46·8	N. 23 15·9	16·0	16·1	16·2	16·2	16·2
2·2		Almak		329 39·6	39·4	39·1	39·0	38·9	39·0	N. 42 07·7	07·8	07·9	08·0	08·1	08·2
3·0	α	Hydri		330 38·3	37·9	37·6	37·5	37·5	37·7	S. 61 46·0	46·0	46·0	46·2	46·3	46·4
2·7		Sheratan		331 54·7	54·5	54·3	54·2	54·1	54·2	N. 20 36·3	36·4	36·5	36·5	36·6	36·6
2·1		Polaris †		330 73·3	63·1	54·9	50·1	50·0	54·9	N. 89 03·9	04·0	04·1	04·3	04·5	04·6
0·6		Achernar	5	335 57·5	57·2	57·0	56·8	56·9	57·0	S. 57 26·5	26·5	26·6	26·7	26·8	26·9
2·8		Ruchbah		339 13·3	13·0	12·7	12·6	12·6	12·7	N. 60 01·0	01·1	01·2	01·4	01·5	01·6
2·4		Mirach		343 08·7	08·5	08·3	08·2	08·2	08·3	N. 35 23·9	24·1	24·2	24·3	24·4	24·4
Var. ‖	γ	Cassiopeiæ		346 26·7	26·4	26·1	26·0	26·1	26·2	N. 60 29·3	29·5	29·6	29·8	29·9	30·0
2·2		Diphda	4	349 37·2	37·0	36·9	36·8	36·9	36·9	S. 18 12·7	12·6	12·6	12·7	12·7	12·8
2·5		Schedar	3	350 27·4	27·1	26·9	26·9	26·9	27·1	N. 56 18·4	18·5	18·7	18·8	19·0	19·0
2·4		Ankaa	2	353 56·3	56·1	55·9	55·9	55·9	56·1	S. 42 31·6	31·6	31·6	31·7	31·8	31·9
2·9	β	Hydri †		354 06·0	05·3	04·9	04·9	05·3	05·9	S. 77 28·9	28·9	29·1	29·2	29·3	29·4
2·9		Algenib		357 13·3	13·1	13·0	12·9	13·0	13·0	N. 14 57·2	57·3	57·4	57·5	57·5	57·5
2·4		Caph		358 15·2	14·9	14·7	14·7	14·8	15·0	N. 58 55·1	55·2	55·4	55·6	55·7	55·7
2·2		Alpheratz	1	358 26·1	25·9	25·8	25·7	25·8	25·9	N. 28 51·7	51·8	51·9	52·0	52·0	52·0

* Formerly Argus   ‡ 0·1—1·2   § 2·3—3·5   ‖ Irregular variable; 1955 mag. 2·8
† Not suitable for use with H.O. 214 (H.D. 486)

THE ELEMENTS OF NAVIGATION AND NAUTICAL ASTRONOMY 413

POLARIS (POLE STAR) TABLES
FOR DETERMINING LATITUDE FROM SEXTANT ALTITUDE AND FOR AZIMUTH

L.H.A. ARIES	0°–9°	10°–19°	20°–29°	30°–39°	40°–49°	50°–59°	60°–69°	70°–79°	80°–89°	90°–99°	100°–109°	110°–119°
	a_0	a_0	a_0	a_0	a_0	a_0	a_0	a_0	a_0	a_0	a_0	a_0
0	0° 10.1′	0° 06.1′	0° 03.8′	0° 03.1′	0° 04.2′	0° 06.9′	0° 11.3′	0° 17.1′	0° 24.2′	0° 32.4′	0° 41.3′	0° 50.8′
1	09.7	05.8	03.6	03.1	04.4	07.3	11.8	17.8	25.0	33.2	42.2	51.7
2	09.2	05.5	03.5	03.2	04.6	07.7	12.3	18.4	25.8	34.1	43.2	52.7
3	08.8	05.3	03.4	03.2	04.8	08.1	12.9	19.1	26.6	35.0	44.1	53.7
4	08.3	05.0	03.3	03.3	05.1	08.5	13.4	19.8	27.4	35.9	45.0	54.6
5	0 07.9	0 04.7	0 03.2	0 03.4	0 05.3	0 08.9	0 14.0	0 20.5	0 28.2	0 36.7	0 46.0	0 55.6
6	07.5	04.5	03.2	03.5	05.6	09.4	14.6	21.2	29.0	37.6	46.9	56.6
7	07.2	04.3	03.1	03.7	05.9	09.8	15.2	22.0	29.8	38.5	47.9	57.6
8	06.8	04.1	03.1	03.8	06.2	10.3	15.8	22.7	30.7	39.5	48.9	58.5
9	06.5	03.9	03.1	04.0	06.6	10.8	16.5	23.5	31.5	40.4	49.8	0 59.5
10	0 06.1	0 03.8	0 03.1	0 04.2	0 06.9	0 11.3	0 17.1	0 24.2	0 32.4	0 41.3	0 50.8	1 00.5

Lat.	a_1	a_1	a_1	a_1	a_1	a_1	a_1	a_1	a_1	a_1	a_1	a_1
0°	0.5′	0.6′	0.6′	0.6′	0.6′	0.5′	0.4′	0.3′	0.2′	0.2′	0.1′	0.1′
10	.5	.6	.6	.6	.6	.5	.4	.4	.3	.2	.2	.1
20	.5	.6	.6	.6	.6	.5	.5	.4	.3	.3	.3	.2
30	.6	.6	.6	.6	.6	.5	.5	.5	.4	.4	.3	.3
40	0.6	0.6	0.6	0.6	0.6	0.6	0.5	0.5	0.5	0.5	0.5	0.4
45	.6	.6	.6	.6	.6	.6	.6	.6	.5	.5	.5	.5
50	.6	.6	.6	.6	.6	.6	.6	.6	.6	.6	.6	.6
55	.6	.6	.6	.6	.6	.6	.6	.7	.7	.7	.7	.7
60	.6	.6	.6	.6	.6	.6	.7	.7	.8	.8	.8	.8
62	0.7	0.6	0.6	0.6	0.6	0.7	0.7	0.8	0.8	0.9	0.9	0.9
64	.7	.6	.6	.6	.6	.7	.7	.8	.9	0.9	1.0	1.0
66	.7	.6	.6	.6	.6	.7	.8	.9	0.9	1.0	1.1	1.1
68	0.7	0.6	0.6	0.6	0.7	0.7	0.8	0.9	1.0	1.1	1.1	1.2

Month	a_2	a_2	a_2	a_2	a_2	a_2	a_2	a_2	a_2	a_2	a_2	a_2
Jan.	0.7′	0.7′	0.7′	0.7′	0.7′	0.7′	0.7′	0.7′	0.7′	0.7′	0.7′	0.7′
Feb.	.6	.7	.7	.7	.8	.8	.8	.8	.8	.8	.8	.8
Mar.	.5	.5	.6	.6	.7	.8	.8	.8	.9	.9	.9	0.9
Apr.	0.3	0.4	0.4	0.5	0.6	0.6	0.7	0.8	0.8	0.9	0.9	1.0
May	.2	.3	.3	.4	.4	.5	.6	.6	.7	.8	.8	0.9
June	.2	.2	.2	.3	.3	.4	.4	.5	.5	.6	.7	.8
July	0.2	0.2	0.2	0.2	0.2	0.3	0.3	0.4	0.4	0.5	0.5	0.6
Aug.	.4	.3	.3	.3	.2	.2	.3	.3	.3	.4	.4	.4
Sept.	.5	.5	.4	.4	.3	.3	.3	.3	.3	.3	.3	.3
Oct.	0.7	0.6	0.6	0.5	0.5	0.4	0.4	0.3	0.3	0.3	0.3	0.3
Nov.	0.9	.8	.8	.7	.7	.6	.5	.5	.4	.4	.3	.3
Dec.	1.0	0.9	0.9	0.9	0.8	0.8	0.7	0.6	0.6	0.5	0.4	0.4

AZIMUTH

Lat.												
0°	0.4°	0.2°	0.1°	359.9°	359.7°	359.6°	359.5°	359.3°	359.2°	359.2°	359.1°	359.1°
20	0.4	0.2	0.1	359.9	359.7	359.6	359.4	359.3	359.2	359.1	359.1	359.0
40	0.5	0.3	0.1	359.9	359.7	359.5	359.3	359.1	359.0	358.9	358.8	358.8
50	0.6	0.4	0.1	359.8	359.6	359.4	359.1	359.0	358.8	358.7	358.6	358.6
55	0.7	0.4	0.1	359.8	359.6	359.3	359.0	358.8	358.6	358.5	358.4	358.4
60	0.8	0.5	0.1	359.8	359.5	359.2	358.9	358.6	358.4	358.3	358.2	358.1
65	0.9	0.6	0.2	359.8	359.4	359.1	358.7	358.4	358.1	358.0	357.9	357.8

Latitude = corrected sextant altitude $-1° + a_0 + a_1 + a_2$

The table is entered with L.H.A. Aries to determine the column to be used; each column refers to a range of 10°. a_0 is taken, with mental interpolation, from the upper table with the units of L.H.A. Aries in degrees as argument; a_1, a_2 are taken, without interpolation, from the second and third tables with arguments latitude and month respectively. a_0, a_1, a_2 are always positive. The final table gives the azimuth of *Polaris*.

POLARIS (POLE STAR) TABLES
FOR DETERMINING LATITUDE FROM SEXTANT ALTITUDE AND FOR AZIMUTH

L.H.A. ARIES	120°–129°	130°–139°	140°–149°	150°–159°	160°–169°	170°–179°	180°–189°	190°–199°	200°–209°	210°–219°	220°–229°	230°–239°
	a_0	a_0	a_0	a_0	a_0	a_0	a_0	a_0	a_0	a_0	a_0	a_0
0	1 00·5	1 10·1	1 19·4	1 28·0	1 35·8	1 42·4	1 47·7	1 51·6	1 53·9	1 54·5	1 53·5	1 50·8
1	01·5	11·1	20·3	28·9	36·5	43·0	48·2	51·9	54·0	54·5	53·3	50·5
2	02·4	12·0	21·2	29·7	37·2	43·6	48·6	52·2	54·1	54·4	53·1	50·1
3	03·4	13·0	22·1	30·5	37·9	44·1	49·0	52·4	54·2	54·4	52·9	49·7
4	04·4	13·9	22·9	31·3	38·6	44·7	49·4	52·7	54·3	54·3	52·6	49·3
5	1 05·3	1 14·8	1 23·8	1 32·0	1 39·2	1 45·2	1 49·8	1 52·9	1 54·4	1 54·2	1 52·4	1 48·9
6	06·3	15·7	24·7	32·8	39·9	45·8	50·2	53·1	54·4	54·1	52·1	48·5
7	07·3	16·7	25·5	33·6	40·5	46·3	50·6	53·3	54·5	54·0	51·8	48·0
8	08·2	17·6	26·4	34·3	41·2	46·8	50·9	53·5	54·5	53·8	51·5	47·6
9	09·2	18·5	27·2	35·1	41·8	47·3	51·3	53·7	54·5	53·7	51·2	47·1
10	1 10·1	1 19·4	1 28·0	1 35·8	1 42·4	1 47·7	1 51·6	1 53·9	1 54·5	1 53·5	1 50·8	1 46·6

Lat.	a_1	a_1	a_1	a_1	a_1	a_1	a_1	a_1	a_1	a_1	a_1	a_1
0	0·1	0·1	0·2	0·3	0·3	0·4	0·5	0·6	0·6	0·6	0·6	0·5
10	·1	·2	·2	·3	·4	·5	·5	·6	·6	·6	·6	·5
20	·2	·3	·3	·4	·4	·5	·5	·6	·6	·6	·6	·5
30	·3	·3	·4	·4	·5	·5	·6	·6	·6	·6	·6	·5
40	0·4	0·5	0·5	0·5	0·5	0·5	0·6	0·6	0·6	0·6	0·6	0·6
45	·5	·5	·5	·5	·6	·6	·6	·6	·6	·6	·6	·6
50	·6	·6	·6	·6	·6	·6	·6	·6	·6	·6	·6	·6
55	·7	·7	·7	·7	·7	·6	·6	·6	·6	·6	·6	·6
60	·8	·8	·8	·8	·7	·7	·6	·6	·6	·6	·6	·6
62	0·9	0·9	0·8	0·8	0·8	0·7	0·7	0·6	0·6	0·6	0·6	0·7
64	1·0	1·0	0·9	·9	·8	·7	·7	·6	·6	·6	·6	·7
66	1·1	1·0	1·0	0·9	·8	·8	·7	·6	·6	·6	·6	·7
68	1·2	1·1	1·1	1·0	0·9	0·8	0·7	0·6	0·6	0·6	0·7	0·7

Month	a_2	a_2	a_2	a_2	a_2	a_2	a_2	a_2	a_2	a_2	a_2	a_2
Jan.	0·6	0·6	0·6	0·6	0·6	0·5	0·5	0·5	0·5	0·5	0·5	0·5
Feb.	·8	·8	·7	·7	·7	·6	·6	·5	·5	·5	·4	·4
Mar.	0·9	0·9	0·9	0·8	0·8	·8	·7	·7	·6	·6	·5	·4
Apr.	1·0	1·0	1·0	1·0	1·0	0·9	0·9	0·8	0·8	0·7	0·6	0·6
May	0·9	1·0	1·0	1·0	1·0	1·0	1·0	0·9	0·9	·8	·8	·7
June	·8	0·9	0·9	1·0	1·0	1·0	1·0	1·0	1·0	0·9	0·9	·8
July	0·7	0·7	0·8	0·9	0·9	0·9	1·0	1·0	1·0	1·0	1·0	0·9
Aug.	·5	·6	·6	·7	·7	·8	0·8	0·9	0·9	0·9	1·0	1·0
Sept.	·4	·4	·5	·5	·6	·6	·7	·7	·8	·8	0·9	0·9
Oct.	0·3	0·3	0·3	0·4	0·4	0·4	0·5	0·6	0·6	0·7	0·7	0·8
Nov.	·2	·2	·2	·2	·3	·3	·3	·4	·4	·5	·5	·6
Dec.	0·3	0·3	0·2	0·2	0·2	0·2	0·2	0·3	0·3	0·3	0·4	0·4

Lat.	AZIMUTH												
0	359·1	359·1	359·2	359·3	359·4	359·5	359·6	359·8	359·9	0·1	0·3	0·4
20	359·0	359·1	359·1	359·2	359·3	359·5	359·6	359·8	359·9	0·1	0·3	0·4
40	358·8	358·8	358·9	359·0	359·2	359·3	359·5	359·7	359·9	0·1	0·3	0·5
50	358·6	358·6	358·7	358·8	359·0	359·2	359·4	359·7	359·9	0·2	0·4	0·6
55	358·4	358·5	358·6	358·7	358·9	359·1	359·4	359·6	359·9	0·2	0·4	0·7
60	358·2	358·2	358·3	358·5	358·7	359·0	359·3	359·6	359·9	0·2	0·5	0·8
65	357·8	357·9	358·1	358·3	358·5	358·8	359·1	359·5	359·9	0·2	0·6	0·9

ILLUSTRATION

January 10 at G.M.T. 22ʰ 17ᵐ 50ˢ in longitude W. 27° 34′ the corrected sextant altitude of *Polaris* was 49° 31′·6.

From the daily pages
G.H.A. Aries (22ʰ)   79 54·9
Increment (17ᵐ 50ˢ)   4 28·2
Longitude (west)   −27 34
L.H.A. Aries   56 49

Corr. Sext. Alt.   49°31′·6
a_0 (argument 56° 49′)   0 09·7
a_1 (lat. 50° approx.)   0·6
a_2 (January)   0·7
Sum − 1° − Lat. −   48 42·6

POLARIS (POLE STAR) TABLES
FOR DETERMINING LATITUDE FROM SEXTANT ALTITUDE AND FOR AZIMUTH

L.H.A. ARIES	240°–249°	250°–259°	260°–269°	270°–279°	280°–289°	290°–299°	300°–309°	310°–319°	320°–329°	330°–339°	340°–349°	350°–359°
	a_0	a_0	a_0	a_0	a_0	a_0	a_0	a_0	a_0	a_0	a_0	a_0
0	1 46·6	1 41·0	1 34·0	1 26·1	1 17·3	1 07·9	0 58·2	0 48·5	0 39·1	0 30·4	0 22·4	0 15·6
1	46·1	40·3	33·3	25·2	16·3	06·9	57·2	47·6	38·2	29·5	21·7	15·0
2	45·6	39·7	32·5	24·4	15·4	06·0	56·3	46·6	37·3	28·7	21·0	14·4
3	45·1	39·0	31·7	23·5	14·5	05·0	55·3	45·7	36·4	27·9	20·2	13·8
4	44·5	38·3	31·0	22·6	13·6	04·0	54·3	44·7	35·5	27·1	19·5	13·2
5	1 43·9	1 37·6	1 30·2	1 21·7	1 12·6	1 03·1	0 53·3	0 43·8	0 34·7	0 26·3	0 18·9	0 12·7
6	43·4	36·9	29·4	20·9	11·7	02·1	52·4	42·8	33·8	25·5	18·2	12·1
7	42·8	36·2	28·6	20·0	10·7	01·1	51·4	41·9	32·9	24·7	17·5	11·6
8	42·2	35·5	27·7	19·1	09·8	1 00·1	50·4	41·0	32·1	23·9	16·9	11·1
9	41·6	34·8	26·9	18·2	08·8	0 59·2	49·5	40·1	31·2	23·2	16·2	10·6
10	1 41·0	1 34·0	1 26·1	1 17·3	1 07·9	0 58·2	0 48·5	0 39·1	0 30·4	0 22·4	0 15·6	0 10·1
Lat.	a_1	a_1	a_1	a_1	a_1	a_1	a_1	a_1	a_1	a_1	a_1	a_1
0	0·4	0·3	0·2	0·2	0·1	0·1	0·1	0·1	0·2	0·3	0·3	0·4
10	·4	·4	·3	·2	·2	·1	·1	·2	·2	·3	·4	·5
20	·5	·4	·3	·3	·3	·2	·2	·3	·3	·4	·4	·5
30	·5	·5	·4	·4	·3	·3	·3	·3	·4	·4	·5	·5
40	0·5	0·5	0·5	0·5	0·5	0·4	0·4	0·5	0·5	0·5	0·5	0·5
45	·6	·6	·5	·5	·5	·5	·5	·5	·5	·5	·6	·6
50	·6	·6	·6	·6	·6	·6	·6	·6	·6	·6	·6	·6
55	·6	·7	·7	·7	·7	·7	·7	·7	·7	·7	·7	·6
60	·7	·7	·8	·8	·8	·8	·8	·8	·8	·8	·7	·7
62	0·7	0·8	0·8	0·9	0·9	0·9	0·9	0·9	0·8	0·8	0·8	0·7
64	·7	·8	·9	0·9	1·0	1·0	1·0	1·0	0·9	·9	·8	·7
66	·8	·9	0·9	1·0	1·0	1·1	1·1	1·0	1·0	0·9	·8	·8
68	0·8	0·9	1·0	1·1	1·1	1·2	1·2	1·1	1·1	1·0	0·9	0·8
Month	a_2	a_2	a_2	a_2	a_2	a_2	a_2	a_2	a_2	a_2	a_2	a_2
Jan.	0·5	0·5	0·5	0·5	0·5	0·5	0·6	0·6	0·6	0·6	0·6	0·7
Feb.	·4	·4	·4	·4	·4	·4	·4	·4	·5	·5	·5	·6
Mar.	·4	·4	·3	·3	·3	·3	·3	·3	·3	·4	·4	·4
Apr.	0·5	0·4	0·4	0·3	0·3	0·2	0·2	0·2	0·2	0·2	0·2	0·3
May	·6	·6	·5	·4	·4	·3	·3	·2	·2	·2	·2	·2
June	·8	·7	·7	·6	·5	·4	·4	·3	·3	·2	·2	·2
July	0·9	0·8	0·8	0·7	0·7	0·6	0·5	0·5	0·4	0·3	0·3	0·3
Aug.	·9	·9	·9	·8	·8	·8	·7	·6	·6	·5	·5	·4
Sept.	·9	·9	·9	·9	·9	·9	·8	·8	·7	·7	·6	·6
Oct.	0·8	0·9	0·9	0·9	0·9	0·9	0·9	0·9	0·8	0·8	0·8	0·8
Nov.	·7	·7	·8	·8	·9	·9	1·0	1·0	1·0	1·0	0·9	0·9
Dec.	0·5	0·6	0·6	0·7	0·8	0·8	0·9	0·9	1·0	1·0	1·0	1·0
Lat.						AZIMUTH						
0°	0·5	0·7	0·8	0·8	0·9	0·9	0·9	0·9	0·8	0·7	0·6	0·5
20	0·6	0·7	0·8	0·9	1·0	1·0	1·0	0·9	0·9	0·8	0·7	0·5
40	0·7	0·9	1·0	1·1	1·2	1·2	1·2	1·2	1·1	1·0	0·8	0·7
50	0·8	1·0	1·2	1·3	1·4	1·4	1·4	1·4	1·3	1·2	1·0	0·8
55	0·9	1·1	1·3	1·5	1·6	1·6	1·6	1·6	1·5	1·3	1·1	0·9
60	1·1	1·3	1·5	1·7	1·8	1·8	1·8	1·8	1·7	1·5	1·3	1·0
65	1·3	1·5	1·8	1·9	2·1	2·2	2·2	2·1	2·0	1·8	1·6	1·3

Latitude = corrected sextant altitude $-1° + a_0 + a_1 + a_2$.

The table is entered with L.H.A. Aries to determine the column to be used; each column refers to a range of 10°. a_0 is taken, with mental interpolation, from the upper table with the units of L.H.A. Aries in degrees as argument; a_1, a_2 are taken, without interpolation, from the second and third tables with arguments latitude and month respectively. a_0, a_1, a_2 are always positive. The final table gives the azimuth of *Polaris*.

ANSWERS TO EXERCISES:

Exercises to Chapter 1 (*page* 9)
1. (i) 0·488;   (ii) 0·469;   (iii) 1·494
2. (i) 55°;   (ii) 44°;   (iii) 37°
3. $\theta = 24°$
   $\cos \theta = 0·9$; $\sec \theta = 1·1$; $\tan \theta = 0·4$;
   $\cot \theta = 2·2$; $\csc \theta = 2·4$
10. $101\frac{1}{2}°$
15. 55° 09′

Exercises to Chapter 2 (*page* 13)
1. (i) 2·393;   (ii) 1·045;   (iii) 0·011
2. (i) 2·133;   (ii) 0·518;   (iii) 3·69
3. (i) 46·97 cm.;   (ii) 273·1 yd.
6. 3,438
7. 6·72 ml.
8. $67\frac{1}{2}°$; $3\pi/8$ or $1·18^c$
9. 6,876 ft.
11. (i) sin 5°=0·872620;   (ii) cosec 2°=28·731669;   (iii) tan 30′=0·0087262;
    (iv) cot 89°=0·0174524
12. (i) 0·01745400;   (ii) 85·940185

Exercises to Chapter 3 (*page* 19)
2. (i) 0·766;   (ii) 2·28;   (iii) 0·726
3. 29° (D. Lat.: Dep.: : 180 : 100)
4. 25°
5. 8°·9
6. 18°   421 lire;   161·9 Fr.
7. 13′·65
8. 94·87 sq. ml.
9. 25 sq. ft.
10. 035°
11. Corr.=48′·41;   Lat.=41° 21′·6
12. 48′·5

Exercises to Chapter 3 (*continued*)

13. (i) 5° E.; (ii) 8°·7 W.
14. (i) 5° E.; (ii) 3°·4 W.
15. 6°·4
16. 3°·25
17. 5·8 ml.
18. 86·6 ft.
19. 419½ ft.
20. 070° 32′
21. 4·216 ml.
22. 21° 48′ N. or S.
23. 43·01 sq. in.
24. 26° 27′

Exercises to Chapter 4 (*page* 24)

5. cos 15°=0·966; sin 75°=0·966
6. tan 15°=0·268; tan 75°=3·732

Exercises to Chapter 5 (*page* 31)

1. 2·873 ml.
2. 4·607 ml.
3. $A=58°\ 24\tfrac{3}{4}'$; $B=96°\ 22\tfrac{3}{4}'$; $C=25°\ 13\tfrac{1}{2}'$
4. 10·57 ml.
5. 2·21 ml.
6. 007° 43′
7. $292\tfrac{1}{2}°$; 3 hr. 11 m.
8. 16·44 ml.
9. 20·1 km.
10. $AD=165\cdot4$ m.; $BD=123\cdot2$ m.; $CD=234\cdot5$ m.
11. $AC=800$ yd.
12. $B=27°\ 04'$ or $78°\ 16'$

Exercises to Chapter 8 (*page* 50)

1. 51° 19′ or 128° 41′
2. $54°\ 01\tfrac{3}{4}'$
3. $168°\ 41\tfrac{1}{2}'$
4. $98°\ 30\tfrac{1}{2}'$
5. 32° 51′

ANSWERS TO EXERCISES

Exercises to Chapter 8 (*continued*)
6. 49° 36′
7. 154° 16¾′
8. $Z=112°\ 28′$; $ZX=57°\ 07¾′$

Exercises to Chapter 9 (*page* 57)
1. $a=32°\ 18¼′$; $b=56°\ 14½′$; $c=48°\ 53½′$
2. $e=27°\ 57′$; $f=77°\ 00¼′$; $G=82°\ 58′$
3. $X=42°\ 04′$; $y=108°\ 51′$; $z=104°\ 13¼′$
4. $Q=72°\ 31¾′$; $p=116°\ 55′$; $q=70°\ 33½′$
5. $Y=110°\ 29½′$; $x=144°\ 52′$; $z=63°\ 33′$
6. $A=60°\ 47½′$; $a=74°\ 23′$; $b=146°\ 31′$
7. $X=52°\ 23′$; $Y=65°\ 19′$; $Z=104°\ 46′$
8. $Z=61°\ 29½′$ or $118°\ 30½′$
9. $\begin{cases} C=58°\ 41½′ \text{ or } 121°\ 18½′ \\ B=123°\ 26′ \text{ or } 22°\ 27′ \end{cases}$
10. 49° 40′

Exercises to Chapter 10 (*page* 67)
12. Lat. 20° N., Long. 165° E.
13. Lat. 46° S., Long. 00°
14. Lat. 00° 00′, Long. 15° 18′ W.
15. (*a*) 04° 09′; 24° 01′
    (*b*) 11° 26′; 176° 34′
19. 59° 50′ N.
20. 6,074·76 ft.

Exercises to Chapter 11 (*page* 75)
9. 263′·0 E.
10. 335½ ml.
11. 58° 37′
12. 17,693·5 ml.
13. 450 kn.
14. 090° × 0·88 kn.
15. 40° 30′ N.; 16° 33½′ W.
16. 178° 04′ W.
17. 090° × 11·1 ml.
18. 25° 17′ N.; 583 ml.
19. Dep.=378·4 ml.; Dist.=462 ml.

420 THE ELEMENTS OF NAVIGATION AND NAUTICAL ASTRONOMY

Exercises to Chapter 11 (*continued*)
20. Dep.=91·2 ml.;   Change in Lat.=05° 01'·6 S.
21. Dist.=354 ml.;   Change in Lat.=04° 39' S.
22. Dist.=100 ml.;   Course=234°
23. Course=212°;   Dist.=324 ml.;   Lat.=46° 39' N.
24. Course=176°;   Dist.=270 ml.;   Lat.=19° 19'·6 S.
25. Course=328½°;   Dist.=625 ml.;   Lat.=25° 29' S.
26. 44·59 ml.
27. Dist.=3,686·2 ml.

Exercises to Chapter 12 (*page* 84)
 9. 33·6
10. 2·388" to 1°
11. 3·40 in.
12. 1·943" to 1' of Long.
13. 2·87" to 1' of Lat.

Exercises to Chapter 13 (*page* 98)
 5. 337·3
 6. Co.=244° 32½';   Dist.=667·1 ml.
 7. Lat.=46° 50' N.;   Long.=23° 44'·3 W.
 8. Co.=248° 20';   Dist.=443·6 ml.
 9. Co.=086° 13½';   Dist.=1,820·1 ml.
10. Dist.=973·1 ml.;   Long.=79° 42'·8 W.
11. Lat.=04° 09' S.;   Dist.=1,711·4 ml.
12. Co.=027° 27';   Dist.=928·4 ml.
13. Bearing=158°;   Dist.=18·3 ml.
14. Lat.=38° 16' N.;   Long.=175° 20' W.
15. Lat.=51° 10'·1 N.;   Long.=08° 32'·5 W.
16. Set=147°;   Drift=14·9 ml.
17. Lat.=32° 21'·6 S.;   Long.=51° 36'·0 E.
18. Set=166°;   Rate=4·6 ml. in 24 hr.

Exercises to Chapter 14 (*page* 109)
 7. 7·694 in.
12. Lat.=36° 25' S.;   Long.=53° 42' W.
14. Within
18. Lat.=29° 00' N.;   Long.=50° 00' E. or 130° 00' W.

ANSWERS TO EXERCISES

Exercises to Chapter 14 (*continued*)

19. Lat.=30° 00' S.;   Long.=80° 00' E.
20. Long.=70° 00' E. or 110° 00' W.
21. Co.=270° 58';   Dist.=2,769¾ ml.
    Points: (i) 49° 24½' N., 16° 31' W.
           (ii) 48° 04½' N., 26° 31' W.
          (iii) 45° 44½' N., 36° 31' W.
           (iv) 42° 10' N., 46° 31' W.
            (v) 37° 18' N., 56° 31' W.
22. Dist.=1,677·5 ml.;   Co.=224° 50'
23. Co.=270° 10';   Dist.=3,396 ml.
    Points: (i) 35° 58½' N., 10° 00' W.
           (ii) 35° 13' N., 20° 00' W.
          (iii) 33° 38' N., 30° 00' W.
           (iv) 31° 09' N., 40° 00' W.
            (v) 27° 42½' N., 50° 00' W.
           (vi) 23° 16' N., 60° 00' W.
24. Dist.=6,188 ml.;   Co.=092° 39'
25. Dist.=5,958 ml.;   Co.=166° 06'
26. Co.=228° 48';   Dist.=4,264½ ml.
27. Co.=285° 35';   Dist.=1,651½ ml.
    Points: (i) 52° 36' N., 20° 00' W.
           (ii) 53° 00' N., 30° 00' W.
          (iii) 53° 00' N., 40° 00' W.
           (iv) 52° 53' N., 50° 00' W.
28. 1,052 ml.

Exercises to Chapter 15 (*page* 119)

5. (*a*) 1 : 729600;   (*b*) 1 : 1216
6. 3·704 cm.

Exercises to Chapter 16 (*page* 129)

9. (i) Error=3° W.;   True Co.=321°
   (ii) Dev.=0°;   Error=5° E.
  (iii) Dev.=24° W.;   True Co.=189°
   (iv) Comp. Co.=088°;   Dev.=2° W.
    (v) Var.=14° E.;   Comp. Co.=247°
   (vi) Error=9° W.;   True Co.=351°
  (vii) Var.=1° W.;   True Co.=266°
 (viii) Var.=14° E.;   True Co.=124°
10. 4° E.
11. Nil
12. 055°

422 THE ELEMENTS OF NAVIGATION AND NAUTICAL ASTRONOMY

Exercises to Chapter 16 (*continued*)

13. 313°
14. (*a*) 117°
    (*b*) 225°
    (*c*) 193¾°
    (*d*) 290°
    (*e*) 283°
15. 330°; 255°; 183°

Exercises to Chapter 18 (*page* 140)

6. 13¾ ml.
7. 9½ ml.; 18° 27′ N.; 63° 25½′ W.
8. 16 ml.; 22 ml.
9. 2·5 ml.
10. 14·63 ml.
11. 13·99 ml.
12. 9·01 ml.
13. 3·5 ml.
14. 9·01 ml.
15. 13·05 ml.

Exercises to Chapter 19 (*page* 146)

6. 0·50 ml.
8. 22·46 ml.
9. 12·82 ml.

Exercises to Chapter 20 (*page* 152)

8. 5·0 ml.; Error=6° W.
9. 3·1 ml.
10. 220° × 2·3 ml.
11. 1·6 ml. or 4·7 ml.
12. 6·3 ml.

Exercises to Chapter 21 (*page* 156)

2. 070°
3. 332°
4. 1·1 ml.
5. Dist.=6·2 ml.; Set=060°; Rate=1·1 kn.
6. Co.=350°; Drift=4·5 ml.

ANSWERS TO EXERCISES

Exercises to Chapter 22 (*page* 161)
 3. 184°
 4. 122°; A/C 23° to starboard
 5. 158°; 13¼ kn.
 6. W. Magnetic
 7. Co.=255°; Current: 288° × 13 ml.
 8. Co.=159°; Time=1 hr. 45 m.; Dist.=17·4 ml.
 9. 280°

Exercises to Chapter 23 (*page* 165)
 6. 090¾°

Exercises to Chapter 24 (*page* 176)
 8. 7 : 3 approx.
 13. 9·8 ft.
 14. 25·9 ft.
 15. 27·0 ft.
 16. 0320 G.M.T.
 17. 0403 G.M.T.
 18. 9·2 ft.
 19. 10·2 ft.
 20. 34·1 ft.
 21. 25·6 ft.
 22. 1856 G.M.T.
 23. 22·7 ft.
 24. 1145 G.M.T.

Exercises to Chapter 25 (*page* 185)
 5. $(2\frac{1}{2})^5$
 6. $(2\frac{1}{2})^{27}$

Exercises to Chapter 27 (*page* 199)
 5. 11° 31½′ N.; 54° 21½′ E.
 6. (*a*) 00° 00′; 90° 00′ E.
    (*b*) 23½° N.; 270°
 7. 46° 16′ N.; 323° 43′·8

424 THE ELEMENTS OF NAVIGATION AND NAUTICAL ASTRONOMY

Exercises to Chapter 31 (*page* 242)

24. 34° 57′·1
25. 36° 39′·8
26. 43° 37′·1
27. 49° 57′·1
28. 52° 35′·2
29. 32° 28′·9
30. 24° 54′·3
31. 54° 29′·5
32. 36° 49′·3
33. 56° 46′·2
34. 34° 22′·9
35. 42° 54′·6
36. 36° 19′·6
37. 47° 40′·6
38. 62° 16′·4
39. 79° 26′·2
40. 64° 09′·2
41. 41° 21′·4
42. 48° 14′·6

Exercises to Chapter 32 (*page* 252)

3. (i) 00° 20′·3 S.;   28° 02′·8 E.
   (ii) 00° 23′·5 N.;   16° 47′·4 W.
   (iii) 00° 10′·5 S.; 178° 05′·0 E.
4. (i) 00° 10′·0 N.;   175° 40′·4 W.
   (ii) 23° 13′·9 N.; 139° 00′·5 E.
5. 23° 20′·5 N.;   00° 07′·2 E.
6. 23° 15′·3 N.;   25° 34′·1 E.
7. 56° 11′·3 N.;   39° 26′·8 E.
8. 23° 15′·9 N.; 175° 10′·5 E.
9. 56° 18′·3 N.;   39° 31′·3 W.
10. 15° 58′·6 S.; 165° 04′·7 W.
11. 14° 34′·8 S.;   6° 53′·7 W.
12. 23° 16′·0 N.; 174° 32′·0 E.
13. 23° 20′·0 S.;   45° 20′·5 W.

ANSWERS TO EXERCISES 425

Exercises to Chapter 33 (*page* 266)

6. (i) 10 h. 00 m.; (ii) 22 h. 01 m.; (iii) 00 h. 33 m.
7. (i) 13 h. 22 m.; (ii) 10 h. 16 m.; (iii) 22 h. 44 m.
8. (i) 15 h. 58 m.; (ii) 00 h. 51 m.
9. (i) 09 h. 44 m.; (ii) 03 h. 59 m.; (iii) 23 h. 31 m.
10. 63° 06′ N.
11. 48° 34′ S.
12. 090°–270° through Lat. 33° 45′·0 S. Long. 78° 45′ E.
13. 090°–270° through Lat. 44° 51′·8 N. Long. 29° 45′ W.
14. 090°–270° through Lat. 50° 00′·3 N. Long. 160° 45′ W.
15. 090°–270° through Lat. 32° 34′·0 S. Long. 156° 00′ E.
16. 090°–270° through Lat. 56° 53′·9 N. Long. 53° 15′ W.
17. 090°–270° through Lat. 50° 32′·9 N. Long. 70° 00′ W.
18. 090°–270° through Lat. 29° 39′·1 S. Long. 75° 00′ E.
19. 090°–270° through Lat. 4° 42′·0 N. Long. 152° 00′ E.
20. 090°–270° through Lat. 68° 25′·0 N. Long. 120° 00′ W.
21. 090°–270° through Lat. 00° 12′·2 S. Long. 23° 00′ W.

Exercises to Chapter 34 (*page* 289)

14. P.L. 000°–180°. Intercept 10·3 ml. *Towards*. Azimuth 090°
15. P.L. 144½°–324½°. Intercept 8·7 ml. *Towards*. Azimuth 234½°
16. P.L. 053°–233°. Intercept 4·4 ml. *Away*. Azimuth 324°
17. P.L. 179°–359°. Intercept 4·6 ml. *Towards*. Azimuth 089°
18. P.L. 159½°–339½°. Intercept 5·8 ml. *Away*. Azimuth 069½°
19. P.L. 073½°–253½°. Intercept 6·8 ml. *Towards*. Azimuth 343½°
20. P.L. 138½°–318½°. Intercept 32·3 ml. *Away*. Azimuth 228½°
21. P.L. 029°–209°. Intercept 2·5 ml. *Away*. Azimuth 299°
22. P.L. 061½°–241½°. Intercept 6·3 ml. *Towards*. Azimuth 151½°
23. P.L. 006½°–186½°. Intercept 7·2 ml. *Away*. Azimuth 096½°
24. P.L. 042°–222°. Intercept 3·5 ml. *Towards*. Azimuth 132°
25. P.L. 136½°–316½°. Intercept 23·4 ml. *Away*. Azimuth 226½°

Exercises to Chapter 35 (*page* 297)

10. 40° 10′ N., 46° 13′ W.
11. 25° 10′ S., 120° 25¼′ W.
12. 44° 20′ N., 155° 26′ W.
13. 60° 29½′ S., 30° 21′ W.
14. 51° 00′ N., 43° 55′ W.

Exercises to Chapter 35 (*continued*)

15. 40° 30′ S., 123° 44½′ W.
16. 39° 52′ N., 00° 58′ W.
17. 36° 25′ N., 09° 18′ W.

Exercises to Chapter 36 (*page* 307)

2. Alioth (32) will cross North
   Gacrux (31) will cross South with a small altitude
   Gienah (29) will cross South
3. Nunki (50) has crossed North
4. Canopus (17) will cross South with a low altitude
   Betelgeuse (16) will cross South
   Alnilam (15) has just crossed South
   Elnath (14) has crossed South
   Bellatrix (13) has crossed South
   Capella (12) has crossed North
   Rigel (11) has crossed North
5. Antares (42) will cross South
   Atria (43) will cross South
6. Eltanin (47) will cross North

Exercises to Chapter 37 (*page* 312)

4. 45° 26′·1 N.
5. 42° 21′·2 N.
6. P.L. 089½°–269½° through Lat. 38° 27′·2 N. Long. 33° 00′ W.
7. P.L. 089°–269° through Lat. 27° 49′·5 N. Long. 75° 00′ W.
8. 7°·5 W.
9. 2°·8 W.
10. 2°·0 W.

Exercises to Chapter 40 (*page* 334)

3. 942′·5 per hour
12. 2 m. 36 s.
13. 1 m. 26 s.

INDEX

	PAGE
ABBREVIATIONS on chart	115
A B C tables	104, 286, 314, 316
Abnormal refraction	146
Absolute magnitude	180
Abscissa	196
Aclinic line	338
Adjustments—	
of compass	340
of sextant	361
Admiralty—	
List of Radio Signals	164, 365
Tide Tables	171
Age of Moon	169, 219, 220
A and *K* Tables	322
Almanac	214
Alternating light	116
Altitude—	
apparent	229
circle of equal	249
corrections	235
maximum	331
meridian	253
observed	229
of Polaris	308
parallel of	197
rate of change of	330
sextant	229
true	229
Ambiguity (180°)	371
Ambiguous case	26
Amphidromic—	
point	170
system	170
Amplitude—	
definition	287
formula	287
tables	287
uses	316
Angle(s)—	
back	240
complementary	4
compound	22
Greenwich hour	199
horizontal	147
horizontal danger	144
hour	198, 207
local hour	198
of cut	133, 149
of elongation	182
sidereal hour	199
spherical	34
vertical sextant	142
Angular acceleration	346

	PAGE
Annular eclipse	223
Anomalistic year	214
Anschutz	343
Antarctic circle	192
Antipodes	33
Aphelion	184, 187, 223
Apogee	184, 218
Apparent—	
altitude	229
annual motion of Sun	188
diurnal motion	201
magnitude	180
solar day	206
time	213
Applied couple	346
Apse line	185
Apsides	185
Arc of sextant	359
Arctic circle	192
Aries, First Point of	196
Arm of couple	344
Arrow up (down)	341
Artificial—	
horizon	242
magnet	337
satellite	181
Ascending node	192, 218
Asteroid	181
Astronomical—	
latitude	65
mean sun	210
position line	227, 245
refraction	230
telescope	360
triangle	199, 268
twilight	325
Astronomy—	
general	177
nautical	277
Atomic—	
clock	214
time	214
Augmentation	232
Autumn	189
Autumnal equinox	189, 219
Axial motion of Earth	187
Axis of great circle	33
Azimuth—	
definition	197, 286
limiting	330
mirror	341
tables: *A B C*	104, 286, 314
tables: Burdwood	286, 318, 323

427

428 THE ELEMENTS OF NAVIGATION AND NAUTICAL ASTRONOMY

	PAGE
Azimuth—	
tables: Davis	286, 318, 323
Azimuthal oscillation	354
BACK angle	354
Ballistic—	
deflection	357
tilt	357
Barycentre	168
Base line	376
Base line extension	376
Bearing(s)—	
and angle	132
and sounding	133
check	131
compass	122
cross	131
definition	197
fixing by	131
four point	138
magnetic	122
Mercatorial	162
open	136
radio	161
relative	132
rhumb line	162
true	122
Bellini-Tosi D.F.	372
Bissextile year	215
Bissextus	215
Bow (doubling angle on)	137
Boyle's Law	367
Brown S. G.	343
Bubble attachment	242, 314
Burdwood azimuth tables	286, 314, 323
CABLE	66
Caesium atom	214
Calculated zenith distance	286
Calendar—	
Christian	216
definition	215
ecclesiastical	216
Gregorian	215
Julian	215
Old Style	215
Roman	215
Calends	215
Calibration of D.F.	373
Cancer, Tropic of	192
Capricorn, Tropic of	192
Cardinal points	62
Cartesian co-ordinates	196
Catalogue of hydrographic charts	114
Cautionary notices	119
C correction	293
Celestial—	
concave	187
equator	188
hemisphere	188
horizon	197, 229

	PAGE
Celestial—	
latitude	198
longitude	196
meridian	198
pole	198
sphere	187
Centrifugal force	184
Centring error	363
Change (New)	183
Chart—	
abbreviations	115
catalogue	114
coastal	113
comparison unit	383
construction	83
corrections	114
cross	
datum	166
description of	113, 115
gnomonic	102
isogonic	338
lattice	375
Mercator	75, 77, 100
miscellaneous	113
plan	113
plotting	294
route	113
scale	113
symbols	117
time zone	213
title	115
triangle	86
wind and current	113
work	112
Charted—	
depth	115
height	175
range	116
Check bearing	131
Chernikeeff	370
Choosing marks for fixing	151
Chosen latitude	270
Chronometer—	
description	227, 364
journal	361
use of	365
winding	365
Circle (of)—	
diurnal	201
equal altitude	249
great	33
illumination	189, 190
latitude	196
position	251
prime vertical	202
small	33
vertical	197
Circular measure	10
Circumpolar—	
body	201, 202
star	202
Civil twilight	325

INDEX

	PAGE
Coastal navigation	113
Coast Radio Station	118
Cocked hat	132
Collimation error	363
Common year	215
Comparison frequency	379
Compass—	
adjustments	340
bearing	122, 340
correction	340
course	122
deviation	124
dry card	338
error	122, 287
gyro	343
instrument	62, 338, 343
liquid	338
magnetic	338
north	121
period	357
Compensator weights	355
Complementary angle	4
Composite great circle	106
Compound angles	15
Confocal hyperbolae	375
Conjunction—	
inferior	182
superior	182
Conservation of angular momentum	346
Consol	118, 377
Constellation	180
Conventional projection	36
Convergency	162
Co-ordinated Universal Time	214
Co-ordinates	196
Co-range lines	170
Correction—	
of altitudes	235–8
of courses	121
of longitude	293
Cosecant	6
Cosine	5
Cosine formula	27
Cotangent	5
Co-tidal lines	170
Couple	345
Course—	
angles	100
compass	122
correction	121
magnetic	122
true	122
Crescent	183
Cross bearings	131
Crossing the equator	94
Culmination	201
Current	74, 95
Current sailing	183
DAMPING—	
definition	354

	PAGE
Damping—	
torque	354
Danger angle	144, 147
Date line	213
Davis' azimuth tables	286, 323
Day—	
apparent solar	206
lunar	205
mean solar	206
sidereal	205
solar	187, 205
Daylight and darkness	190
Day's run	95
Day's work	96
Dead reckoning—	
definition	73
position	157
"d" correction	248
Decca—	
fix	380
lane	380
lattice chart	379
navigator	379
zone	380
Declination	198
Decometer	380
Deep sea lead line	367
Departure	69
Departure position	73
Descending node	192, 219
Detector light	118
Deviation—	
card	125
definition	124
table	125
D.F. station	118
Diaphone	118
Difference of—	
latitude	63
longitude	63, 69
meridional parts	80
Dip—	
effect of refraction on	231
magnetic	338
needle	337
of sea horizon	229
Dipping range	145
Direction finder	162
Direct motion	195
Distance—	
great circle	100
of sea horizon	229
off by vertical angle	142
on Mercator chart	78
rhumb line	68
zenith	249
Distance difference systems	375
Diurnal—	
circles	201
tides	166
Dominical letter	216

Doppler shift	386
Doubling the angle on the bow	137
Drift (current)	74
Drift (gyro)	349
Drifting	350
D.R. position	73, 157
Duration of tide	166
Dynamical Mean Sun	210
EARTH—	
annual motion	187
axial motion	187
magnetism	337
-Moon system	346
radius	61
revolution	187
rotation	61, 187
satellite	385
shape	61
size	61
true shape	64
Easter	216
Ebb tide	171
Ecclesiastical calendar	216
Echo-sounder	368
Eclipse—	
annular	223
of Moon	222
of Sun	222
partial	223
Ecliptic—	
definition	188
obliquity of	188
system	196
Ellipticity of the Earth	64
Elongation angle	182, 183
Engine distance	95
Epact	221
Ephemeris time	214
Equation of time	208, 209
Equator—	
crossing the	94
definition	62
magnetic	338
Equatorial system	198
Equilibrium theory of the tide	167
Equinoctial point	189
Equinox—	
autumnal	189
definition	189
spring	189
vernal	189
Equisignal	377
Eratosthenes	61
Erecting telescope	360
Error(s)—	
compass	122, 287
gyro compass	354
in altitude	293
in longitude	292
latitude	354

Error(s)—	
latitude, course and speed	355
rolling	355
Estimated position	73, 157
Euler, L.	385
Evening star	182
Ex-meridian—	
altitude	301
by intercept method	302
by Napier's rules	303
formula	302
observation	301
problem	306
tables	304
Extra-galactic nebulae	179
Extreme range	145
FALSE echoes	383
Family of confocal hyperbolae	375
Figure drawing	45
First Point of Aries	192, 196, 205
Fix (by)—	
bearing and angle	132
bearing and sounding	133
cross bearings	131
Decca	380
four point bearing	138
horizontal angles	147
reliability of	149
running	136
transit bearings	134
vertical angles	142
Fixed light	116
Fixing, choosing marks for	133
Flamsteed, J.	385
Flashing light	116
Flinders' bar	340
Flinders, M.	340
Flood stream	171
Fog—	
bell	118
detector light	118
diaphone	118
gong	118
gun	118
horn	118
nautophone	118
whistle	118
Force	344
Foucault's pendulum	343
Four parts formula	49
Four point bearing	138
Free gyroscope	351
Frigid zone	191
GALACTIC arch	179
Galaxy	179
Geocentric latitude	65
Geographical—	
latitude	65

INDEX

	PAGE
Geographical—	
mile	67
position	245
Geometrical projection	36
Geometry of sphere	33
Gibbous	183, 218
Gnomonic—	
chart	102
projection	114
Golden Number	216
Goniometer	372
Graduation error	363
Gravity	344
Gravity control	351
Great circle—	
course	104
definition	33
distance	104
sailing	100
vertex of	103
Greenwich—	
hour angle	199
meridian	62
Gregorian calendar	215
Gregory (Pope)	215
Ground radiation	373
Gymbals	339
Gyroscope—	
compass	343
free	351
Gyroscopic—	
errors of compass	354
inertia	188, 344, 347
HALF cardinal points	62
Half convergency	163, 373
Halley, E.	385
Hand lead line	367
Hard iron	339
Harrison, J.	227
Harvest Moon	221
Haversine—	
definition	4
formula	28
H.D. *486*	319
Head-up display	383
Height of tide	172
Hemisphere	62
Hesperus	182
H.O. *214*	318
Horizon—	
artificial	242
bubble	243
celestial	197, 229
dip of sea	229
glass	359
north point of	197
sensible	229
system	197
theoretical	144, 230
visible	229

	PAGE
Horizontal—	
angle	147
danger angle	149
parallax	234
Hour angle—	
definition	198, 207
Greenwich	199
local	198
sidereal	199
Hour circle—	
definition	198
6 o'clock	202
Hunter's Moon	221
Hydrographic Department	114
Hyperbola	375
Hyperbolic navigation	375
IDES	215
Incremental table	248
Index—	
definition	359
error	362
maps	114
mirror	362
Inertia, law of	344
Inertial navigation	385, 387
Inferior—	
conjunction	182
passage	201, 257
planet	182
transit	202
Information on chart	114
Inspection tables	271, 318
Instruments—	
azimuth mirror	341
Bellini-Tosi D.F.	372
chronometer	364
Consol	377
Decca navigator	379
echo-sounder	368
gyroscopic compass	343
lead line	367
liquid compass	338
logs	369
Loran	378
magnetic compass	338
mechanical sounding machine	367
pitometer log	370
quartz chronometer	365
Racon	383
Radar	382
radio D.F.	371
Ramark	383
sextant	359
SINS	388
taffrail log	370
TRANSIT	386, 387
two-day chronometer	365
Intercept	279
Intercept method	269, 278
Intermediate points	62

	PAGE
International date line	213
International System (S.I.) Units	214
Interpolation tables	248
Interval between maximum and meridian altitudes	332
Irradiation	235
Island universe	179
Isoclinic lines	338
Isogonal	123
Isogonic chart	123, 338
JULIAN calendar	25
KAUFMAN, G.	77
Kelvin—	
Lord	340, 385
sounding machine	367
sounding tube	367
spheres	340
Kepler, J.	183
Kepler's Laws	183
Knot	66
LAGGING of tides	170
Lalande, J. J.	385
Land effect	373
Latitude—	
astronomical	65
by meridian altitude	253, 255
by Polaris	308
celestial	196
circle of	196
error	354
geocentric	65
geographical	65
mean	90
middle	90
of Geographical Position	245
parallel of	62
reduced	65
reduction to	65
Lattice charts	375
Lead line—	
deep sea	367
hand	367
Leap second	214
Leap year	215
Leeway	127
Libra, First Point of	196
Librations	222
Light—	
alternating	116
characteristics	116
dipping	145
fixed	116
flashing	116
fog detector	118
group flashing	116
Morse code	116
occulting	116

	PAGE
Light—	
rising	145
sector	116
year	179
Limiting azimuth	330
Line of apsides	185
Lines of force	337
Liquid compass	338
Liquid gravity control	352
Local hour angle	198
Locus	147
Log—	
clock	369
definition	369
Longitude—	
and time	212
celestial	196
definition	62
general remarks	271
method of sight reduction for	269
of Geographical Position	245
Looming	146
Loop aerial	371
Loran	378
Loxodromic curve	68
Lubber line	339
Lucifer	182
Lunar—	
day	205
method	227
month	220
tide	169
Lunation	220
Luni-solar tide	169
MACROBIUS	215
Magnet—	
artificial	337
corrector	340
Magnetic—	
bearing	122
compass	338
course	122
dip	337
equator	338
field	337
lines of force	337
meridian	121
north	121
poles	338
variation	123
Magnetism—	
Earth's	121
indices	339
permanent	339
ship's	339
terrestrial	121, 337
Magnitude—	
absolute	180
apparent	180

INDEX

	PAGE
Main stream of flood	170
Map projection—	
gnomonic	101
Mercator	77
stereographic	36
Marcq St. Hilaire method	278
Mass	344
Mayer, T.	385
Maximum altitude	331
Mean—	
H.W.S.	116, 142
latitude	90
solar day	187, 205
Sun	206, 210
tide level	166
to middle latitude correction	90
Mercator chart—	
advantages	77
defects	78
description	75, 77, 100
distortion	78
exaggeration of meridians	79
exaggeration of parallels	78
principle	81
projection	36
sailing	75, 86
Mercatorial bearing	162
Meridian—	
altitude	253
celestial	197
celestial bodies near the	299
convergency of	162
definition	62
magnetic	121
observer's inferior	197
observer's superior	197
passage	201, 257
virtual	355
zenith distance	253
Meridional parts	79
Mer. Pts. for spheroid	82
Meton	216
Metonic cycle	216
Micrometer	359
Middle latitude sailing	75, 86, 90
Midnight Sun	327
Mile—	
geographical	67
nautical	65
Milky Way	179
Minor planet	181
Miscellaneous charts	113
Modified formula	283
Moment of couple	345
Momentum	344
Moon—	
age of	220
altitude corrections for	235
augmentation	232
change	219

	PAGE
Moon—	
crescent	220
declination of	221
distance of	218
eclipse of	222
first quarter	220
full	220
gibbous	220
harvest	221
horizontal parallax of	234
hunter's	221
last quarter	220
librations of	222
new	219
orbit of	218
phases of	216
retardation of	218
rise (set)	324
waxing and waning	220
winter and summer full	221
Morning star	182
Morse code light	116
Motion—	
direct	195
retrograde	195
Myerscough and Hamilton tables	322
NADIR	197
Napier's Rules—	
for ex-meridian problem	303
for oblique spherical triangles	56
for quadrantal spherical triangles	55
for right angled spherical triangles	51
Natural scale	113, 115
Nautical—	
Almanac 170, 213, 227, 232, 242, 246, 250, 308, 324	
astronomy	227
mile	65
twilight	325
Nautophone	118
Navigational—	
instruments (*see* Instruments)	
planets	183
satellites	385
stars	179
Neap tide	169
New Moon	183, 219
Newton's Laws of Motion	344
Night effect	373
Non-adjustable errors of sextant	361
Node	192, 218
Noon position	95
North—	
compass	121
magnetic	121
pole	121
true	121
North-up display	383
Notices to mariners	114
Number (Golden)	216

	PAGE
OBLATE spheroid	64
Oblique triangles	25
Obliquity of the ecliptic	188
Observed—	
altitude	229
position	73, 157
zenith distance	302
Observer's celestial meridian	197
Occultation	224
Ocean chart	113
Ogura's table	320
Old Style calendar	215
Open bearings	136
Opposition	182
Orbit of—	
Earth	187
Moon	218
Ordinate	196
Omega	380
PARALLAX—	
definition	232
formula for	233
horizontal	233
-in-altitude	234
Parallel (of)—	
altitude	197
declination	198
latitude	62
sailing	69
Parallelogram of velocities	159
Pendulum	343
Penumbra	223
Performing disc	346
Perigee	184, 185, 218
Perihelion	184, 185, 187
Period—	
of compass	357
sidereal	218
synodic	220
Perpendicularity error	362
Perpetual day (or night)	191
Perspective projection	36
Perturbation	184
Phases of—	
Moon	216, 219
Venus	183
Pilots	113
Pitometer log	370
Pitot tube	370
Plan	113
Plan Position Indicator (P.P.I.)	382
Plane—	
sailing	71
sailing formula	69
triangle	86
Planet—	
inferior	182
minor	181
navigational	183
superior	182

	PAGE
Plotting—	
chart	294
sheet	291
Polar distance	203
Polar variation	214
Polaris	308
Pole—	
altitude of	308
celestial	198
Earth's	61
of ecliptic	196
of great circle	33
of magnet	337
Position—	
chosen	270
circle	251
D.R.	157
estimated	157
geographical	245
line	131, 227, 262, 291
observed	157
on celestial sphere	196
on Earth	62
Position line—	
hyperbolic	376
transferred	135
Precession	344, 348
Primary great circle	33
Primary radar	382
Prime meridian	62
Prime vertical circle	201, 203
Priming	170
Projection—	
gnomonic	101
Mercator	77
stereographic	36
Pulse repetition frequency	382
PZX-triangle	268
QUADRANTAL—	
error	373
system	62
triangle	55
Quadrature	182
Quartz chronometer	365
Quick flashing light	116
RADAR—	
beacon	383
description	382
horizon	382
navigation	382
primary	383
responder beacon	383
secondary	383
Racon	383
Radian	10
Radio—	
beacon	383
bearing	162
direction finder	371

INDEX

	PAGE
Radio—	
time signal	228
Radius—	
of curvature	65
of Earth	64
of gyration	346
vector	184
Ramark	383
Range of tide	166
Rates of change	328
Reckoning	73
Recording fix by Horizontal Angles	150
Reduced latitude	66
Reduction to the latitude	66
Reduction to the meridian	304
Reduction to soundings	171
Reed	118
Refraction	146, 230
Relative bearing	132
Relative display	383
Retardation	218
Retrograde motion	195
Retrogressive curve	195
Rhumb line—	
definition	68, 100
sailing	68
Rising range	146
Rising (setting) phenomena	323
Rising piece	359
Rolling error	355
Rotational times	214
Running fix	136
SAILINGS—	
approximate great circle	100
composite great circle	106
current	158
direction	114
great circle	100
Mercator	75, 86
Middle latitude	75, 86, 90
parallel	69
plane	71
traverse	73
Satellites	181
Scale—	
natural	113, 115
on Mercator chart	78
Schuler tuning	388
Sea clutter	383
Sea horizon—	
dip of	229
distance of	144
Search coil	372
Seasons	189
Secant	6
Secondary great circle	33, 196
Sector light	116
Semi-diameter	232
Semi-diurnal tide	166
Sense aerial	372

	PAGE
Sense finding unit	372
Set (of current)	74
Sextant—	
adjustments	361
altitude	229
care of	363
description of	359
errors	362
principles	360
shades	359
telescopes	360
use of	363
Shade error	363
Ship magnetism	339
Shore effect	373
Short-method tables	320, 372
Side error	362
Sidereal—	
day	205
hour angle	199
period	218
time	207
year	214
Sine	5
Sine formula	25
Six o'clock hour circle	202
Sight	268
Sight reduction	268
Sight reduction tables	319
Slip	95
Small angles	12
Small circle	33
Small corrections	118
Soft iron	339
Solar—	
day	187, 205
system	176
tide	168
time	205
Solstice	189
Sosigenes	215
Sound, speed in water	368
Sounding machine	367
Special angles	7, 139
Speed error	355
Sperry E. B.	343
Sphere—	
celestial	187
geometry of	33
Spherical—	
angle	34
cosine formula	47
haversine	48
hyperbolae	375
sine formula	46
triangle	34
trigonometry	35
Spinning top	348
Spring—	
equinox	218
tide	169

	PAGE
Standard—	
formulae	7
nautical mile	66
standard port	171
time	213
Standing wave theory of the tide	170
Star magnitudes	180
Stars	179
Station pointer	151
Steaming time	95
Stereographic projection	36
Summer—	
season	189
solstice	189
Sun—	
astronomical mean	210
dynamical mean	210
mean	206
rise (set)	323
true	206
Sunday letter	216
Superior transit	202
Synchronism	170
Synodic period	220
TABLES—	
A and K	322
$A\,B\,C$	286, 314
amplitude	287
azimuth	302
Burdwood's	286
Comrie's	322
Davis'	286
Dreisonstok's	322
Gingrich's	322
longitude correction	293
Myerscough and Hamilton	322
nautical astronomical	314
Ogura's	320
Pole Star	308, 309, 311
sight reduction	319
Smart and Shearme	322
traverse	15, 70
Taffrail log	370
Tangent	5
Tangent formula	30
Tangent screw	359
Temperate zone	192
Theoretical horizon	144
Theoretical sunrise (set)	323
Three bearing problem	154
Three letter points	62
Tidal curve	166
Tidal streams	171
Tide—	
diurnal	166
ebb	171
effect of Earth's rotation on	167
equilibrium theory of	167
height of	172
lagging of	170

	PAGE
Tide—	
luni-solar	167
Moon's effect on	168
neap	169
priming of	170
progressive wave theory of	170
semi-diurnal	166
standing wave theory of	170
Sun's effect on	168
Tilt	349
Tilting	350
Tilt oscillations	354
Time—	
and longitude	211
apparent solar	207
at any instant	207
at sea	212
atomic	214
comparison of	211
difference systems	375
ephemeris	214
equation of	208
mean solar	207
rotational	214
sidereal	207
signals	228
standard	213
units	205
zones	213
Tipsy	365
Torrid zone	192
Total correction table—	
Moon	239
star	235
Sun	237
Tracing paper for fix by horizontal angles	151
Transferred position line	135
Transit (marks in)	132, 194, 201
TRANSIT (satellite navigation)	386
Traverse sailing	70, 73
Traverse table	15, 70
Trigonometrical—	
function	3
ratio	3
Trigonometry	3
Tropics	192
Tropical year	214
True altitude	229
True latitude	65
True motion display	383
Twilight—	
astronomical	325
civil	325
definition	325
nautical	325
Tyfon	118
UNIVERSAL law of gravitation	184
Universal time	214
Universe	179

INDEX

	PAGE
Unwanted echoes	383
U.T.*1*	214
U.T.*2*	214
U.T.C.	214
Umbra	223
VARIATION—	
magnetic	123, 338
polar	214
"v" correction	248
Vector	344
Velocity	344
Venus, phases of	183
Vernal equinox	189
Versine	4, 6
Vertex	102
Vertical—	
angle	142
danger angle	144
definition	65
Vertical circle—	
definition	197
prime	201
Virtual meridian	355
Visible—	
horizon	229
sunrise (set)	323

	PAGE
WANING	220
Waxing	220
Weight	144
Winter	189
Winter solstice	189
Wright, E.	77, 81
YEAR—	
anomalistic	214
bissextile	215
civil	215
common	215
leap	215
of confusion	213
sidereal	213
tropical	214, 215
ZENITH	197
Zenith distance	249
Zodiacal belt	193
Zodiac, signs of	193
Zone—	
climatic	191
frigid	191
of perpetual daylight (darkness)	190, 191
temperate	191
time	213
torrid	191

N.S.C.C. NAUTICAL INSTITUTE
P.O. Box 1225
Port Hawkesbury, N.S. B0E 2V0

WITHDRAWN